# Molecular Biology of the Cell

# Molecular Biology of the Cell

Editor: Erik Pierre

R CALLISTO
REFERENCE

www.callistoreference.com

**Callisto Reference,**
118-35 Queens Blvd., Suite 400,
Forest Hills, NY 11375, USA

Visit us on the World Wide Web at:
www.callistoreference.com

ISBN: 978-1-64116-061-2 (Hardback)

**Cataloging-in-Publication Data**

Molecular biology of the cell / edited by Erik Pierre.
     p. cm.
Includes bibliographical references and index.
ISBN 978-1-64116-061-2
1. Cytology--Molecular aspects. 2. Molecular biology. I. Pierre, Erik.
QH581.2 .M65 2019
571.6--dc23

# Table of Contents

# Preface

It is often said that books are a boon to mankind. They document every progress and pass on the knowledge from one generation to the other. They play a crucial role in our lives. Thus I was both excited and nervous while editing this book. I was pleased by the thought of being able to make a mark but I was also nervous to do it right because the future of students depends upon it. Hence, I took a few months to research further into the discipline, revise my knowledge and also explore some more aspects. Post this process, I begun with the editing of this book.

Molecular biology studies fundamental theories and concepts related to interactions between RNA and DNA, their biosynthesis and functions. As a sub-branch of biochemistry, this scientific field studies the interactions and activities concerned with biomolecules present in the cell. This book is a valuable compilation of topics, ranging from the basic to the most complex advancements in the field of molecular biology. The various sub-fields of molecular biology along with technological progress that have future implications are glanced at in this book. It is a vital tool for all researching or studying molecular biology as it gives incredible insights into emerging trends and concepts.

I thank my publisher with all my heart for considering me worthy of this unparalleled opportunity and for showing unwavering faith in my skills. I would also like to thank the editorial team who worked closely with me at every step and contributed immensely towards the successful completion of this book. Last but not the least, I wish to thank my friends and colleagues for their support.

Editor

# Conserved and highly expressed tRNA derived fragments in zebrafish

Ana Raquel Soares[1*], Noémia Fernandes[1], Marisa Reverendo[1], Hugo Rafael Araújo[2], José Luís Oliveira[2], Gabriela M. R. Moura[1] and Manuel A. S. Santos[1*]

## Abstract

**Background:** Small non-coding RNAs (sncRNAs) are a class of transcripts implicated in several eukaryotic regulatory mechanisms, namely gene silencing and chromatin regulation. Despite significant progress in their identification by next generation sequencing (NGS) we are still far from understanding their full diversity and functional repertoire.

**Results:** Here we report the identification of tRNA derived fragments (tRFs) by NGS of the sncRNA fraction of zebrafish. The tRFs identified are 18–30 nt long, are derived from specific 5′ and 3′ processing of mature tRNAs and are differentially expressed during development and in differentiated tissues, suggesting that they are likely produced by specific processing rather than random degradation of tRNAs. We further show that a highly expressed tRF (5′tRF-Pro$^{CGG}$) is cleaved in vitro by Dicer and has silencing ability, indicating that it can enter the RNAi pathway. A computational analysis of zebrafish tRFs shows that they are conserved among vertebrates and mining of publicly available datasets reveals that some 5′tRFs are differentially expressed in disease conditions, namely during infection and colorectal cancer.

**Conclusions:** tRFs constitute a class of conserved regulatory RNAs in vertebrates and may be involved in mechanisms of genome regulation and in some diseases.

**Keywords:** tRNA derived fragments, Zebrafish, Small non coding RNAs, tRNAs

## Background

Small non-coding regulatory RNAs (sncRNAs) play fundamental roles in many aspects of biology and their classification is based on their size, structure, biogenesis and function [1]. MicroRNAs (miRNAs) constitute the most extensively studied class of sncRNAs and are known to regulate the expression of target genes at the translational level [2]. The advent of next generation sequencing (NGS) allowed for the identification and production of miRNA profiles [3–6], identification of piRNAs [7], 21-U RNAs [8] and rasiRNAs [9]. More recently, sncRNAs up to 30 nucleotides derived from tRNAs—the tRNA derived fragments (tRFs)—have also been identified [10, 11]. Fragments of tRNA have been retrieved by many computational analysis of NGS datasets, but they

were initially considered random degradation products of tRNAs and were discarded from further analysis [12]. However, recent experimental data showed that they are a stable and functional class of sncRNAs produced by specific cleavage of certain tRNAs [10, 11, 13–15], and can be classified according to their origin, namely 5′tRFs, also known as tRF-5 series, which derive from 5′ processing of the mature tRNAs; 3′tRFs, also known as tRF-3 series, which derive from 3′ processing of the mature tRNA and 3′U-tRFs, or tRF-1 series, which derive from pre-tRNA processing by RNAse Z cleavage [11, 16]. Different studies have also demonstrated that some 5′- and 3′tRFs are generated by Dicer cleavage, similarly to miRNAs [10, 17], and that some of them are incorporated into Argonaute (Ago) proteins [13, 18] or are associated with piwi proteins [15]. Indeed, deep sequencing of HeLa cells identified 5′tRFs generated by Dicer cleavage at the D-loop, both in vitro and in vivo [10]. Dicer-dependent generation of 3′tRFs also occurs in a human embryonic

*Correspondence: ana.r.soares@ua.pt; msantos@ua.pt
[1] Department of Medical Sciences and Institute for Biomedicine–iBiMED, University of Aveiro, 3810-193 Aveiro, Portugal
Full list of author information is available at the end of the article

kidney 293 cell line and luciferase reporter assays demonstrated that they silence target genes, suggesting that they are involved in gene regulation [18]. Moreover, a Dicer dependent 3′tRF is down-regulated in B cell lymphoma and represses the RPA1 gene, which is involved in DNA repair, similarly to miRNAs [17]. This functional similarity between tRFs and miRNAs may happen because some tRNAs have the potential to form a long hairpin as an alternative to the typical tRNA cloverleaf secondary structure, thus functioning as a Dicer substrate, as is the case of the tRNA$^{Ile}$ in mouse embryonic stem cells [19].

Besides tRFs, longer tRNA fragments (30–38 nt) corresponding to tRNA halves are also produced by cleavage at the 5′ or 3′ sides of the tRNA anticodon by specific endonucleases, rather than Dicer, in response to nutritional deprivation and other stress conditions [20, 21], as is the case in *Tetrahymena thermophila* during amino acid starvation [22]. The 5′ derived tRNA halves induced by stress (tiRNAs) in human cells have the ability to inhibit translation initiation [23] and trigger the formation of stress granules [24], due to a terminal oligoguanine motif [25], indicating that these molecules are involved in gene expression regulation.

We have previously applied NGS to the discovery of zebrafish sncRNAs and identified novel miRNAs in this organism [26]. A new analysis of the sequencing datasets described herein identified 10 new tRFs that originate from specific cleavage of tRNAs. Expression analysis by northern blot shows that these tRFs are differentially expressed at different developmental stages and in certain tissues and their abundance is higher than the corresponding mature tRNA. Our data show that a 5′tRF, namely 5′tRF-Pro$^{CGG}$ can be generated by Dicer and has trans-silencing ability, indicating that it may enter the RNAi pathway, controlling gene expression of specific targets. Northern blot and computational analysis also demonstrate that those tRFs are conserved in vertebrates and are differentially expressed in some disease states, namely during infection and cancer, suggesting their involvement in the mechanisms underlying diseases and their potential use as disease biomarkers.

## Results

### Identification of tRFs in zebrafish

The Roche 454 NGS platform (max nr reads = 100,000) was used previously by our group to identify miRNA molecules in zebrafish adult tissues and developmental stages [26]. In order to identify sequences corresponding to other non-coding RNAs besides miRNAs, the retrieved reads were aligned against a database of known small RNAs extracted from Biomart/Ensembl, including snRNAs, snoRNAs, rRNAs and tRNAs, as described in the "Methods" (Additional file 1: Figure S1). 8 % of total sequencing reads matched the selected ncRNAs, where 61 % of them matched known tRNA loci. The majority of these reads matched to one particular structural domain of tRNAs, suggesting specific processing rather than random tRNA degradation, as described previously [10, 12]. We have considered specific cleavage whenever a given tRF sequence appeared more than three times in the cDNA libraries and the overall tRF alignments with a mature tRNA were dominated by that specific tRF. This methodology identified a total of ten tRFs, which aligned with the 3′ end of tRNAs (six 3′tRFs) and with the 5′end (four 5′tRFs) (Table 1; Fig. 1).

The tRFs identified in this study were 18–30 nt long and half of the 3′tRFs had the trinucleotide signature CCA at the 3′end, which was similar to previous reported cases [10, 11].

In zebrafish 12292 genes code for the different tRNAs, according to the genomic tRNA database. Those with highest copy numbers are: Lysine (1478 genes), Glycine (1162 genes), Leucine (1132 genes) and Serine (1065

**Table 1 Zebrafish tRFs identified by deep sequencing**

| tRF ID | Sequence (5′–3′) | Nr of reads | Expression |
|---|---|---|---|
| 5′tRF-Lys$^{TTT}$ | GCCCGGATAGCTCAGTCGGTAGAGCATCAG | 4 | Adult tissues |
| 5′tRF-Val$^{CAC}$ | GTTTCCGTAGTGTAGTGGTTATCACGTTCG | 4 | Adult tissues |
| 5′tRF-Glu$^{CTC}$ | TCCCTGGTGGTCTAGTGGTTAGGATTCGGC | 26 | Development, adult tissues |
| 5′tRF-Pro$^{CGG}$ | TAGGGGTATGATTCTCGC | 12 | Development, adult tissues |
| 3′tRF-Ala$^{AGC}$ | AGAGGTAGCGGGATCGTTGCCC | 16 | Adult tissues |
| 3′tRF-Lys$^{CTT}$ | CAGGGTCGTGGGTTCGAGCCCC | 4 | Adult tissues |
| 3′tRF-Ile$^{AAT}$ | CAAGGTGCGCGGGTTCGTTCCCC | 6 | Adult tissues |
| 3′tRF-Glu$^{CTC}$ | TCGATTCCCGGTCAGGGAACCA | 24 | Development, adult tissues |
| 3′tRF-Pro$^{AGG}$ | TCCCGGACGAGCCCCCA | 13 | Development, adult tissues |
| 3′tRF-Arg$^{TCG}$ | GTCCCTTCGTGGTCGCCA | 7 | Adult tissues |

tRF identification, sequence, number of reads and expression are shown

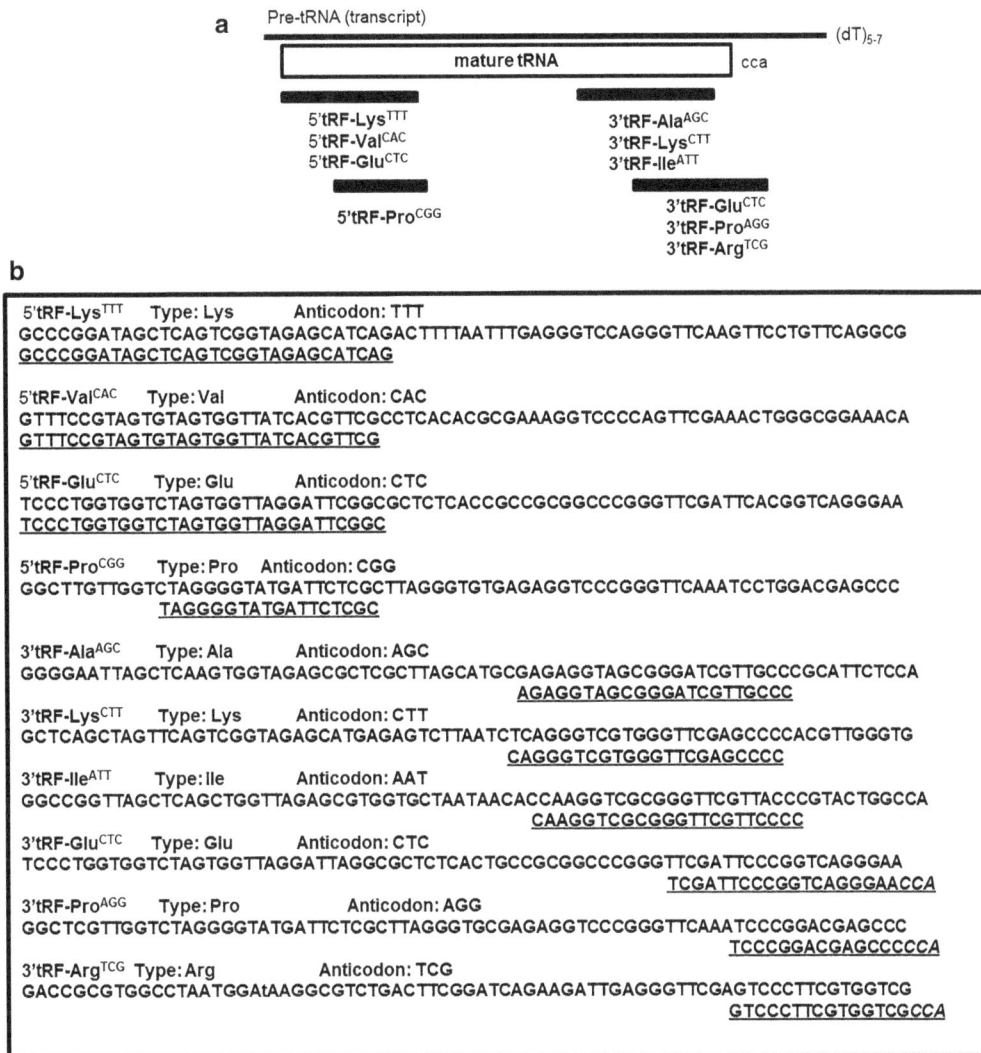

**Fig. 1** Classification of the zebrafish tRNAfs identified by deep sequencing. **a** Schematic view of the localization of the tRFs in the linear structure of mature tRNAs. Six tRFs align in the 3′ end of the mature tRNAs and 4 tRFs align to the 5′ region of the tRNA. **b** Alignment of the 12 tRFs in the corresponding mature tRNA

genes). Although the 10 tRFs identified derived from different classes of tRNAs, there seems to be no correlation between the number of tRNA genes and the generation of tRFs, as most of these abundant tRNAs did not generate tRFs, with exception of Lysine (Fig. 1b). There was a slight enrichment in tRFs generated from Lysine, Glutamine and Proline tRNAs (Fig. 1b), suggesting that the generation of tRFs was not dependent on tRNA expression levels. 5′tRF-Lys$^{TTT}$ and 3′tRF-Lys$^{CTT}$ derived from different tRNA$^{Lys}$ isoacceptors, 5′tRF-Glu$^{CTC}$ and 3′tRF-Glu$^{CTC}$ derived from tRNA$^{Glu}$, whereas 5′tRF-Pro$^{CGG}$ and 3′tRF-Pro$^{AGG}$ derived from tRNA$^{Pro}$ isoacceptors.

### tRF profiling

In order to confirm the deep sequencing data, the expression patterns of four tRFs, namely 5′tRF-Lys$^{TTT}$, 5′tRF-Glu$^{CTC}$, 5′tRF-Pro$^{CGG}$ and 3′tRF-Pro$^{AGG}$, were studied in different developmental stages [24 h post fertilization (hpf), 48, 72 and 96 hpf] and in different zebrafish adult tissues (brain, fins, bone/muscle, skin) using northern blot analysis, as described in the "Methods".

There was poor correlation between the abundance of tRFs and mature tRNAs (Fig. 2a–d). During zebrafish development (from 24 to 96 hpf), only 5′tRF-Glu$^{CTC}$ and 5′tRF-Pro$^{CGG}$ were detected and at low levels (less

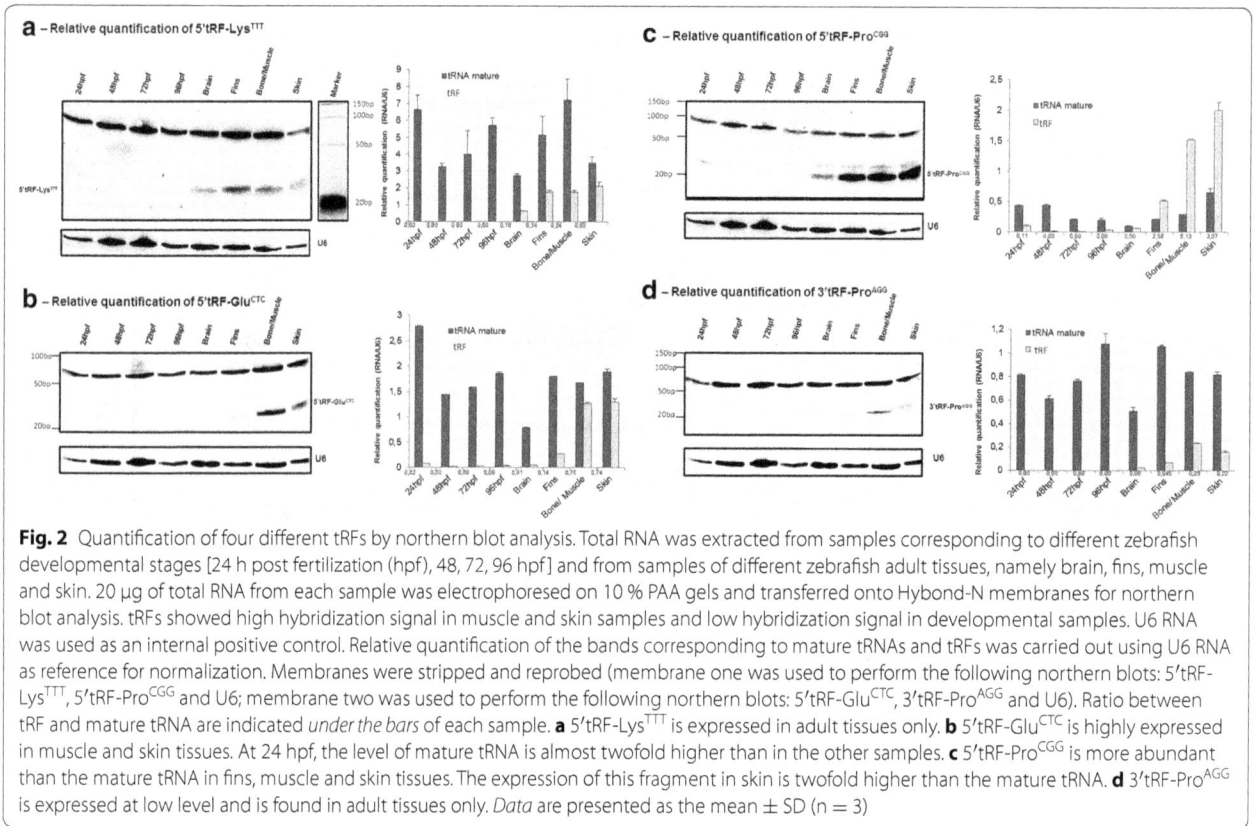

**Fig. 2** Quantification of four different tRFs by northern blot analysis. Total RNA was extracted from samples corresponding to different zebrafish developmental stages [24 h post fertilization (hpf), 48, 72, 96 hpf] and from samples of different zebrafish adult tissues, namely brain, fins, muscle and skin. 20 μg of total RNA from each sample was electrophoresed on 10 % PAA gels and transferred onto Hybond-N membranes for northern blot analysis. tRFs showed high hybridization signal in muscle and skin samples and low hybridization signal in developmental samples. U6 RNA was used as an internal positive control. Relative quantification of the bands corresponding to mature tRNAs and tRFs was carried out using U6 RNA as reference for normalization. Membranes were stripped and reprobed (membrane one was used to perform the following northern blots: 5'tRF-Lys$^{TTT}$, 5'tRF-Pro$^{CGG}$ and U6; membrane two was used to perform the following northern blots: 5'tRF-Glu$^{CTC}$, 3'tRF-Pro$^{AGG}$ and U6). Ratio between tRF and mature tRNA are indicated *under the bars* of each sample. **a** 5'tRF-Lys$^{TTT}$ is expressed in adult tissues only. **b** 5'tRF-Glu$^{CTC}$ is highly expressed in muscle and skin tissues. At 24 hpf, the level of mature tRNA is almost twofold higher than in the other samples. **c** 5'tRF-Pro$^{CGG}$ is more abundant than the mature tRNA in fins, muscle and skin tissues. The expression of this fragment in skin is twofold higher than the mature tRNA. **d** 3'tRF-Pro$^{AGG}$ is expressed at low level and is found in adult tissues only. *Data are presented as the mean $\pm$ SD (n = 3)*

than 0.2 tRF/U6 relative expression). These tRFs were expressed at high levels in bone/muscle and skin (>1 tRF/U6 relative expression), regardless of the levels of mature tRNAs in those tissues (Fig. 2b, c). Remarkably, the abundance of 5'tRF-Pro$^{CGG}$ was higher than that of its corresponding mature tRNA in most tissue samples (fins, bone/muscle and skin, with a tRF/tRNA ratio of 2.58, 5.13 and 3.07, respectively), suggesting that it may have a functional role (Fig. 2c).

The other tRFs tested were not detected by northern blotting during development (Fig. 2a, d), but were detected in adult tissues. The abundance of the 5'tRF-Lys$^{TTT}$ was similar in fins, muscle and skin (~1.5 tRF/U6 relative expression), whereas the abundance of the corresponding mature tRNA was higher in muscle (7 tRNA/U6 relative expression) than in other tissues (Fig. 2a). 3'tRF-Pro$^{AGG}$ was also detected in adult tissues only, at low levels—maximum 0.3 tRF/U6 relative expression (Fig. 2d).

The lack of correlation between the mature tRNAs and the respective tRFs abundance in different tissues suggested that tRF generation is a regulated process, rather than a random degradation process. Moreover, almost no bands of intermediate molecular weight were detected in the northern blots, indicating that tRNA cleavage occurs

at specific cleavage sites. In some tissues, namely brain, fins and skin a band of intermediate size ~30 nt was detected for 5'tRF-Pro$^{CGG}$ (Fig. 2c), however the smaller band, corresponding to 5'tRF-Pro$^{CGG}$ was always the most prominent one, indicating preferential accumulation of this tRF.

Since northern blot analysis revealed that both 5'tRF-Glu$^{CTC}$ and 5'tRF-Pro$^{CGG}$ were highly abundant tRFs, we have focused our attention on these two molecules and studied them in more detail. We have extended our previous developmental analysis up to 2 months post fertilization (mpf) (Fig. 3a). The levels of both 5'tRF-Pro$^{CGG}$ and 5'tRF-Glu$^{CTC}$ were highest at 2 mpf and for this reason the tRF expression was considered 100 % at this particular stage. 5'tRF-Glu$^{CTC}$ was barely detected at 24 and 48 hpf, but its levels increased steadily during development (Fig. 3a). 5'tRF-Pro$^{CGG}$ was detected already at 24 hpf and there was an increased generation of this tRF at 72 hpf (30 % increase), however at 10 days post fertilization (dpf) the 5'tRF-Pro$^{CGG}$ levels decreased to levels similar to those observed at 24 and 48 hpf and increased again reaching high abundance at 2 mpf.

These tRFs are also differentially produced in a variety of tissues (Fig. 3b). The expression of both tRFs in bone is approximately twofold higher than in muscle. In fact,

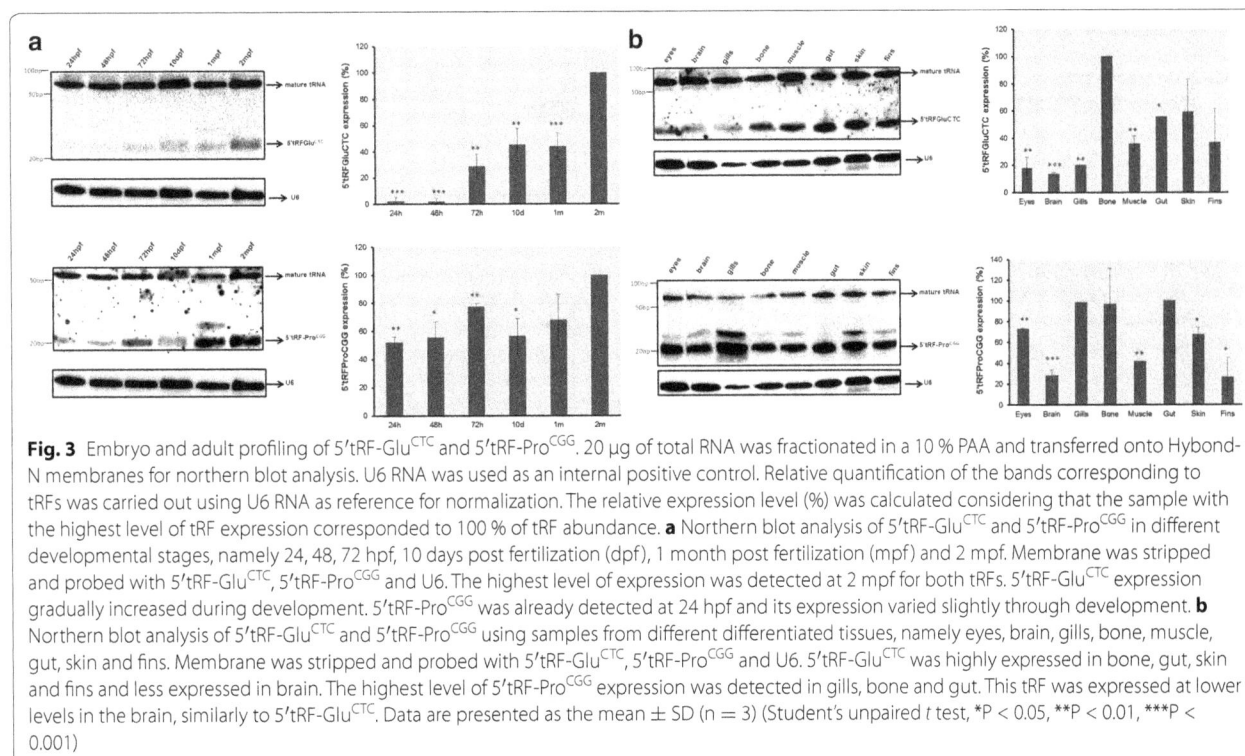

**Fig. 3** Embryo and adult profiling of 5′tRF-Glu$^{CTC}$ and 5′tRF-Pro$^{CGG}$. 20 μg of total RNA was fractionated in a 10 % PAA and transferred onto Hybond-N membranes for northern blot analysis. U6 RNA was used as an internal positive control. Relative quantification of the bands corresponding to tRFs was carried out using U6 RNA as reference for normalization. The relative expression level (%) was calculated considering that the sample with the highest level of tRF expression corresponded to 100 % of tRF abundance. **a** Northern blot analysis of 5′tRF-Glu$^{CTC}$ and 5′tRF-Pro$^{CGG}$ in different developmental stages, namely 24, 48, 72 hpf, 10 days post fertilization (dpf), 1 month post fertilization (mpf) and 2 mpf. Membrane was stripped and probed with 5′tRF-Glu$^{CTC}$, 5′tRF-Pro$^{CGG}$ and U6. The highest level of expression was detected at 2 mpf for both tRFs. 5′tRF-Glu$^{CTC}$ expression gradually increased during development. 5′tRF-Pro$^{CGG}$ was already detected at 24 hpf and its expression varied slightly through development. **b** Northern blot analysis of 5′tRF-Glu$^{CTC}$ and 5′tRF-Pro$^{CGG}$ using samples from different differentiated tissues, namely eyes, brain, gills, bone, muscle, gut, skin and fins. Membrane was stripped and probed with 5′tRF-Glu$^{CTC}$, 5′tRF-Pro$^{CGG}$ and U6. 5′tRF-Glu$^{CTC}$ was highly expressed in bone, gut, skin and fins and less expressed in brain. The highest level of 5′tRF-Pro$^{CGG}$ expression was detected in gills, bone and gut. This tRF was expressed at lower levels in the brain, similarly to 5′tRF-Glu$^{CTC}$. Data are presented as the mean ± SD (n = 3) (Student's unpaired $t$ test, *P < 0.05, **P < 0.01, ***P < 0.001)

the levels of the 5′tRF-Glu$^{CTC}$ were highest in bone and for comparative purposes its relative expression was considered 100 % in this tissue. Its relative expression was also high in skin (60 %) and gut (58 %) and lower in eyes (18 %), brain (15 %) and gills (20 %). This data confirmed the initial profiling data which showed that 5′tRF-Glu$^{CTC}$ generation was low in brain and high in skin (Fig. 2). The 5′tRF-Pro$^{CGG}$ showed the maximal abundance in gills and for this reason we considered its relative expression as 100 % in this sample. The relative abundance of this tRF was also variable between tissues: eyes (75 %), bone (98 %), skin (70 %) and gut (98 %) and lower in brain and fins (<30 %), confirming the initial profiling data (Figs. 2c, 3b). Therefore, different tRFs are differentially produced and accumulated in the various tissues of zebrafish, in particular both 5′tRF-Glu$^{CTC}$ and 5′tRF-Pro$^{CGG}$ are generated at lower levels in the brain than in any other tissues tested.

## Biogenesis of tRFs

Previous studies have implicated Dicer (RNAse III family member) in tRF generation [10, 18]. To test whether Dicer could be responsible for 5′tRF-Glu$^{CTC}$ and 5′tRF-Pro$^{CGG}$ biogenesis we incubated total RNA from 72 hpf zebrafish embryos with this enzyme and performed northern blot analysis. Both tRFs were detected after 30 min of incubation with Dicer and their abundance

increased over time (Fig. 4a), suggesting that this enzyme was involved in tRNA$^{Glu}$ and tRNA$^{Pro}$ cleavage and in the biogenesis of 5′tRF-Glu$^{CTC}$ and 5′tRF-Pro$^{CGG}$, respectively. Dicer did not produce non-specific tRNA cleavage products, suggesting that this enzyme is directly involved in the biogenesis of both 5′tRF-Glu$^{CTC}$ and 5′tRF-Pro$^{CGG}$ in vitro. As Dicer has also been implicated in the generation of 3′tRFs [17, 18], we tested its ability to generate 3′tRF-Ala$^{AGC}$, however, we have only observed bands corresponding to the mature tRNA(Ala)$^{AGC}$, indicating that Dicer is not involved in the generation of this particular 3′tRF in vitro (Fig. 4a).

Angiogenin, an RNAseA family member has also been implicated in the generation of some tRNA derived fragments, namely tRNA halves, in response to stress [20, 23, 24]. Another study also shows that angiogenin is involved in the production of 20–25 nt tRFs in vitro in HEK293 cells [27]. Our experiments confirmed that angiogenin is not involved in the generation of the tRFs tested, (Fig. 4b), reinforcing the role of Dicer in the biogenesis of 5′tRF-Glu$^{CTC}$ or 5′tRF-Pro$^{CGG}$ and indicating that 3′tRF-Ala$^{AGC}$ generation is probably dependent on the action of alternative ribonucleases. Besides we did not detect any increase in 5′tRF-Glu$^{CTC}$ or 5′tRF-Pro$^{CGG}$ production upon exposure to stress (data not shown), but the tRNAs were cleaved into short non-specific tRNA fragments after 10 min of incubation with Angiogenin (Fig. 4b).

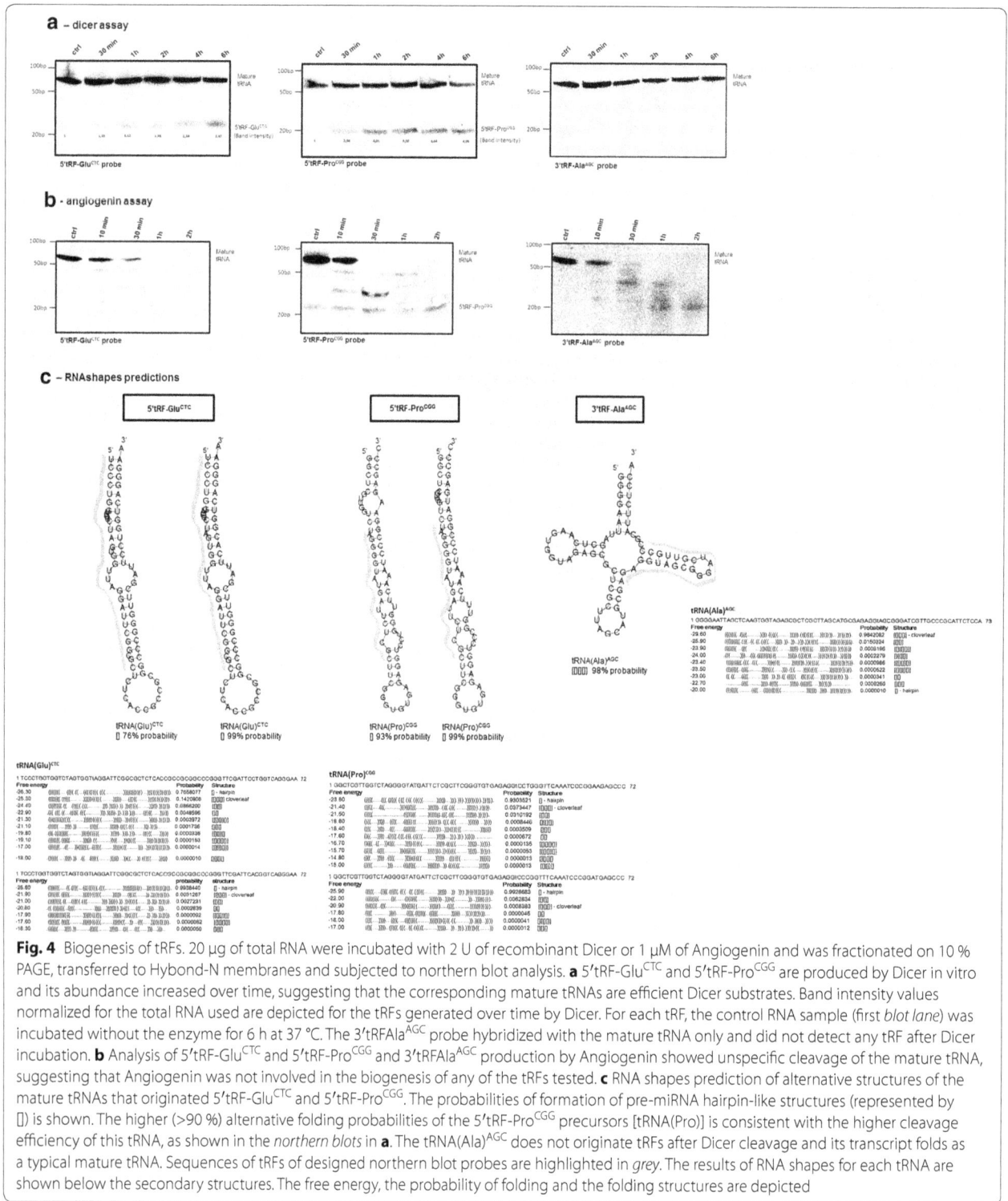

**Fig. 4** Biogenesis of tRFs. 20 µg of total RNA were incubated with 2 U of recombinant Dicer or 1 µM of Angiogenin and was fractionated on 10 % PAGE, transferred to Hybond-N membranes and subjected to northern blot analysis. **a** 5'tRF-Glu^CTC and 5'tRF-Pro^CGG are produced by Dicer in vitro and its abundance increased over time, suggesting that the corresponding mature tRNAs are efficient Dicer substrates. Band intensity values normalized for the total RNA used are depicted for the tRFs generated over time by Dicer. For each tRF, the control RNA sample (first *blot lane*) was incubated without the enzyme for 6 h at 37 °C. The 3'tRFAla^AGC probe hybridized with the mature tRNA only and did not detect any tRF after Dicer incubation. **b** Analysis of 5'tRF-Glu^CTC and 5'tRF-Pro^CGG and 3'tRFAla^AGC production by Angiogenin showed unspecific cleavage of the mature tRNA, suggesting that Angiogenin was not involved in the biogenesis of any of the tRFs tested. **c** RNA shapes prediction of alternative structures of the mature tRNAs that originated 5'tRF-Glu^CTC and 5'tRF-Pro^CGG. The probabilities of formation of pre-miRNA hairpin-like structures (represented by []) is shown. The higher (>90 %) alternative folding probabilities of the 5'tRF-Pro^CGG precursors [tRNA(Pro)] is consistent with the higher cleavage efficiency of this tRNA, as shown in the *northern blots* in **a**. The tRNA(Ala)^AGC does not originate tRFs after Dicer cleavage and its transcript folds as a typical mature tRNA. Sequences of tRFs of designed northern blot probes are highlighted in *grey*. The results of RNA shapes for each tRNA are shown below the secondary structures. The free energy, the probability of folding and the folding structures are depicted

Longer incubation times increased the degradation of tRNA and after 2 h the complete fraction of mature tRNA was degraded. Moreover, angiogenin degraded completely the fragments of the tRNA(Glu), but was not able to degrade the 5'tRF-Pro^CGG, indicating that these tRFs are not angiogenin targets (Fig. 4b).

Recent computational analysis also suggested that Dicer cleavage of tRNAs may occur when the primary transcripts form long hairpins, as an alternative fold to the standard tRNA cloverleaf secondary structure [19]. To test this hypothesis, we determined the probability of the tRNA transcripts forming pre-miRNA like folds using RNAshapes [28]. The Shape Probability option was used to calculate the shape probabilities based on the partition function where the probability of a shape is the sum of the probabilities of all structures that fall into this shape. $3'$tRF-Ala$^{AGC}$ can only derive from one tRNA(Ala) locus, while $5'$tRF-Glu$^{CTC}$ and $5'$tRF-Pro$^{CGG}$ can be assigned to different tRNA(Glu) and tRNA(Pro) locus in zebrafish (>100 tRNA locus), as found in the genomic tRNA database [29]. We have tested the folding of all of them using RNAShapes and verified that 5 % of tRNA(Glu) and 50 % of tRNA(Pro) transcripts are more prone to acquire a hairpin like folding than any other folding. In most of these cases, the probability of hairpin-like folding is higher than 70 %. Figure 4c shows some of the examples of alternative folds of tRNA(Glu) and tRNA(Pro) obtained, supporting the hypothesis that $5'$tRF-Glu$^{CTC}$ and $5'$tRF-Pro$^{CGG}$ can be produced by Dicer. On the other hand the mature tRNA(Ala)$^{AGC}$ showed 98 % probability of forming a typical cloverleaf tRNA-like structure (Fig. 4c).

## $5'$tRF-Pro$^{CGG}$ has the ability to silence gene expression

A dual fluorescence reporter system (DFRS) consisting of a GFP-Reporter/mRFP-Sensor plasmid was injected into one cell stage zebrafish embryos to evaluate the silencing ability of endogenously available $5'$tRF-Glu$^{CTC}$ and $5'$tRF-Pro$^{CGG}$. This reporter expresses both RFP and GFP under the control of the same promoter. The RFP contained a $3'$UTR cassette complementary to the tRF of interest, which functions as a silencing sensor; the GFP lacks complementary sites functioning as an internal control reporter. While the GFP is always expressed in cells that incorporate the plasmid, the expression of the RFP reporter is repressed if the endogenous tRF of interest has trans-silencing activity. We have observed a decrease in the expression of the RFP signal relative to the control at 24 hpf upon injection of the DFRS-$5'$tRF-Pro$^{CGG}$ (Fig. 5a–f). The silencing was still observed at 72 hpf (Additional file 2: Figure S2D–F) indicating that the endogenous $5'$tRF-Pro$^{CGG}$ binds to its complementary sites and induces silencing. To further confirm this observation, we have engineered mutations in the $5'$ and $3'$ ends of the RFP sites that were complementary to the $5'$tRF-Pro$^{CGG}$ to disrupt binding and silencing (Fig. 5p). We have assumed that any mutation would affect silencing if full complementarity between the tRF and its target was required or that silencing would be affected by specific mutations only if partial complementarity, similar to that used by miRNAs, would be required. We have

observed that the mutations that affected the binding of the $5'$end of $5'$tRF-Pro$^{CGG}$ (corresponding to nucleotides 2, 3, 5 and 6 of the tRF) with the reporter (DFRS-$5'$tRF-Pro$^{CGG}$-Mut5) derepressed RFP expression, indicating that these sites are essential for endogenous $5'$tRF-Pro$^{CGG}$ silencing activity (Fig. 5g–i). Injection of DFRS-$5'$tRF-Pro$^{CGG}$-Mut3, which affects binding of $5'$tRF-Pro$^{CGG}$ $3'$end to complementary target sites (corresponding to nucleotides 13, 14, 16 and 17 of the tRF), did not affect RFP expression, indicating that tRF $3'$end nucleotides are not essential for target recognition and silencing (Fig. 5j–l).

There was a slight decrease in RFP intensity levels after microinjection of DFRS-$5'$tRF-Glu$^{CTC}$ relative to GFP, but the RFP signal was not abolished (Fig. 5m–o). Since $5'$tRF-Glu$^{CTC}$ levels are low at 24 and 48 hpf we checked for RFP expression after 72 hpf, but no considerable alterations were observed (Additional file 2: Figure S2G–I), indicating that the endogenous $5'$tRF-Glu$^{CTC}$ does not silence gene expression as efficiently as $5'$tRF-Pro$^{CGG}$. Approximately 50 embryos were microinjected per condition and biological replicate. Three biological replicates were performed for each condition tested.

## tRFs are conserved between zebrafish and humans

Since tRNAs are well conserved across vertebrates, it is plausible that the same is true for tRFs. To validate the conservation hypothesis, total RNA was extracted from different zebrafish, human and mouse cell lines and analyzed by northern blots. ZFb1 and ZFb2 cells derived from jaw, vertebra and branquial arch tissues of zebrafish, HeLa cells derived from human epithelial cervical cancer, HEK293 cells derived from human embryonic kidney, AGS cells derived from human stomach adenocarcinoma and mouse NIH3T3 cells derived from mouse embryonic fibroblasts, were analyzed. RNA from zebrafish bone/muscle was used as a positive control, as it corresponds to a tissue where these tRFs are abundant, as shown previously (Fig. 2). $5'$tRF-Glu$^{CTC}$ and $5'$tRF-Pro$^{CGG}$ were detected in both human and zebrafish cell lines, and to less extent in mouse NIH3T3 cells (Fig. 6). Both $5'$tRF-Glu$^{CTC}$ and $5'$tRF-Pro$^{CGG}$ were slightly shorter in the Zfb1 cell line, suggesting slightly different processing in different organisms.

## $5'$tRF levels are affected during infection and in colorectal cancer

We have also analysed the presence of tRFs in human sequencing datasets deposited in the GEO database. From the different datasets analysed, we were able to identify $5'$tRF-Lys$^{TTT}$, $5'$tRF-Val$^{CAC}$, $5'$tRF-Glu$^{CTC}$, $5'$tRF-Pro$^{CGG}$, $3'$tRF-Glu$^{CTC}$ and $3'$tRF-Pro$^{AGG}$ in 5 of them, namely GSM1293576 [30], GSE33584 [31], GSE29173

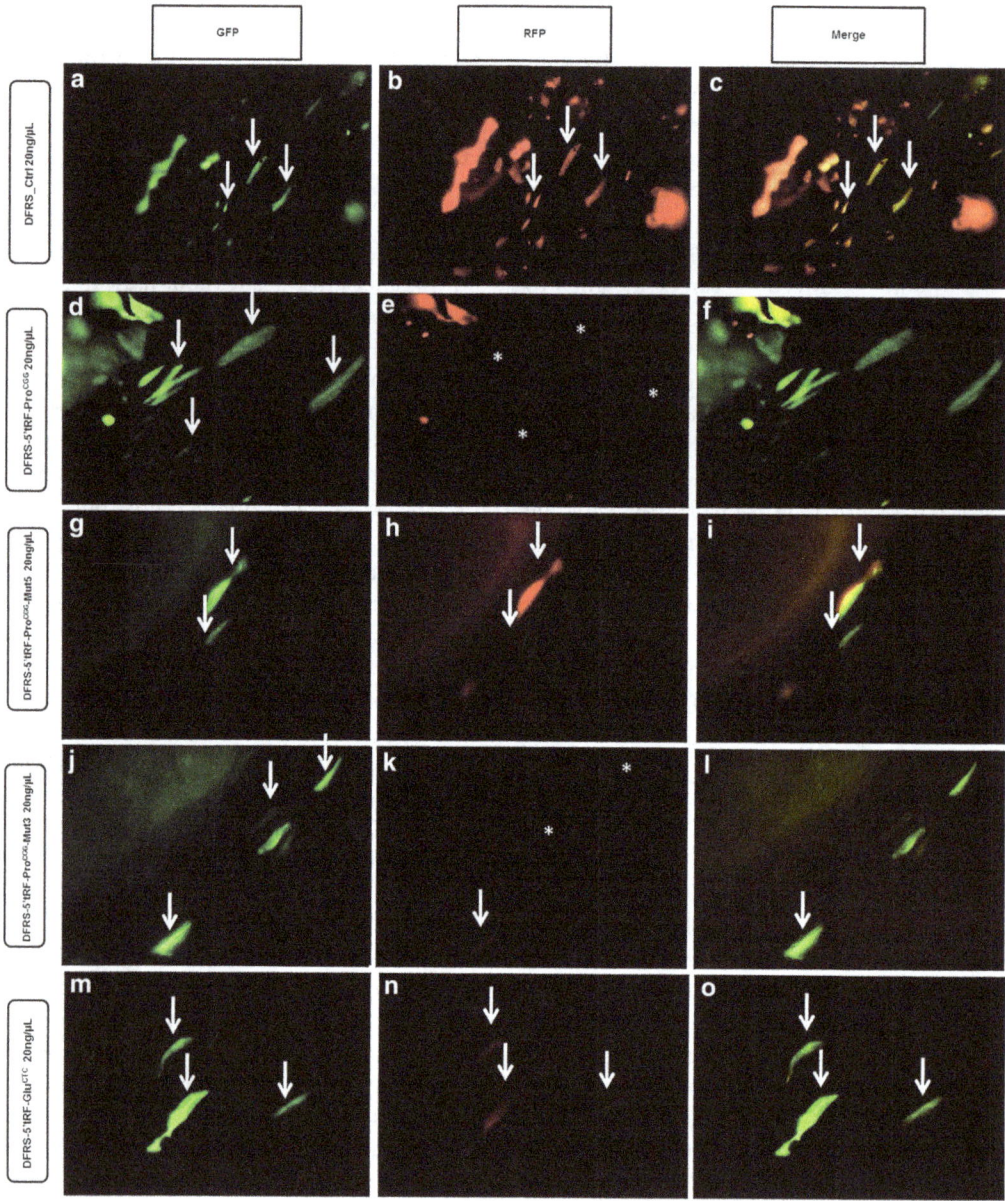

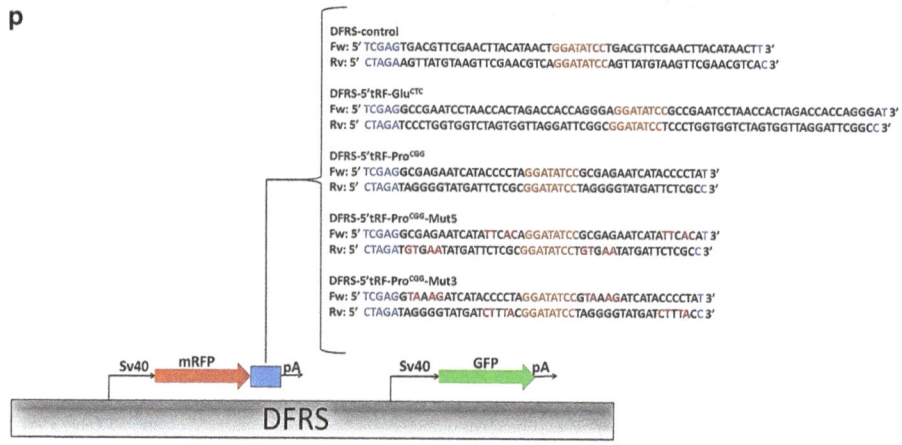

(See figure on previous page.)

**Fig. 5** Silencing ability of tRFs using a dual reporter plasmid. Endogenous 5′tRF-Pro$^{CGG}$ has silencing ability. Embryos injected with 20 ng/µL of DFRS control plasmid (**a, b, c**), show GFP and mRFP signal, while the RFP signal is lost after microinjection of 20 ng/µL DFRS-5′tRF-Pro$^{CGG}$ plasmid (**d, e, f**). 5′ portion of 5′tRF-Pro$^{CGG}$ is necessary for target silencing, as mutations in the reporter that affect binding at the 5′ end (DFRS-5′tRF-Pro$^{CGG}$_Mut5) result in recovery of RFP fluorescence (**g, h, i**) when compared to non-mutated reporter (**d, e, f**). Mutations that affect binding of the 3′ end of the tRF do not affect silencing ability as RFP is repressed (**j, k, l**), similarly to the non-mutated reporter. There is only a slight decrease in RFP signal after microinjection of 20 ng/µL DFRS-5′tRF-Glu$^{CTC}$ (**m, n, o**) plasmid, indicating that the endogenous 5′tRF-Glu$^{CTC}$ does not have trans-silencing ability. *Arrows* indicate cells containing both GFP-reporter and mRFP-sensor. *Asterisks* indicate muscle fibers that lost mRFP fluorescence. Orientation of embryos: caudal, *left*; ventral, *up* ×20 magnification. **p** DFRS plasmid scheme. The DFRS plasmid bears two fluorescent proteins, namely GFP (*green*) and mRFP (*red*), controlled by SV40 promoters. The mRFP contains a 3′UTR cassette (*blue box*) complementary to the tRF of interest. The sequences inserted in the DFRS to obtain the different reporters (DFRS-control, DFRS-5′tRF-Glu$^{CTC}$, DFRS-5′tRF-Pro$^{CGG}$, DFRS-5′tRF-Pro$^{CGG}$-Mut5 and DFRS-5′tRF-Pro$^{CGG}$-Mut3) are depicted. Restriction sites are shown in *blue*, *Eco*RV cleavage site is shown in *orange* and the mutated nucleotides are highlighted in *red*

[32], GSE22918 [33] and GSE43550. Low number of reads of 5′tRF-Val$^{CAC}$ and 5′tRF-Glu$^{CTC}$ (maximum 185 reads) were identified in serum of old and young individuals (GSM1293576), with no particular age preference (Additional file 3: Table S1). Both tRFs identified were 4 nt longer than the original zebrafish tRFs, probably indicating variation in processing of these molecules in human serum.

3′tRF-Pro$^{AGG}$ was the only zebrafish tRF identified in breast tissue (normal and tumour tissue)—GSE29173—, but no significant change in read numbers was found between samples (Additional file 3: Table S1). Only 5′tRF-Glu$^{CTC}$ reads were found in the nucleus and in the cytoplasm of 5-8F cells—a nasopharyngeal carcinoma cell line (GSE22918). Reads corresponding to other tRFs, namely 5′tRF-Lys$^{TTT}$, 5′tRF-Pro$^{CGG}$, 3′tRF-Glu$^{CTC}$ and 3′tRF-Pro$^{AGG}$ were only found in the cytoplasm of 5-8F cells, indicating that most tRFs are present in the cytoplasm rather than in the nucleus (Additional file 3: Table S1). This is not surprising as mature cytoplasmic tRNAs are the precursors of most tRFs, namely 3′ and 5′tRFs [34]. The fact that 5′tRF-Glu$^{CTC}$ was found in the nucleus indicates that tRFs migrate to this organelle, similarly to

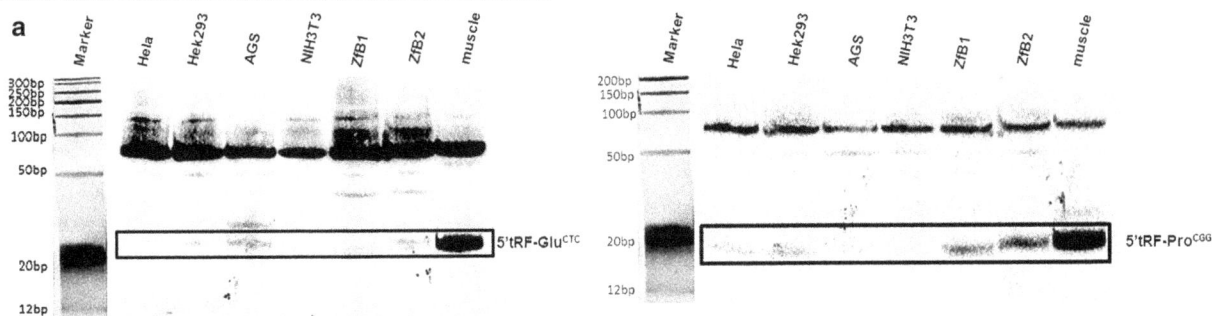

**Fig. 6** tRFs are conserved among vertebrate species. **a** Northern blots showing the expression of 5′tRF-Glu$^{CTC}$ and 5′tRF-Pro$^{CGG}$ in human cell lines (HeLa, HEK293, AGS), a mouse cell line (NIH3T3) and zebrafish cell lines derived from bone (ZFb1, ZFb2). RNA extracted from adult zebrafish muscle was used as a positive control. 5′tRF-Glu$^{CTC}$ and 5′tRF-Pro$^{CGG}$ were detected in human and zebrafish cell lines, but barely detected in the mouse embryonic fibroblast cell line (NIH3T3). **b** ClustalW alignments of mature tRNA and tRFs showing high conservation in different species, namely human, mouse and zebrafish

what has already been found for other sncRNAs, namely miRNAs [35, 36].

Analysis of the sequencing data from colorectal tumour samples (GSE43550) revealed the presence of 5′tRF-Lys$^{TTT}$, 5′tRF-Val$^{CAC}$, 5′tRF-Glu$^{CTC}$ and 5′tRF-Pro$^{CGG}$, both in adjacent normal and in the tumour tissue. These tRFs, in particular 5′tRF-Glu$^{CTC}$ showed sequence variations at the 3′ end, indicating that the cleavage events are not precise even in the same tissue, similarly to miRNAs [37] (Fig. 7a; Additional file 3: Table S1). 5′tRF-Pro$^{CGG}$ was only detected in tissues (normal and tumour) without transcatheter arterial infusion chemotherapy. 5′tRF-Lys$^{TTT}$ reads decreased in tumour samples without arterial infusion chemotherapy, but no significant difference in read number was detected between normal and tumour samples with treatment. A slight decrease (~7000 reads difference) was detected for 5′tRF-Val$^{CAC}$ in tumour tissue without transcatheter arterial infusion chemotherapy relative to the corresponding normal tissue. The expression of this tRF seemed to be affected by transcatheter arterial infusion chemotherapy as the total number of reads decreased from ~100,000 (normal tissue with transcatheter arterial infusion) to

~35,000 (tumour tissue with transcatheter arterial infusion). 5′tRF-Glu$^{CTC}$ was also down regulated in tumour tissues, independently of transcatheter arterial infusion chemotherapy. The total number of 5′tRF-Glu$^{CTC}$ reads decreased from ~30,000 (normal adjacent tissue with infusion chemotherapy) to ~8000 (tumour tissue with infusion chemotherapy) and decreased to about half that number in tumour tissue without infusion chemotherapy compared to its corresponding normal tissue (Fig. 7b). The decrease in read number was always observed for all the isoforms (different sequences with 3′end variations) and some of them were absent in tumour samples. This data show that 5′tRF-Val$^{CAC}$ and especially 5′tRF-Glu$^{CTC}$ may have a role in colorectal cancer and have potential to be used as cancer biomarkers.

5′tRF-Lys$^{TTT}$, 5′tRF-Val$^{CAC}$ and 5′tRF-Glu$^{CTC}$ were identified in 24 and 72 h non-infected and cytomegalovirus infected primary human fibroblasts (GSE33584). The three identified tRFs were always 2 nt longer when compared to the original zebrafish tRF. Read numbers were relatively low for 5′tRF-Lys$^{TTT}$ and 5′tRF-Val$^{CAC}$, but still 5′tRF-Lys$^{TTT}$ reads were fourfold higher in infected

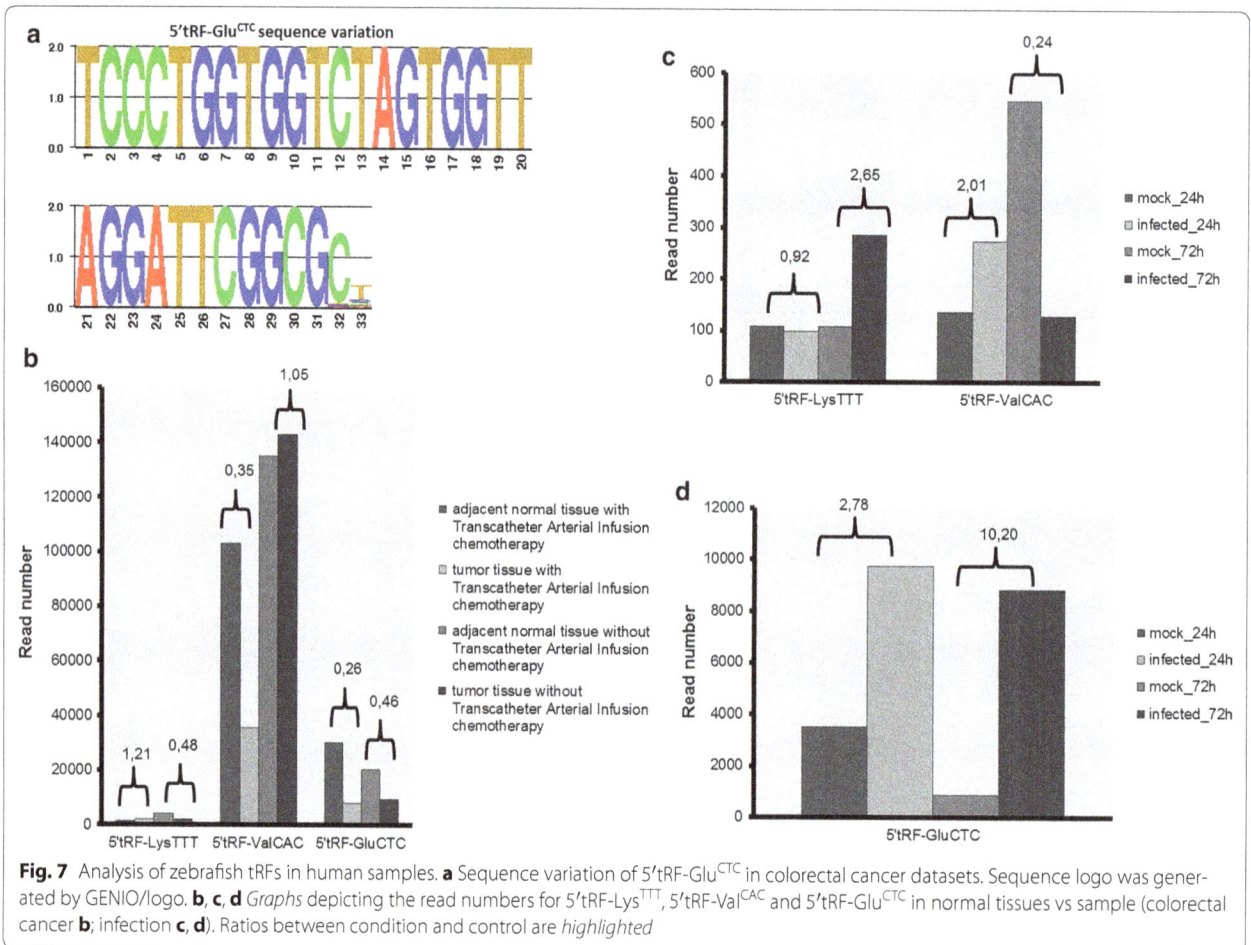

**Fig. 7** Analysis of zebrafish tRFs in human samples. **a** Sequence variation of 5′tRF-Glu$^{CTC}$ in colorectal cancer datasets. Sequence logo was generated by GENIO/logo. **b, c, d** *Graphs* depicting the read numbers for 5′tRF-Lys$^{TTT}$, 5′tRF-Val$^{CAC}$ and 5′tRF-Glu$^{CTC}$ in normal tissues vs sample (colorectal cancer **b**; infection **c, d**). Ratios between condition and control are *highlighted*

cells 72 h post-infection while 5′tRF-Val$^{CAC}$ reads were fourfold lower in 72 h post-infected cells. The number of 5′tRF-Glu$^{CTC}$ reads was ~threefold higher 24 h post-infection and ~tenfold higher 72 h post-infection, indicating a role for this tRF during infection (Additional file 3: Table S1; Fig. 7c, d). This particular tRF and 5′tRF-Lys$^{TTT}$, among others, were up-regulated upon respiratory syncytial virus infection in airway epithelial cells (A549 cells) and were implicated in RSV replication [14]. In this case, 5′tRF-Glu$^{CTC}$ was only 1 nt longer than the zebrafish 5′tRF-Glu$^{CTC}$ and, consequently, 1 nt shorter than the 5′tRF-Glu$^{CTC}$ detected in human fibroblasts. Since 5′tRF-Glu$^{CTC}$ variability was found in different cell lines it is likely that the 3′end variability is tissue specific and reflect tissue specific variation in processing. 5′tRF-Lys$^{TTT}$ had no variation in A549 cells relative to the zebrafish tRF and 5′tRF-Glu$^{CTC}$ was abundant after rickettsia infection in mice and exhibited no sequence variation when compared to the equivalent zebrafish tRF [38].

## Discussion

### tRFs are highly expressed in adult zebrafish

In this study we have identified 10 zebrafish tRFs—4 belonging to the tRF-5 series and 6 belonging to the tRF-3 series—distributed throughout development and adult tissues. Since our deep sequencing experiment was designed to identify miRNAs with a NGS read length cutoff between 15 and 30 nt, it is likely that longer tRFs that are generally induced by nutritional deprivation and stress conditions were missed [20, 22, 39]. Studies are needed to clarify this question.

Several tRFs were previously identified in zebrafish early developmental stages, up to 24 hpf [37]. Sequencing of sncRNAs in different developmental stages, namely 256-cell, sphere, shield and 24 hpf showed that tRFs are expressed at low level during early development, but there is a slight enrichment at the 256 cell stage and at 24 hpf [37]. Our analysis started at 24 hpf, when most tRFs are not expressed and progressed to zebrafish tissues where tRFs are abundantly expressed. Most tRFs detected in previous studies [37] were not identified by us, probably due to the higher sequencing coverage used [26, 37]. Alternatively, those tRFs are present during early development and were missed because we did not analyze the early developmental stages. It is also possible that those tRFs are expressed at exceptionally low levels and were missed by our NGS experiment, as we used a chip that retrieves a maximum of 100,000 reads. Nevertheless, 3′tRF-Pro$^{AGG}$ was present in 256-cell, sphere stages and 24 hpf. 5′tRF-Pro$^{CGG}$ was also present at 24 hpf [35]. Furthermore, analysis of zebrafish sequencing datasets available in GEO showed that these two tRFs are present in endothelial cells of 24 hpf embryos [40].

We have found that the cellular concentration of tRFs is not correlated with the abundance of the corresponding tRNAs, which is an indication that these molecules are not products of random tRNA degradation, as previously shown [10]. However, in some cases, namely 5′tRF-Pro$^{CGG}$, an alternative band of approximately 30 nt was detected in adult ZF tissues, indicating that tRF processing may differ across different tissues or developmental stages and thus, different regulation mechanisms of tRF generation may exist and should be further explored in the future.

Northern blot analysis of the tRFs identified by NGS showed that they are highly expressed in mature tissues, such as bone and skin. One of the most striking results was obtained for 5′tRF-Pro$^{CGG}$ whose higher abundance in the skin and bone/muscle relative to the corresponding mature tRNA may suggest a role in these particular tissues that needs to be further explored. Its in vitro cleavage by Dicer supports this hypothesis as it allows for incorporation into the RNAi pathway [10, 17, 18], but in vivo cleavage should be tested in the future with zebrafish dicer mutants [41]. How Dicer recognizes these tRNAs still needs to be clarified. One possibility is that the pre-tRNA transcripts may have alternative folds with long hairpins that are recognized and cleaved by Dicer [13, 19]. Coincidently, part of the mature tRNAs that originate 5′tRF-Glu$^{CTC}$ and 5′tRF-Pro$^{CGG}$ can form long hairpins recognizable by Dicer. It will be interesting to experimentally validate these folding predictions to clarify the role of tRNA folding in the biogenesis of tRFs. Moreover, the folding of tRNAs into alternative secondary and tertiary structures may also explain the existence of tRNA gene duplications in the genomes of vertebrates and the lack of correlation between the abundance of mature tRNAs and their respective tRFs.

### Zebrafish tRFs are conserved in vertebrates

We have shown by northern blot analysis that 5′tRF-Glu$^{CTC}$ and 5′tRF-Pro$^{CGG}$ are conserved in vertebrates, namely in mouse and in human, which is supported by published studies. Indeed, the 5′tRF-Glu$^{CTC}$ is described in a human fetus hepatic tissue [20], a human monocytic cell line (THP-1) [13] and in adenocarcinomic human alveolar basal epithelial cells (A549) [14]. The tRNA$^{Glu}$ fragment cloned from human fetus hepatic tissue is 3 nt longer than the zebrafish 5′tRF-Glu$^{CTC}$, whereas the 5′tRF-Glu$^{CTC}$ identified in THP-1 cells and A549 is 2 nucleotides and 1 nucleotide shorter, respectively, suggesting that the 3′-end of this fragment is heterogeneous, similarly to miRNAs. This may be due to mismatches between the retrieved sequences and the genomic loci, but the cleavage position may also vary with the cell type or tissue [4, 37] due to differential processing [5]. Besides, our analysis of publically available NGS datasets identified several of the zebrafish tRFs described in our study, namely 5'tRF-Lys$^{TTT}$, 5'tRF-Val$^{CAC}$, 5'tRF-Glu$^{CTC}$, 5'tRF-Pro$^{CGG}$, 3'tRF-Glu$^{CTC}$ and 3'tRF-Pro$^{AGG}$. Taken together, our data confirms the conservation of tRFs in vertebrates, suggests conserved tRF processing pathways and tissue specific expression of the tRFs.

## 5'tRF-Pro$^{CGG}$ has silencing ability

Dicer-dependent tRFs share some features of the RNAi pathway and may represent a class of sncRNA molecules with specific functions, including gene silencing. Silencing has been demonstrated with standard reporter assays for a Dicer-dependent tRF [18] and a recent study showed that a Dicer dependent 3'tRF can repress expression of a set of endogenous genes, including RPA1, by binding to target sites on the mRNA 3'UTR, similarly to miRNAs [17]. Our study shows that 5'tRF-Pro$^{CGG}$ functions in the silencing pathway, as the endogenous 5'tRF-Pro$^{CGG}$ is able to induce silencing of a RFP reporter by binding to complementary target sites present on its 3'UTR. This miRNA-like feature was further confirmed by loss of silencing of a target gene mutated on the seed region of the tRF (between nucleotides 2 and 8). Since this tRF is highly expressed in bone and skin it will be interesting to validate putative gene targets to understand its functional role. We have predicted computationally putative targets for this tRF, as described before [26], by maintaining the seed match between nucleotides 2 and 7, and allowing up to 6 mismatches in the remaining sequence. Some of the predicted target genes are involved in embryonic patterning, cartilage and skeletal development, namely sec23b and myst3, which is consistent with the expression pattern of 5'tRF-Pro$^{CGG}$ (Additional file 4: Table S2).

Endogenous 5'tRF-Glu$^{CTC}$ did not silence efficiently the reporter used in this study, despite its apparent Dicer dependent production in vitro. This could be explained by low levels of endogenous 5'tRF-Glu$^{CTC}$ at 24 hpf, however even at 72 hpf there was no obvious RFP silencing, meaning that this fragment does not have miRNA-like trans-silencing ability. Recently 5'tRF-Glu$^{CTC}$ was identified in A549 cells after respiratory syncytial virus infection and the authors demonstrated that silencing of a reporter occurred through a mechanism distinct from the miRNA/siRNA pathway [14]. The silencing mechanism was not investigated, but other studies also showed that 5'tRFs can inhibit translation by alternative mechanisms. For example, Sobala and colleagues demonstrated that translational inhibition by 5'tRFs is dependent on a conserved dinucleotide (GG) motif present at position 19 and does not need complementary target sites in the mRNA [16]. These authors also found that endogenous 5'tRFs were not able to silence reporters transfected into

cells and that silencing was only observed when a duplex mimic of those tRFs was co-transfected, which is not surprising as the synthetic duplex functions as a siRNA and induces miRNA-like-silencing. General inhibition of translation was achieved when different synthetic 5'tRFs bearing the GG dinucleotide, not in the duplex form, were transfected into cells [16]. A synthetic tRF similar to 5'tRF-Glu$^{CTC}$ was the only 5'tRF that did not exhibit silencing ability, despite the GG motif [16].

## tRFs are differentially expressed during infection and in cancer

5'tRF-Glu$^{CTC}$ is up-regulated after respiratory syncytial virus [14] and cytomegalovirus infection [31] and is down regulated in colon rectal cancer, as shown in the results section. Other tRFs are down regulated in cancer. For example, a 3'tRF (also called CU1276) is down-regulated in B cell lymphoma, similarly to 5'tRF-Val$^{CAC}$, 5'tRF-Lys$^{TTT}$ and 5'tRF-Glu$^{CTC}$ in colorectal cancer [17]. These data show a trend for down regulation of tRFs in cancer and up regulation during infection, but further validation experiments are needed to clarify whether these molecules can be used as disease biomarkers. For example, it would be interesting to analyze the tRF profile in different cancers and under different types of infections and also in different populations and experimentally validate the sequencing data by northern blot and qPCR techniques.

## Conclusions

Our data show that tRFs are conserved in vertebrates and confirms that tRFs are expressed in a cell and tissue specific manner. For example, in zebrafish the identified tRFs are mostly expressed in muscle, bone and skin and less expressed in the brain and during development, which is probably related to its biological function. Moreover, our data shows that 5'tRF-Pro$^{CGG}$ is more expressed in these tissues than its corresponding mature tRNA and that it can play a role in gene expression regulation as it exhibits silencing ability. It is now important to experimentally validate the biological targets of these molecules and determine if those targets are correlated with its high expression. Our data also shows that besides being conserved in vertebrates, tRF expression is affected in specific disease conditions, namely infection and cancer. The differential expression of tRFs in these conditions indicates that tRFs have the potential to be used as disease biomarkers. It would be interesting to analyze the available NGS datasets of human diseases to identify all the tRFs present in specific samples and identify situations where these molecules are deregulated. Their similarity with miRNAs may allow them to recruit the miRNA machinery, but they may have their own machinery or cooperate with other sncRNAs to control important biological processes.

# Methods

## Zebrafish husbandry

Wild type AB zebrafish strain was obtained from the fish facility at the Department of Biology, University of Aveiro and maintained at 28 °C on a 14 h-light/10 h-dark cycle. Zebrafish maintenance followed the Portuguese law for animal experimentation (Regulatory Guideline no 113/2013, 7th August, 2013) and the experiments were approved by the National Food and Veterinary Authority (DGAV) in Portugal and by the committee for animal experimentation and well-being of the Biology Department, University of Aveiro.

## Computational analysis of sequencing reads

Base calling and quality trimming of sequence reads was carried out as described before [42]. For the identification of tRFs, reads that did not match miRNAs were aligned against a small RNA database extracted from Biomart/Ensembl. Up to two mismatches were allowed in alignments to ensure that sequences with sequencing errors or post-transcriptional modifications, which could produce reverse transcription errors during cDNA library construction, were not discarded. tRFs identified were then aligned against their mature tRNA to verify if the sequences were not randomly distributed through the mature sequence.

## RNA analysis

Total RNA was extracted from zebrafish embryos, adult tissues and cultured cells with TRIZOL® (Invitrogen). Twenty micrograms of total RNA was fractionated on 10 % denaturing polyacrylamide (PAA) gels. RNA was then transferred during 30 min to Hybond-N membranes (GE Healthcare) using a semidry transfer system and was UV-crosslinked. Antisense oligonucleotides complementary to predicted tRF candidates were radio labeled with $[^{32}P]$-ATP and T4 polynucleotide kinase (Takara) and were used as hybridization probes. Membranes were pre-hybridized for 4 h in pre-hybridization/hybridization buffer containing 5× Denhardt's solution, 1 % SDS and 6.6× SSPE at 64 °C (5′tRF-Lys$^{TTT}$ and 5′tRF-Glu$^{CTC}$), 40 °C (5′tRF-Pro$^{CGG}$), 57 °C (3′tRF-Pro$^{AGG}$ and 3′tRF-Ala$^{AGC}$ probe) and 56 °C (U6). Membranes were stripped twice and probed with a maximum of three probes. For the stripping, membranes were incubated with a solution of 50 % formamide and 2xSSPE at 65 °C for 1 h, followed by 15 min incubation with TE buffer to neutralize the membrane. $[^{32}P]$-ATP labeled probes were added to the hybridization chamber and incubated with the membranes overnight at the mentioned temperatures. Membranes were then washed twice with washing solution, containing 2xSSPE and 0.1 % SDS at room temperature and twice with washing solution at 57 °C for 3 min, and were exposed to a phosphor screen (Biorad®) overnight and scanned using a Molecular Imager® FX (Biorad), equipped with Quantity One FX software.

Probes:

5′tRF-Lys$^{TTT}$:  5′-CTGATGCTCTACCGACTGAGCTATCCGGGC-3′
5′tRF-Glu$^{CTC}$:  5′-GCCGAATCCTAACCACTAGACCACCAGGGA-3′
5′tRF-Pro$^{CGG}$: 5′-GCGAGAATCATACCCCTA-3′
3′tRF-Pro$^{AGG}$: 5′-TGGGGGCTCGTCCGGGA-3′
3′tRF-Ala$^{AGC}$: 5′-GGGCATCGATCCCGCTACCTCT-3′
U6:5′-AATATGGAACGCTTCACGAATTTGCGTGTC-3′

## Angiogenin and dicer in vitro assays

For the Angiogenin cleavage assay, 20 μg of total RNA extracted from 72 h post fertilization (hpf) zebrafish embryos were incubated with 1 μM of the enzyme in PBS + 0.1 % BSA for 10 min, 30 min, 1 and 2 h at 37 °C. The cleaved products were recovered using phenol/chlorophorm extraction followed by ethanol precipitation. Approximately 20 μg of each sample were fractionated on 10 % PAA gels and transferred onto Hybond-N membranes for northern blot analysis as described previously.

For the Dicer assay, recombinant human Dicer enzyme kit from Genlantis was used according to the manufacturer's instructions with minor changes. Briefly, 20 μg of total RNA extracted from 72 hpf zebrafish embryos were incubated with 2 U of recombinant Dicer in 20 μL reactions, for 30 min, 1, 2, 4 and 6 h. Reactions were stopped with Dicer Stop Solution and were electrophoresed on 10 % PAA gels and transferred onto Hybond-N membranes for northern blot analysis, as described previously.

## Dual reporter assay

To test tRFs silencing ability, a dual fluorescence reporter system (DFRS)—pDSV2-eGFP-mRFP was used. This reporter bears two fluorescent proteins, GFP and RFP, controlled by the same promoter. GFP identifies the tissues expressing the plasmid and the RFP, which contains a 3′UTR cassette complementary to the tRF of interest, functions as a silencing sensor. DFRS plasmids were obtained from DR. Wieland Huttner [43] and slightly modified. Briefly, primers containing multiple cloning sites (XhoI-SalI_NheI), as described in [44] were annealed and inserted into ECORI-NotI sites of the DFRS plasmid. Next, two different DFRS plasmids containing complementary sites for 5′tRF-Glu$^{CTC}$ and 5′tRF-Pro$^{CGG}$ and a control DFRS were generated. DFRS-5′tRF-Glu$^{CTC}$, DFRS-5′tRF-Pro$^{CGG}$ and DFRS-control were each cloned into XhoI-NheI site by annealing nucleotides A–B, C–D and E–F, respectively.

To study the 5′tRF-Pro$^{CGG}$ functional domains DFRS-5′tRF-Pro$^{CGG}$ bearing mutations on the 5′ and on the 3′ end were generated. 5′ end mutations of DFRS-5′tRF-Pro$^{CGG}$

were obtained after annealing primers G–H. 3′ end mutations were obtained after annealing primers I–J. Mutated sequences were cloned into *XhoI-NheI* site.

Primers (written in direction 5′–3′):

A-TCGAGGCCGAATCCTAACCACTAGACCACC
AGGGAGGATATCCGCCGAATCCTAACCACTAG
ACCACCAGGGAT
B-CTAGATCCCTGGTGGTCTAGTGGTTAG
GATTCGGCGGATATCCTCCCTGGTGGTCTAGTG
GTTAGGATTCGGCC
C-TCGAGGCGAGAATCATACCCCTAGGATATC
CGCGAGAATCATACCCCTAT
D-CTAGATAGGGGTATGATTCTCGCGGATATC
CTAGGGGTATGATTCTCGCC
E-TCGAGTGACGTTCGAACTTACATAACTGGA
TATCCTGACGTTCGAACTTACATAACTT
F-CTAGAAGTTATGTAAGTTCGAACGTCAGGA
TATCCAGTTATGTAAGTTCGAACGTCAC
G-TCGAGGCGAGAATCATATTCACAGGATATC
CGCGAGAATCATATTCACAT
H-CTAGATGTGAATATGATTCTCGCGGATATC
CTGTGAATATGATTCTCGCC
I-TCGAGGTAAAGATCATACCCCTAGGATATCCG
TAAAGATCATACCCCTAT
J-CTAGATAGGGGTATGATCTTTACGGATATC
CTAGGGGTATGATCTTTACC

## Zebrafish embryos microinjection

One cell zebrafish eggs were microinjected with ~1000 pL of a solution containing 20 ng/μL of the reporter plasmid, phenol red and 0.9 M KCl. Embryos were kept at 28 °C and analyzed at 24 and 72 hpf under an epifluorescence microscope Imager.Z1 (Zeiss), a GFP and mRFP filter, AxioCam HRm camera (Zeiss) and AxioVision software (Zeiss). Approximately 50 embryos were used per replica and per condition. Three biological replicates were performed.

## Target prediction

The 3′UTR sequences of zebrafish mRNAs were extracted from Biomart (http://www.biomart.org) and blasted against the antisense 5′tRF-Pro$^{CGG}$ sequence. Sequences with perfect seed match between nucleotides 2 and 7, and no more than six mismatches in the remaining sequence were retained for further analysis. Targets were considered positive whenever RNAhybrid confirmed them thermodynamically. Targets were discarded when RNAhybrid did not retrieve the targets obtained in the first approach.

## Computational analysis of publicly available sequencing datasets

Publicly available NGS sequencing datasets deposited in GEO were analyzed to identify the tRFs uncovered in zebrafish. Briefly, datasets were downloaded from GEO and a blast search for each tRF was performed. Lists of identified tRFs and respective number of reads were generated for each sample. Only sequences with more than 50 reads were kept for further analysis. All data used corresponded to the processed sequence along with the cloning frequency that was available in all datasets used. Normalized data against the total number of reads corresponding to sncRNAs for each sample was used to compare the relative levels of tRFs in different samples in the same experiment (sample vs control).

## Additional files

**Additional file 1: Figure S1.** Sequencing data analysis pipeline. Pipeline describing the protocol used for identifying miRNAs and other small noncoding RNAs present in pyrosequencing datasets. All reads that were not considered as being miRNAs were blasted against other small RNAs and were identified. Potential tRFs were catalogued.

**Additional file 2: Figure S2.** Sensor plasmid at 72 hpf. At 72 hpf GFP and RFP expression is equivalent after the injection of the control reporter (A, B, C). Silencing by endogenous 5′tRF-Pro$^{CGG}$ is still present at 72hpf (D, E, F), as RFP expression is repressed. Endogenous 5′tRF-Glu$^{CTC}$ does not efficiently silence putative targets even at 72hpf, as no alterations in RFP expression is observed when compared to GFP expression (G, H, I). Arrows indicate cells containing both GFP-reporter and mRFP-sensor. Asterisks indicate muscle fibers that lost mRFP fluorescence. Orientation of embryos: caudal, left; ventral, up. 20× magnification.

**Additional file 3: Table S1.** Sequencing data relative to zebrafish tRFs found in publicly available GEO datasets. For each experiment the sequences of conserved tRFs identified and corresponding number of reads are displayed. ID corresponds to the tRF identification; "Query" corresponds to the zebrafish tRF that was used as a template to search for similar molecules in the datasets; "Sequence" corresponds to the sequence found in the datasets that is related to the "query"; "Count" gives the total number of reads retrieved for each "sequence"; "Source tag" is the GEO dataset sample ID number; "description" depicts the type of sample where the tRF was found.

**Additional file 4: Table S2.** Target predictions for 5′tRF-Pro$^{CGG}$. Seed match between nucleotides 2 and 7 were maintained, and up to 6 mismatches in the remaining sequence were allowed. Two of the target genes predicted play roles in embryonic patterning, cartilage and skeletal development, namely Sec23b and Myst3, which correlates with the expression pattern of 5′tRF-Pro$^{CGG}$ (high expression in bone).

## Authors' contributions

ARS and MASS conceived and design the study. ARS, NF and MR performed the experiments. ARS, HRA, JLO and GM performed the computational analysis of NGS datasets. ARS wrote the manuscript. All authors read and approved the final manuscript.

## Author details

[1] Department of Medical Sciences and Institute for Biomedicine–iBiMED, University of Aveiro, 3810-193 Aveiro, Portugal. [2] IEETA, University of Aveiro, 3810-193 Aveiro, Portugal.

## Acknowledgements
The authors are most grateful to the Portuguese Foundation for Science and Technology (FCT) for funding our work through project COMPETE/FEDER FCT-ANR/IMI-MIC/0041/2012 and Grant SFRH/BPD/77528/2011 to A.R.S. The funders had no role in study design, data collection and analysis, decision to publish, or preparation of the manuscript.

## Competing interests
The authors declare that they have no competing interests.

## References

1. Kim VN, Han J, Siomi MC. Biogenesis of small RNAs in animals. Nat Rev Mol Cell Biol. 2009;10(2):126–39.
2. Bartel DP. MicroRNAs: genomics, biogenesis, mechanism, and function. Cell. 2004;116(2):281–97.
3. Burnside J, Ouyang M, Anderson A, Bernberg E, Lu C, Meyers BC, Green PJ, Markis M, Isaacs G, Huang E et al. Deep sequencing of chicken microR-NAs. Bmc Genom. 2008;9:185.
4. Landgraf P, Rusu M, Sheridan R, Sewer A, Iovino N, Aravin A, Pfeffer S, Rice A, Kamphorst AO, Landthaler M, et al. A mammalian micro-RNA expression atlas based on small RNA library sequencing. Cell. 2007;129(7):1401–14.
5. Morin RD, O'Connor MD, Griffith M, Kuchenbauer F, Delaney A, Prabhu AL, Zhao Y, McDonald H, Zeng T, Hirst M, et al. Application of massively parallel sequencing to microRNA profiling and discovery in human embryonic stem cells. Genome Res. 2008;18(4):610–21.
6. Sunkar R, Zhou XF, Zheng Y, Zhang WX, Zhu JK. Identification of novel and candidate miRNAs in rice by high throughput sequencing. Bmc Plant Biol. 2008;8:25.
7. Houwing S, Kamminga LM, Berezikov E, Cronembold D, Girard A, van den Elst H, Filippov DV, Blaser H, Raz E, Moens CB, et al. A role for Piwi and piRNAs in germ cell maintenance and transposon silencing in Zebrafish. Cell. 2007;129(1):69–82.
8. Ruby JG, Jan C, Player C, Axtell MJ, Lee W, Nusbaum C, Ge H, Bartel DP. Large-scale sequencing reveals 21U-RNAs and additional microRNAs and endogenous siRNAs in C. elegans. Cell. 2006;127(6):1193–207.
9. Aravin AA, Lagos-Quintana M, Yalcin A, Zavolan M, Marks D, Snyder B, Gaasterland T, Meyer J, Tuschl T. The small RNA profile during Drosophila melanogaster development. Dev Cell. 2003;5(2):337–50.
10. Cole C, Sobala A, Lu C, Thatcher SR, Bowman A, Brown JW, Green PJ, Barton GJ, Hutvagner G. Filtering of deep sequencing data reveals the existence of abundant Dicer-dependent small RNAs derived from tRNAs. RNA. 2009;15(12):2147–60.
11. Lee YS, Shibata Y, Malhotra A, Dutta A. A novel class of small RNAs: tRNA-derived RNA fragments (tRFs). Genes Dev. 2009;23(22):2639–49.
12. Kawaji H, Nakamura M, Takahashi Y, Sandelin A, Katayama S, Fukuda S, Daub CO, Kai C, Kawai J, Yasuda J, et al. Hidden layers of human small RNAs. BMC Genom. 2008;9:157.
13. Burroughs AM, Ando Y, Hoon ML, Tomaru Y, Suzuki H, Hayashizaki Y, Daub CO. Deep-sequencing of human Argonaute-associated small RNAs provides insight into miRNA sorting and reveals Argonaute association with RNA fragments of diverse origin. RNA Biol. 2011;8(1):158–77.
14. Wang Q, Lee I, Ren J, Ajay SS, Lee YS, Bao X. Identification and functional characterization of tRNA-derived RNA fragments (tRFs) in respiratory syncytial virus infection. Mol Ther J Am Soc Gene Ther. 2013;21(2):368–79.
15. Keam SP, Young PE, McCorkindale AL, Dang TH, Clancy JL, Humphreys DT, Preiss T, Hutvagner G, Martin DI, Cropley JE, et al. The human Piwi protein Hiwi2 associates with tRNA-derived piRNAs in somatic cells. Nucleic Acids Res. 2014;42(14):8984–95.
16. Sobala A, Hutvagner G. Small RNAs derived from the 5' end of tRNA can inhibit protein translation in human cells. RNA Biol. 2013;10(4):553–63.
17. Maute RL, Schneider C, Sumazin P, Holmes A, Califano A, Basso K, Dalla-Favera R. tRNA-derived microRNA modulates proliferation and the DNA damage response and is down-regulated in B cell lymphoma. Proc Natl Acad Sci USA. 2013;110(4):1404–9.
18. Haussecker D, Huang Y, Lau A, Parameswaran P, Fire AZ, Kay MA. Human tRNA-derived small RNAs in the global regulation of RNA silencing. RNA. 2010;16(4):673–95.
19. Babiarz JE, Ruby JG, Wang Y, Bartel DP, Blelloch R. Mouse ES cells express endogenous shRNAs, siRNAs, and other Microprocessor-independent, Dicer-dependent small RNAs. Genes Dev. 2008;22(20):2773–85.
20. Fu H, Feng J, Liu Q, Sun F, Tie Y, Zhu J, Xing R, Sun Z, Zheng X. Stress induces tRNA cleavage by angiogenin in mammalian cells. FEBS Lett. 2009;583(2):437–42.
21. Thompson DM, Parker R. The RNase Rny1p cleaves tRNAs and promotes cell death during oxidative stress in Saccharomyces cerevisiae. J Cell Biol. 2009;185(1):43–50.
22. Lee SR, Collins K. Starvation-induced cleavage of the tRNA anticodon loop in Tetrahymena thermophila. J Biol Chem. 2005;280(52):42744–9.
23. Yamasaki S, Ivanov P, Hu GF, Anderson P. Angiogenin cleaves tRNA and promotes stress-induced translational repression. J Cell Biol. 2009;185(1):35–42.
24. Emara MM, Ivanov P, Hickman T, Dawra N, Tisdale S, Kedersha N, Hu GF, Anderson P. Angiogenin-induced tRNA-derived stress-induced RNAs promote stress-induced stress granule assembly. J Biol Chem. 2010;285(14):10959–68.
25. Ivanov P, Emara MM, Villen J, Gygi SP, Anderson P. Angiogenin-induced tRNA fragments inhibit translation initiation. Mol Cell. 2011;43(4):613–23.
26. Soares AR, Pereira PM, Santos B, Egas C, Gomes AC, Arrais J, Oliveira JL, Moura GR, Santos MAS. Parallel DNA pyrosequencing unveils new zebrafish microRNAs. Bmc Genom. 2009;10:195.
27. Li Z, Ender C, Meister G, Moore PS, Chang Y, John B. Extensive terminal and asymmetric processing of small RNAs from rRNAs, snoRNAs, snRNAs, and tRNAs. Nucleic Acids Res. 2012;40(14):6787–99.
28. Steffen P, Voss B, Rehmsmeier M, Reeder J, Giegerich R. RNAshapes: an integrated RNA analysis package based on abstract shapes. Bioinformatics. 2006;22(4):500–3.
29. Chan PP, Lowe TM. GtRNAdb: a database of transfer RNA genes detected in genomic sequence. Nucleic Acids Res. 2009;37(Database issue):D93–97.
30. Noren Hooten N, Fitzpatrick M, Wood WH, 3rd, De S, Ejiogu N, Zhang Y, Mattison JA, Becker KG, Zonderman AB, Evans MK. Age-related changes in microRNA levels in serum. Aging. 2013;5(10):725–40.
31. Stark TJ, Arnold JD, Spector DH, Yeo GW. High-resolution profiling and analysis of viral and host small RNAs during human cytomegalovirus infection. J Virol. 2012;86(1):226–35.
32. Farazi TA, Horlings HM, Ten Hoeve JJ, Mihailovic A, Halfwerk H, Morozov P, Brown M, Hafner M, Reyal F, van Kouwenhove M, et al. MicroRNA sequence and expression analysis in breast tumors by deep sequencing. Cancer Res. 2011;71(13):4443–53.
33. Liao JY, Ma LM, Guo YH, Zhang YC, Zhou H, Shao P, Chen YQ, Qu LH. Deep sequencing of human nuclear and cytoplasmic small RNAs reveals an unexpectedly complex subcellular distribution of miRNAs and tRNA 3' trailers. PLoS One. 2010;5(5):e10563.
34. Phizicky EM, Hopper AK. tRNA biology charges to the front. Genes Dev. 2010;24(17):1832–60.
35. Jeffries CD, Fried HM, Perkins DO. Nuclear and cytoplasmic localization of neural stem cell microRNAs. RNA. 2011;17(4):675–86.
36. Huang V, Li LC. miRNA goes nuclear. RNA Biol. 2012;9(3):269–73.
37. Wei C, Salichos L, Wittgrove CM, Rokas A, Patton JG. Transcriptome-wide analysis of small RNA expression in early zebrafish development. RNA. 2012;18(5):915–29.
38. Gong B, Lee YS, Lee I, Shelite TR, Kunkeaw N, Xu G, Lee K, Jeon SH, Johnson BH, Chang Q, et al. Compartmentalized, functional role of angiogenin during spotted fever group rickettsia-induced endothelial barrier dysfunction: evidence of possible mediation by host tRNA-derived small noncoding RNAs. BMC Infect Dis. 2013;13:285.
39. Thompson DM, Lu C, Green PJ, Parker R. tRNA cleavage is a conserved response to oxidative stress in eukaryotes. RNA. 2008;14(10):2095–103.
40. Nicoli S, Knyphausen CP, Zhu LJ, Lakshmanan A, Lawson ND. miR-221 is required for endothelial tip cell behaviors during vascular development. Dev Cell. 2012;22(2):418–29.
41. Giraldez AJ, Cinalli RM, Glasner ME, Enright AJ, Thomson JM, Baskerville S, Hammond SM, Bartel DP, Schier AF. MicroRNAs regulate brain morphogenesis in zebrafish. Science. 2005;308(5723):833–8.

42. Soares AR, Pereira PM, Santos B, Egas C, Gomes AC, Arrais J, Oliveira JL, Moura GR, Santos MA. Parallel DNA pyrosequencing unveils new zebrafish microRNAs. BMC Genom. 2009;10:195.

43. De Pietri Tonelli D, Calegari F, Fei JF, Nomura T, Osumi N, Heisenberg CP, Huttner WB. Single-cell detection of microRNAs in developing vertebrate embryos after acute administration of a dual-fluorescence reporter/sensor plasmid. Biotechniques. 2006;41(6):727–32.

44. Mishima Y, Abreu-Goodger C, Staton AA, Stahlhut C, Shou C, Cheng C, Gerstein M, Enright AJ, Giraldez AJ. Zebrafish miR-1 and miR-133 shape muscle gene expression and regulate sarcomeric actin organization. Genes Dev. 2009;23(5):619–32.

45. Edgar R, Domrachev M, Lash AE. Gene expression omnibus: NCBI gene expression and hybridization array data repository. Nucleic Acids Res. 2002;30(1):207–10.

2

# Properties of STAT1 and IRF1 enhancers and the influence of SNPs

Mohamed Abou El Hassan[1,2,3†], Katherine Huang[1†], Manoja B. K. Eswara[1], Zhaodong Xu[1], Tao Yu[1], Arthur Aubry[1], Zuyao Ni[1,6], Izzy Livne-bar[1], Monika Sangwan[1], Mohamad Ahmad[1] and Rod Bremner[1,4,5*]

## Abstract

**Background:** STAT1 and IRF1 collaborate to induce interferon-γ (IFNγ) stimulated genes (ISGs), but the extent to which they act alone or together is unclear. The effect of single nucleotide polymorphisms (SNPs) on in vivo binding is also largely unknown.

**Results:** We show that IRF1 binds at proximal or distant ISG sites twice as often as STAT1, increasing to sixfold at the MHC class I locus. STAT1 almost always bound with IRF1, while most IRF1 binding events were isolated. Dual binding sites at remote or proximal enhancers distinguished ISGs that were responsive to IFNγ versus cell-specific resistant ISGs, which showed fewer and mainly single binding events. Surprisingly, inducibility in one cell type predicted ISG-responsiveness in other cells. Several dbSNPs overlapped with STAT1 and IRF1 binding motifs, and we developed methodology to rapidly assess their effects. We show that in silico prediction of SNP effects accurately reflects altered binding both in vitro and in vivo.

**Conclusions:** These data reveal broad cooperation between STAT1 and IRF1, explain cell type specific differences in ISG-responsiveness, and identify genetic variants that may participate in the pathogenesis of immune disorders.

## Background

IFNγ is a pleiotropic cytokine that plays essential roles in antiviral and anticancer immune responses (reviewed in [1, 2]). IFNγ binds to its receptor complex and activates receptor-associated JAK kinases, which phosphorylate a substantial fraction of cytoplasmic signal transducer and activator of transcription 1 (STAT1). Phosphorylated STAT1 forms homodimers that translocate to the nucleus and bind IFNγ activation sites (GAS). STAT1 recruits histone acetyltransferases (HATs) and other transcriptional co-activators to acetylate chromatin and facilitate transcription. Genomic studies showed that STAT1 binds at promoter proximal and distal sites, suggesting a role in remote gene regulation [3–6]. Indeed, IFNγ induces long range interactions between STAT1-bound enhancers and target promoters [7–9].

Interferon regulatory factor 1 (*IRF1*) is a primary target gene of STAT1. Like STAT1, IRF1 also acts as a transcription factor (TF), binding to IRF-E motifs and interferon-stimulated response elements (ISRE) [10, 11]. Access of both STAT1 and IRF1 to target enhancers requires the SWI/SNF chromatin remodeling complex to counter PRC2, which uses the histone methyl transferase EZH2 to deposit H3K27me3 and block the induction of many other cytokine and cytokine responsive loci [7, 12, 13]. IRF1 functions at the transcription initiation level by facilitating RNA Pol II recruitment to ISGs promoters [14, 15]. IRF1 also binds to remote enhancers of the *CIITA* locus that loop together to form a 3D interconnected hub with the promoter [7]. Indeed, ChIP-chip and ChIP-seq studies show that IRF1 binds many remote enhancers [6, 16–18], and analysis of 128 transcription factors in K562 cells revealed that STAT1-IRF1 co-binding is a recurring pattern in IFNγ treated cells [19]. Notably, STAT1 is essential but not sufficient for gene induction [11], and both STAT1 and IRF1 are required for the IFNγ-induced expression of *CIITA*, *GBP1*, and gp19 [14, 15, 20]. In addition, STAT1 complexes with

*Correspondence: bremner@lunenfeld.ca
†Mohamed Abou El Hassan and Katherine Huang contributed equally to this work
[1] Lunenfeld Tanenbaum Research Institute, Mt Sinai Hospital, Toronto, ON, Canada
Full list of author information is available at the end of the article

IRF1 at the *LMP2* promoter and maintains its constitutive expression [21].

Here, we studied the extent of STAT1 and IRF1 cooperation in HeLa cells within ISG-rich chromosomal segments encompassing ~10% of all known ISGs. Most of these loci responded to IFNγ in HeLa cells, leaving ~20% resistant ISGs. IRF1 binding sites outnumbered STAT1 sites 2 to 1. A large fraction of STAT1/IRF1 binding occurred at remote sites and looping studies confirmed the functional role of putative enhancers at the *SOCS1* locus. Most STAT1 binding occurred at or near to IRF1 sites (dual binding), but IRF1 often bound isolated from STAT1. Dual STAT1 and IRF1 but not isolated IRF1 or STAT1 binding was linked to ISG responsiveness. Finally, several variants affecting STAT1/IRF1 motifs induce or impair binding.

## Results

### Diverse gene responses to IFNγ

To define patterns of TF binding around ISGs, we employed tiling arrays to focus on 16 Mb distributed across 11 distinct chromosomal segments with a high density of ISGs (Fig. 1a, Additional file 1: Table S1). Nine segments were 1 Mb genomic regions on six chromosomes centered on specific IFNγ target genes (e.g. 1 Mb around *IRF1* etc.). Two others included a 2 Mb segment centered on *CIITA*, and a 5 Mb segment covering the complete classical 3.6 Mb MHC locus and an additional 1.4 Mb 5′ region including much of the so-called extended MHC class I region (Fig. 1a; Additional file 1: Table S1). Within these regions 25% (95/375) of the genes are known ISGs, ~5× more than the genome-wide ISG frequency (1167/24996 ISGs) and ~15-fold above the average ISGs density per Mb. The total number of Refseq genes and UCSC Known Genes in the 16 Mb regions is 394 (Additional file 1: Table S1). Of these, 95% (375) were represented on the Illumina-12 Human WG-6v3 array used to assess gene expression (see below). The frequency of Refseq genes across the genome is ~6/Mb, but most of the 11 chromosomal segments in our study were gene dense (average 24/Mb), especially at the MHC (35/Mb), PSME (39/Mb) and IFITM clusters (45/Mb) (Additional file 1: Table S1). There are also 126 pseudogenes across the 16 Mb, with most (93) located at the MHC cluster (Additional file 1: Table S1). Pseudogenes are not represented on the Illumina genome wide array we used to study expression.

Signaling pathway target loci show cell-specific responsiveness, but the exact TF binding patterns that distinguish induction versus resistance in a specific cell type are unclear. Thus, we compared the pattern of STAT1 and IRF1 binding at different gene types. For this we compiled a database of ~ all known ISGs using our own and

prior transcriptome data (Additional file 1: Table S2). As summarized in Fig. 1b, ISGs fell into 8 classes depending on whether IFNγ caused induction, no effect (resistant ISGs in HeLa cells), or repression, and whether induction/repression were early (detected at 6 h) or late (24 or 48 h), and strong (differential score ≥13, and ≥twofold change) or weak (differential score ≥13, <twofold). The microarray expression data was validated using reverse transcription and quantitative PCR (RT-qPCR), which confirmed 83% (20/24) of indISGs and 87% (21/24) of resISGs (Fig. 1c). Of all 95 known ISGs on the array, 31 (33%) genes were es-indISGs, 15 (16%) were ls-indISGs, 29 (31%) were ew-indISGs or lw-indISGs, and 20 were resISGs (Fig. 1b). Es-indISGs were distributed at an average density of 1.9/Mb within the studied regions (Additional file 1: Table S1). The highest density was observed at the IFIT and GBP clusters with an average of 4.0 es-indISGs/Mb.

No genes were repressed (IFNγ repressed genes, IRGs) at the early 6 h time point, while 19 and 23 were strongly or weakly repressed at later times, respectively (ls-IRGs and lw-IRGs; Fig. 1b), suggesting indirect regulation of IRGs (perhaps through activation of a repressor). The remaining genes that were not IFNγ-responsive either in this or any prior study were termed "Other Genes". In summary, known ISGs fall into induced and resistant subclasses in HeLa cells, providing a useful system to define STAT1 and IRF1 binding patterns linked to responsiveness.

### Validation of STAT1 and IRF1 ChIP-chip analyses

ChIP-chip was used to locate STAT1 and IRF1 sites at promoter proximal and distal sites of the genes of each category. STAT1 and IRF1 ChIPs were performed on chromatin from HeLa cells that were either untreated or exposed to IFNγ for 6 h. Hybridization intensities were normalized to internal standards and values from quadruplicate spots were averaged. Significantly different intensities between ChIP DNA and input DNA samples in three biological replicates (p < 0.0001) were determined using the Wilcoxon rank sum test. Peaks representing the significantly enriched DNA regions (p < 0.0001) where the ratio of ChIP to input DNA was ≥ 1.5-fold were visualized in the UCSC browser and plotted on a log2 scale. Only 2 STAT1 and 28 IRF1 peaks were identified in untreated cells, rising to 92 and 196 post-IFNγ treatment, respectively. Browser views are shown in Additional file 2: Figure S1 and can be visualized at http://research.lunenfeld.ca/IFNγ. ChIP-qPCR validated 91% (20/22) and 96% (23/24) of STAT1 and IRF1 ChIP-chip peaks, respectively (Fig. 2). We compared STAT1 binding at 6 h (this study) with IFNγ-induced STAT1 binding after 30 min [22], also assessed in HeLa

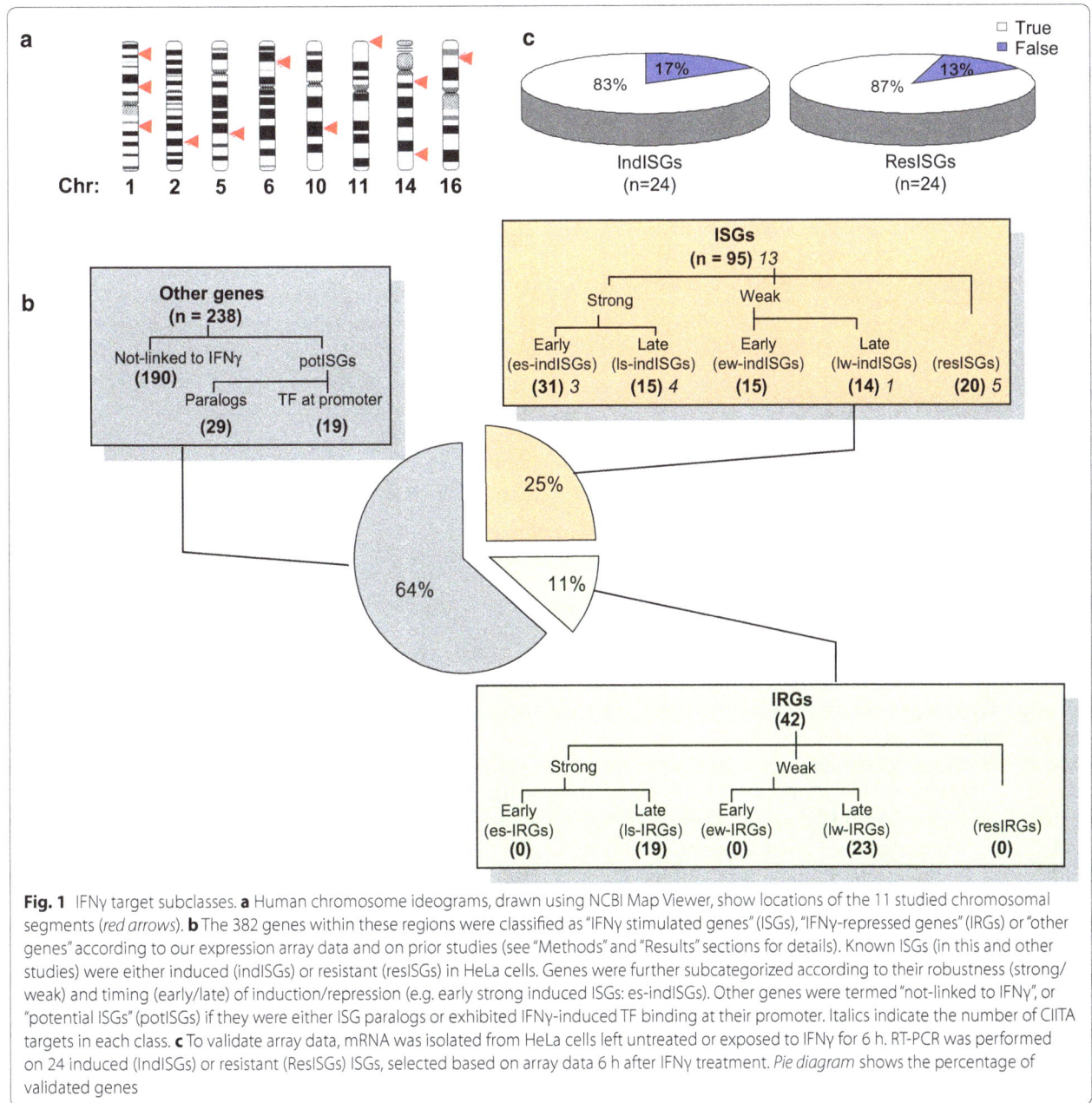

**Fig. 1** IFNγ target subclasses. **a** Human chromosome ideograms, drawn using NCBI Map Viewer, show locations of the 11 studied chromosomal segments (*red arrows*). **b** The 382 genes within these regions were classified as "IFNγ stimulated genes" (ISGs), "IFNγ-repressed genes" (IRGs) or "other genes" according to our expression array data and on prior studies (see "Methods" and "Results" sections for details). Known ISGs (in this and other studies) were either induced (indISGs) or resistant (resISGs) in HeLa cells. Genes were further subcategorized according to their robustness (strong/weak) and timing (early/late) of induction/repression (e.g. early strong induced ISGs: es-indISGs). Other genes were termed "not-linked to IFNγ", or "potential ISGs" (potISGs) if they were either ISG paralogs or exhibited IFNγ-induced TF binding at their promoter. Italics indicate the number of CIITA targets in each class. **c** To validate array data, mRNA was isolated from HeLa cells left untreated or exposed to IFNγ for 6 h. RT-PCR was performed on 24 induced (IndISGs) or resistant (ResISGs) ISGs, selected based on array data 6 h after IFNγ treatment. *Pie diagram* shows the percentage of validated genes

cells. In the 16 Mb of DNA assessed here, the latter study detected 26 STAT1 sites, of which 21 overlapped with the 92 STAT1 sites we detected.

### Basal TF binding

Unphosphorylated STAT1 has roles in regulating ISGs days after IFN treatment [23, 24], but its role in untreated cells is less clear, although STAT1 nuclear cytoplasmic shuttling occurs even in untreated cells [25–27]. Basal STAT1 binding is linked to the nuclear localization of unphosphorylated STAT1 and contributes to the constitutive expression of some targets [21, 28]. IRF1 is also expressed to low levels in unstimulated HeLa cells [7] and it cooperates with STAT1 to maintain low basal expression levels of LMP2 [21]. In addition, there is also some STAT1 phosphorylation (below detectable levels) in untreated cells that contributes to basal activity, as shown elegantly by knockin studies in mice [29]. We detected 2 STAT1 and 28 IRF1 binding sites in untreated cells, accounting for 2.2 and 14.3% of induced sites, respectively. Our data accords with another ChIP-chip analysis of STAT1 binding which reported that 6.5% of IFNγ-induced STAT1 sites in HeLa cells treated for 30 min with IFNγ (as opposed to 6 h in our case) are occupied in uninduced cells [22].

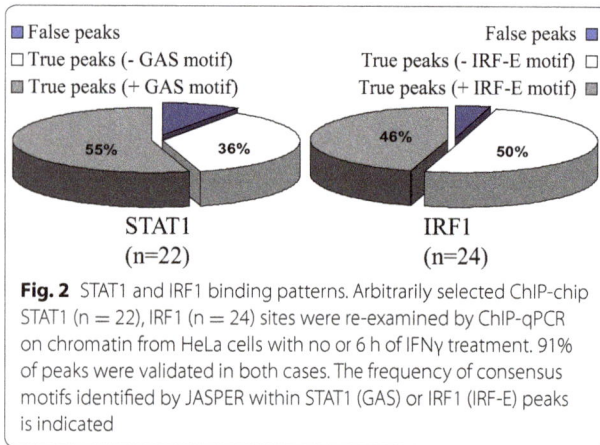

■ False peaks
□ True peaks (- GAS motif)
■ True peaks (+ GAS motif)

False peaks ■
True peaks (- IRF-E motif) □
True peaks (+ IRF-E motif) ■

STAT1
(n=22)

IRF1
(n=24)

**Fig. 2** STAT1 and IRF1 binding patterns. Arbitrarily selected ChIP-chip STAT1 (n = 22), IRF1 (n = 24) sites were re-examined by ChIP-qPCR on chromatin from HeLa cells with no or 6 h of IFNγ treatment. 91% of peaks were validated in both cases. The frequency of consensus motifs identified by JASPER within STAT1 (GAS) or IRF1 (IRF-E) peaks is indicated

Further analysis suggests that basal TF binding detected here is physiologically relevant. Of the genes with basal STAT1 or IRF1 binding, we assessed 21 by microarray and/or RT-PCR and all were expressed in untreated cells (Additional file 1: Table S3). In contrast, of 26 randomly selected ISGs that lacked basal TF binding, only 13 were basally expressed. Indeed, constitutive expression of PSMB9 and TAP2 requires constitutive IRF1 binding [21, 30]. In addition, several loci with basal TF binding are in paralogous gene clusters suggesting conservation of high affinity binding sites during gene duplication (e.g. PSMB8 and PSMB9, GBP2 and GBP3, and IFIT1, IFIT2 and IFIT3). A high fraction (82%) of the 28 IRF1 basally occupied sites possessed IRF1 binding motifs. Thus, our data supports the notion that basal binding of STAT1 and IRF1 is physiologically relevant.

### Remote IFNγ activated enhancers are common at ISGs

In IFNγ treated cells, 54% (50/92) of STAT1 and 44% (87/196) of IRF1 peaks were within 5 kb of the transcription start site (TSS) of all 394 Known Genes on the array (Fig. 3a; Additional file 1: Table S4). Adding other

databases, including predicted genes, raised the fraction to 64% for STAT1 and 57% for IRF1 (Additional file 2: Figure S2A). Of equal numbers of randomly generated sites, the proportion at <5 kb from gene starts was much lower (Fig. 3a; Additional file 2: Figure S2A). Thus STAT1 binding is slightly skewed to promoter proximal sites, while IRF1 binding is slightly biased toward remote sites (Fig. 3b).

Prior analysis of chromatin modification and looping at *CIITA*, and partial looping analysis at *1* locus support the idea that remote sites are functionally important [7, 8]. To further test this notion we performed additional assessment of *SOCS1*, a key negative regulator of IFNγ signaling that is responsive to IFNs and other immune signaling pathways [31, 32]. *SOCS1* responds to IFNγ in HeLa cells (Fig. 4a). ChIP-chip data exposed 6 IFNγ-induced STAT1 and/or IRF1 peaks ± 100 kb of the *SOCS1* TSS (Fig. 4b). ChIP-qPCR analysis verified binding at the *SOCS1* promoter (*pSOCS1; −0.1 kb*), and at +50, +15, −3, −55, −68 and −72 kb (Fig. 4c). A negative region at −63 kb was also validated. In favor of functional relevance of proximal and remote sites, we detected constitutive histone H3 acetylation (H3ac) and/or H4ac at pSOCS1 (−0.1 kb), −3 and −55 kb (Fig. 4c), and IFNγ induced acetylation at pSOCS1 and the 6 remote sites but not at the irrelevant −63 kb site (Fig. 4c). These constitutive and inducible events paralleled recruitment of the HATs CBP and/or p300 (Fig. 4c). H3K4me2 also marks enhancers [33], and constitutive H3K4me2 was detected at *pSOCS1*, −3 kb, −55 kb, matching constitute histone acetylation, and also at the −72 kb enhancer (Fig. 4c), which contacts the promoter (see below). IFNγ treatment did not further increase methylation at these sites, but did induce H3K4me2 at +50 kb, +15 kb and −72 kb (Fig. 4c). Finally, we detected constitutive Pol II recruitment at *pSOCS1* and −55 kb but not at the other TF binding or negative control sites (Fig. 4c). After IFNγ treatment, Pol II recruitment increased at *pSOCS1*,

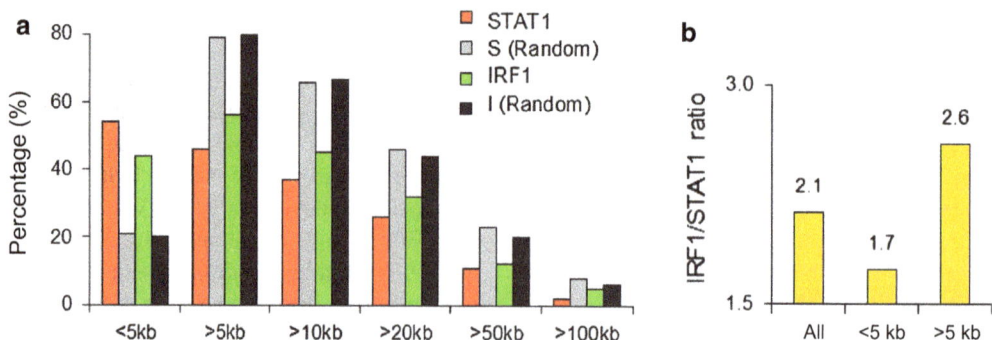

**Fig. 3** Proximal *vs.* distal STAT1 and IRF1 binding. **a** The percentage of IFNγ-induced STAT1 or IRF1 binding sites (*top*) or randomly generated controls at proximal (≤5 kb) or distal (>5 kb) sites relative to the TSS of Known genes. **b** IRF1 to STAT1 ratio at proximal and remote sites

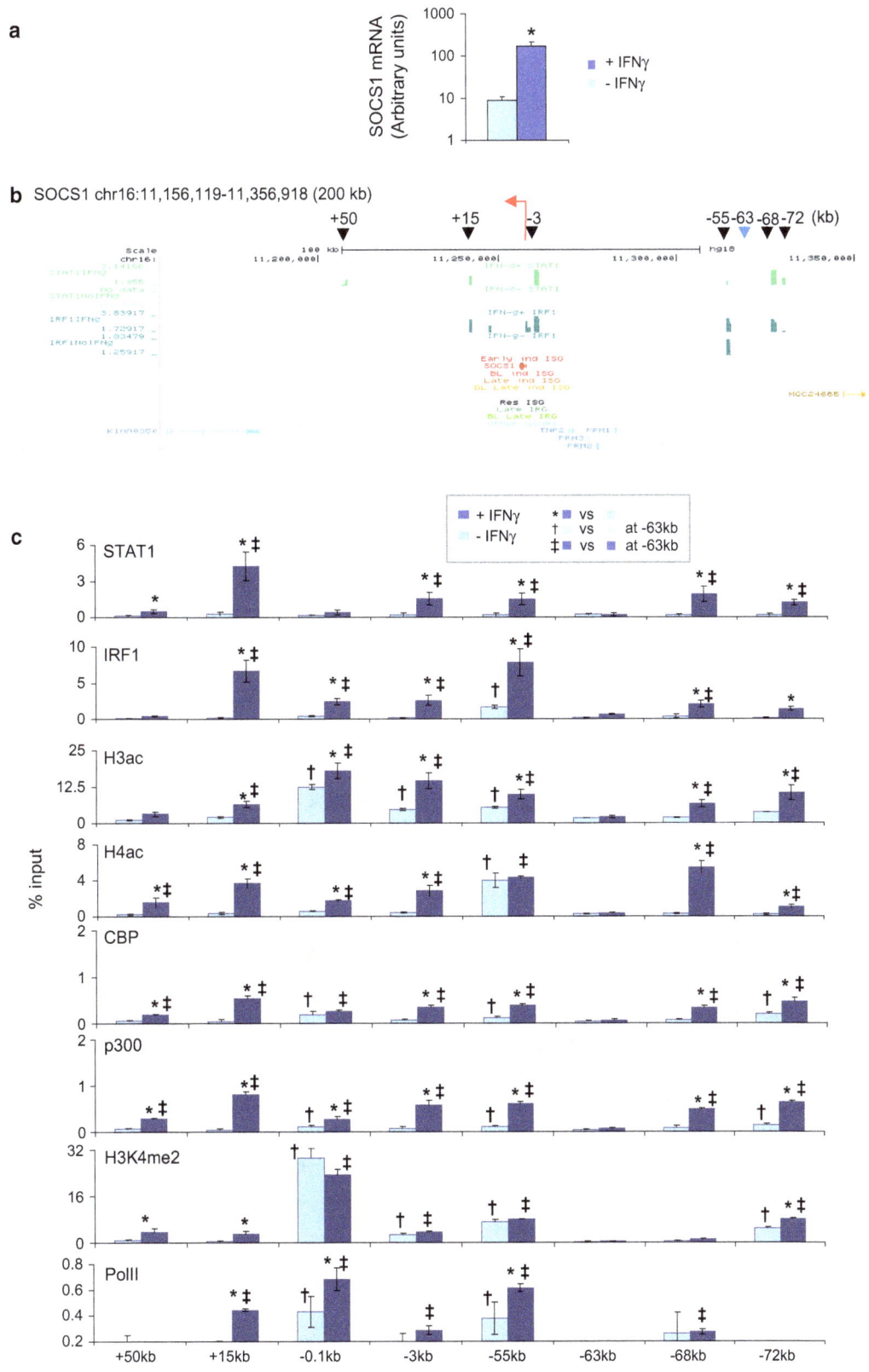

(See figure on previous page.)
**Fig. 4** STAT1 and IRF1 binding at *SOCS1*. **a** RT-qPCR for SOCS1 mRNA in HeLa cells treated with IFNγ for 0 or 6 h. Data are in arbitrary units relative to β-actin levels (log scale). **b** ChIP-chip maps of STAT1 and IRF1 binding across the *SOCS1* locus treated as in **a**. *Black arrowheads* indicate TF binding sites of interest, with distances from the TSS (*red arrow*) in kb. The TF-free −63 kb site (*blue*) is used as a negative control in **c**. **c** ChIP-qPCR analysis of the basal and IFNγ-induced histone modifications or recruitment of the indicated factors. Marks (*, †, ‡) show significant differences (p < 0.05, ANOVA followed by Fisher test) in the indicated comparisons (mean ± SD, n = 3)

+15 kb and −55 kb (Fig. 4c). Association with the +15 kb element may reflect IFNγ-induced promoter looping (see below).

Chromosome conformation capture (3C) revealed both constitutive and IFNγ-induced contacts between the promoter and remote STAT1 and/or IRF1 sites at *CIITA* [7]. To examine looping at *SOCS1*, we studied six EcoRI (I–VI) or four NcoI (VII–X) fragments (Fig. 5a). Of a total of 22 possible interactions, we studied 4 previously ([8]; underlined in Fig. 5b) and assessed an additional 7 in this study. As expected, no interaction was observed between fragments containing the promoter and irrelevant sites at −63, −6, or +70 kb, either before or after cytokine exposure. However, between suspected functional elements, we detected a total of 3 loops of all 11 putative interactions in the basal state, and each of these loops was enhanced after IFNγ treatment and was accompanied by a new interaction between the +50 and −72 kb enhancers that lie 122 kb apart (Fig. 5b, c). Our data suggest that the SOCS1 locus is basally present in a mega looping complex that becomes more compact and involves more inter-element interactions after IFNγ treatment. Together, the ChIP and 3C data show that STAT1 and IRF1 binding is linked to extensive chromatin modifications and looping.

## Unusual IRF1 distribution at MHC loci
As noted earlier, IRF1 exceeded STAT1 sites by ~twofold, but this varied at some regions, most notably at the MHC class I locus where the ratio was 3.5:1 (56 IRF1:16 STAT1 sites; Additional file 1: Table S4). The ratio was particularly skewed at the extended (6:1) versus classic (2.9:1) MHC class I region. 26 of all MHC class I IRF1 sites were within 5 kb of Known Gene starts and 17 within 5 kb of pseudogenes (Additional file 1: Tables S4, S5), giving a total of 77% (43/56) promoter proximal sites, which is

higher than the 44% at all loci (Fig. 3a; Additional file 1: Table S4). However, whereas 66% (25/38) of IRF1 sites were promoter-proximal in the classical MHC class I region, this dropped to only 6% (1/18) at the extended MHC class I region, and was low even after including pseudogenes (4/18; Additional file 1: Tables S4, S5), leaving an unusually high fraction of remote IRF1 sites (78%). Thus, IRF1 seems to play a broader role than STAT1 at the MHC class I cluster, primarily at proximal elements in the classic region, but at remote elements in the extended region.

STAT1 and IRF1 induce *CIITA*, the master regulator of MHC class II expression (reviewed in [34]). The number of STAT1 and IRF1 sites was typically very low in the MHC class II region (Additional file 1: Table S4). Out of 13 MHC class II genes, 5 (DRB5, DQB1, DQB2, DQA2, and DOA) were resistant to IFNγ in HeLa cells, 5 (DOB, DRB1, DQA1, DPA1, DPB1) responded only after 24 h, a time of maximum production of CIITA [35], and 3 (DRA, DMB and DMA) were es-indISGs. With the exception of DOB, none of the resistant or late-induced genes exhibited STAT1 or IRF1 promoter binding. However, of the 3 es-indISGs, two had promoter proximal IRF1 binding while DMA had IRF1 binding fairly near (~8 kb) its promoter. Thus IRF1 may cooperate with CIITA at a subset of MHC class II promoters. Others reported CIITA-independent induction of MHC class II genes [36–39], which may, therefore, involve IRF1.

## STAT1 and IRF1 binding is enriched at robustly induced ISGs
As discussed, ISGs fell into 8 classes depending on whether IFNγ caused induction, no effect (resistant ISGs in HeLa cells), or repression, and whether induction/repression were early or late, and strong or weak (Fig. 1b). We plotted the distribution of STAT1 and IRF1

(See figure on next page.)
**Fig. 5** Basal and IFNγ-induced looping at *SOCS1*. **a** A schematic view of the *SOCS1* locus. *Circles* indicate the *SOCS1* promoter (*purple*), putative remote enhancers (*red*), and negative control sites (*blue*), with distances from the TSS (*red arrow*) indicated above in kb, while fragments used in 3C assays with primers (*black arrowheads*) are shown below. **b** Cross linking frequencies between the promoter and remote sites across the *SOCS1* locus. Quantitative 3C was performed with chromatin from HeLa cells left untreated or exposed to IFNγ for 6 h. Bar graphs show the crosslinking frequency of a selected number of interactions. *Underlined* interactions were published previously [8]. Marked interactions (*, †, ‡) are significantly different at the indicated comparisons (p < 0.05, ANOVA followed by Fisher test, mean ± SD, n = 3). **c** Summary of looping events. Interacting sites and DNA strands are colored as in **a**. STAT1/IRF1 (*green/red dots*) and Hac/H3K4me2 (*green/gray diamonds*; data from Fig. 6c) are also depicted

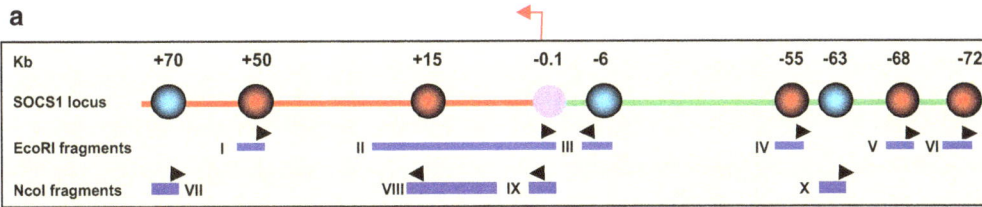

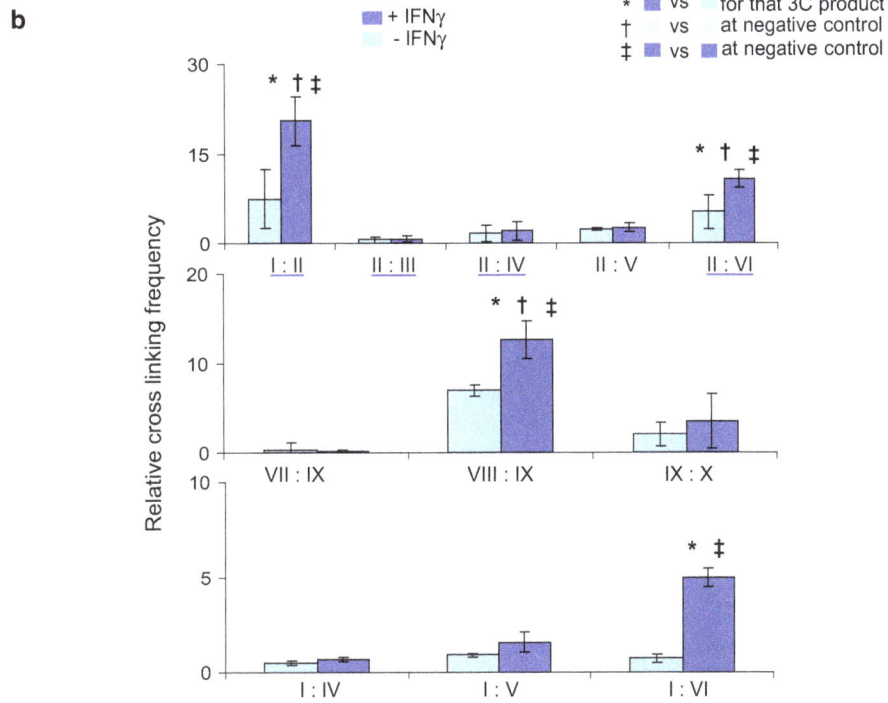

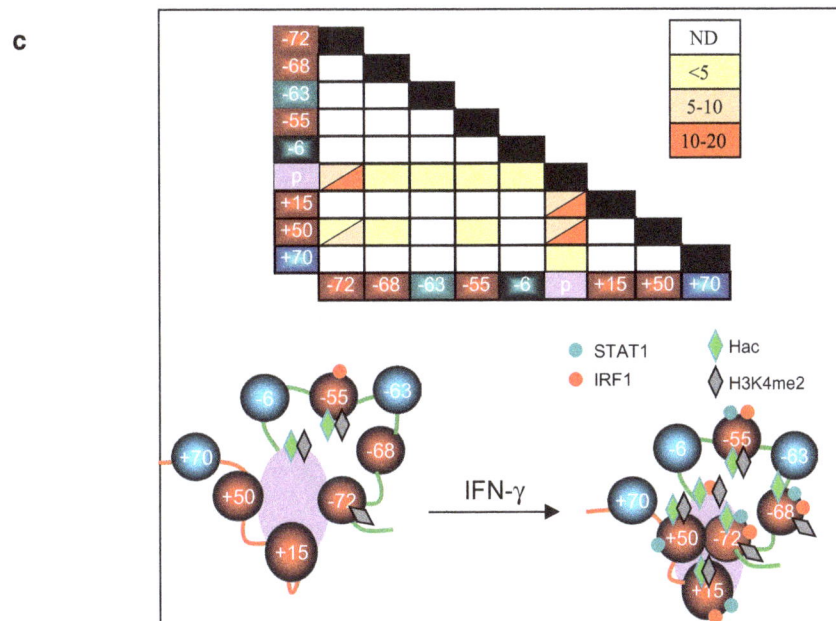

binding sites relative to all 8 gene classes. Binding sites were assigned to the nearest gene class, designated as proximal or distal when ≤5 or >5 kb from the TSS, respectively, and were compared to 288 randomly chosen sites equaling number of STAT1 + IRF1 sites (Fig. 6a). We also calculated the TF enrichment ratio (TER) in which the % distribution of TFs at proximal and distal locations was normalized to the % distribution of random sites (Fig. 6b). A binding frequency twice that of random sites (TER = 2) was assigned as an arbitrary minimum threshold.

STAT1 and IRF1 binding sites were most highly associated with robustly induced IFNγ targets (es-indISGs; Fig. 6). This applied when STAT1 or IRF1 were considered together, separately, and at proximal or remote locations (Fig. 6b). Consistent with this finding, weakly induced genes (ew-indISGs) had fewer binding events

and lower TERs (Fig. 6). Of 236 Other genes (never classified as an ISG in any study), a total of only 17 had 9 STAT1 and 16 IRF1 proximal peaks, mostly (13/17) located at the MHC and RT-qPCR confirmed no induction at 10/10 of these genes (Additional file 1: Table S6). IFNγ enhancers loop over large distances at *CIITA* [7] and *SOCS1* (Fig. 5), so proximal and distal enhancers nearest to Other Genes may target neighboring ISGs. In summary, the data indicate a clear bias of STAT1 and IRF1 binding at rapidly and robustly induced ISGs, but not other gene classes.

### Isolated or dual STAT1 and IRF1 recruitment is directed by binding motifs

Next we compared the fraction of isolated or dual STAT1/IRF1 binding events. Of a total of 230 discrete

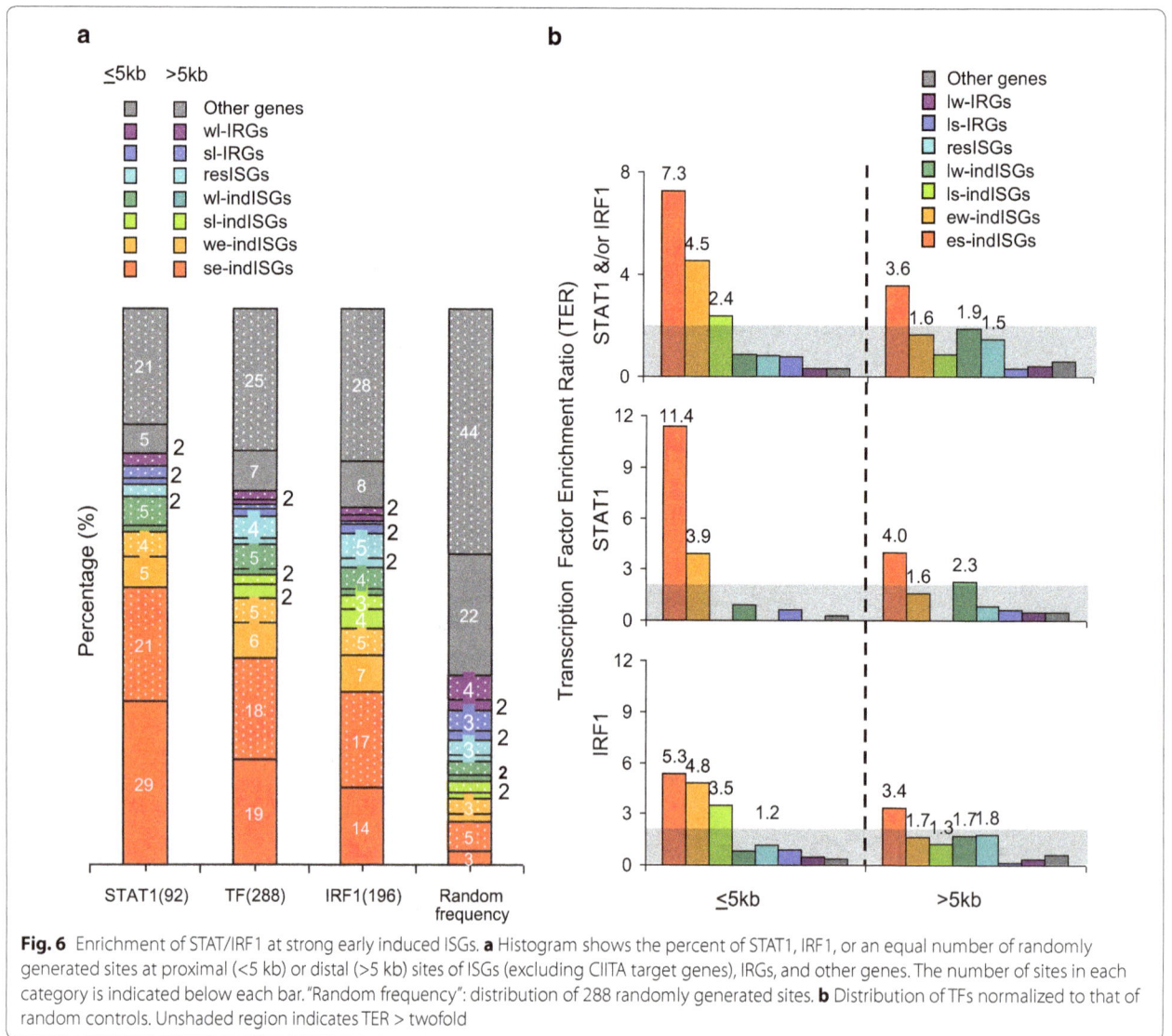

**Fig. 6** Enrichment of STAT/IRF1 at strong early induced ISGs. **a** Histogram shows the percent of STAT1, IRF1, or an equal number of randomly generated sites at proximal (<5 kb) or distal (>5 kb) sites of ISGs (excluding CIITA target genes), IRGs, and other genes. The number of sites in each category is indicated below each bar. "Random frequency": distribution of 288 randomly generated sites. **b** Distribution of TFs normalized to that of random controls. Unshaded region indicates TER > twofold

TF binding regions, 16% (36/230) exhibited STAT1 binding alone (isolated STAT1), of which almost half (17/36) were proximal; 61% (140/230) exhibited only IRF1 binding (isolated IRF1), of which 42% (59/140) were proximal; and 23% (54/230) showed overlap (dual STAT1/IRF1), of which slightly more than half (31/54) were proximal (Additional file 1: Table S4). Randomly generated sites showed negligible overlap (Additional file 2: Figure S2B), but dual STAT1/IRF1 binding represented more than half (54/92; 59%) of all STAT1 sites and about a quarter (54/196; 28%) of IRF1 peaks (Fig. 7a). Fewer overlapping IRF1 peaks reflect their twofold excess relative to STAT1 peaks. Thus, STAT1 preferentially binds with IRF1 at IFNγ enhancers, whereas most IRF1 sites are not co-localized with STAT1.

JASPER analysis of IRF1 and STAT1 peak regions revealed that the cognate binding motif was observed at a statistically significant level relative to equal numbers of random peaks (Fig. 7b). 60% of isolated STAT1 peaks had a STAT1 motif, and only 30% had an IRF1 motif, while 70% of isolated IRF1 peaks possessed an IRF1 motif, but only 20% had a STAT1 motif. A strong correlation existed between STAT1/IRF1 binding and the presence of the corresponding motifs (Fig. 7c). Indeed 40% of dual STAT1/IRF1 sites had both binding motifs, whereas there were none at equal numbers of randomly generated sites (Fig. 7b). Dual sites which have only a STAT1 or IRF1 binding motif may reflect protein–protein interaction or DNA looping as seen at the *SOCS1* and *CIITA* loci (Fig. 4c) [7, 8]. In summary, DNA sequence directs isolated or dual STAT1/IRF1 binding in IFNγ treated cells.

### Dual STAT1 and IRF1 targeted enhancers distinguish responsive from resistant ISGs

Comparing inducible ISGs in our array study with ~all known ISGs in a large database (Additional file 1: Table S2) revealed resistant ISGs (res-ISGs) in HeLa cells (Fig. 1b). There were far fewer STAT1/IRF1 binding events at res-ISGs vs ind-ISGs, and the TER (ratio of actual TF binding to random sites) at res-ISGs was low, and similar to that at other genes (Fig. 6). There was near ubiquitous association of both STAT1 and IRF1 at es-indISG promoters, but they were virtually absent at res-ISG promoters (Fig. 6). To quantify the types of TF binding events (isolated, dual, etc.), we plotted the frequency of genes with at least one binding event within or beyond 5 kb (Fig. 8b), and the density of each type of binding event per gene (Fig. 8c). Isolated TF binding did not discriminate the two gene classes, whereas there was significantly more dual STAT1 and IRF1 binding at esISGs, at both proximal and distal sites (Fig. 8b, c). Thus, cooperation between STAT1 and IRF1 plays a central role in mediating IFNγ responsiveness.

### Degree of TF binding and responsiveness in HeLa predicts ISG responsiveness in other cell types

Many studies have analyzed IFN-gene responsiveness, but a comprehensive analysis of which ISGs show broad or cell-type specific expression and, more importantly, the mechanism underlying such variability, has not been attempted. To assess variability in ISG induction, we compiled expression data on ISGs from 7 different human cell lines or primary cells, including 5 listed in Additional file 1: Table S2, plus HeLa cells (this work) and BRG1-reconstituted SW13 cells [13]. Across all 7 cell lines there were a total of 312 ISGs, the majority (61%) were exclusively induced in only one cell type, 28% were induced in 2–4 cell types, and 11% were induced in most (5–7) cell types (Fig. 9a; Additional file 1: Table S8). Only 9 genes were induced in every context and these included STAT1 and IRF1, in line with their apical role in IFNγ signaling.

We assessed the relationship between broad responsiveness, degree of induction, and STAT1/IRF1 binding. HeLa ChIP-chip data provided STAT1 and IRF1 binding information for 24 es-indISGs present in all 7 expression array datasets. Of these, 3/24 were induced exclusively in HeLa, 10/24 were induced in 2–4 lines and 11/24 were induced in 5–7 lines (Fig. 9b). Of note, genes induced exclusively in HeLa were up-regulated to a much lower extent than ubiquitously IFNγ-responsive targets (Fig. 9c). Greater induction of ubiquitously responsive loci was paralleled by a higher density of TF binding at promoter proximal sites (Fig. 9d). Thus, the level of induction is linked to the degree of STAT1 and IRF1 recruitment, and there is an unexpected link between the strength of ISG induction in one context (HeLa in this case) and competency to respond to IFNγ in other contexts.

### SNPs modulate STAT1 and IRF1 binding in vitro

Defects in IFNγ signaling are linked to a wide range of disorders [40–44]. Several studies focused on the association between genetic variants and the risk of IFNγ related disorders, but at gene promoters or coding regions of ISGs rather than IFNγ responsive enhancers. Within the 16 Mb of DNA around ISGs studied here, there are a total of $7.1 \times 10^5$ dbSNPs [hg19; SNPs (141)]. Of these, 6648 dbSNPs lay within the 230 STAT1/IRF1 peaks. Only 7 of these 6648 dbSNPs were listed on the GWAS database. GWAS SNPs do not define all disease associated SNPs (DA-SNPs) because GWAS genotyping arrays provide low genomic coverage [45] and therefore the 6648 dbSNPs may encompass other DA-SNPs not mapped yet. None of the 7 DA-SNPs (GWAS database) overlapped with a STAT1/IRF1 motif, but 80 of the 6648 dbSNPs overlapped with 27 STAT1 and 47 IRF1 motifs (Additional file 1: Table S9).

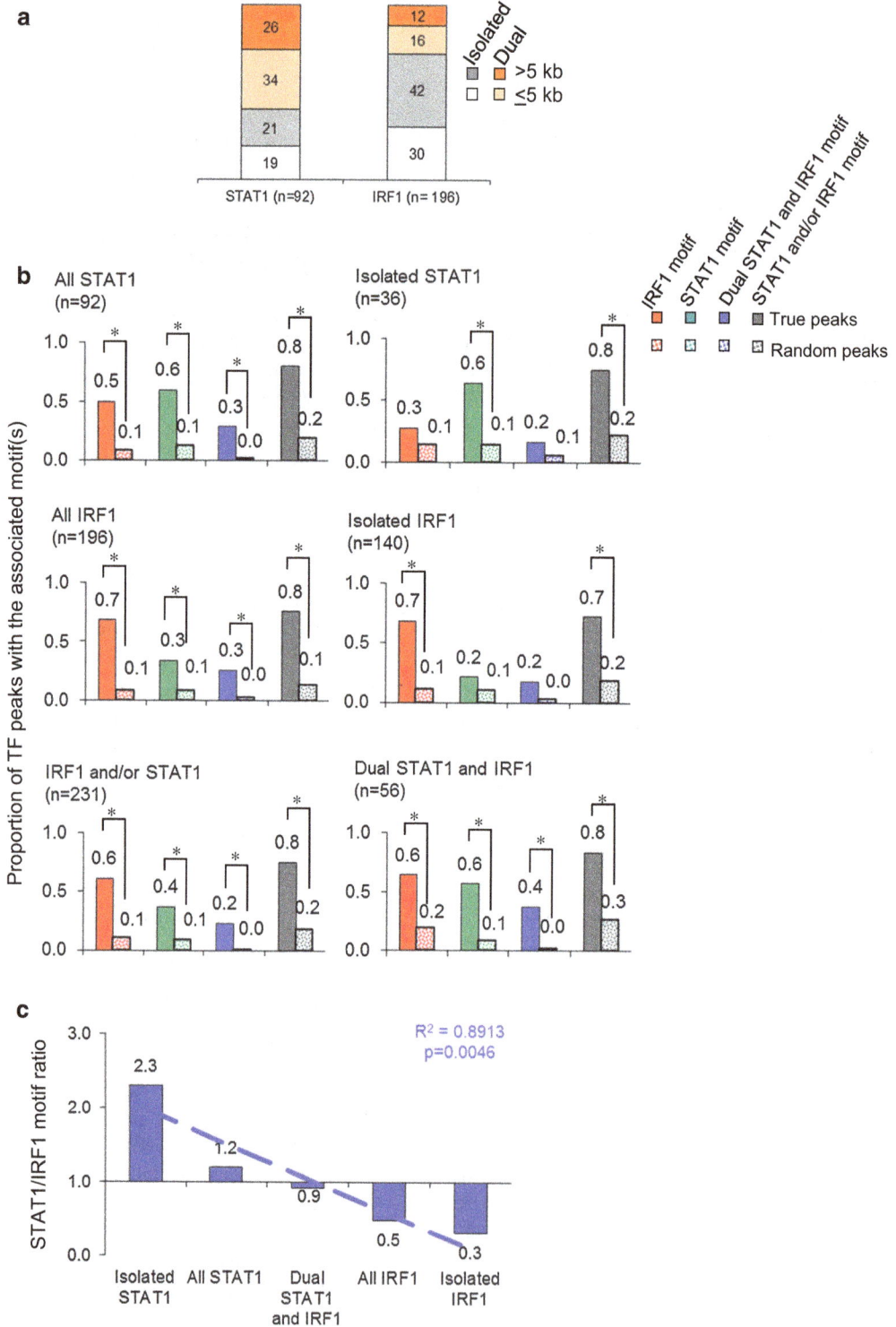

**Fig. 7** STAT1 and IRF1 binding correlates with binding motifs. **a** Percent distribution of isolated STAT1 or IRF1 or dual STAT1 + IRF1 binding at proximal (≤5 kb) or remote (>5 kb) sites of Known gene promoters. **b** TF binding sites were classified into 6 subclasses, then mapped motifs using CisGenome's "Known Motif Mapping" program (see "Methods" section for details). Sets of equal numbers of randomly generated "peaks" were used to define the background occurrence of STAT1 and IRF1 motifs. *Asterisk* indicates significant difference between true and random peaks (p < 0.00005, two-sided probability test in R). **c** Ratio of STAT1/IRF1 motifs at different categories of peaks

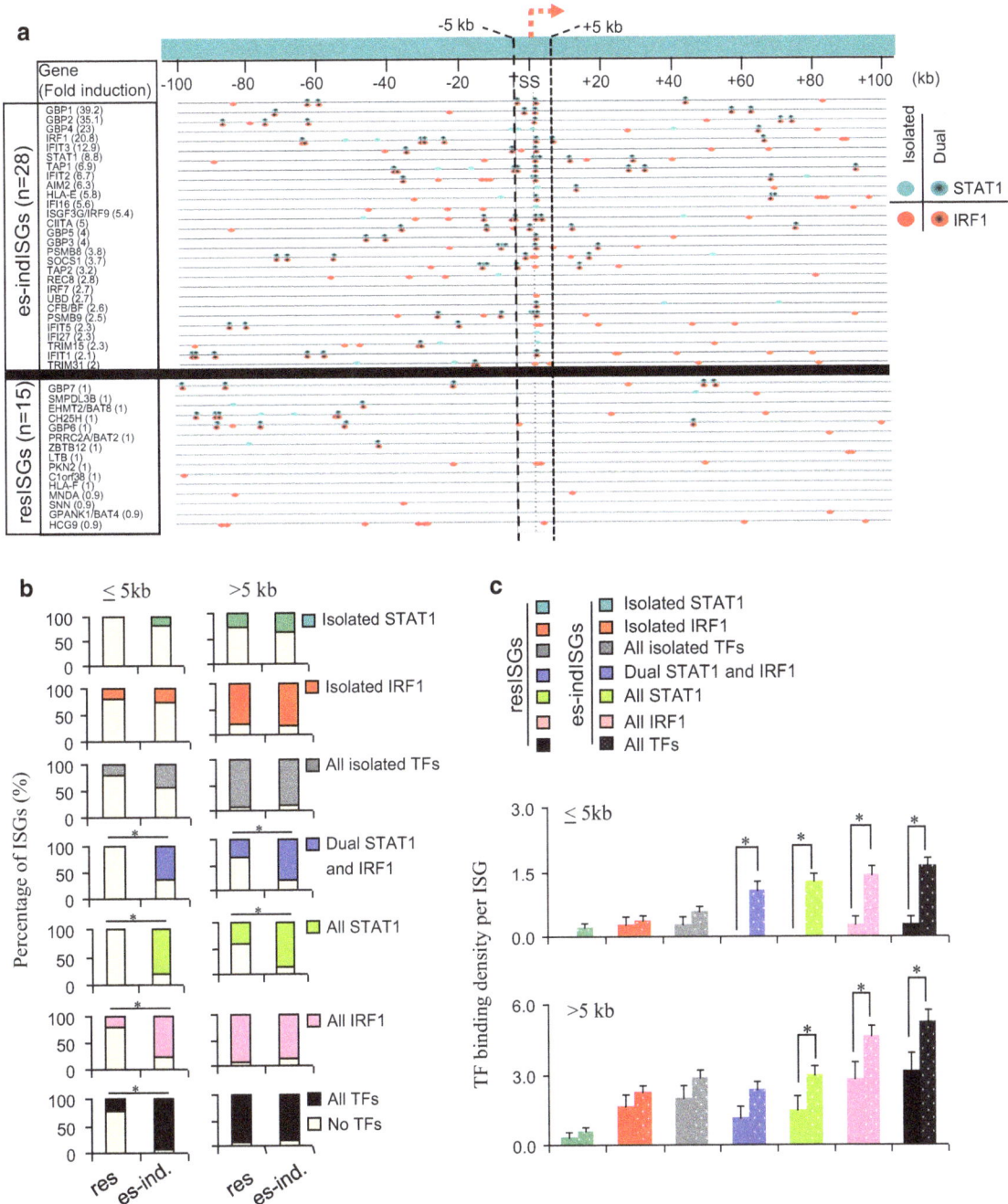

**Fig. 8** STAT/IRF1 recruitment at es-indISGs vs res-ISGs. **a** Map of isolated and dual STAT1 and IRF1 binding ± 100 kb of the TSS of es-indISGs (*top*) and resISGs (*bottom*), after removing CIITA targets, in HeLa cells treated for 6 h with IFNγ. *Red arrow* indicated TSS and direction of transcription. Genes are ranked according to fold induction, indicated in *brackets*. **b** Histograms show the percentage of es-indISGs or resISGs with proximal (≤5 kb) or distal (>5 kb) binding of STAT1 and/or IRF1. *Asterisk* p < 0.05, Fisher exact test. **c** Average number of TF binding events at proximal and remote sites at resISGs or es-indISGs. Error bar: SEM; *Asterisk* p < 0.05, ANOVA followed by Fisher test

We studied which of these 80 SNPs affect STAT1/IRF1 binding. First, we utilized the CisGenome "Known Motif Mapping" program to predict which of the variants may modulate STAT1/IRF1 binding motifs (see "Methods" section). CisGenome compares the position weight matrix (PWM) in the JASPAR CORE database and creates likelihood scores for the reference or variant allele. We calculated the fold change in likelihood scores (variant/reference allele) to assess the predicted relative effect. At a cutoff of 1.5-fold, the variant alleles of 34/80

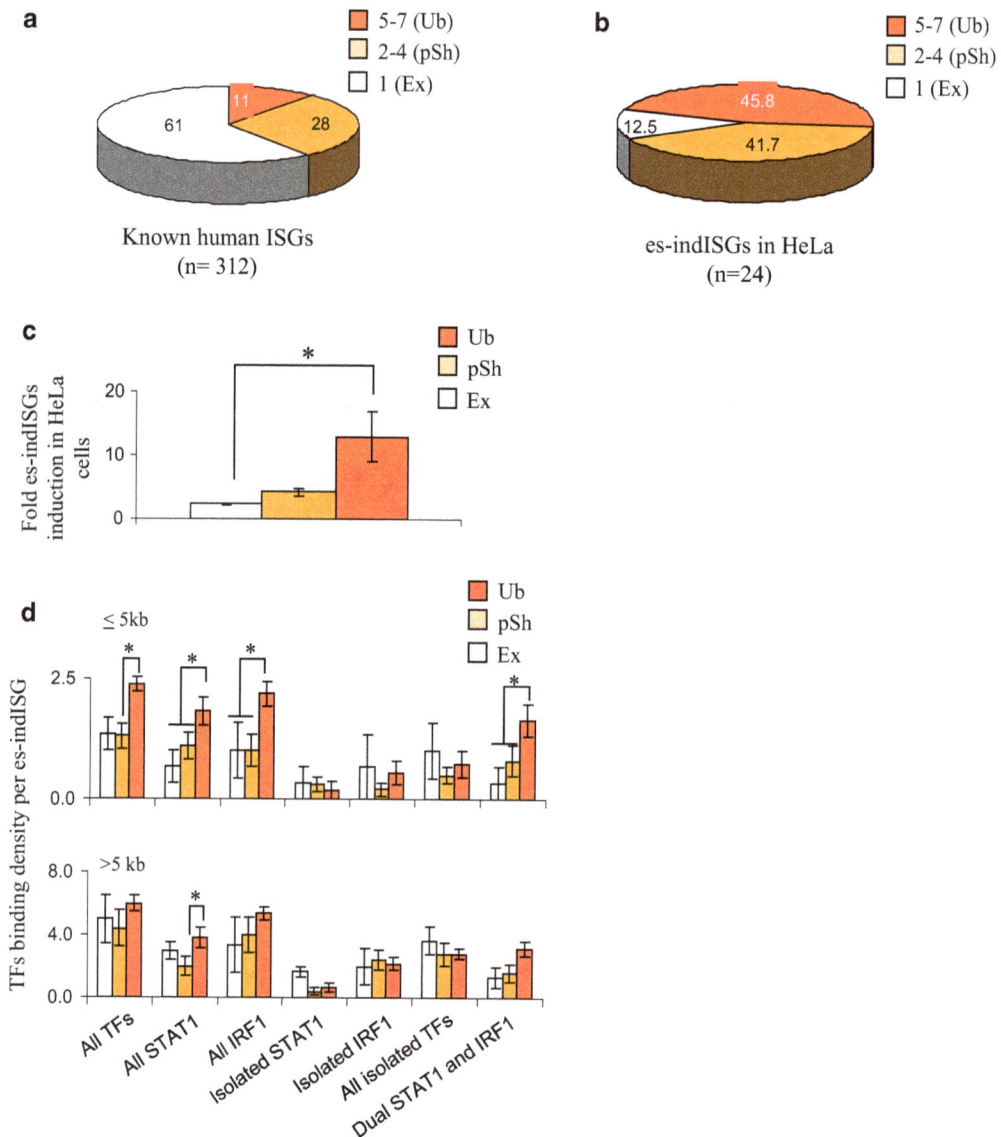

**Fig. 9** Link between ISG responsiveness in HeLa cells, responsiveness in other cell types, and STAT1/IRF1 binding. **a** Percentage of ubiquitous (Ub), partially shared (pSh), or exclusive (Ex) human ISGs based on their responsiveness to IFNγ in the indicated number of cell lines. **b** Percentage of HeLa es-indISGs which respond in only HeLa or in more cell types (as in **a**). For full lists of ISGs and es-indISGs see Additional file 1: Tables S7, S8. **c** Level to which Ub, pSh or Ex es-indISGs are induced in HeLa cells. **d** Average number of TF binding at promoter (≤5 kb) and remote (>5 kb) sites of the indicated types of es-indISGs. Error bar indicates SEM; *Asterisk* p < 0.05, ANOVA followed by Fisher test

dbSNPs were predicted to modulate the binding affinity of 24 IRF1 motifs and 10 STAT1 motifs (Additional file 1: Table S9).

To test these predictions in vitro, we developed an ELISA-based DNA affinity assay (see "Methods" section). Canonical STAT1 or IRF1 motif-containing biotinylated 33-mers were immobilized on streptavidin-coated 96-well plates. Cell lysates from HeLa cells exposed to IFNγ for 6 h were mixed with either no or various amounts of Wt (positive control), mutated (negative control), or dbSNP (test) competitor probes, then added to

the immobilized biotinylated probe, and the amount of bound TF determined using anti-STAT1 or anti-IRF1 antibody. We tested 4 or 1 SNPs affecting IRF1 or STAT1 sites, respectively (Fig. 10; Additional file 1: Tables S9, S10). Wt IRF1 and STAT1 probes exhibited strong binding with low $IC_{50}$s of $9.6 \pm 1.5$ or $4.2 \pm 0.9$ pmol/well, respectively, whereas control mutated probes had minimal/no effect (Fig. 10; Additional file 1: Table S10). 3/6 of the IRF1 SNPs decreased affinity (rs365393, rs9262216, rs34494346) and 1/6 created a strong IRF1 site (rs9260102) (Fig. 10a; Additional file 1: Table S9). The

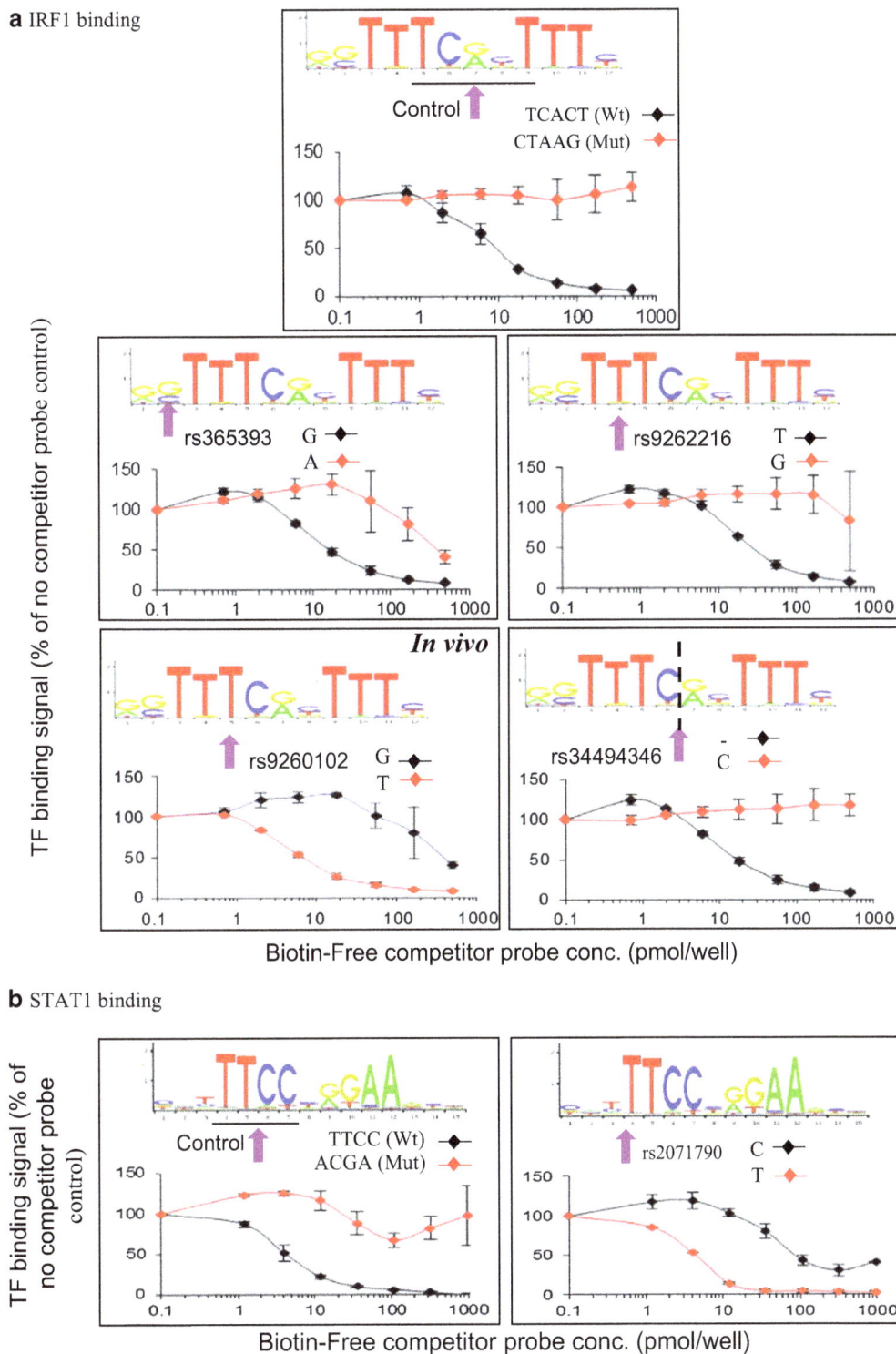

**Fig. 10** SNPs modulate STAT1 and IRF1 binding in vitro. **a** IRF1, and: **b** STAT1 binding assays. Graphs show STAT1 and IRF1 binding signal to immobilized probes in the presence of different concentrations of competitor probes with either the variant or reference allele. 100% binding is that obtained in the absence of competitor. *Arrows* highlight the affected base (or 4 bases in the control mutated probe). As indicated, rs9260102 was also assessed in vivo (Fig. 11)

single STAT1 SNP that we tested created a putative binding site, and indeed the T allele of rs2071790 showed high affinity binding (Fig. 10b; Additional file 1: Table S10). Our ChIP-chip data indicated that this SNP lies within an isolated remote IRF1 peak, suggesting that the T allele would convert this regulatory element to a dual STAT1/IRF1 enhancer. In summary, these data show close concordance between the predicted and actual effects of SNPs on STAT1 and IRF1 binding. Thus, it is likely that most of the 34 predicted functional SNPs do in fact alter binding.

### rs9260102 affects IRF1 binding in vivo

Next we asked if the T allele of rs9260102, which creates an IRF1 site in vitro (Fig. 10a), has this effect in vivo. This SNP lies ~1 kb upstream of the HLA-A locus, within an IFNγ-responsive IRF1 ChIP-chip peak in HeLa cells

(Fig. 11a). To test whether it affects IRF1 binding in vivo we employed the EBV-transformed lymphoblastic cell line GM18857, which is heterozygous for rs9260102 (G/T), implying that IRF1 should only bind to one (the T) allele. Treatment with IFNγ for 6 h induced a 1.8-fold increase in the total IRF1 ChIP-qPCR signal (Fig. 11b). Snapshot sequencing revealed that this IFNγ-dependent increase was due solely to elevated binding to the T allele (Fig. 11c). Thus, in silico prediction, an in vitro binding assay, and in vivo allele specific ChIP all show that the G to T switch creates an IRF1 binding site (Fig. 10a; Additional file 1: Table S10).

### Discussion
STAT1 and IRF1 drive the induction of IFN induced genes, but the extent to which they act collectively is unclear. We report that most STAT1 binding (60%)

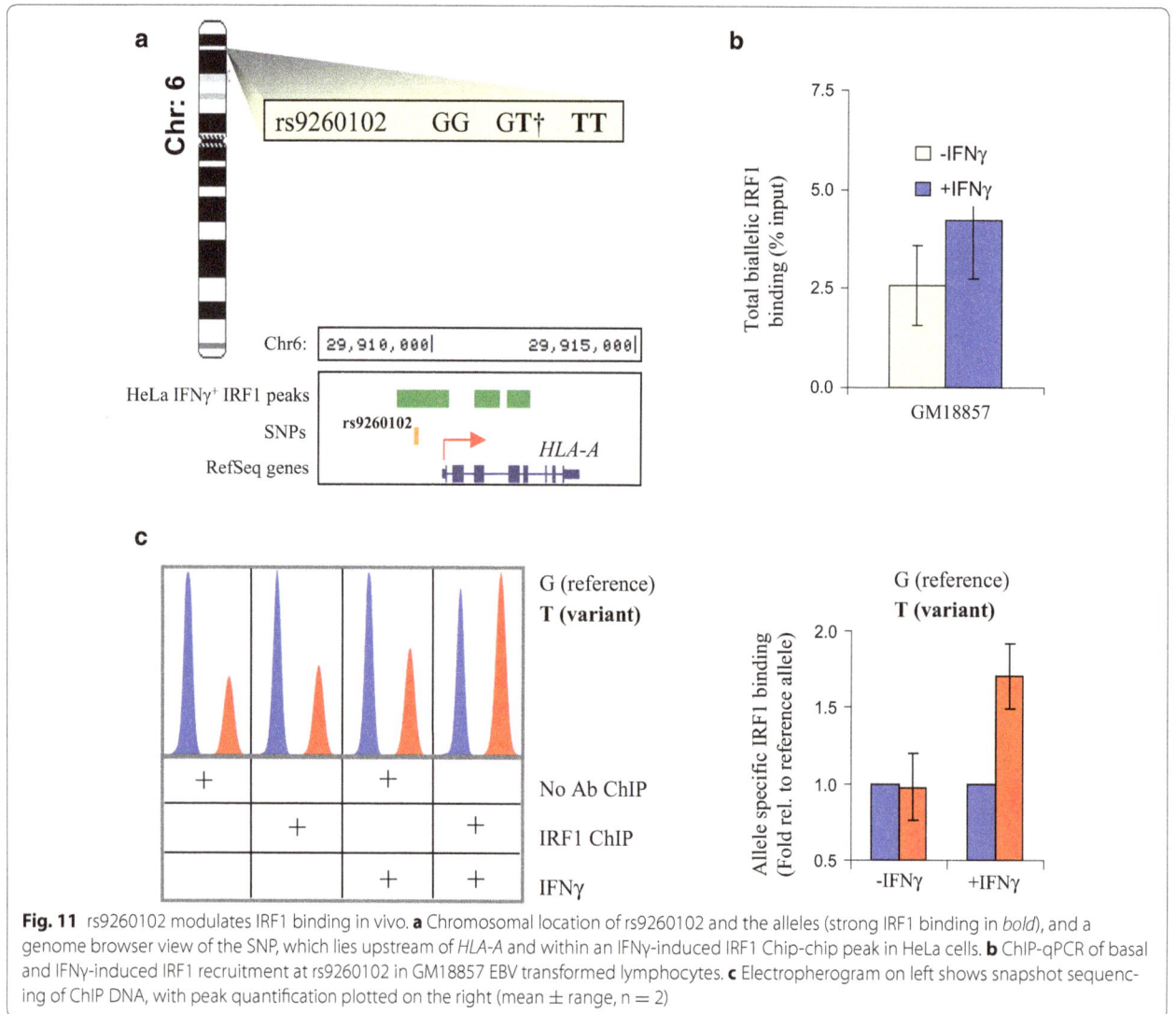

**Fig. 11** rs9260102 modulates IRF1 binding in vivo. **a** Chromosomal location of rs9260102 and the alleles (strong IRF1 binding in *bold*), and a genome browser view of the SNP, which lies upstream of *HLA-A* and within an IFNγ-induced IRF1 Chip-chip peak in HeLa cells. **b** ChIP-qPCR of basal and IFNγ-induced IRF1 recruitment at rs9260102 in GM18857 EBV transformed lymphocytes. **c** Electropherogram on left shows snapshot sequencing of ChIP DNA, with peak quantification plotted on the right (mean ± range, n = 2)

occurs together with IRF1, but most IRF1 binding (72%) is isolated (Fig. 7a). Binding occurs where there are cognate binding motifs (Fig. 7), suggesting that most ChIP signals reflect direct recruitment. Both proximal and remote STAT1 and IRF1 binding is observed at robustly induced ISGs, but not at other loci (Fig. 6). In line with the importance of TF occupancy for responsiveness [46], every responsive locus exhibits a mixture of STAT1 and IRF1 bound enhancers (Fig. 8a). Moreover, dual bound enhancers distinguish induced vs resistant ISGs, whereas single bound enhancers are found with similar frequency at responsive or non-responsive ISGs (Fig. 8). This is not to say, however, that single bound enhancers are irrelevant. For example, while multiple remote *SOCS1* enhancers recruit both TFs, the +50 kb element or promoter are targeted only by STAT1 or IRF1, respectively, yet both are involved in 3D looping (Figs. 4, 5). Similarly, while dual STAT1/IRF1 binding occurs at the active *CIITA* promoter, the −50 kb and +59 kb enhancers recruit only STAT1 or IRF1, respectively, yet contribute to 3D looping and are essential for responsiveness [7]. Indeed, all the responsive genes we surveyed exhibit a mix of STAT1-only, IRF1-only, and STAT1/IRF1 dual enhancers (Fig. 8). Together, these results suggest that IFNγ-responsiveness requires cooperation between enhancers that bind both or either TF, but that STAT1- or IRF1-only enhancers are insufficient for gene induction. Irrespective, it is clear that responsive ISGs integrate information from both STAT1 and IRF1.

Previously, we showed that there is a pre-existing 3D structure at the silent *CIITA* locus, generated through looping between enhancers that subsequently recruit STAT1 and IRF1 upon IFNγ treatment [7]. This was true even in the absence of BRG1, a chromatin remodeling enzyme that is critical to allow stable TF recruitment and thus IFNγ-responsiveness. Subsequent genome-wide analyses indicate that enhancer looping in the poised but silent state is common at inducible loci [47]. We observed the same phenomenon at the IFNγ responsive *SOCS1* locus (Fig. 5). Potentially, these contacts are mediated by pioneer factors that mark responsive enhancers, but their identity at ISGs is unknown. The data here and other studies show that STAT1 and IRF1 can bind some sites in the basal state [21, 28], so in theory, low/unstable binding (undetectable by ChIP) could poise ISG enhancers. It would thus be interesting to perform looping studies at ISGs in STAT1/IRF1 deficient cells. It is of note that the degree to which ISGs were induced in HeLa cells predicted whether they were likely to respond to IFNγ in other cells (Fig. 9). Thus, the chromatin at these genes is accessible in many contexts. The factors that mediate this broad poised, open state may also initiate the basal looping at ISGs.

Over 90% of the disease markers identified in GWAS studies lie within the non-protein-coding regions of the genome [48]. These markers correlate with gene expression [49–52], and lie within gene regulatory regions [53–56]. There is thus considerable interest in identifying SNPs that influence TF binding and, therefore, gene regulation. We identified 80 SNPs within STAT1 or IRF1 motifs, and in silico assessment predicted that 34 may alter binding. In vitro quantification confirmed these predictions in 5/5 cases, arguing that most of/all the other predictions are accurate. The availability of a cell line heterozygous for one such SNP allowed us to test whether the prediction held up in vivo. Indeed, the T allele of rs9260102, which lies just upstream of the HLA-A locus, bound IRF1 whereas the G allele did not, as observed in silico and in vitro. These data serve as proof of principle that in silico prediction is a reliable tool to anticipate the effect of SNPs on STAT1 and IRF1 binding.

## Conclusions
This study provides strong evidence for widespread cooperation between STAT1 and IRF1 at ISGs, and suggests that in silico predictions reliably predict the effect of nucleotide variants on binding in vivo.

## Methods
### Custom oligonucleotide ChIP Tiling array design
A custom oligonucleotide tiling array was designed to cover 11 genomic regions spanning a total of 16 Mb of human genomic DNA in 8 chromosomes (Additional file 1: Table S1). Regions covered from 1 to 5 Mb genomic sequences. Arrays consisted of 50 mers, in quadruplicate, with median probe spacing of 80 bp within non-repetitive DNA regions.

### ChIP on tiled genome arrays (ChIP-chip) and ChIP–quantitative PCR
Details of primers and antibodies used in ChIP assays are in Additional file 1: Tables S11, S12. HeLa-ini1-11 cells (HeLa), were grown as described [12]. Cells were left untreated or exposed to 300 units/ml of human IFN-γ for 6 h (PHC4834, BioSource International, Camarillo, CA, USA). Crosslinked chromatin was sonicated to an average size of about 500 base pairs and was incubated with STAT1 or IRF1 antibody. Bound fragments were purified by ChIP and amplified by ligation-mediated PCR, as described [7], then labeled and hybridized to the arrays. Hybridization intensities were normalized to internal standards and values from quadruplicate spots were averaged. Significantly different intensities between ChIP DNA and input DNA samples in three biological replicates (p < 0.0001) were determined with the Wilcoxon rank-sum test. Peaks representing significantly enriched

DNA regions (p < 0.0001) where the ratio of ChIP to input DNA was 1.5-fold or more were visualized with the University of California at Santa Cruz Human (Homo sapiens) Genome Browser (Phast-Cons) and are plotted on a log2 scale. Peaks in a sliding window of 500 base pairs were merged with an in-house Perl script pipeline. ChIP—quantitative PCR was done as described [7], and in all cases, the low background signal obtained with a no-antibody control was subtracted.

## Custom oligonucleotide ChIP tiling array data analysis

Raw intensities from three independent biological replicates, quality assessed by Nimbelgen SignalMap software, were quantile normalized [57], and averaged for each quadruplicate 50 mer. We developed a Wilcoxon Rank Sum test [58] based software to studying the difference between the intensities of the ChIP signal compared to the input DNA signal for each probe within a 500 bp sliding window. Genomic positions with statistically higher intensities (>1.5-fold, $p < 10^{-4}$) from input DNA were merged to form a peak. ChIP-chip data were imported into UCSC genome browser (assembly hg17, NCBI build 35) as two sets of separate tracks for each antibody before and 6 h after IFNγ treatment (http://research.lunenfeld.ca/IFNγ).

## STAT1 and IRF1 motif analysis

We mapped STAT1 and IRF1 consensus motifs to the STAT1 and IRF1 ChIP-chip binding regions by using CisGenome "Known Motif Mapping" program. Motif occurrences were determined by using position frequency matrices (PFMs) of STAT1 (ID: MA0137.2) and IRF1 (ID: MA0050.1) from the JASPAR CORE database. The PFMs were converted to pseudo-count matrix for CisGenome's input. A motif mapping location is selected by the cut-off of a likelihood ratio (LR) > 500. The LR is determined by comparing the motif's PFM with a background model estimated from input ChIP-chip regions in the $3^{rd}$ order. Seven sets of ChIP-chip peak regions were mapped and compared with background control regions: STAT1 peaks (92), IRF1 peaks (196), merged STAT1 and/or IRF1 peaks (231), dual STAT1 and IRF1 peaks (56), STAT1 peaks isolated from IRF1 peaks (isolated-STAT1, 56), IRF1 peaks isolated from STAT1 peaks (isolated-IRF1, 140), and basal IRF1 peaks. For each set of peaks, the same number and size of control background regions were randomly sampled from blank regions without any ChIP-chip bindings following the same frequency distribution as the real binding peaks within each cluster segments on the chromosome of ChIP-chip data. The frequencies of regions mapped with motifs were compared between ChIP-chip peaks and random control sites by using the two-sided probability test in R for each paired set of peaks.

## RNA extraction, expression microarray analysis, and Reverse transcriptase-qPCR

RNA extraction and reverse transcription were done from HeLa cells left untreated or at 6, 24 or 48 h after IFNγ treatment as described previously [12]. RNA quality was checked using both Nanodrop (Thermo Fischer Scientific; 260/280 ratio was ≥1.8) and Bioanalyzer (Agilent Inc.; RNA Integrity Number, RIN, ≥ 9.4, range 9.4–9.9). RNA samples were converted to cDNA, followed by a second strand synthesis, and cRNA was prepared using the Ambion kit (Applied Biosystems). The cRNA was column purified and quality was checked using Bioanalyzer (Agilent Inc.). A total of 1.5 μg of cRNA was hybridized to human whole-genome expression arrays (HumanRef-6 Expression BeadChip, Illumina, Inc.) using standard Illumina protocols. Slides were scanned on an Illumina Beadstation and analyzed using BeadStudio (Illumina, Inc). Genes induced by ≥twofold compared to controls and that achieved a differential score of ≥13 were classified as strongly induced ISGs. ISGs which achieved a differential score of ≥13 but fold induction less than 2 were considered weakly induced. Genes reduced by ≥twofold compared to control and achieved a differential score of ≤ −13 were considered strongly reduced. Genes that had a differential score of ≤ −13 but the fold reduction was less than 2 were considered weakly reduced. Three biological replicates were included for each treatment group.

RT-qPCR was performed much as described [59]. Briefly, RNA was extracted from HeLa cells left untreated or at 6, 24 or 48 h after IFNγ treatment using Trizol (Invitrogen), and quality assessed by RIN and OD260/280 as above. cDNA was prepared from 1 mg RNA using random primers and SuperScript RT (Invitrogen). Amplification of cDNA was performed using gene specific primer pairs and SYBER Green Mix (ABI). PCR was ran on Applied Biosystems PRISM 7900HT. Primers were designed in the coding region of each gene (Additional file 1: Table S12). Human genomic DNA was used to prepare calibrators for the quantification of cDNAs. Dissociation curves were inspected to ensure a single product and all PCR products were also tested on a gel to confirm amplification specificity. In addition, no template controls (NTC) were included to ensure the absence of DNA contamination. Gene expression was normalized to multiple house-keeping/reference genes to control for the total amount of RNA. All experiments were done in triplicate.

## Chromosome conformation capture (3C)

The 3C assay was conducted as described [7, 8]. Primer sequences are provided in Additional file 1: Table S12.

## Assessment of TF binding distribution around different classes of ISGs

We compared the distribution of TFBS in the vicinity of es-indISGs and resISGs in STAT1 peaks, IRF1 peaks, and merged STAT1 and IRF1 peaks. We first aligned the TFBS within a range of 400 kb region around the TSS of each gene in a 1 kb resolution, which means that we divided each region into 1 kb windows with the window 0 centered at the TSS and others line up to the 200 kb end upstream and 200 kb end downstream, and then scored the frequency of TFBS at each window as the number of peaks whose center was within the window. If 400 kb extended beyond the ChIP-chip segments, the binding frequencies along the truncated regions were regarded as missing data. Then we plotted the average binding frequencies per 1 kb window versus the relative distance of each to the TSS. Missing values were discarded for averaging the frequencies.

## Defining STAT1 and IRF1 functional SNPs

First, we queried dbSNPs located within the 230 ChIP-chip peaks (UCSC, Build hg19; Track, All SNPs(141); Table, snp141). We defined a total of 6648 dbSNPs. Next we defined SNPs that overlap with STAT1 or IRF1 binding motifs within the 230 ChIP-chip peaks. Then we selected a region of ±50 bp around the SNPs that overlapped with STAT1/IRF1 binding motifs and recovered the DNA sequence of these regions using CisGenome. Then we computationally evaluated the binding affinity of the reference or variant sequence using the likelihood scores obtained from the Cisgenome "known motif mapping" program with the sequences as input to map the STAT1 and IRF1 motif matrix. In some cases the introduction of the variant SNP renders the motif unidentifiable and in this case the sequence of the motif was indicated as "NULL" and the likelihood score was considered as zero (Additional file 1: Table S9). The cutoff value of affinity change was set at 1.5-fold.

## ELISA-based DNA binding affinity assay and ChIP coupled with DNA sequencing

We designed 33-mers with either the reference or variant alleles (Additional file 1: Table S13). Control probes with wild type or dead mutant STAT1 or IRF1 motifs were also included. Probes were ordered biotinylated or biotin-free (competitors). Two pmol of biotinylated probes were immobilized per well of 96-well streptavidin-coated plates. Cell lysates where incubated with different concentrations of the competitor probes (probe-lysate mix) at 4 °C for 3 h to allow STAT1 or IRF1 binding. The probe-lysate mix was then added to streptavidin-coated plates with immobilized biotin probes and incubated overnight at 4 °C. To quantify bound TFs, wells were washed and probed with STAT1 or IRF1 primary antibodies, followed by IR-800 conjugated secondary antibodies. Excess antibodies were washed thoroughly and plates were scanned and quantified using Odyssey Infrared imaging system (LICOR). Signal from no-competitor well is considered as 100% and the % antibody signal was plotted against competitor probe concentration. IC50 values were calculated using Graphpad PRISM 5.2.

For in vivo studies, EBV-transformed lymphoblastic GM18857 cells, cultured as recommended by the supplier (Coriell Biorepositories), were treated with IFNγ for 6 h, fixed and harvested for ChIP analysis. Chromatin was immunoprecipitated using IRF1 antibody and isolated DNA was sequenced using Snapshot sequencing.

### Abbreviations

3C: chromosome conformation capture; GAS: IFNγ activation site; IRF1: interferon regulatory factor 1; ISRE: interferon-stimulated response element; HAT: histone acetyltransferase; IFNγ: interferon-γ; ISGs: IFNγ stimulated genes; esISGs: early strong induced ISGs; lsISGs: late strong induced ISGS; ewISGs: early weak induced ISGs; lwISGs: late weak induced ISGs; resISGs: resistant ISGs; potISGs: potential ISGS; IRGs: IFNγ repressed genes; lsIRGs: late strong IRGs; lwIRGs: late weak IRGs; SNPs: single nucleotide polymorphisms; STAT1: signal transducer and activator of transcription 1; TF: transcription factor; TSS: transcription start site.

### Authors' contributions

MAEH generated and analyzed most of the data and KH performed most of the bioinformatics. MBKE and MAEH assessed TF binding in vitro, and were helped by AA. ZX helped with the bioinformatics. TY and ZN helped MAEH with ChIP. MS, MA, and IL helped generate reagents and analyze data. RB obtained funding, analyzed data, and wrote the paper with MAEH and KH. All authors read and approved the manuscript.

### Author details

[1] Lunenfeld Tanenbaum Research Institute, Mt Sinai Hospital, Toronto, ON, Canada. [2] Clinical Chemistry Division, Provincial Laboratory Services, Queen Elizabeth Hospital, Charlottetown, PE, Canada. [3] Department of Pathology, Faculty of Medicine, Dalhousie University, Halifax, NS, Canada. [4] Department of Lab Medicine and Pathobiology, University of Toronto, Toronto, ON, Canada. [5] Department of Ophthalmology and Vision Science, University of Toronto, Toronto, ON, Canada. [6] Present Address: Donnelly Centre, University of Toronto, Toronto, ON, Canada.

### Acknowledgements

We thank Malcolm Macleod for designing the software to create TF binding maps.

Funding was provided by Canadian Institutes of Health Research (Grant No. MOP 111004); Canadian Cancer Society Research Institute (Grant No. 703079).

### Competing interests

The authors declare that they have no competing interests.

**Funding**
This work was funded by grants to R.B. from the Canadian Cancer Society Research Institute (CCSRI), the Canadian Health Research Institutes (CIHR), and the Krembil Foundation.

**References**

1. Vesely MD, Kershaw MH, Schreiber RD, Smyth MJ. Natural innate and adaptive immunity to cancer. Annu Rev Immunol. 2011;29:235–71.
2. Maher SG, Romero-Weaver AL, Scarzello AJ, Gamero AM. Interferon: cellular executioner or white knight? Curr Med Chem. 2007;14:1279–89.
3. Hartman SE, Bertone P, Nath AK, Royce TE, Gerstein M, Weissman S, et al. Global changes in STAT target selection and transcription regulation upon interferon treatments. Genes Dev. 2005;19:2953–68.
4. Robertson G, Hirst M, Bainbridge M, Bilenky M, Zhao Y, Zeng T, et al. Genome-wide profiles of STAT1 DNA association using chromatin immunoprecipitation and massively parallel sequencing. Nat Methods. 2007;4:651–7.
5. Robertson AG, Bilenky M, Tam A, Zhao Y, Zeng T, Thiessen N, et al. Genome-wide relationship between histone H3 lysine 4 mono- and tri-methylation and transcription factor binding. Genome Res. 2008;18:1906–17.
6. Au-Yeung N, Mandhana R, Horvath CM. Transcriptional regulation by STAT1 and STAT2 in the interferon JAK-STAT pathway. JAK-STAT. 2013;2:e23931.
7. Ni Z, Abou El Hassan M, Xu Z, Yu T, Bremner R. The chromatin-remodeling enzyme BRG1 coordinates CIITA induction through many interdependent distal enhancers. Nat Immunol. 2008;9:785–93.
8. Abou El Hassan M, Bremner R. A rapid simple approach to quantify chromosome conformation capture. Nucleic Acids Res. 2009;37:e35.
9. Harismendy O, Notani D, Song X, Rahim NG, Tanasa B, Heintzman N, et al. 9p21 DNA variants associated with coronary artery disease impair interferon-γ signalling response. Nature. 2011;470:264–8.
10. Ning S, Huye LE, Pagano JS. Regulation of the transcriptional activity of the IRF7 promoter by a pathway independent of interferon signaling. J Biol Chem. 2005;280:12262–70.
11. Schroder K, Hertzog PJ, Ravasi T, Hume DA. Interferon-gamma: an overview of signals, mechanisms and functions. J Leukoc Biol. 2004;75:163–89.
12. Pattenden SG, Klose R, Karaskov E, Bremner R. Interferon-gamma-induced chromatin remodeling at the CIITA locus is BRG1 dependent. EMBO J. 2002;21:1978–86.
13. Abou El Hassan M, Yu T, Song L, Bremner R. Polycomb repressive complex 2 confers BRG1 dependency on the CIITA locus. J Immunol. 2015;194:5007.
14. Morris AC, Beresford GW, Mooney MR, Boss JM. Kinetics of a gamma interferon response: expression and assembly of CIITA promoter IV and inhibition by methylation. Mol Cell Biol. 2002;22:4781–91.
15. Ramsauer K, Farlik M, Zupkovitz G, Seiser C, Kröger A, Hauser H, et al. Distinct modes of action applied by transcription factors STAT1 and IRF1 to initiate transcription of the IFN-gamma-inducible gbp2 gene. Proc Natl Acad Sci USA. 2007;104:2849–54.
16. Shi L, Perin JC, Leipzig J, Zhang Z, Sullivan KE. Genome-wide analysis of interferon regulatory factor I binding in primary human monocytes. Gene. 2011;487:21–8.
17. Frontini M, Vijayakumar M, Garvin A, Clarke N. A ChIP-chip approach reveals a novel role for transcription factor IRF1 in the DNA damage response. Nucleic Acids Res. 2009;37:1073–85.
18. Rettino A, Clarke NM. Genome-wide identification of IRF1 binding sites reveals extensive occupancy at cell death associated genes. J Carcinog Mutagen. 2013;(Spec Iss Apoptosis):S6–009.
19. Xie D, Boyle AP, Wu L, Zhai J, Kawli T, Snyder M. Dynamic trans-acting factor colocalization in human cells. Cell. 2013;155:713–24.
20. Kumatori A, Yang D, Suzuki S, Nakamura M. Cooperation of STAT-1 and IRF-1 in interferon-gamma-induced transcription of the gp91(phox) gene. J Biol Chem. 2002;277:9103–11.
21. Chatterjee-Kishore M, Wright KL, Ting JP, Stark GR. How Stat1 mediates constitutive gene expression: a complex of unphosphorylated Stat1 and IRF1 supports transcription of the LMP2 gene. EMBO J. 2000;19:4111–22.
22. Heintzman ND, Hon GC, Hawkins RD, Kheradpour P, Stark A, Harp LF, et al. Histone modifications at human enhancers reflect global cell-type-specific gene expression. Nature. 2009;459:108–12.
23. Sung PS, Cheon H, Cho CH, Hong S-H, Park DY, Seo H-I, et al. Roles of unphosphorylated ISGF3 in HCV infection and interferon responsiveness. Proc Natl Acad Sci USA. 2015;112:10443–8.
24. Cheon H, Stark GR. Unphosphorylated STAT1 prolongs the expression of interferon-induced immune regulatory genes. Proc Natl Acad Sci USA. 2009;106:9373–8.
25. Reich NC, Liu L. Tracking STAT nuclear traffic. Nat Rev Immunol. 2006;6:602–12.
26. Lödige I, Marg A, Wiesner B, Malecová B, Oelgeschläger T, Vinkemeier U. Nuclear export determines the cytokine sensitivity of STAT transcription factors. J Biol Chem. 2005;280:43087–99.
27. Vinkemeier U. Getting the message across, STAT! Design principles of a molecular signaling circuit. J Cell Biol. 2004;167:197–201.
28. Meyer T, Marg A, Lemke P, Wiesner B, Vinkemeier U. DNA binding controls inactivation and nuclear accumulation of the transcription factor Stat1. Genes Dev. 2003;17:1992–2005.
29. Majoros A, Platanitis E, Szappanos D, Cheon H, Vogl C, Shukla P, et al. Response to interferons and antibacterial innate immunity in the absence of tyrosine-phosphorylated STAT1. EMBO Rep. 2016;17:367–82.
30. White LC, Wright KL, Felix NJ, Ruffner H, Reis LF, Pine R, et al. Regulation of LMP2 and TAP1 genes by IRF-1 explains the paucity of CD8+ T cells in IRF-1−/− mice. Immunity. 1996;5:365–76.
31. Dimitriou ID, Clemenza L, Scotter AJ, Chen G, Guerra FM, Rottapel R. Putting out the fire: coordinated suppression of the innate and adaptive immune systems by SOCS1 and SOCS3 proteins. Immunol Rev. 2008;224:265–83.
32. Wu J, Ma C, Wang H, Wu S, Xue G, Shi X, et al. A MyD88-JAK1-STAT1 complex directly induces SOCS-1 expression in macrophages infected with Group A Streptococcus. Cell Mol Immunol. 2015;12:373–83.
33. Heintzman ND, Stuart RK, Hon G, Fu Y, Ching CW, Hawkins RD, et al. Distinct and predictive chromatin signatures of transcriptional promoters and enhancers in the human genome. Nat Genet. 2007;39:311–8.
34. Wright KL, Ting JP. Epigenetic regulation of MHC-II and CIITA genes. Trends Immunol. 2006;27:405–12.
35. Ni Z, Karaskov E, Yu T, Callaghan SM, Der S, Park DS, et al. Apical role for BRG1 in cytokine-induced promoter assembly. Proc Natl Acad Sci USA. 2005;102:14611–6.
36. Chou SD, Khan AN, Magner WJ, Tomasi TB. Histone acetylation regulates the cell type specific CIITA promoters, MHC class II expression and antigen presentation in tumor cells. Int Immunol. 2005;17:1483–94.
37. Collinge M, Pardi R, Bender JR. Class II transactivator-independent endothelial cell MHC class II gene activation induced by lymphocyte adhesion. J Immunol. 1998;161:1589–93.
38. Zhou H, Su HS, Zhang X. Douhan J rd, Glimcher LH. CIITA-dependent and -independent class II MHC expression revealed by a dominant negative mutant [In Process Citation]. J Immunol. 1997;158:4741–9.
39. Magner WJ, Kazim AL, Stewart C, Romano MA, Catalano G, Grande C, et al. Activation of MHC class I, II, and CD40 gene expression by histone deacetylase inhibitors. J Immunol. 1950;2000(165):7017–24.
40. Guerra SG, Vyse TJ. Cunninghame Graham DS. The genetics of lupus: a functional perspective. Arthritis Res Ther. 2012;14:211.
41. Kim H-J, Eom C-Y, Kwon J, Joo J, Lee S, Nah S-S, et al. Roles of interferon-gamma and its target genes in schizophrenia: proteomics-based reverse genetics from mouse to human. Proteomics. 2012;12:1815–29.
42. Kantarci OH, Hebrink DD, Schaefer-Klein J, Sun Y, Achenbach S, Atkinson EJ, et al. Interferon gamma allelic variants: sex-biased multiple sclerosis susceptibility and gene expression. Arch Neurol. 2008;65:349–57.
43. Cooke GS, Campbell SJ, Sillah J, Gustafson P, Bah B, Sirugo G, et al. Polymorphism within the interferon-gamma/receptor complex is associated with pulmonary tuberculosis. Am J Respir Crit Care Med. 2006;174:339–43.
44. Thye T, Burchard GD, Nilius M, Müller-Myhsok B, Horstmann RD. Genomewide linkage analysis identifies polymorphism in the human interferon-gamma receptor affecting Helicobacter pylori infection. Am J Hum Genet. 2003;72:448–53.

45. Grant SFA, Hakonarson H. Microarray technology and applications in the arena of genome-wide association. Clin Chem. 2008;54:1116–24.
46. Dogan N, Wu W, Morrissey CS, Chen K-B, Stonestrom A, Long M, et al. Occupancy by key transcription factors is a more accurate predictor of enhancer activity than histone modifications or chromatin accessibility. Epigenetics Chromatin. 2015;8:16.
47. Jin F, Li Y, Dixon JR, Selvaraj S, Ye Z, Lee AY, et al. A high-resolution map of the three-dimensional chromatin interactome in human cells. Nature. 2013;503:290–4.
48. Andersson R, Gebhard C, Miguel-Escalada I, Hoof I, Bornholdt J, Boyd M, et al. An atlas of active enhancers across human cell types and tissues. Nature. 2014;507:455–61.
49. Ellinghaus D, Ellinghaus E, Nair RP, Stuart PE, Esko T, Metspalu A, et al. Combined analysis of genome-wide association studies for Crohn disease and psoriasis identifies seven shared susceptibility loci. Am J Hum Genet. 2012;90:636–47.
50. Nicolae DL, Gamazon E, Zhang W, Duan S, Dolan ME, Cox NJ. Trait-associated SNPs are more likely to be eQTLs: annotation to enhance discovery from GWAS. PLoS Genet. 2010;6:e1000888.
51. Zhong H, Beaulaurier J, Lum PY, Molony C, Yang X, Macneil DJ, et al. Liver and adipose expression associated SNPs are enriched for association to type 2 diabetes. PLoS Genet. 2010;6:e1000932.
52. Franke A, McGovern DPB, Barrett JC, Wang K, Radford-Smith GL, Ahmad T, et al. Genome-wide meta-analysis increases to 71 the number of confirmed Crohn's disease susceptibility loci. Nat Genet. 2010;42:1118–25.
53. International Genetics of Ankylosing Spondylitis Consortium (IGAS), Cortes A, Hadler J, Pointon JP, Robinson PC, Karaderi T, et al. Identification of multiple risk variants for ankylosing spondylitis through high-density genotyping of immune-related loci. Nat Genet. 2013;45:730–8.
54. Ricaño-Ponce I, Wijmenga C. Mapping of immune-mediated disease genes. Annu Rev Genomics Hum Genet. 2013;14:325–53.
55. Maurano MT, Humbert R, Rynes E, Thurman RE, Haugen E, Wang H, et al. Systematic localization of common disease-associated variation in regulatory DNA. Science. 2012;337:1190–5.
56. Schaub MA, Boyle AP, Kundaje A, Batzoglou S, Snyder M. Linking disease associations with regulatory information in the human genome. Genome Res. 2012;22:1748–59.
57. Bolstad BM, Irizarry RA, Astrand M, Speed TP. A comparison of normalization methods for high density oligonucleotide array data based on variance and bias. Bioinform Oxf Engl. 2003;19:185–93.
58. Hollander M, Wolfe DA. Nonparametric statistical methods, solutions manual. 2nd ed. New York: Wiley. http://www.wiley.com/WileyCDA/WileyTitle/productCd-047132986X.html. Accessed 6 Dec 2016.
59. Chen D, Pacal M, Wenzel P, Knoepfler PS, Leone G, Bremner R. Division and apoptosis of E2f-deficient retinal progenitors. Nature. 2009;462:925–9.

# 3

# MicroRNA 433 regulates nonsense-mediated mRNA decay by targeting *SMG5* mRNA

Yi Jin[1,2†], Fang Zhang[1†], Zhenfa Ma[1] and Zhuqing Ren[1,2*]

## Abstract

**Background:** Nonsense-mediated mRNA decay (NMD) is a RNA quality surveillance system for eukaryotes. It prevents cells from generating deleterious truncated proteins by degrading abnormal mRNAs that harbor premature termination codon (PTC). However, little is known about the molecular regulation mechanism underlying the inhibition of NMD by microRNAs.

**Results:** The present study demonstrated that miR-433 was involved in NMD pathway via negatively regulating *SMG5*. We provided evidence that (1) overexpression of miR-433 significantly suppressed the expression of *SMG5* (P < 0.05); (2) Both mRNA and protein expression levels of *TBL2* and *GADD45B*, substrates of NMD, were increased when *SMG5* was suppressed by siRNA; (3) Expression of *SMG5*, *TBL2* and *GADD45B* were significantly increased by miR-433 inhibitor (P < 0.05). These results together illustrated that miR-433 regulated NMD by targeting *SMG5* mRNA.

**Conclusions:** Our study highlights that miR-433 represses nonsense mediated mRNA decay. The miR-433 targets 3'-UTR of *SMG5* and represses the expression of *SMG5*, whereas NMD activity is decreased when SMG5 is decreased. This discovery provides evidence for microRNA/NMD regulatory mechanism.

**Keywords:** miR-433, *SMG5*, NMD

## Background

Nonsense-mediated mRNA decay (NMD) recognizes and degrades mRNAs that contain premature termination codon (PTC). This surveillance mechanism could prevent from generating deleterious truncated proteins [1, 2]. UPF1 is an essential factor for NMD, which can recognize abnormal translation termination, and activate NMD [3]. Previous studies have indicated that suppressor with morphogenetic effect on genitalia 5 (SMG5) was involved in de-phosphorylation of UPF1 [4, 5]. SMG5 and 7 shared a conserved 14-3-3 like domain that could recognize the phosphorylated UPF1 [6]. When UPF1 was de-phosphorylated, *SMG5* and *7* were recruited to

the P-bodies [7]. Phosphorylated serine 1096 of UPF1 could bind the heterodimer formed by SMG5–7 [8]. In addition this heterodimer cold increase affinity combining to UPF1 [9]. SMG5–7 and protein phosphatase 2A (PP2A) plays an important role in the recycling for UPF1 [10, 11]. Furthermore, SMG5–7 interacts with deadenylases CCR4-NOT complex [12], and SMG5 was associated to the decapping enhancer Prorich nuclear receptor co-activator 2 (PNRC2) recruiting the general decapping complex [13]. As the consequence, it is very important to investigate the expression and regulation of SMG5 for further understanding of NMD. One recent study indicated that DEAD box protein Ddx5 regulated *SMG5* expression [14]. However, few studies have reported that microRNAs could regulate *SMG5* expression. Here we demonstrated that a microRNA, miR-433, regulated *SMG5* expression. Shih et al indicated that miR-433 was associated to poorer survival in ovarian cancer [15]. MiR-433 expression was altered in gastric cancer [16], lung

*Correspondence: renzq@mail.hzau.edu.cn
†Yi Jin and Fang Zhang contributed equally to this work
[2] The Cooperative Innovation Center for Sustainable Pig Production, Huazhong Agricultural University, Wuhan 430070, Hubei, People's Republic of China
Full list of author information is available at the end of the article

dysplasia and myeloproliferative neoplasms [17, 18]. MiR-433 could also regulate the expression of *SFRP2*, *GRB2*, *CREB1*, and *HDAC6*, all of which were closely associated to cancer [19–23].

Two recent studies reported that miR-128, miR-125a, and miR-125b regulated NMD through silencing *UPF1* and *SMG1* [24, 25]. In present study, we demonstrated that miR-433 repressed the expression of *SMG5*, thereby suppressed NMD activity. Additionally, there may be a miR-433/NMD regulatory circuit in eukaryotic cells.

## Results

### MiRNA-433 targets 3′-UTR of SMG5

We selected the possible miRNAs targeting *SMG5* through miRNA algorithm TargetScan (http://www.targetscan.org/), miRanda (http://www.microrna.org/microrna/home.do), RNAhydrid (http://www.bibiserv.techfak.uni-bielefeld.de/rnahybrid/), and FINDTAR (http://www.bio.sz.tsinghua.edu.cn/). The prediction of all the softwares indicates that miR-433 targets *SMG5* (Fig. 1a). The 3′-UTR of *SMG5* holds a sequence motif (AUCAUGA) that is identical to the seed sequence UAGUACU (UAGUACU) of *miR-433* (Fig. 1a). We performed the dual-luciferase assay to investigate miR-433 targeting *SMG5*. The wild report gene vector pmirGLO-SMG5-WT and miR-433 mimic were co-transfected into HEK-293T cells. Then we detected the normalized luciferase values, which showed 50 % reduction compared with controls (P < 0.01, Fig. 1b). However, this regulation was abrogated when a four-nucleotide mutation (AUGA–UACU) was introduced in the miR-433 seed sequence in the 3′-UTR of *SMG5* (Fig. 1b). Additionally, the expression level of luciferase in SMG5 mutant was increased compared to SMG5 WT in the presence of NC (Fig. 1d). Furthermore, the dual-luciferase vectors and miR-433 inhibitor were co-transfected into Hela cells. The result showed *SMG5* expression was increased (Fig. 1c). This regulation was also abrogated when the pmirGLO-SMG5-MUT was transfected (Fig. 1c). In general, the results together showed that miR-433 targets 3′-UTR of *SMG5* directly.

### The suppression of SMG5 repressed NMD activity

The previous studies have indicated that SMG5 was an important NMD factor, so we are going to suppresse *SMG5* expression by RNAi to confirm this function. The siRNA was transfected into C2C12 cells for 24–48 h. The results showed *SMG5* was decreased in both mRNA and protein expression level (Fig. 2a, b). According to the NMD mechanism, if NMD activity reduced, the expression of NMD substrates would be reduced. The NMD activity thereby was indicated by the expression level of NMD substrates, *TBL2* and *GADD45B*. We detected the

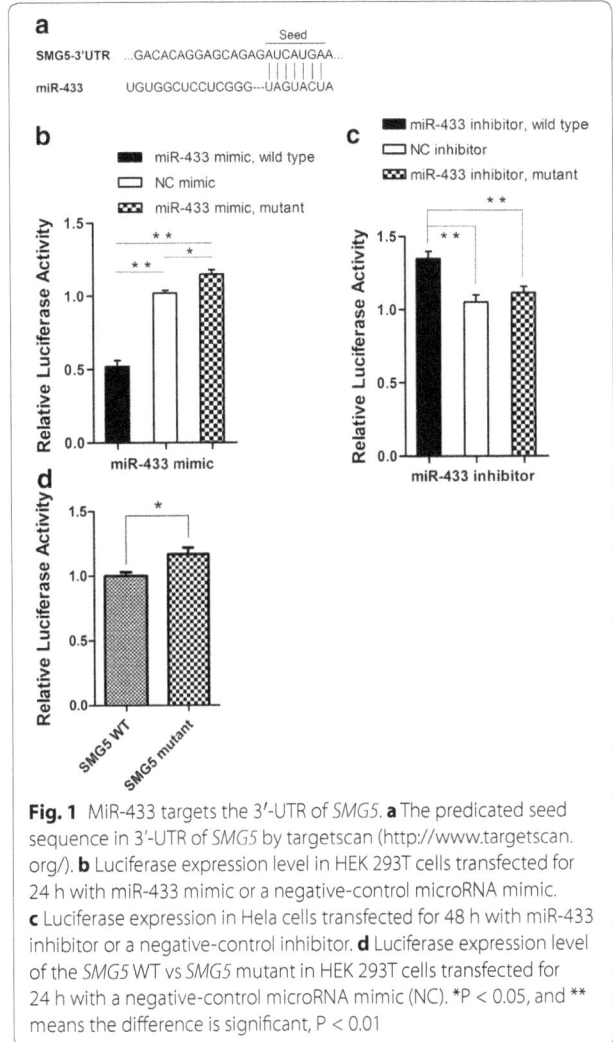

**Fig. 1** MiR-433 targets the 3′-UTR of *SMG5*. **a** The predicated seed sequence in 3′-UTR of *SMG5* by targetscan (http://www.targetscan.org/). **b** Luciferase expression level in HEK 293T cells transfected for 24 h with miR-433 mimic or a negative-control microRNA mimic. **c** Luciferase expression in Hela cells transfected for 48 h with miR-433 inhibitor or a negative-control inhibitor. **d** Luciferase expression level of the *SMG5* WT vs *SMG5* mutant in HEK 293T cells transfected for 24 h with a negative-control microRNA mimic (NC). *P < 0.05, and ** means the difference is significant, P < 0.01

expression level of *TBL2* and *GADD45* and found that both of them were increased (P < 0.05, Fig. 2c–f). Therefore the RNAi experiment indicated that suppression of SMG5 repressed NMD activity.

### MiR-433 repressed SMG5 expression

When we transfected miR-433 mimic into BHK cells, the expression level of *SMG5* was decreased, (P < 0.01, Fig. 3). Furthermore, the expression level of *SMG5* was increased significantly when the miR-433 inhibitor was transfected into the BHK cells (P < 0.01, Fig. 3). As the consequence, the *SMG5* expression was down-regulated by miR-433.

### MiR-433 repressed NMD activity

Since we have demonstrated that miR-433 repressed *SMG5* expression and suppression of *SMG5* repressed NMD activity, miR-433 would repress NMD activity. In

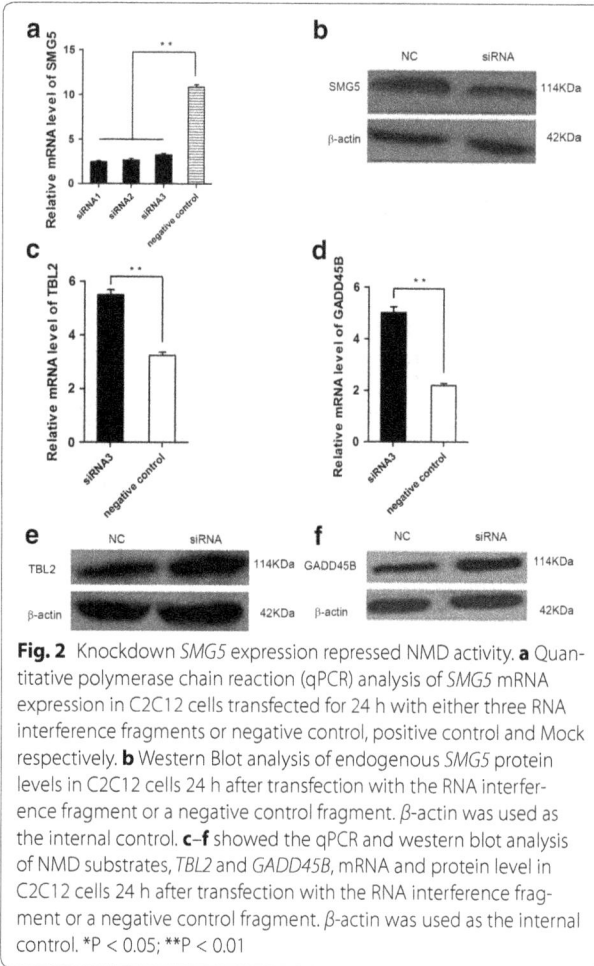

**Fig. 2** Knockdown *SMG5* expression repressed NMD activity. **a** Quantitative polymerase chain reaction (qPCR) analysis of *SMG5* mRNA expression in C2C12 cells transfected for 24 h with either three RNA interference fragments or negative control, positive control and Mock respectively. **b** Western Blot analysis of endogenous *SMG5* protein levels in C2C12 cells 24 h after transfection with the RNA interference fragment or a negative control fragment. β-actin was used as the internal control. **c–f** showed the qPCR and western blot analysis of NMD substrates, *TBL2* and *GADD45B*, mRNA and protein level in C2C12 cells 24 h after transfection with the RNA interference fragment or a negative control fragment. β-actin was used as the internal control. *P < 0.05; **P < 0.01

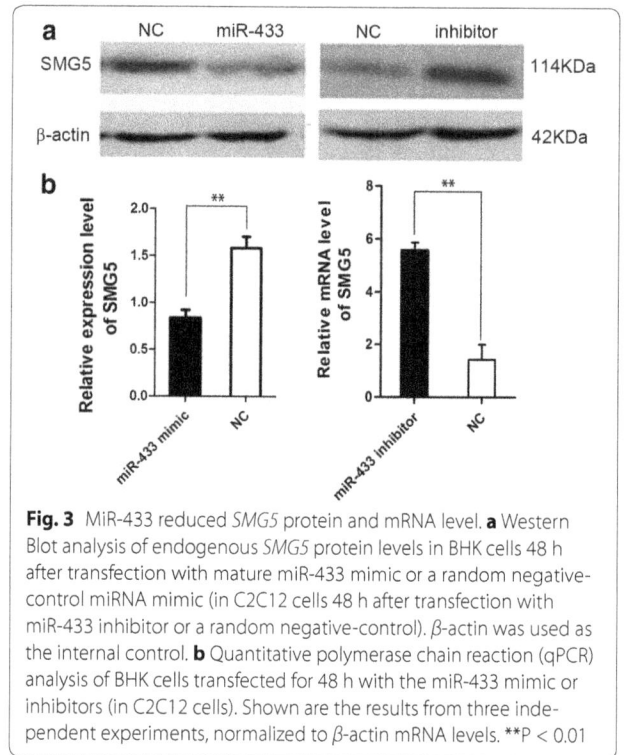

**Fig. 3** MiR-433 reduced *SMG5* protein and mRNA level. **a** Western Blot analysis of endogenous *SMG5* protein levels in BHK cells 48 h after transfection with mature miR-433 mimic or a random negative-control miRNA mimic (in C2C12 cells 48 h after transfection with miR-433 inhibitor or a random negative-control). β-actin was used as the internal control. **b** Quantitative polymerase chain reaction (qPCR) analysis of BHK cells transfected for 48 h with the miR-433 mimic or inhibitors (in C2C12 cells). Shown are the results from three independent experiments, normalized to β-actin mRNA levels. **P < 0.01

this section, the results showed that the NMD substrates *TBL2* and *GADD45B* were increased in BHK cells when transfected with miR-433 mimic (P < 0.05, Fig. 4). In addition, we chose another cell line (C2C12) to transfect miR-433 inhibitor. The expression level of *TBL2* and *GADD45B* were also decreased by miR-433 inhibitor (P < 0.05, Fig. 4).

## Discussion

Nonsense-mediated mRNA decay, a surveillance system, which recognizes mRNAs with translation termination codons positioned in abnormal contexts and degrades aberrant mRNAs, scrutinizes mRNA quality in all mammalian cells [26]. In this intricate process, a set of NMD factors are influential to the recognition and degradation of aberrant mRNAs. UPF1, the master regulator of NMD, is considered to determine the NMD process [27–29]. The ATPase and helicase activities as well as the phosphorylation of N- and C- terminal domains bestow on UPF1 ability that recognizes premature termination codon containing mRNAs selectively [30–33]. The phosphorylation and de-phosphorylation of UPF1 were

very important for NMD. SMG5, an important NMD factor, played an important role in UPF1 de-phosphorylation process. As our results showed, NMD activity was repressed when the expression of *SMG5* was suppressed. So the regulation for *SMG5* could affect NMD activity. We selected a microRNA, miR-433, which targets 3'-UTR of *SMG5* by prediction. The regulation of miR-433 targeted *SMG5* was detected by dual-luciferase report assay. The result indicated that *SMG5* expression was reduced by 50 % when we transfected with miR-433 mimic. It has demonstrated that the NMD activity was repressed when *SMG5* was suppressed (Fig. 3). Hence, miRNA-433 repressed NMD activity by suppressing *SMG5* expression.

As a conserved mRNA surveillance system in eukaryotic cells, NMD is crucial for many physiological processes, and these crucial function of NMD have been published by many researches. Numbers of NMD factors were important for embryo development. When *UPF1*, *UPF2*, *SMG1* or *SMG6* was knockout or knockdown technologically, all mice died during an early embryonic stage [34–37]. A study implied that NMD has the potency that affects cell type diversification [34]. NMD activity was identified to be reduced in neuronal stem cells differentiation process. And this regulation of NMD activity is essential for nervous system development [38]. Additionally, NMD also regulates neural

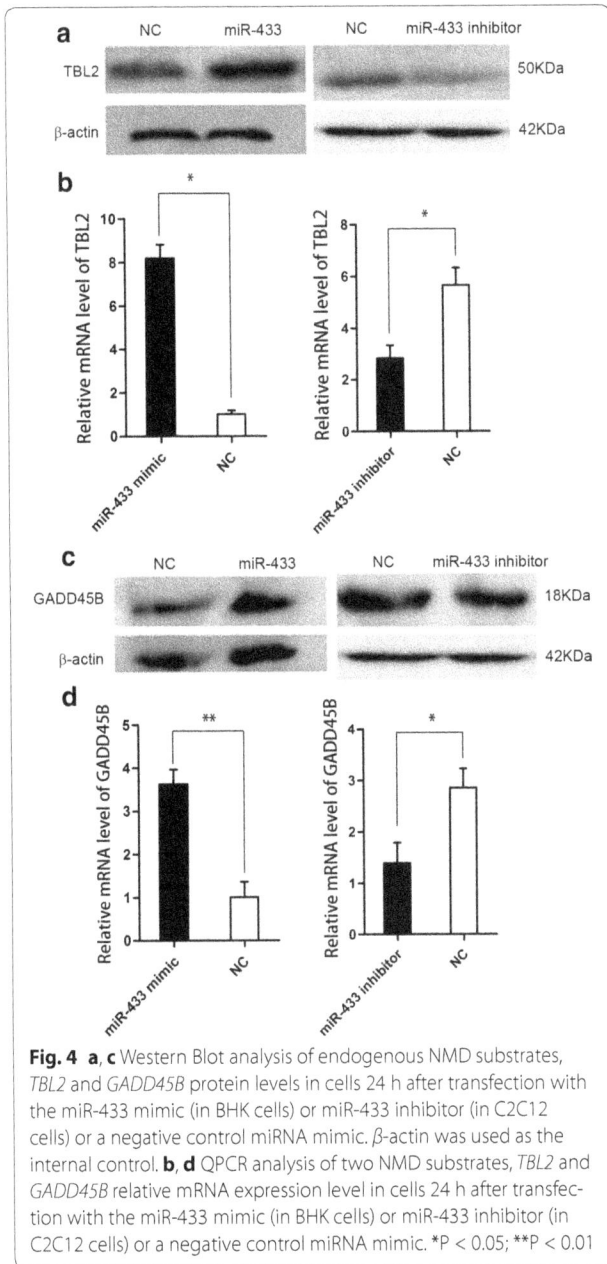

Fig. 4 a, c Western Blot analysis of endogenous NMD substrates, TBL2 and GADD45B protein levels in cells 24 h after transfection with the miR-433 mimic (in BHK cells) or miR-433 inhibitor (in C2C12 cells) or a negative control miRNA mimic. β-actin was used as the internal control. b, d QPCR analysis of two NMD substrates, TBL2 and GADD45B relative mRNA expression level in cells 24 h after transfection with the miR-433 mimic (in BHK cells) or miR-433 inhibitor (in C2C12 cells) or a negative control miRNA mimic. *P < 0.05; **P < 0.01

development related genes [39, 40]. Here, our experiment indicated that miR-433 repressed SMG5 expression, and therefore suppressed NMD activity. This result indicated a circuit regulation of miR-433 and NMD. Hence, miR-433 could participate in the process regulated by NMD. For the circuit regulation, previous study identified miR-128 repressed NMD by UPF1 and MLN51 [24]. MiR-125 and 128 were involved in many nervous system disorder diseases such as autism [41], prion-induced neuron degeneration [42], Huntington's disease [43], Parkinson's disease [44], as well as Alzheimer's disease [45]. Some studies indicated that NMD

played an important role in brain development and embryonic development [46, 47], and three essential embryonic brain vesicles formation disrupted by knockout of SMG1 [48]. Therefore, we assumed that there may be a miR-433/NMD regulatory circuit in early embryo development, especially in early nervous and brain development. This study would provide evidence to make clear the microRNA/NMD regulatory mechanism and give advice to neurological disease and related cancers therapies.

## Conclusion

It is well known that nonsense mediated mRNA decay is an important mRNA quality surveillance system. However, the microRNA/NMD regulatory mechanism is not fully clear. Our study claim that NMD activity is repressed by a microRNA, miR-433. The miR-433 targets 3'UTR of SMG5 and decreases the expression of SMG5. NMD activity is suppressed when expression of SMG5 is reduced. This discovery provides evidence for microRNA/NMD regulatory mechanism and therapeutic advice for NMD-related diseases.

## Methods

### Cell culture

The cell lines, HEK-293T, Hela, BHK-21, and C2C12, were purchased from China for type culture collection (CCTCC). Cells expect for Hela were maintained in Dulbecco's modified Eagle's medium (DMEM, Hyclone, Logan, Utah, USA) supplemented with 10 % fetal bovine serum (FBS, Hyclone, Logan, Utah, USA) at 37 °C in a humidified atmosphere of 5 % $CO_2$. Hela cells were maintained in DMEM (Hyclone, Logan, Utah, USA) containing 20 % FBS (Hyclone, Logan, Utah, USA) at 37 °C in a humidified atmosphere of 5 % $CO_2$. Medium was changed twice weekly until the fibroblasts migrated out to cover the dishes. Passage 6–9 cells were used for all experiments.

### Isolation of 3'-UTR sequences of SMG5

C2C12 genome DNA was used as template to amplify 3'-UTR of SMG5. To isolate the 464-bp 3'-UTR of 2.6-kb genome, the following primer pairs were used: SMG5-3'UTR-F and SMG5-3'UTR-R [5'-cgagctcCTG TACTGGATAAGGGGTGCC-3' and 5'-ccgctcgagCA ACTGTTCCCTGGTTTTCCC, with uppercase bases corresponding to the mouse 2.6-kb SMG5 genome and lowercase bases indicating 5' extensions with restriction enzyme sites (underline) for SacI and XhoI, respectively]. The protocol was as follows: pre-denaturation at 94 °C for 2 min, denaturation at 94 °C for 30 s, annealing at 64 °C for 30 s, elongation at 72 °C for 30 s per 1 kb, thirty-five cycles, elongation at 72 °C for 2 min, and end.

PCR products were investigated by 1–2 % agarose gel electrophoresis. The restriction enzyme, SacI and XhoI were purchased from Fermentas (Thermo Fisher Scientific, Rockford, IL, USA).

### Wild type report gene constructs

The 2.2. PCR product was purified and cloned into the SacI and XhoI sites of pmirGLO dual-luciferase miRNA target expression vector (Promega, Madison, WI, USA). The enzyme digestion system contained 10× buffer 4 μL, SacI 4 μL, XhoI 4 μL, pmirGLO vector 12 μL, and the ddH$_2$O up to 40 μL. The enzyme digestion system was incubated at 37 °C for 2 h. The T$_4$ DNA ligase (TaKaRa) interaction system contained enzyme digestion fragment 3 μL, digested pmirGLO plasmid 2 μL, Solution I 5 μL. The system was incubated at 16 °C for 2 h. The accuracy of inserted 3'-UTR of SMG5 were confirmed by sequencing. Then the plasmid was isolated and purified using E.Z.N.A.® endo-free plasmid mini kit II (OMEGA, Bio-Tek, Norcross, GA, USA). The constructed wild type report gene recombinants named pmirGLO-SMG5-WT was stored at −20 °C.

### Introducing four-nucleotide mutagenesis into SMG5 3'-UTR

Two primer pairs, SMG5-3'UTR-F (5'-cgagctcCTGTA CTGGATAAGGGGTGCC-3') and SMG5-3'UTR-R (5'-ccgctcgagCAACTGTTCCCTGGTTTTCCC), SMG5-3'UTR-MF (5'-AGCAGAGATC**TACT**ATACTC AG-3') and SMG5-3'UTR-MR (5'-GCCCCTGAGT AT**AGTA**GATCTCTG-3'), were designed (uppercase bases corresponding to the mouse 2.6-kb SMG5 genome and lowercase bases indicating 5' extensions with restriction enzyme sites [underline] for SacI and XhoI, respectively. And the bold bases mean four-nucleotide mutation sites of SMG5 3'-UTR). The former was used to construct pmirGLO-SMG5-WT, whereas the latter to amplify the mutation sites and its upstream sequence (production I), and the mutation sites and its downstream sequence (production II) respectively, the pmirGLO-SMG5-WT plasmid was the template. Then using fusion PCR combined these two products. The PCR system contained 2× taq PCR mix 11 μL, recovered production I 2 μL, recovered production II 2 μL, ddH$_2$O 10 μL. PCR conditions: an initial denaturation at 95 °C for 4 min, followed by 30 cycles of 95 °C for 30 s, 40 °C for 30 s, 45 °C for 5 s, 50 °C for 5 s, 55 °C for 5 s, 60 °C for 5 s and 72 °C for 30 s, with a final extension of 5 min at 72 °C. PCR products were visible after electrophoresis of 25-μL reaction using an agarose gel (1.5 %). Then the fusion PCR production was recovered and used as a template.

### Mutant report gene constructs

The 464 bp mutant SMG5 3'-UTR was amplified, purified, and cloned into the SacI and XhoI sites of pmirGLO dual-luciferase miRNA target expression vector (Promega, Madison, WI). The enzyme digestion system contained 10× buffer 4 μL, SacI 4 μL, XhoI 4 μL, pmirGLO vector 12 μL, and ddH$_2$O up to 40 μL. The enzyme digestion system was incubated at 37 °C for 2 h. The T$_4$ DNA ligase (TaKaRa) interaction system contained enzyme digestion fragment 3 μL, digested pmirGLO plasmid 2 μL, Solution I 5 μL. This system was incubated at 16 °C for 2 h. The accuracy of inserted SMG5 3'-UTR was confirmed by sequencing. Then the plasmid was isolated and purified using E.Z.N.A.® endo-free plasmid mini kit II (OMEGA, Bio-Tek, Norcross, GA, USA). The mutant report gene vector,named pmirGLO-SMG5-MUT conserved at −20 °C.

### Cell transfection and dual-luciferase assay

HEK-293T cells were seeded onto 24-well plates at 30–40 % confluence. The next day, medium was changed to Opti-MEM I Reduced Serum Medium (Hyclone, Logan, Utah, USA), and cells were transfected with pmirGLO-SMG5-WT constructs (200 ng per well). After 4–6 h, medium was replaced by MEM containing 10 % FBS (Hyclone, Logan, Utah, USA). Transfected cells were washed twice with cold PBS (Hyclone, Logan, Utah, USA), lysed using 1× passive lysis buffer (Promega) for 30 min, and assayed for firefly and Renilla luciferase activities by the dual-luciferase assay (Promega) in a PerkimELmer 2030 Microplate Reader (Lenovo, China) according to the manufacturer's instructions. At least three independent experiments were performed for each assay, each time with a minimum of n = 4. For co-transfection, cells were grown on 24-well plates. For each well, a transfection mixture (200 μL) consisting of 0.2-μg reporter construct, 1.25 μL miR-433 mimic or negative control, and 2 μL Lipofectamine 2000 Reagent (Invitrogen) was prepared. Opti-MEM I Reduced Serum Medium was used as diluent. Hella cells were treated as same as the HEK-293T, but the medium was MEM containing 20 % FBS.

### RNA isolation and mRNA/miRNA quantitation

The RNA isolation reagent was TRizol® Reagent (Invitrogen, Carlsbad, CA, USA). RNA was quantified spectroscopically (ND-1000 Spectrophotometer; NanoDrop, Wilmington, DE, USA), and integrity was assessed by agarose gel electrophoresis. According to the manufacturer's protocol, RNA was reverse-transcribed using RNA PCR kit (Takara Bio Inc.). The gene expression was investigated by RT-PCR. Primers for RT-PCR were as follows: SMG5-qF (5′-TACCTCATCCCTGACACCCA-3′) and SMG5-qR (5′-GCCCCTGGCTGTTCTTTCT-3′), miR-433-qF (CTGGTAGGATCATGATGGGAT) and miR-433-qR (TCAACTGGTGTCGTGGAGT), U6-qF (CTGGTAGGGTGCTCGCTTCGGCAG) and U6-qR (CAACTGGTGTCGTGGAGTCGGC), TBL2-qF (5′-AAGTATCTGGCCACCTGTGC-3′) and TBL2-qR (5′-GGCCAGCCAAACAATGAAGG-3′), GADD45B-qF (5′-AAGGCGGCCAAACTGATGAAT-3′) and GADD45B-qR (5′-ATTGTCGCAGCAGAACGACT-3′), β-actin-F (5′-GCCTCACTGTCCACCTTCCA-3′) and β-actin-R (5′-AGCCATGCCAATGTTGTCTCTT-3′). All the primer pairs for each gene were used for gene-specific amplification. The expression of β-actin was the internal control. For quantification, the comparative threshold cycle method was used to assess relative changes in mRNA levels [49].

## MicroRNA overexpression assay

MiR-433 mimic and inhibitor were purchased from GenePharma (shanghai, China). MiR-433 mimic could enhance endogenous miR-433 function, and miR-433 inhibitor could repress the endogenous miR-433 function. HK-21 cells and C2C12 cells were seeded onto 6-well plates in DMEM supplemented with 10 % FBS at 37 °C in a humidified atmosphere of 5 % $CO_2$. The next day, when the cells confluence get 70–80 %, the medium was changed to Opti-MEM I Reduced Serum Medium (Hyclone, Logan, Utah, USA), and cells were transfected with 10 µL miRNA mimic (0.02 nmol/µL) using Lipofectamine 2000 Reagent (Invitrogen). After 4–6 h, medium was replaced by MEM containing 10 % FBS (Hyclone, Logan, Utah, USA). Transfected cells were washed twice with cold PBS (Hyclone, Logan, Utah, USA). At least three independent experiments were performed for each assay, each time with a minimum of n = 4. Opti-MEM I Reduced Serum Medium was used as diluent. The mimic was added fluorescent tag so that it could be detected under fluorescence microscope to calculate the transfection efficiency.

## Small interfering RNA design and cell transfection

The musSMG5 mRNA sequence and internal control (β-actin) mRNA sequences were obtained from NCBI database (http://www.ncbi.nlm.nih.gov/). The siRNAs were purchased from Genpharma (Inc. Shanghai, China). The sequences if siRNAs were as follows: siRNA1 5′-GCCGCUUCAUCAUCAUCAUTT-3′ for sense and 5′-AUGAUGAUGAUGAAGCGGCTT-3′ for antisense; siRNA2 5′-GGAGUGUGAAAGUGGAUAUTT-3′ for sense and 5′-AUAUCCACUUUCACACUCCTT-3′ for antisense; siRNA3 5′-GCAGGCAGCAA

GUAUUACATT-3′ for sense and 5′-UGUAAUACUUGCUGCCUGCTT-3′ for antisense; NC (negative control) 5′-UUCUCCGAACGUGUCACGUTT-3′ for sense and 5′-ACGUGACACGUUCGGAGAATT-3′ for antisense. C2C12 cells were incubated in 5 % CO2 at 37 °C in DMEM (Hyclone, Logan, Utah, USA) supplemented with 10 % FBS (Hyclone, Logan, Utah, USA). One day before transfection, $10^5$ cells were plated in 2 mL DMEM containing 10 % FBS per well of a 6-well plate. The next day, when the cells confluence get 50 %, the medium was changed to Opti-MEM I Reduced Serum Medium (Hyclone, Logan, Utah, USA), and cells were transfected with 10 µL siRNA using 10 µL Lipofectamine 2000 Reagent (Invitrogen). After 4–6 h, medium was replaced by MEM containing 10 % FBS (Hyclone, Logan, Utah, USA). The cells were incubated for 24–48 h. Transfected cells were washed twice with cold PBS (Hyclone, Logan, Utah, USA). At least three independent experiments were performed for each assay, each time with a minimum of n = 4. Opti-MEM I Reduced Serum Medium was used as diluent.

## Western blotting

The whole-cell protein of BHK-21 and C2C12 were collected using RIPA and PMSF, and conserved at −20 °C. The concentration of protein was detected using the bicinchoninic acid (BCA) method (Beyotime Biotechnology, Jiangsu, China). Subsequently, samples were separated by SDS–PAGE (10 %) and electroblotted onto polyvinylidene fluoride (PVDF) membrane (Millipore, USA). Membrane was blocked with 1× TBST supplemented with 5 % skimmed milk powder (servicebio, Wuhan, China). Membranes were incubated with the primary antibody, anti-SMG5 (P-14) (goat. #SC50980, Santa Cruz Biotechnology, Inc.). For normalization of the results, membranes were reblotted for β-actin (anti-β-actin, servicebio, Wuhan). Proteins were detected with secondary antibody (HRP-rabbit, anti-goat, 1:1000; boster, Wuhan, China). For the TBL2 and GADD45B western blotting analysis, the primary antibodies were anti-TBL2 (L-15) (goat. #SC-104692, Santa Cruz Biotechnology, Inc.), and anti-GADD45B (rabbit, #ab128920, abcam, Inc.). Bands on the X-ray films were quantified with WCIF ImageJ software for the densitometry analysis.

## Statistical analysis

All data were displayed as the mean ± SD of three independent experiments, in which each assay was performed in triplicate. Statistically significant differences between two groups were determined by Student's t test. $P < 0.05$ was considered statistically significant.

## Abbreviations

miR-433: microRNA-433; NMD: nonsense mediated mRNA decay; SMG5: suppressor with morphological effect on genitalia 5; PTC: premature termination codon; TBL2: transducin beta-like protein 2; GADD45B: growth arrest and DNA damage-inducible protein GADD45 beta; UTR: untranslated regions; UPF1: up-framshift 1; SMG7: suppressor with morphological effect on genitalia 7; PP2A: protein phosphatase 2A; PNRC2: prorich nuclear receptor co-activator 2; SFRP2: secreted frizzled-related protein 2; GRB2: growth factor receptor-bound protein 2; CREB1: CAMP-response element binding protein; HDAC6: histone deacetylase 6.

## Authors' contributions

YJ sorted and analyzed data, wrote the article. FZ designed this experiment, and performed the experiment. ZM purchased the regents and materials for the experiment, and performed some part of the experiment. ZR designed this work and participated in the revision of this article. All authors read and approved the final manuscript.

## Author details

[1] Key Laboratory of Swine Genetics and Breeding of Ministry of Agriculture & Key Laboratory of Agriculture Animal Genetics, Breeding and Reproduction of Ministry of Education, College of Animal Science, Huazhong Agricultural University, Wuhan 430070, Hubei, People's Republic of China. [2] The Cooperative Innovation Center for Sustainable Pig Production, Huazhong Agricultural University, Wuhan 430070, Hubei, People's Republic of China.

## Acknowledgements

None.

## Competing interests

The authors declare that they have no competing interests.

## Funding

This work was supported by the Fundamental Research Funds for the Central Universities (No. 2662015PY103, No. 2013PY032), the Natural Science Foundation of Hubei Province of China (No. 2014CFB265) and the National Key Technology Support Program of China (No. 2015BAI09B06).

## References

1. Nagy E, Maquat LE. A rule for termination-codon position within intron-containing genes: when nonsense affects RNA abundance. Trends Biochem Sci. 1998;23(6):198–9.
2. Peltz SW, Brown AH, Jacobson A. mRNA destabilization triggered by premature translational termination depends on at least three cis-acting sequence elements and one trans-acting factor. Genes Dev. 1993;7(9):1737–54.
3. Hilleren P, Parker R. mRNA surveillance in eukaryotes: kinetic proofreading of proper translation termination as assessed by mRNP domain organization? RNA. 1999;5(6):711–9.
4. Chiu SY, Serin G, Ohara O, Maquat LE. Characterization of human Smg5/7a: a protein with similarities to *Caenorhabditis elegans* SMG5 and SMG7 that functions in the dephosphorylation of Upf1. RNA. 2003;9(1):77–87.
5. Ohnishi T, Yamashita A, Kashima I, Schell T, Anders KR, Grimson A, Hachiya T, Hentze MW, Anderson P, Ohno S. Phosphorylation of hUPF1 induces formation of mRNA surveillance complexes containing hSMG-5 and hSMG-7. Mol Cell. 2003;12(5):1187–200.
6. Fukuhara N, Ebert J, Unterholzner L, Lindner D, Izaurralde E, Conti E. SMG7 is a 14-3-3-like adaptor in the nonsense-mediated mRNA decay pathway. Mol Cell. 2005;17(4):537–47.
7. Unterholzner L, Izaurralde E. SMG7 acts as a molecular link between mRNA surveillance and mRNA decay. Mol Cell. 2004;16(4):587–96.
8. Okada-Katsuhata Y, Yamashita A, Kutsuzawa K, Izumi N, Hirahara F, Ohno S. N- and C-terminal Upf1 phosphorylations create binding platforms for SMG-6 and SMG-5:SMG-7 during NMD. Nucleic Acids Res. 2012;40(3):1251–66.
9. Jonas S, Weichenrieder O, Izaurralde E. An unusual arrangement of two 14-3-3-like domains in the SMG5-SMG7 heterodimer is required for efficient nonsense-mediated mRNA decay. Genes Dev. 2013;27(2):211–25.
10. Anders KR, Grimson A, Anderson P. SMG-5, required for *C. elegans* nonsense-mediated mRNA decay, associates with SMG-2 and protein phosphatase 2A. EMBO J. 2003;22(3):641–50.
11. Durand S, Cougot N, Mahuteau-Betzer F, Nguyen CH, Grierson DS, Bertrand E, Tazi J, Lejeune F. Inhibition of nonsense-mediated mRNA decay (NMD) by a new chemical molecule reveals the dynamic of NMD factors in P-bodies. J Cell Biol. 2007;178(7):1145–60.
12. Loh B, Jonas S, Izaurralde E. The SMG5-SMG7 heterodimer directly recruits the CCR4-NOT deadenylase complex to mRNAs containing nonsense codons via interaction with POP2. Genes Dev. 2013;27(19):2125–38.
13. Cho H, Han S, Choe J, Park SG, Choi SS, Kim YK. SMG5-PNRC2 is functionally dominant compared with SMG5-SMG7 in mammalian nonsense-mediated mRNA decay. Nucleic Acids Res. 2013;41(2):1319–28.
14. Geissler V, Altmeyer S, Stein B, Uhlmann-Schiffler H, Stahl H. The RNA helicase Ddx5/p68 binds to hUpf3 and enhances NMD of Ddx17/p72 and Smg5 mRNA. Nucleic Acids Res. 2013;41(16):7875–88.
15. Shih KK, Qin LX, Tanner EJ, Zhou Q, Bisogna M, Dao F, Olvera N, Viale A, Barakat RR, Levine DA. A microRNA survival signature (MiSS) for advanced ovarian cancer. Gynecol Oncol. 2011;121(3):444–50.
16. Ueda T, Volinia S, Okumura H, Shimizu M, Taccioli C, Rossi S, Alder H, Liu CG, Oue N, Yasui W, et al. Relation between microRNA expression and progression and prognosis of gastric cancer: a microRNA expression analysis. Lancet Oncol. 2010;11(2):136–46.
17. Del Vescovo V, Meier T, Inga A, Denti MA, Borlak J. A cross-platform comparison of affymetrix and Agilent microarrays reveals discordant miRNA expression in lung tumors of c-Raf transgenic mice. PLoS ONE. 2013;8(11):e78870.
18. Lin X, Rice KL, Buzzai M, Hexner E, Costa FF, Kilpivaara O, Mullally A, Soares MB, Ebert BL, Levine R, et al. miR-433 is aberrantly expressed in myeloproliferative neoplasms and suppresses hematopoietic cell growth and differentiation. Leukemia. 2013;27(2):344–52.
19. Gotanda K, Hirota T, Matsumoto N, Ieiri I. MicroRNA-433 negatively regulates the expression of thymidylate synthase (TYMS) responsible for 5-fluorouracil sensitivity in HeLa cells. BMC Cancer. 2013;13:369.
20. Luo H, Zhang H, Zhang Z, Zhang X, Ning B, Guo J, Nie N, Liu B, Wu X. Down-regulated miR-9 and miR-433 in human gastric carcinoma. J Exp Clin Cancer Res. 2009;28:82.
21. Simon D, Laloo B, Barillot M, Barnetche T, Blanchard C, Rooryck C, Marche M, Burgelin I, Coupry I, Chassaing N, et al. A mutation in the 3'-UTR of the HDAC6 gene abolishing the post-transcriptional regulation mediated by hsa-miR-433 is linked to a new form of dominant X-linked chondrodysplasia. Hum Mol Genet. 2010;19(10):2015–27.
22. Snyder CM, Rice AL, Estrella NL, Held A, Kandarian SC, Naya FJ. MEF2A regulates the Gtl2-Dio3 microRNA mega-cluster to modulate WNT signaling in skeletal muscle regeneration. Development. 2013;140(1):31–42.
23. Yang Z, Tsuchiya H, Zhang Y, Hartnett ME, Wang L. MicroRNA-433 inhibits liver cancer cell migration by repressing the protein expression and function of cAMP response element-binding protein. J Biol Chem. 2013;288(40):28893–9.
24. Bruno IG, Karam R, Huang LL, Bhardwaj A, Lou CH, Shum EY, Song HW, Corbett MA, Gifford WD, Gecz J, et al. Identification of a MicroRNA that activates gene expression by repressing nonsense-mediated RNA decay. Mol Cell. 2011;42(4):500–10.
25. Wang G, Jiang B, Jia C, Chai B, Liang A. MicroRNA 125 represses nonsense-mediated mRNA decay by regulating SMG1 expression. Biochem Biophys Res Commun. 2013;435(1):16–20.
26. Lykke-Andersen SJT. Nonsense-mediated mRNA decay: an intricate machinery that shapes transcriptomes. Nat Rev Mol Cell Biol. 2015;16:665–7.
27. Kervestin S, Jacobson A. NMD: a multifaceted response to premature translational termination. Nat Rev Mol Cell Biol. 2012;13(11):700–12.

28. Schweingruber C, Rufener SC, Zund D, Yamashita A, Muhlemann O. Nonsense-mediated mRNA decay - mechanisms of substrate mRNA recognition and degradation in mammalian cells. Biochim Biophys Acta. 2013;1829(6–7):612–23.

29. Siwaszek A, Ukleja M, Dziembowski A. Proteins involved in the degradation of cytoplasmic mRNA in the major eukaryotic model systems. RNA Biol. 2014;11(9):1122–36.

30. Franks TM, Singh G, Lykke-Andersen J. Upf1 ATPase-dependent mRNP disassembly is required for completion of nonsense-mediated mRNA decay. Cell. 2010;143(6):938–50.

31. Kashima I, Yamashita A, Izumi N, Kataoka N, Morishita R, Hoshino S, Ohno M, Dreyfuss G, Ohno S. Binding of a novel SMG-1-Upf1-eRF1-eRF3 complex (SURF) to the exon junction complex triggers Upf1 phosphorylation and nonsense-mediated mRNA decay. Genes Dev. 2006;20(3):355–67.

32. Kurosaki T, Li W, Hoque M, Popp MW, Ermolenko DN, Tian B, Maquat LE. A post-translational regulatory switch on UPF1 controls targeted mRNA degradation. Genes Dev. 2014;28(17):1900–16.

33. Weng Y, Czaplinski K, Peltz SW. Genetic and biochemical characterization of mutations in the ATPase and helicase regions of the Upf1 protein. Mol Cell Biol. 1996;16(10):5477–90.

34. Li T, Shi Y, Wang P, Guachalla LM, Sun B, Joerss T, Chen YS, Groth M, Krueger A, Platzer M, et al. Smg6/Est1 licenses embryonic stem cell differentiation via nonsense-mediated mRNA decay. EMBO J. 2015;34(12):1630–47.

35. McIlwain DR, Pan Q, Reilly PT, Elia AJ, McCracken S, Wakeham AC, Itie-Youten A, Blencowe BJ, Mak TW. Smg1 is required for embryogenesis and regulates diverse genes via alternative splicing coupled to nonsense-mediated mRNA decay. Proc Natl Acad Sci USA. 2010;107(27):12186–91.

36. Medghalchi SM, Frischmeyer PA, Mendell JT, Kelly AG, Lawler AM, Dietz HC. Rent1, a trans-effector of nonsense-mediated mRNA decay, is essential for mammalian embryonic viability. Hum Mol Genet. 2001;10(2):99–105.

37. Weischenfeldt J, Damgaard I, Bryder D, Theilgaard-Monch K, Thoren LA, Nielsen FC, Jacobsen SE, Nerlov C, Porse BT. NMD is essential for hematopoietic stem and progenitor cells and for eliminating by-products of programmed DNA rearrangements. Genes Dev. 2008;22(10):1381–96.

38. Lou CH, Shao A, Shum EY, Espinoza JL, Huang L, Karam R, Wilkinson MF. Posttranscriptional control of the stem cell and neurogenic programs by the nonsense-mediated RNA decay pathway. Cell Rep. 2014;6(4):748–64.

39. Colak D, Ji SJ, Porse BT, Jaffrey SR. Regulation of axon guidance by compartmentalized nonsense-mediated mRNA decay. Cell. 2013;153(6):1252–65.

40. Nguyen LS, Jolly L, Shoubridge C, Chan WK, Huang L, Laumonnier F, Raynaud M, Hackett A, Field M, Rodriguez J, et al. Transcriptome profiling of UPF3B/NMD-deficient lymphoblastoid cells from patients with various forms of intellectual disability. Mol Psychiatry. 2012;17(11):1103–15.

41. Abu-Elneel K, Liu T, Gazzaniga FS, Nishimura Y, Wall DP, Geschwind DH, Lao K, Kosik KS. Heterogeneous dysregulation of microRNAs across the autism spectrum. Neurogenetics. 2008;9(3):153–61.

42. Saba R, Goodman CD, Huzarewich RL, Robertson C, Booth SA. A miRNA signature of prion induced neurodegeneration. PLoS ONE. 2008;3(11):e3652.

43. Lee ST, Chu K, Im WS, Yoon HJ, Im JY, Park JE, Park KH, Jung KH, Lee SK, Kim M, et al. Altered microRNA regulation in Huntington's disease models. Exp Neurol. 2011;227(1):172–9.

44. Kim J, Inoue K, Ishii J, Vanti WB, Voronov SV, Murchison E, Hannon G, Abeliovich A. A MicroRNA feedback circuit in midbrain dopamine neurons. Science. 2007;317(5842):1220–4.

45. Lukiw WJ. Micro-RNA speciation in fetal, adult and Alzheimer's disease hippocampus. NeuroReport. 2007;18(3):297–300.

46. Wittkopp N, Huntzinger E, Weiler C, Sauliere J, Schmidt S, Sonawane M, Izaurralde E. Nonsense-mediated mRNA decay effectors are essential for zebrafish embryonic development and survival. Mol Cell Biol. 2009;29(13):3517–28.

47. Anastasaki C, Longman D, Capper A, Patton EE, Caceres JF. Dhx34 and Nbas function in the NMD pathway and are required for embryonic development in zebrafish. Nucleic Acids Res. 2011;39(9):3686–94.

48. McIlwain DR, Pan Q, Reilly PT, Elia AJ, McCracken S, Wakeham AC, Itie-Youten A, Blencowe BJ, Mak TW. Smg1 is required for embryogenesis and regulates diverse genes via alternative splicing coupled to nonsense-mediated mRNA decay. P Natl Acad Sci USA. 2010;107(27):12186–91.

49. Livak KJ, Schmittgen TD. Analysis of relative gene expression data using real-time quantitative PCR and the 2(-Delta Delta C(T)) method. Methods. 2001;25(4):402–8.

# 4

# 3C-digital PCR for quantification of chromatin interactions

Meijun Du and Liang Wang[*]

## Abstract

**Background:** Chromosome conformation capture (3C) is a powerful and widely used technique for detecting the physical interactions between chromatin regions in vivo. The principle of 3C is to convert physical chromatin interactions into specific DNA ligation products, which are then detected by quantitative polymerase chain reaction (qPCR). However, 3C-qPCR assays are often complicated by the necessity of normalization controls to correct for amplification biases. In addition, qPCR is often limited to a certain cycle number, making it difficult to detect fragment ligations with low frequency. Recently, digital PCR (dPCR) technology has become available, which allows for highly sensitive nucleic acid quantification. Main advantage of dPCR is its high precision of absolute nucleic acid quantification without requirement of normalization controls.

**Results:** To demonstrate the utility of dPCR in quantifying chromatin interactions, we examined two prostate cancer risk loci at 8q24 and 2p11.2 for their interaction target genes *MYC* and *CAPG* in LNCaP cell line. We designed anchor and testing primers at known regulatory element fragments and target gene regions, respectively. dPCR results showed that interaction frequency between the regulatory element and *MYC* gene promoter was 0.7 (95% CI 0.40– 1.10) copies per 1000 genome copies while other regions showed relatively low ligation frequencies. The dPCR results also showed that the ligation frequencies between the regulatory element and two *EcoRI* fragments containing *CAPG* gene promoter were 1.9 copies (95% CI 1.41–2.47) and 1.3 copies per 1000 genome copies (95% CI 0.76–1.92), respectively, while the interaction signals were reduced on either side of the promoter region of *CAPG* gene. Additionally, we observed comparable results from 3C-dPCR and 3C-qPCR at 2p11.2 in another cell line (DU145).

**Conclusions:** Compared to traditional 3C-qPCR, our results show that 3C-dPCR is much simpler and more sensitive to detect weak chromatin interactions. It may eliminate multiple and complex normalization controls and provide accurate calculation of proximity-based fragment ligation frequency. Therefore, we recommend 3C-dPCR as a preferred method for sensitive detection of low frequency chromatin interactions.

**Keywords:** Chromatin interaction, Chromosome conformation capture, Digital PCR, Quantitative PCR

## Background

Chromosome conformation capture (3C) has been widely used for detecting the physical interactions of chromosomal regions in vivo [1]. In general, 3C library is first built by three basic steps involving fixation of chromatin spatial configuration by formaldehyde, digestion of cross-linked chromatin with restriction enzymes, and intra-molecular ligation of digested fragments that favors proximity. Chromatin interactions are then detected by measuring ligation frequency of two interacting fragments by polymerase chain reaction (PCR) [2–4]. Initial 3C assays estimate ligation frequency based on intensity of ethidium bromide-stained PCR products separated by agarose gel electrophoresis [3]. The gel-based assays, however, are hardly quantitative, making it difficult to differentiate subtle difference or detect weak signals. With advent of real-time quantitative PCR (qPCR), quantification of 3C ligation frequency becomes more accurate by monitoring the signal strength after each amplification cycle [4]. Due to relatively low ligation frequency in 3C library [1, 5], the qPCR assay

*Correspondence: liwang@mcw.edu
Department of Pathology and MCW Cancer Center, Medical College of Wisconsin, Milwaukee, WI 53226, USA

usually detects amplification signals at high cycle threshold (Ct) (such as Ct ≥ 35), which significantly reduces the assay's sensitivity. In addition, current 3C-qPCR is complex because randomly ligated control is needed to normalize the amplification efficiency of different primer pairs.

Recently, digital PCR (dPCR) has been emerged as a powerful tool for nucleic acid quantification, in particular, for rare molecule detection [6]. The technology detects number of targeted nucleic acids for absolute quantification by molecular counting. During dPCR, DNA samples are partitioned into thousands or millions of individual PCR reactions. Due to significant dilution, each reaction partition contains zero or one target molecule, sometimes multiple copies if dilution is not sufficient. After PCR amplification, each independent reaction can be defined as positive or negative for the target molecule by intensity of its recorded fluorescence signal [6]. Characterized by high sensitivity and specificity, the dPCR is increasingly being used for various applications such as absolute nucleic acid quantification, rare mutation detection, and copy number variation [7–9]. Here we reported a 3C-dPCR assay by incorporating dPCR technology into 3C assay [10]. We tested this assay at two prostate cancer risk regions of 8q24 and 2p11.2 for their interaction target genes *MYC* and *CAPG* [11, 12]. Our results show that 3C-dPCR is easier to use and more sensitive in determining chromatin interactions. The 3C-dPCR is likely to offer a valuable alternative method for accurate quantification of low frequency chromatin interactions.

## Result

### 3C-dPCR workflow

To identify a chromatin interaction through looping structure, it is necessary to show that the two interaction fragments have higher contact frequency than randomly ligated fragments. The first step of the procedure is to design primer and probe. In principle, a series of primers covering both pre-defined regions should be selected. In this study, we examined one fixed anchor primer (interaction hot spot) in combination with a series of test primers (covering target region). TaqMan probe was located downstream of the anchor primer (Fig. 1A). The second step is to build a 3C library including chromatin crosslinking, restriction enzyme digestion and intramolecular fragment ligation (Fig. 1B). The third step is to measure interaction (ligation) frequency using primers specific for the restriction fragments of interest. After PCR amplification in a digital PCR system, positive and negative reactions were determined by the fluorescence signal intensity. The number of the concentration of ligation product was reported as copies/μL (Fig. 1C).

## Characterization of dPCR for detection of 3C product

dPCR assay provides a convenient and straightforward approach to run up to millions of PCR reactions in parallel. In this study, we applied 3D Digital PCR system and performed duplex dPCR by including both target and genome copy control. Figure 2a and b displayed the representative plot showing digestion efficiency and self-ligation rate, where we observed 197 copies undigested *EcoRI* fragments and 44 self-ligated copies per 1000 genome copies. Figure 2c indicated moderate adjacent fragment ligation with 3.6 copies per 1000 genome copies. Figure 2d showed the representative plot of long-range chromatin interaction with 1.7 copies per 1000 genome copies. For each plot, signals in the lower left quadrant were negative (yellow) for both targets, in the lower right quadrant were positive for genome copy number (red), and in the upper left quadrant were positive for target fragment ligation (blue). The green signals between red and blue were positive for both the target and genome copy control. The intensity of fluorescence signals reflected target copy numbers after PCR amplification. The signals were specific to each primer/probe set.

## Detection of chromatin interactions at selected target regions

To determine the ligation frequency between the two restriction fragments in 3C libraries, we first tested a previous reported interaction between 8q24 region 1 and *MYC* gene [11], we examined an anchor primer 8L at *EcoRI* site (chr8: 128537495) paired with five test primers at *MYC* gene locus (*MYC*1–5). The anchor primer was also paired with three other test primers at 9, 85 and 92 kb downstream as nearby ligation and random ligation controls (Fig. 3a). The highest ligation frequency (approximately 2%) was observed for the fragment located directly upstream of the anchor fragment (Fig. 3b). The interaction frequency between the regulatory element and *MYC* gene promoter fragment (*MYC*3), ~200 kb away from the anchor fragment, was 0.7 copies (95% CI 0.40–1.10) per 1000 genome copies while the regions (−85 and −92 kb regions) assumed to be looped out from the hub [11] showed ligation frequencies of 0.18 (95% CI 0.08–0.29) and 0.16 (95% CI 0.07–0.30), respectively. Moreover, there are no obvious signals for ddH$_2$O and random ligated genomic DNA negative control (data not shown).

We further tested the ligation frequency between prostate cancer risk locus 2p11.2 and gene *CAPG*, which was described in our previous study [12]. The anchor primer at 2q11.3 was designed at the position (chr2: 85778503) named 2L. Eleven test primers were designed on 2q11.3 covering twelve *EcoRI* cutting sites from chr2: 85619044

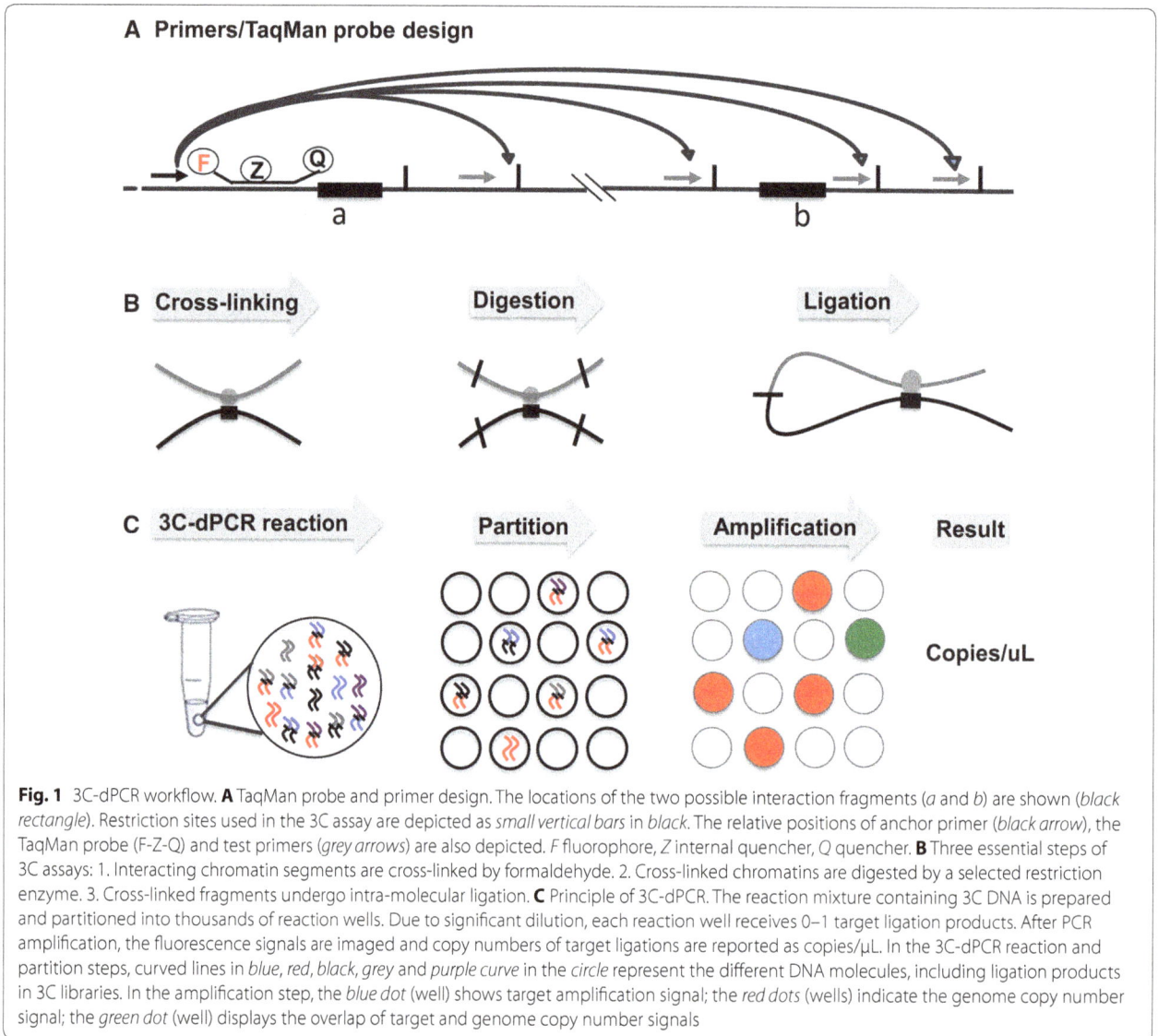

**Fig. 1** 3C-dPCR workflow. **A** TaqMan probe and primer design. The locations of the two possible interaction fragments (*a* and *b*) are shown (*black rectangle*). Restriction sites used in the 3C assay are depicted as *small vertical bars* in *black*. The relative positions of anchor primer (*black arrow*), the TaqMan probe (F-Z-Q) and test primers (*grey arrows*) are also depicted. *F* fluorophore, *Z* internal quencher, *Q* quencher. **B** Three essential steps of 3C assays: 1. Interacting chromatin segments are cross-linked by formaldehyde. 2. Cross-linked chromatins are digested by a selected restriction enzyme. 3. Cross-linked fragments undergo intra-molecular ligation. **C** Principle of 3C-dPCR. The reaction mixture containing 3C DNA is prepared and partitioned into thousands of reaction wells. Due to significant dilution, each reaction well receives 0–1 target ligation products. After PCR amplification, the fluorescence signals are imaged and copy numbers of target ligations are reported as copies/µL. In the 3C-dPCR reaction and partition steps, curved lines in *blue*, *red*, *black*, *grey* and *purple curve* in the *circle* represent the different DNA molecules, including ligation products in 3C libraries. In the amplification step, the *blue dot* (well) shows target amplification signal; the *red dots* (wells) indicate the genome copy number signal; the *green dot* (well) displays the overlap of target and genome copy number signals

(T1) to chr2: 85679686 (T12), which corresponded to the promoter and nearby region of *CAPG* (Fig. 4a). We observed strong interaction signals at *Eco*RI fragments containing primer T7 and T6 with the ligation frequency 1.9 copies (95% CI 1.41–2.47) and 1.3 copies (95% CI 0.76–1.92) per 1000 genome copies, respectively. The interaction signals were reduced on either side of this *Eco*RI site. Another interaction peak was with primer T11, the interaction frequency was 1.2 copies (95% CI 0.84–1.65) molecule per 1000 genomes. The lowest interaction signal was 0.45 copies (95% CI 0.27–0.72) per 1000 genomes (Fig. 4b). We also examined the frequency of self-ligation in the 3C library by pairing the anchor primer with a primer on the same fragment. A primer pair across the *Eco*RI site was used to test the enzyme

digestion efficiency. We found 44 copies (95% CI 34–58) per 1000 genomes for the frequency of self-ligation and 197 copies (95% CI 172–226) per 1000 genomes for the undigested *Eco*RI fragments (Fig. 2a, b).

### Comparison of 3C-dPCR with 3C-qPCR
To investigate the potential precision of 3C-dPCR and compare it with the established technique of 3C-qPCR, a comparison of dPCR and qPCR was performed to detect the interaction frequency between 2p11.2 and the cluster of *Eco*RI fragments on *CAPG* locus in another 3C library made from cell line DU145. The 3C product was run separately by both dPCR and standard TaqMan qPCR to directly compare the interaction frequency of different primer pairs. Figure 5a shows strong signals between

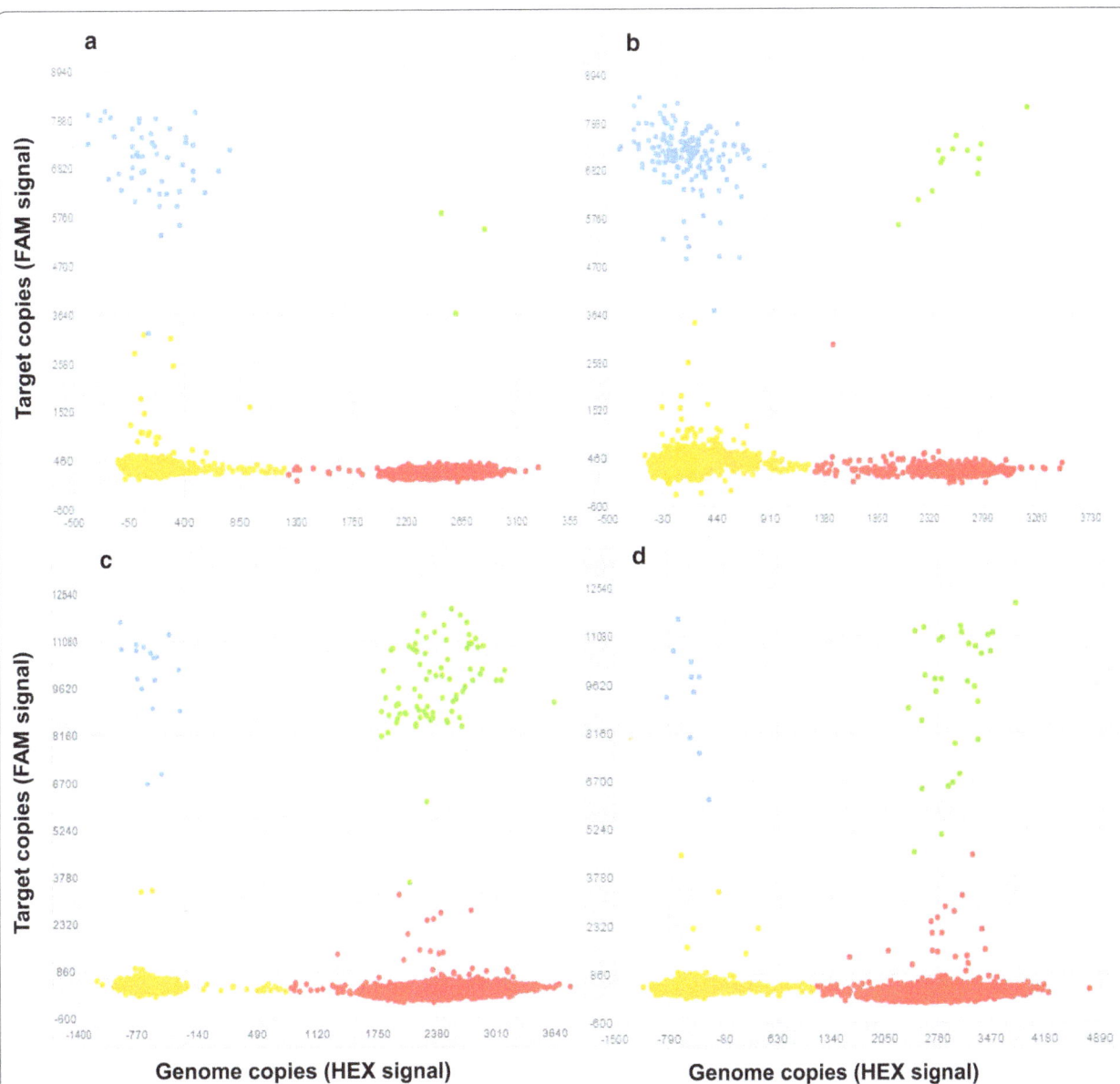

**Fig. 2** Representative duplex 3C-dPCR plots in cell line DU-145. **a** *Eco*RI digestion efficiency test, 197 copies of undigested *Eco*RI fragment per 1000 genome using a primer pair across the *Eco*RI site 2L at 2q11.2 loci. **b** Strong interaction signal for self-ligation fragment by the primer designed on the same fragment with anchor primer and paired with the anchor primer at 2q11.2, 44 copies per 1000 genome. **c** Moderate interaction signals for nearby ligation by the adjacent primer designed on the fragment next to the anchor primer at 2q11.2, 3.6 copies per 1000 genome. **d** Relative weak interaction signal for long-distance interaction between 2L with *Eco*RI fragment T7 covering the promoter of gene *CAPG*, 1.7 copies per 1000 genome. The x-axis displays the amplitude of genome copy number control (labeled by HEX, *red*) and the y- axis is signal strength of target ligation products (labeled by FAM, *blue*). The signals in the *lower left quadrant* are negative for both targets (*yellow*). The signals in the *upper right quadrant* are positive for both targets (*green*)

the *Eco*RI fragments covering primer T7, T6, T11 and anchor primer with 1.7 (95% CI 1.26–2.32), 1.2 (95% CI 0.82–1.66) and 1.0 copy (95% CI 0.66–1.46) molecules per 1000 genomes, respectively. However, fragments near this interaction were 2–5 folds lower than the active interaction fragments. Figure 5b showed the corresponding 3C-qPCR results. Although the interaction peak was slightly different between dPCR and qPCR, the overall trend from two results were highly consistent.

## Discussion

To identify a chromatin interaction, it is necessary to demonstrate higher ligation frequency between two restriction fragments than randomly ligated fragments.

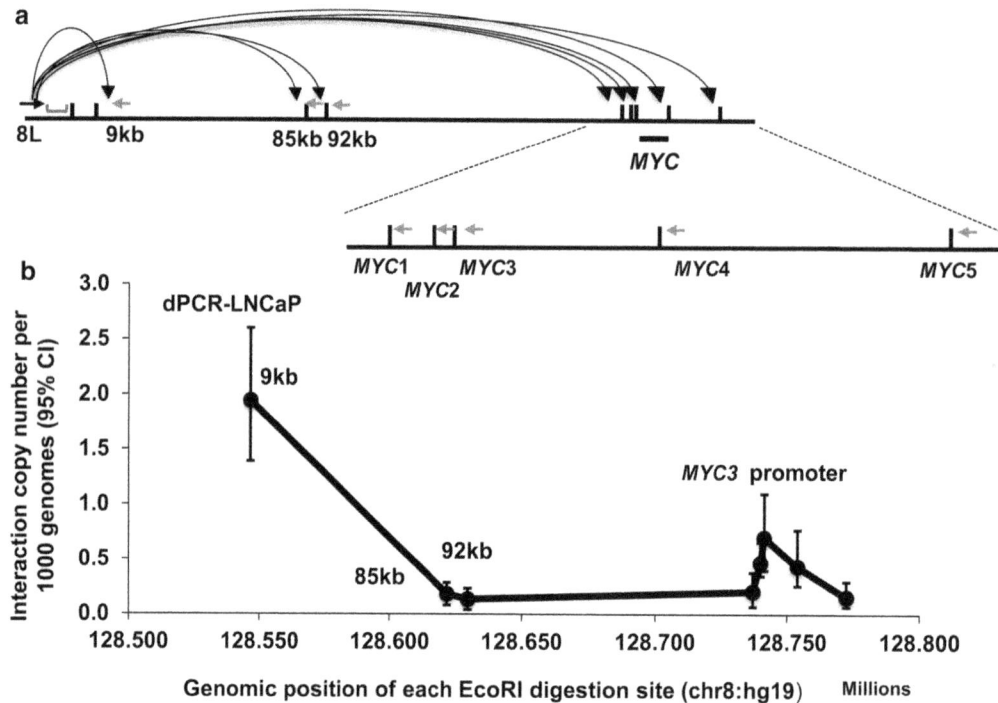

**Fig. 3** Interactions between prostate cancer risk region 1 and *MYC* gene locus at 8q24. **a** The anchor primer 8L, TaqMan probe; five target test primers (*MYC1–MYC5*) and three control test primers (9, 85, and 92 kb) are designed for the detection of the *cis*-acting interactions. *Small vertical bars* in *black* represent *Eco*RI digestion sites. *Black* and *grey arrows* show the anchor primer and test primers, respectively. The TaqMan probe is depicted as *grey bar*. **b** The copy number of ligation products at each selected restriction site. The highest interaction is at *MYC*-3 fragment, which contains the *MYC* promoter region. The y-axis displays the ligation copy numbers at each *Eco*RI site per 1000 genome. The x-axis is the genomic position of each *Eco*RI site. The *error bars* represent 95% CIs

Because ligation frequency is generally low between any two non-adjacent fragments [1, 5], a meaningful 3C analysis critically relies on the accurate quantification of different ligation products. In this study, we evaluated dPCR, the latest DNA quantification technology, for sensitive detection of chromatin interactions. Our results show that the 3C-dPCR is user-friendly and able to detect all previously reported interactions. Its simplicity and accuracy make it ideal for low copy number analysis such as low ligation frequency between chromatin interactions.

Currently, the most commonly used qPCR-based 3C assay has its own limitations. First, the assay requires preparation of randomly ligated control template DNA to normalize the amplification efficiency differences among different primer pairs [13]. Second, this assay generates relative quantification of ligated fragments [4, 6]. Third, the assay may not be sensitive enough to detect low frequency ligation products. Low concentration of ligation products in standard 3C library often leads to high Ct value, sometimes beyond the limitation of qPCR detection. In contrast, by sub-dividing a reaction mix into thousands of individual replicates, the dPCR assay

significantly reduces the total number (hence diversity) of DNA molecules in any given partition effectively enriched for the sequences of interest and diluted out other background sequences. Therefore, dPCR assay is more sensitive and more specific than qPCR assay [14]. It also effectively overcomes qPCR biases due to primer amplification efficiency differences. In addition, the 3C-dPCR is able to generate absolute numbers of ligated target fragments and genome copies in one reaction without requirement of normalization controls. Therefore, the 3C-dPCR is simpler and more sensitive in determining low interaction frequency at the target regions of interest.

The dPCR may also simplify quality control during 3C library preparation. For example, dPCR can be used to determine efficiency of restriction enzyme digestion and proximity-based ligation. In current 3C protocol, internal control primer pair is required to accurately calculate percentage of digested fragments and ligated fragments among all available genome copies. The dPCR, however, does not have amplification bias and can accurately calculate digestion and ligation efficiency. For the low frequency ligations that are close to the lower limit

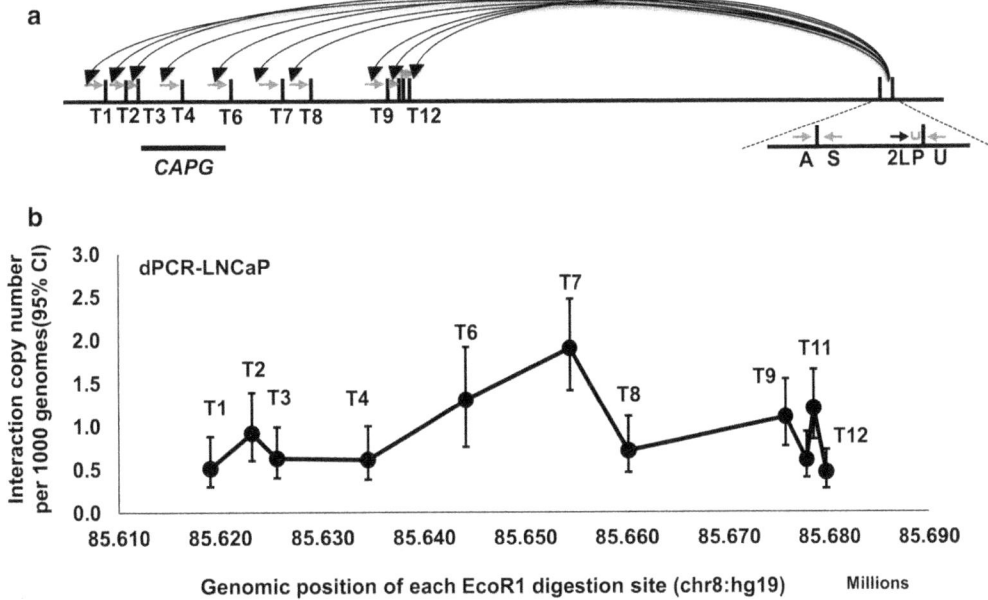

**Fig. 4** Interactions between 2L and the cluster of *EcoRI* fragments on *CAPG* gene locus at 2p11.2. **a** Anchor primer 2L, TaqMan probe on prostate cancer SNP risk region and eleven test primers (from T1 to T12) around the *CAPG* gene locus are selected for the detection of the long-range interactions. *Small vertical bars* in *black* represent *EcoRI* digestion sites. *Black* and *grey arrows* show the anchor primer and test primers, respectively. The TaqMan probe is depicted as *grey bar*. **b** The copy number of ligation products at each selected restriction site. Relative strong interaction signals were found with the *EcoRI* fragments covering primer T6, T7 and T11. The y-axis displays the ligation copy numbers at each *EcoRI* site per 1000 genome. The x-axis is the genomic position of each *EcoRI* site. *Error bars* represent standard deviation of triplicate dPCR results. *A* adjacent ligation primer, *S* self-ligation primer, *U* undigested control primer, *P* TaqMan probe. The *error bars* represent 95% CIs

of detection, dPCR system allows increasing the 3C DNA concentration in the PCR mix to provide more target ligations available for detection. The system also allows running a larger volume of the same sample on multiple chips and pooling the data into one larger "virtual" chip for low frequency ligation detection.

## Conclusion

Over the years 3C-based technologies have been evolved from single PCR assay to massive parallel sequencing assay [15–17]. Although the sequencing assays have significantly extended the scope of chromatin loop mediated long distance interaction and facilitated understanding biological mechanisms underlying gene regulation, most studies still rely on PCR-based assay to evaluate interactions at pre-defined genomic regions. By introducing dPCR into 3C assay, we show that this digital technology not only eliminates the potential variations of PCR amplification efficiency but also provides more accurate measurement of proximity-based fragment ligation frequency. The 3C-dPCR is a preferred method for sensitive and specific quantification of chromatin interactions.

## Methods

### Selection of chromatin interaction loci and primers/probes design

Previous study showed that prostate cancer risk loci at 8q24 were interacted with *MYC* region [11]. To test the feasibility of dPCR in detection of such chromatin interactions, we designed an anchor primer that was located upstream of the *EcoRI* site at chr8: 128537495 on 8q24 named as 8L. This site was shown to have an interaction peak with *MYC* in a previous report [11, 18]. Five test primers were selected downstream of each *EcoRI* site around the *MYC* gene region from chr8:128737079 to chr8:128772550 (named as *MYC*1 to *MYC*5, respectively). *MYC* gene promoter was in *MYC*3 fragment (Fig. 3a). One test primer, 9 kb downstream from the anchor primer, was used as positive control (nearby ligation). Two other test primers, 85 and 92 kb away from the anchor primer were used as long-distance random ligation (negative) controls. We named the corresponding primers as 9, 85 and 92 kb accordingly. Each of these test primers was paired with the anchor primer. One pair of primer within *EcoRI* fragment (Chr8:128521424–128537496) was used to normalize genome copy number.

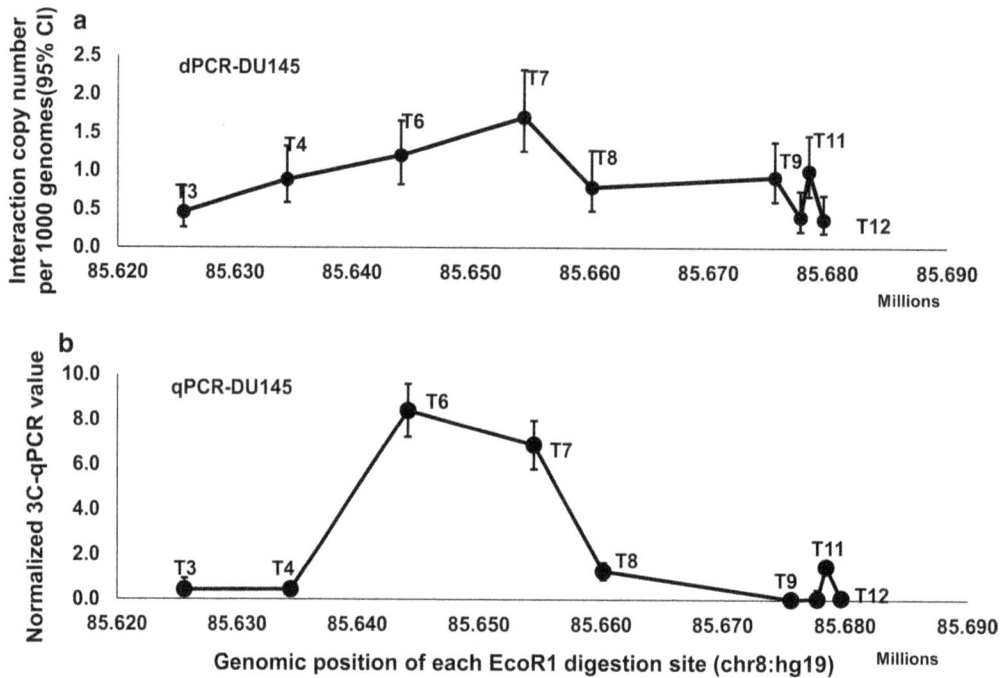

**Fig. 5** Comparison of 3C-dPCR with 3C-qPCR. **a** Copy number of ligation products at each selected restriction site detected by 3C-dPCR. Relative strong interaction signals were found with the *Eco*RI fragments covering primer T6, T7 and T11. The y-axis displays the ligation copy numbers at each *Eco*RI site per 1000 genome. The x-axis is the genomic position of each *Eco*RI site. The *error bars* represent 95% CIs. **b** 3C qPCR-based interaction signals between anchor primers and test primers. Higher interaction signals were detected with the *Eco*RI fragments covering primer T6, T7 and T11. The y-axis displays normalized 3C-qPCR values. The x-axis is the genomic position of each *Eco*RI site. *Error bars* represent standard deviation of triplicate qPCR results. The TaqMan assay at T3, T4 T8, T9, T12 fragments are beyond detection limitation of qPCR (>45Ct). Quantification values may not reflect fragment ligation frequency

TaqMan probes were located downstream of the anchor primer and labeled with 5′ FAM (targets) or HEX (copy number control) (Fig. 1A). The primers and probes were synthesized by Integrated DNA Technologies (Coralville, IA, USA). TaqMan probes were dissolved in TE pH 8.0 and stored at −20 °C as 2.5 µM aliquots. The sequences of the primers and probes were listed in Additional file 1: Table S1.

For the interaction between 2q11.3 and gene *CAPG*, primers and probe were designed as previously reported [12]. In brief, the anchor primer at 2q11.3 was designed near the cutting site chr2: 85778503 named 2L. Eleven test primers were spread twelve *Eco*RI cutting sites from chr2: 85619044 (T1) to chr2: 85679686 (T12), which covered the promoter and nearby region of gene *CAPG*. Adjacent ligation primer was designed on the fragment next to 2L. Each test primer was paired with the anchor primer. Self-ligation primer was designed on the same fragment with anchor primer and paired with the anchor primer to test self-ligated DNA circles. Undigested control primer was across the *Eco*RI site 2L and paired with the anchor primer (Fig. 4a). The sequences of the primers are listed in Additional file 1: Table S1.

### 3C library preparation

3C libraries were prepared as previously described [4]. Briefly, $1 \times 10^7$ cells were cross-linked with 1% formaldehyde for 10 min, and quenched with a final concentration of 0.125 mM glycine for 5 min at room temperature. Cells were counted and placed into aliquots $5 \times 10^6$ cells. Each aliquot of cells was lysed with 500 µL 1× cold lysis buffer (10 Mm Tris–HCl Ph 8.0, 10 Mm NaCl, 0.2% Ige cal CA630) including 1× protease inhibitor (Roche, Indianapolis, IN, USA) for at least 15 min on ice. Cell nuclei were pelleted, washed twice with 500 µL ice cold 1× *Eco*RI buffer (NEB, Ipswich, MA, USA), and then re-suspended in 500 µL 1× *Eco*RI buffer with 0.3% SDS and incubated for 1 h at 37 °C, followed by adding 1% Triton X-100 and incubated for another 1 h to sequester the SDS. Each sample was digested overnight with 600 U restriction enzyme at 37 °C. To stop the restriction digestion, 1.6% SDS (final concentration) was added, and samples were incubated at 65 °C for 20 min. Ligation were performed at 16 °C for 4 h in 15 mL tubes containing 745 µL 10× T4 ligase Buffer, 10% Triton-X 100, 80 µL 10 mg/ mL BSA, 6 mL water, 575 µL of cell lysate, 10 µL 1U/ µL T4 ligase (Invitrogen, Grand Island, NY, USA). The crosslinks were reversed with Proteinase K (Invitrogen) at 65 °C overnight. 3C samples were then purified using phenol–chloroform extraction and quantified by Qubit dsDNA HS Assay (Life Technologies).

## Digital PCR

QuantStudio 3D Digital PCR System (Life Technology, Carlsbad, CA, USA) was used for the dPCR. For each Chip, reactions were performed in 18ul volume using 9 μL of 2× 3D Digital PCR master mix, 500 nM of target primer pairs, 250 nM of probes, 80 ng of 3C template DNA examining long-distance interaction copies and 8 ng of 3C samples testing the enzyme digestion efficiency and self-ligation copies. A copy number control primer/probe mix was added in the same concentration as target primer/probe mix for duplex dPCR for genome copy number determination. To exclude the false positive result caused by high level non-specific background signal from PCR amplification, ddH$_2$O and random ligation control genomic LNCaP DNA after *Eco*RI digestion and T4 ligation were used as dPCR negative controls. Reaction mix was evenly loaded into a Digital PCR 20 K Chip containing 20,000 partitions. After sealing, the Chip was loaded into the Dual Flat Block GeneAmp PCR System 9700. Reactions were performed under universal cycling conditions: 96 °C for 10 min, followed by 45 cycles at 58 °C/60 °C for 2 min and 98 °C for 30 s with final extension at 60 °C for 2 min.

The Chip signal image was captured by the QuantStudio 3D Digital PCR system. Data analysis was performed using the AnalysisSuite Software (Life technology), which provided the copy number per μL reaction mix. Thresholds were determined based on results from negative control wells containing no template DNA; only wells above a minimum amplitude threshold were counted as positive. As template DNAs were randomly distributed among the all partitions, a Poisson correction was applied to correct for potential multiple copies per well. The confidence interval (CI) calculations for the absolute quantity (AQ) accounted for the Poisson error and resulted in a CI that was consistent with the random distribution of molecules across the chip, assuming that the deposition of the molecules follows a Poisson process. For the CI around the relative quantity (RQ), the absolute quantity of each target was first determined along with the CI around the AQ. The RQ was then calculated along with a CI around the RQ, consistent with the CI expected for the ratio of two types of target molecules distributed by two independent Poisson processes. For replicate chips, the combined RQ result across the replicate chips was calculated using a weighted average of the RQ result from individual chips, where the weighting factor was derived from the inverse of the CI around the RQ value from each chip [19]. The interaction frequency (=target copy number per 1000 genomes) was calculated: 1000× target copies/μL divided by genome copies/μL.

## Real-time quantitative PCR

To confirm the dPCR data, TaqMan qPCR technology was used to quantify the ligation frequency of 2p11.2 risk locus and the cluster of *Eco*RI fragments on *CAPG* locus. All PCR reactions were performed using Taqman Universal Master Mix II (Applied Biosystems, Foster City, CA, Cat# 4440038). Each 10 μL reaction consisted of 1× Taqman Universal MasterMix II, 1 μL 5uM anchor primer, 1 μL test primer, 1 μL Taqman probe (2.5 μM), and 100 ng 3C DNA. PCR cycles were as follows: an initial step 2 min at 50 °C, 10 min at 95 °C, 50 cycles of 15 s at 95 °C, 60 s at 58–60 °C. Each PCR reaction was performed in triplicate, and the data presented were the average of at least two independent experiment results for all PCR reactions. The contact frequency of each interaction pair was normalized using a 3C-control library prepared from pooled PCR products that contained 16 *Eco*RI-digested and T4 ligase-ligated fragments covering target *Eco*RI cutting sites and primer binding sites [12]. Adjacent fragment ligation frequency was used to normalize the different loading, fixation and ligation efficiencies between different cell lines.

### Abbreviations

3C: chromosome conformation capture; PCR: polymerase chain reaction; qPCR: quantitative PCR; dPCR: digital PCR; Ct: cycle threshold; CI: confidence interval; AQ: absolute quantity; RQ: relative quantity (RQ).

### Authors' contributions

Conceived and designed the experiments: MD, LW; Performed the experiments: MD; Wrote the manuscript: MD; Read and revised the manuscript: LW. Both authors read and approved the final manuscript.

### Acknowledgements

None.

### Competing interests

The authors declare that they have no competing interests.

### Funding

This work was supported by Medical College of Wisconsin Cancer Center Seed Fund [3305738 to LW] and National Institute of Health [R01CA157881 to LW].

**References**

1. Simonis M, Kooren J, de Laat W. An evaluation of 3C-based methods to capture DNA interactions. Nat Methods. 2007;4(11):895–901.
2. Miele A, Dekker J. Mapping cis- and trans- chromatin interaction networks using chromosome conformation capture (3C). Methods Mol Biol. 2009;464:105–21.
3. Dekker J, Rippe K, Dekker M, Kleckner N. Capturing chromosome conformation. Science. 2002;295(5558):1306–11.
4. Hagege H, Klous P, Braem C, Splinter E, Dekker J, Cathala G, de Laat W, Forne T. Quantitative analysis of chromosome conformation capture assays (3C-qPCR). Nat Protoc. 2007;2(7):1722–33.
5. Gavrilov AA, Golov AK, Razin SV. Actual ligation frequencies in the chromosome conformation capture procedure. PLoS ONE. 2013;8(3):e60403.
6. Hindson BJ, Ness KD, Masquelier DA, Belgrader P, Heredia NJ, Makarewicz AJ, Bright IJ, Lucero MY, Hiddessen AL, Legler TC, et al. High-throughput droplet digital PCR system for absolute quantitation of DNA copy number. Anal Chem. 2011;83(22):8604–10.
7. Nadauld L, Regan JF, Miotke L, Pai RK, Longacre TA, Kwok SS, Saxonov S, Ford JM, Ji HP. Quantitative and sensitive detection of cancer genome amplifications from formalin fixed paraffin embedded tumors with droplet digital PCR. Transl Med. 2012;2(2).
8. Reid AL, Freeman JB, Millward M, Ziman M, Gray ES. Detection of BRAF-V600E and V600K in melanoma circulating tumour cells by droplet digital PCR. Clin Biochem. 2014.
9. Li N, Ma J, Guarnera MA, Fang H, Cai L, Jiang F. Digital PCR quantification of miRNAs in sputum for diagnosis of lung cancer. J Cancer Res Clin Oncol. 2014;140(1):145–50.
10. Link N, Kurtz P, O'Neal M, Garcia-Hughes G, Abrams JM. A p53 enhancer region regulates target genes through chromatin conformations in cis and in trans. Genes Dev. 2013;27(22):2433–8.
11. Du M, Yuan T, Schilter KF, Dittmar RL, Mackinnon A, Huang X, Tschannen M, Worthey E, Jacob H, Xia S, et al. Prostate cancer risk locus at 8q24 as a regulatory hub by physical interactions with multiple genomic loci across the genome. Hum Mol Genet. 2015;24(1):154–66.
12. Du M, Tillmans L, Gao J, Gao P, Yuan T, Dittmar RL, Song W, Yang Y, Sahr N, Wang T, et al. Chromatin interactions and candidate genes at ten prostate cancer risk loci. Sci Rep. 2016;6:23202.
13. Miele A, Gheldof N, Tabuchi TM, Dostie J, Dekker J. Mapping chromatin interactions by chromosome conformation capture. Curr Protoc Mol Biol. 2006.
14. Whale AS, Huggett JF, Cowen S, Speirs V, Shaw J, Ellison S, Foy CA, Scott DJ. Comparison of microfluidic digital PCR and conventional quantitative PCR for measuring copy number variation. Nucleic Acids Res. 2012;40(11):e82.
15. de Wit E, de Laat W. A decade of 3C technologies: insights into nuclear organization. Genes Dev. 2012;26(1):11–24.
16. Dekker J, Marti-Renom MA, Mirny LA. Exploring the three-dimensional organization of genomes: interpreting chromatin interaction data. Nat Rev Genet. 2013;14(6):390–403.
17. Sajan SA, Hawkins RD. Methods for identifying higher-order chromatin structure. Annu Rev Genomics Hum Genet. 2012;13:59–82.
18. Ahmadiyeh N, Pomerantz MM, Grisanzio C, Herman P, Jia L, Almendro V, He HH, Brown M, Liu XS, Davis M, et al. 8q24 prostate, breast, and colon cancer risk loci show tissue-specific long-range interaction with MYC. Proc Natl Acad Sci USA. 2010;107(21):9742–6.
19. Majumdar N, Wessel T, Marks J. Digital PCR modeling for maximal sensitivity, dynamic range and measurement precision. PLoS ONE. 2015;10(3):e0118833.

# 5

# Identification of microRNAs and their response to the stress of plant allelochemicals in *Aphis gossypii* (Hemiptera: Aphididae)

Kang-Sheng Ma, Fen Li, Ying Liu, Ping-Zhuo Liang, Xue-Wei Chen and Xi-Wu Gao*

## Abstract

**Background:** MicroRNAs (miRNAs) are a group of short non-coding RNAs involved in the inhibition of protein translation or in mRNA degradation. Although the regulatory roles of miRNAs in various biological processes have been investigated, there is as yet an absence of studies about the regulatory roles of miRNAs involved in the metabolism of plant allelochemicals in insects.

**Results:** We constructed five small RNA libraries from apterous *Aphis gossypii* adults that had fed on an artificial diet containing various allelochemicals. Using Illumina sequencing, a total of 73.27 million clean reads was obtained, and 292 miRNAs were identified from *A. gossypii*. Comparative analysis of read counts indicated that both conserved and novel miRNAs were differently expressed among the five libraries, and the differential expression was validated via qRT-PCR. We found that the transcript levels of several miRNAs were increased or decreased in all of the allelochemical treatment libraries compared to the control. The putative target genes of the miRNAs were predicted using in silico tools, and the target genes of several miRNAs were presumed to be involved in the metabolism of xenobiotic compounds. Furthermore, the target prediction results were confirmed using dual luciferase reporter assay, and Ago-miR-656a-3p was demonstrated to regulate the expression of *CYP6J1* post-transcriptionally through binding to the 3' UTR of *CYP6J1*.

**Conclusion:** Our research results indicate that miRNAs may be involved in the metabolism of plant allelochemicals in *A. gossypii*, and these results also represent an important new small RNA genomics resource for further studies on this topic.

**Keywords:** MicroRNAs, Plant allelochemicals, *Aphis gossypii*, Illumina sequencing, Regulatory roles

## Background

MicroRNAs (miRNAs) are small non-coding RNAs (18-24 nucleotides in length) that regulate gene expression at the post-transcriptional level [1, 2]. Through binding complementarily to 3' untranslated regions (UTRs), coding sequences, or 5' UTRs, miRNA suppress the translation of target mRNA molecules, and thereby silence target gene expression [3–5]. miRNAs are generated in all eukaryotes and viruses [1, 6], and many miRNAs are conserved among related species [7]. Since the first miRNA was reported to regulate the timing of development in *Caenorhabditis elegans* [8], numerous miRNAs have been identified from animals, plants, and viruses. In the last two decades, thousands of miRNAs have been isolated from insect species, including *Drosophila melanogaster* [9], *Bombyx mori* [10, 11], *Manduca sexta* [12–14], *Plutella xylostella* [15, 16], and *Helicoverpa armigera* [17, 18]. Functional studies carried out in these insect species have demonstrated that insect miRNAs play very important regulatory roles in various biological processes, such as development, immune responses, metabolism, and host-pathogen interactions [1, 6, 19, 20].

The cotton aphid, *Aphis gossypii* Glover (Hemiptera: Aphididae), is an important insect pest in cotton and cucurbit crops that causes economic damage both

*Correspondence: gaoxiwu@263.net.cn
China Agricultural University, Beijing, China

through direct feeding and through the transmission of viruses [21, 22]. Given that *A. gossypii* has a very wide host range that encompasses least 300 species [23, 24], this pest encounters multiple plant toxic chemicals that are produced by host plants to defend against herbivores. These compounds have strong deleterious effects on herbivorous insects by affecting the growth and development or even by directly causing mortality [25, 26]. In humans and large mammals, increasing evidence suggests that miRNAs play very important roles in the metabolism of xenobiotic compounds [5, 27]. Such miRNAs can mediate the detoxification metabolism of xenobiotics by regulating the expression of xenobiotic-metabolizing enzymes and nuclear receptors [5]. For example, human P450 CYP1A1, which is involved in the metabolism of carcinogenic metabolites, was found to be post-transcriptionally regulated by miR-892a [28]. In addition, previous studies carried out in mosquitoes *Culex pipiens* suggested that miRNAs participate in the resistance to pyrethroid insecticides by mediating the expression levels of P450 genes [29, 30]. While there are a lot known about the miRNAs that participate in regulating the detoxification of xenobiotics in animals and miRNAs likely have essential roles in insecticide resistance, less is understood about the regulatory roles of miRNAs in the metabolism of plant allelochemicals in insects.

In the present study, five small RNA libraries were built from apterous *A. gossypii* adults fed on artificial diets that contained various plant allelochemicals (gossypol, 2-tridecanone, quercetin and tannic acid respectively, which are toxic chemicals found naturally in cotton plant or other host plants of the cotton aphid) and control. A total of 73.27 million clean reads was obtained by deep sequencing, and 292 miRNAs were identified from the five sample libraries. In order to identify putative allelochemical metabolism-related miRNAs, the expression levels of both conserved and novel miRNAs were compared among the five libraries, and the targets of the newly identified miRNAs were predicted. The results of this study deepen our understanding of the regulatory roles of miRNAs in *A. gossypii* and indicate that miRNAs are likely involved in the insect metabolism of plant allelochemicals.

## Results
### Deep sequencing of *A. gossypii* small RNA libraries
In order to examine the potential role of small RNAs in *A. gossypii* responses to plant allelochemicals, we collected apterous *A. gossypii* adults that had been fed artificial diet containing various allelochemicals for 24 h for small RNA sequencing. Five *A. gossypii* small RNA libraries were constructed and sequenced using the Illumina sequencing platform. A total of 76.96 million raw reads

was obtained from the five libraries, and after filtering out sequences shorter than 18 nt and filtering the lower quality reads from the raw data, 73.27 million quality reads were obtained. The number of clean reads differed among the five libraries; more clean reads (16.09 million) were obtained from the 2-tridecanone-fed library than from any of the other four libraries (14.48, 13.96, 14.01, and 14.74 million, in the tannic acid, quercetin, gossypol, and control libraries respectively) (Table 1). The length of these small RNAs ranged from 18 to 30 nt. In all five libraries, the highest peak for nucleotide length distribution was that for 22 nt (Fig. 1; Additional file 1).

### Identification of conserved and novel miRNAs, and analysis of their features
To identify potential candidate miRNAs from the *A. gossypii* sample libraries, the raw sequencing data were analyzed with miRDeep2 software. The *Acyrthosiphon pisum* genome was used as a reference because the *A. gossypii* genome was not available at the time of the analysis. Total of 292 unique miRNA candidates were identified from among the five libraries, including 246 conserved

**Table 1 Total number of reads obtained from the small RNA libraries from aphids fed artificial diet containing various plant allechemicals**

| Allechemicals | Raw reads | Clean reads |
|---|---|---|
| Control | 15,193,417 | 14,483,649 |
| 2-tridecanone | 16,819,205 | 16,090,539 |
| Tannic acid | 14,630,011 | 13,956,679 |
| Quercetin | 14,743,194 | 14,006,538 |
| Gossypol | 15,575,107 | 14,736,872 |
| Total | 76,960,934 | 73,274,277 |

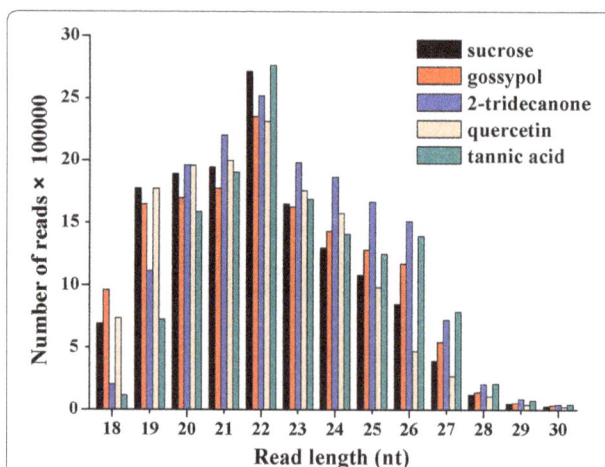

**Fig. 1** Length distribution of small RNAs from *A. gossypii* identified by deep sequencing. This length distribution was assessed using clean reads after filtering out the redundant small RNAs

miRNAs and 46 potentially novel miRNAs. The newly identified miRNAs of *A. gossypii* were named by prefixed with "Ago", where "Ago" means *A. gossypii*. The length and copy number distribution analysis of these newly identified miRNAs indicated that the miRNAs of 22 nt in length were the most abundant (47.46%), and the miR-NAs of 21–22 nt in length accounted for more than 81% of reads (Table 2). Further, we analyzed the common and unique distribution of the newly identified miRNAs from among the five libraries. The results demonstrated that 221 miRNAs were common among the five libraries; only a few of the miRNAs were uniquely expressed in a particular library (Additional file 2: Figure S1).

Since the dominance of uracil at the first position of the 5′ terminus terminal is considered to be one of the conserved features of mature miRNAs [31], and given that the first base toward the 5′ end of miRNAs is known to play very important roles in the interaction between miRNAs and argonaute complexes [32], the position-specific nucleotide occurrence of the candidate miRNA sequences was analyzed. *A. gossypii* miRNAs showed a nucleotide bias towards uracil (U) at the first nucleotide position (Fig. 2). In addition, the base composition of the miRNAs at each position was analyzed, and the nucleotide U was the most abundant nucleotide at most of the positions; this was especially pronounced at positions 1, 17, 22, 23, 25, and 30 (Fig. 2).

### A. gossypii miRNAs differentially expressed following allelochemical treatments

To identify the miRNAs that may play important roles in the responses of *A. gossypii* against plant allelochemicals, the differential expression of the identified miRNAs were analyzed using edgeR software. The expression values as assessed by miRDeep2 were used to analyze differential expression of *A. gossypii* miRNAs, and the miRNAs that had read counts of more than 10 in all five libraries were

**Table 2 Length distribution and copy number of *A. gossypii* miRNAs**

| miRNA length | Number of miR-NAs | Copy number | Percentage (%) |
|---|---|---|---|
| 18 | 65 | 11,291 | 0.32 |
| 19 | 64 | 62,839 | 1.79 |
| 20 | 40 | 100,111 | 2.85 |
| 21 | 37 | 1,183,236 | 33.69 |
| 22 | 54 | 1,666,826 | 47.46 |
| 23 | 17 | 309,952 | 8.82 |
| 24 | 11 | 178,019 | 5.07 |
| 25 | 4 | 31 | 0.00 |

selected for differential expression analysis. Compared to the aphids fed on a 0.5 M sucrose solution (control), 134 miRNAs were found to be differentially expressed in the aphids fed on diet containing allelochemicals (Fig. 3). Of these, following for aphids feeding on gossypol, 2-tridecanone, quercetin, and tannic acid for 24 h, there were 33, 73, 59, and 42 differentially expressed miRNAs, respectively (Table 3). Interestingly, we found that most of the differentially expressed miRNAs were up-regulated in the aphids treated with gossypol, quercetin, and 2-tridecanone, but that the miRNAs from the tannic acid treated aphids possessed the tendency of down-regulation (Table 3).

The expression levels of these miRNAs were dysregulated among five libraries. For instance, Ago-miR-2997-5p expression was low in the control and quercetin treatment libraries and was not detected in the 2-tridecanone or tannic acid treatment, but was up-regulated in the gossypol library (Fig. 3). Although the expression patterns of most of the miRNAs differed among the various allelochemical treatments, several miRNAs were consistently up/down-regulated in all four of the allelochemical treatment libraries. For instance, compared to the control, the expressions of some miRNAs (Ago-miR-8798a, Ago-miR-331-3p, Ago-miR-3191-3p, Ago-miR-1773-5p, Ago-miR-2179-5p, Ago-miR-92b-5p, Ago-miR-9083-2, and Ago-miR-719) were up-regulated in all four of the allelochemical treatments (Table 4). While the expression levels of Ago-let-7-5p, Ago-miR-100-5p, Ago-miR-44b-3p, Ago-miR-7054-3p, Ago-miR-4021-3p, Ago-miR-656a-3p, Ago-miR-4661a-3p, and Ago-miR-2238j-3p were down-regulated in all four treatment libraries in compare with the control (Table 5). It is worth nothing that Ago-miR-7475a-5p expression was not detected in the control library, but was expressed in each of the allelochemical treatment libraries (Table 4).

To verify the expression analysis results from the sequencing experiments (read count), 6 differentially expressed miRNAs were selected, and their expression levels were measured via qRT-PCR analysis. Five of the selected miRNAs showed similar expression patterns as those assessed using sequencing read counts (Figs. 3, 4). In the sequencing results, the expression of Ago-miR-2179-5p was up-regulated, and Ago-let-7-5p was down-regulated in all of the allelochemicals treatment libraries, and the qRT-PCR analysis showed similar results (Fig. 4; Tables 4, 5). The qRT-PCR results for Ago-miR-3051-2 agreed with the sequencing results: it was highly expressed when apterous *A. gossypii* adults fed on an artificial diet containing 0.2% tannic acid (Figs. 3, 4). The expression level of Ago-miR-5468a as measured by qRT-PCR was different from that as assessed by the

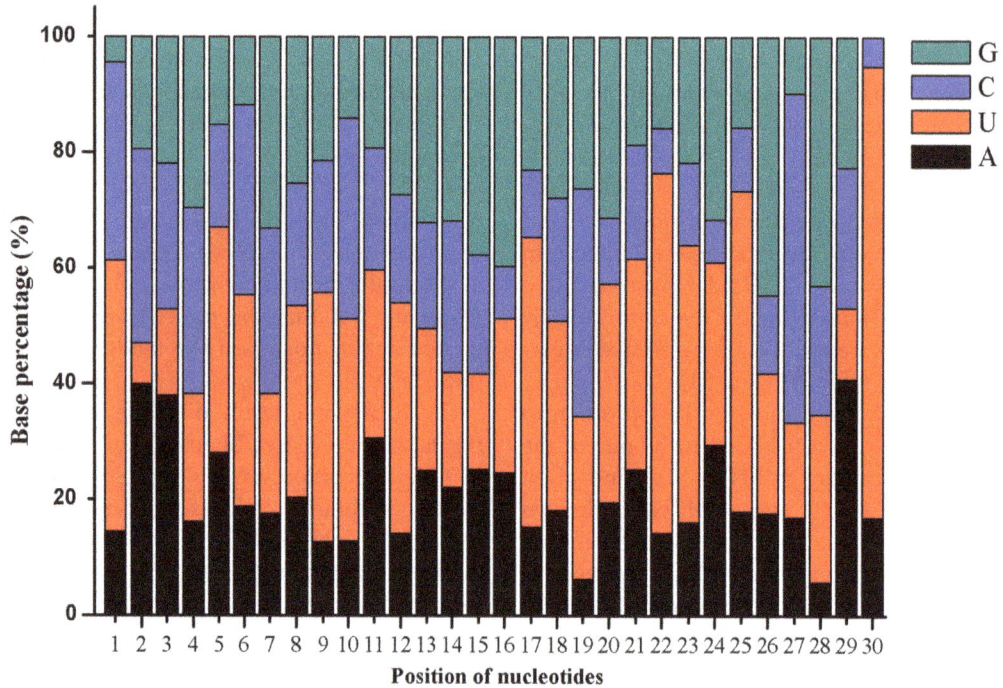

**Fig. 2** The position-specific nucleotide occurrence of *A. gossypii* mature miRNAs. Uracil dominated the first nucleotide position towards the 5′ end of miRNAs

sequencing read count, the reason for this discrepancy is unknown.

### Prediction of the putative target genes of the miRNAs

To further understand the role of the various miRNAs identified in *A. gossypii*, the putative targets of these identified miRNAs were computationally predicted. The miRanda and RNAhybrid programs were used to identify the targets of these miRNAs from an *A. gossypii* transcriptome database. Combining the results of these two target prediction programs, a total of 5929 genes were putatively targeted by 236 miRNAs from *A. gossypii*. GO annotation of these target genes included the following categories: cellular components, molecular function, and biological process. About 49.8% of the predicted target genes were classified into the biological process category, including cellular process (19.4%), metabolic process (17.6%), and biological regulation (10.8%) (Additional file 2: Figure S2).

To characterize the potential roles of *A. gossypii* miRNAs in the defense responses against allelochemicals, we focused our attention to the predicted target genes that were likely to be involved in xenobiotic metabolism. Interestingly, we found that several miRNAs were targeted to genes that are known to play very important roles in insect responses to xenobiotic stress, including cytochrome P450s, acetylcholinesterases, glutathione

S-transferases, sodium channel proteins, etc. (Table 6). Several miRNAs were predicted to have many target genes, and numerous genes were putatively targeted by multiple miRNAs. For instance, Ago-miR-4467a-1 and Ago-miR-4973-5p-11 were found have many predicted target genes, and CYP6J1 was putatively targeted by Ago-miR-656a-3p, Ago-miR-669c-5p, and Ago-miR-4172-3p (Table 6). In addition, some of these miRNAs, which were putatively targeted to xenobiotic metabolism-related genes, were differently expressed following allelochemical treatment, including Ago-miR-3191-3p, Ago-miR-8798a, and Ago-miR-656a-3p (Table 6; Fig. 3).

### Validation of the target prediction

The target prediction results were validated by selecting a cascading of Ago-miR-656a-3p and *CYP6J1*. To determine whether or not Ago-miR-656a-3p could bind to the 3′ UTR of *CYP6J1* and suppress the expression of *CYP6J1*, the 3′ UTR of *CYP6J1* containing the target site of Ago-miR-656a-3p was inserted into a pmirGLO vector to yield a recombined vector, pmirGLO-CYP6J1-UTR. The results of dual luciferase reporter assay showed that the firefly luciferase activity normalized to Renilla was significantly reduced after pmirGLO-CYP6J1-UTR was co-transfected with Ago-miR-656a-3p agomir in comparison with the negative mimic control; while the co-transfection of Ago-miR-656a-3p agomir with control

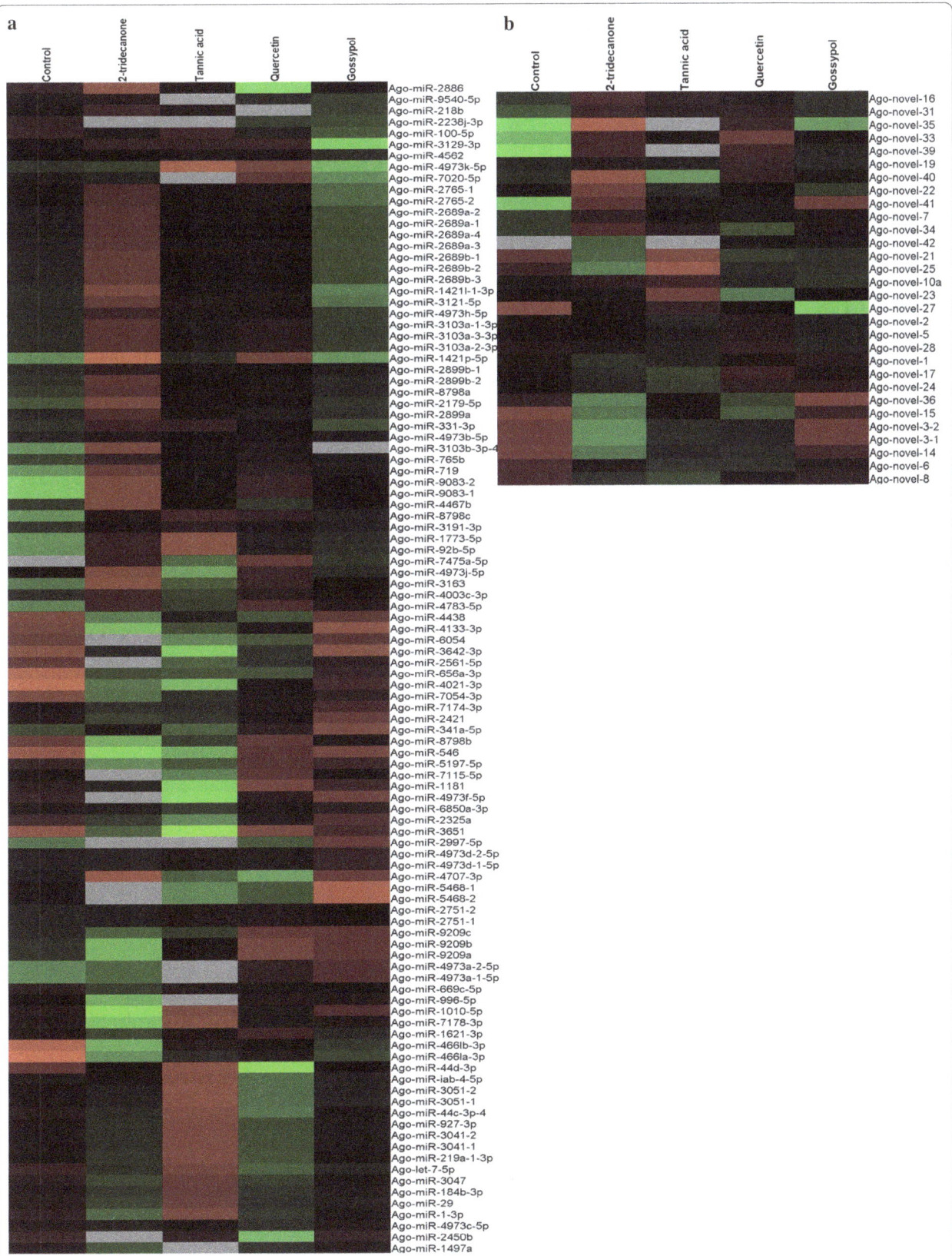

**Fig. 3** Differential expressions of *A. gossypii* miRNAs after feeding on various allelochemicals for 24 h. **a** Expression of conserved miRNAs. **b** Expression of novel miRNAs. *Green color* represents low expression levels and *red color* represents the high expression levels of the miRNAs

**Table 3** Number of miRNAs that were significantly differentially regulated after the different allelochemical treatments

| Allelochemicals | Number of up-regulated miRNAs | Number of down-regulated miRNAs |
| --- | --- | --- |
| 2-tridecanone | 42 | 31 |
| Tannic acid | 4 | 38 |
| Quercetin | 49 | 10 |
| Gossypol | 29 | 4 |

vector (pmirGLO) did not decrease the relative activity of luciferase (Fig. 5).

## Discussion

Allelochemicals are very important plant natural products that are known to play essential roles in plant defense responses to herbivores. Previous studies have shown that allelochemicals have great impacts on herbivorous insects. For example, detrimental effects were observed when *H. armigera* larvae were exposed to a high concentration of gossypol [33, 34]. Similarly, naringenin and quercetin have been reported to cause detrimental effects in the pea aphid, *A. pisum*, by effecting

development, fecundity, and mortality [26]. As a very important polyphagous pest, cotton aphids encounter multiple allelochemicals in their life cycles, including gossypol, quercetin, and tannic acid. There is no doubt that allelochemicals have great effects on *A. gossypii*; for example, Gao et al. found that higher levels of gossypol adversely affected the development, longevity, and reproduction of *A. gossypii* [35]. miRNAs have been demonstrated to play very important regulatory roles in many biological processes over last few years. Therefore, it is reasonable to conjecture that small RNA molecules potentially function in regulatory roles in cotton aphid responses to plant allelochemicals. The identification and analysis of the expression profiles of miRNAs in allelochemical treated *A. gossypii* can potentially provide insight into the regulatory mechanisms underlying insect detoxification of plant allelochemicals. The present study was undertaken to identify the conserved and novel miRNAs of *A. gossypii* and to investigate the potential regulatory roles of these miRNAs in the metabolism of allelochemicals.

With the development of high throughput sequencing technology, small RNA sequencing has become a popular experimental approach to identifying miRNAs

**Table 4** The read counts and sequences of *A. gossypii* miRNAs that had increased expression following plant allelochemical treatment

| miRNA name | miRNA sequence | Control | 2-tridecanone | Tannic acid | Quercetin | Gossypol |
| --- | --- | --- | --- | --- | --- | --- |
| Ago-miR-8798a | CGCGGTCGCCGCGCCGCC | 116 | 199 | 117 | 154 | 162 |
| Ago-miR-331-3p | GCCCCTGGTGGTCATGTTGGA | 39 | 58 | 56 | 57 | 42 |
| Ago-miR-3191-3p | GGGGGACGAGGTGGCCGAGCGGT | 37 | 39 | 45 | 60 | 54 |
| Ago-miR-1773-5p | GGGGGGAGGAGGAGGAGGA | 19 | 41 | 75 | 36 | 39 |
| Ago-miR-2179-5p | ATGCAAAATACATTTGTGTACT | 27 | 66 | 32 | 45 | 35 |
| Ago-miR-92b-5p | CGGGACGGCGAGGGTTGGGG | 16 | 36 | 64 | 30 | 31 |
| Ago-miR-719 | CTCTCGGCCGTCGGCGCGGC | 2 | 6 | 3 | 6 | 5 |
| Ago-miR-9083-2 | GGCCACGCCGCGCCGTCGG | 1 | 5 | 2 | 5 | 4 |
| Ago-miR-7475a-5p | CGCCACCGCCGCGCCGTCGT | 0 | 6 | 2 | 13 | 5 |

**Table 5** The read counts and sequences of *A. gossypii* miRNAs that had decreased expression following plant allelochemical treatment

| miRNA name | miRNA sequence | Control | 2-tridecanone | Tannic acid | Quercetin | Gossypol |
| --- | --- | --- | --- | --- | --- | --- |
| Ago-let-7-5p | TGAGGTAGTTGGTTGTATAGT | 2943 | 673 | 2378 | 811 | 1586 |
| Ago-miR-100-5p | GACCCGTAGATCCGAACTTGTG | 2501 | 803 | 1521 | 1171 | 962 |
| Ago-miR-44b-3p | TGACTAGAGTTTATACTACCGA | 1475 | 497 | 1375 | 606 | 835 |
| Ago-miR-7054-3p | CCAACTTGGCAGCTTCTGA | 158 | 17 | 25 | 52 | 95 |
| Ago-miR-4021-3p | TAAGTATTTGGCTCTTGG | 52 | 3 | 2 | 10 | 22 |
| Ago-miR-656a-3p | CATATTATGGTCGTGAGTA | 24 | 2 | 2 | 3 | 10 |
| Ago-miR-466la-3p | TATAAATATTGTAGGTACC | 23 | 1 | 2 | 5 | 3 |
| Ago-miR-2238j-3p | TATGACGAGAGGGCAAAT | 16 | 0 | 0 | 6 | 7 |

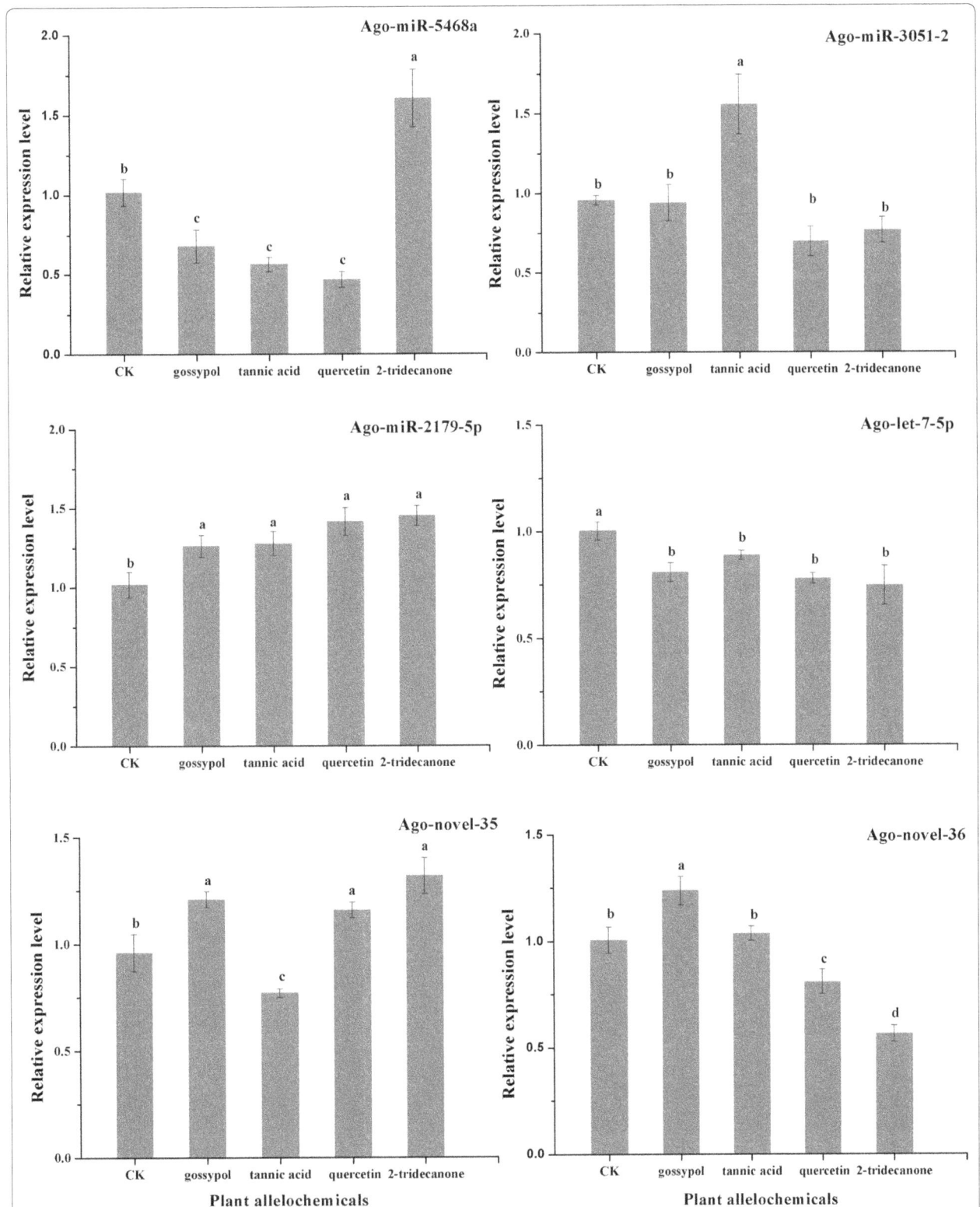

**Fig. 4** Differential expressions of miRNAs following plant allelochemical treatment. The results are presented as mean ± SD for three independent replicates. *Different letters* on the *bars* of the histogram indicate significant differences based on ANOVA followed by Tukey's HSD multiple comparison test ($P < 0.05$)

**Table 6  Putative xenobiotic metabolism-related target genes of *A. gossypii* miRNAs**

| miRNA | Xenobiotic metabolism related target genes |
| --- | --- |
| Ago-miR-1-3p | G-protein coupled receptor |
| Ago-miR-341b-5p | Acetylcholinesterase, CYP6A8, CYP6A14 |
| Ago-miR-656a-3p | CYP6J1, glutathione S-transferase sigma 1, carboxylesterase |
| Ago-miR-669c-5p | CYP6J1, glutathione S-transferase sigma 1 |
| Ago-miR-1181 | CYP6A13 |
| Ago-miR-1332-3p | UDP-glucuronosyltransferase, CYP6A8, CYP6K1, CYP4C1, CYP6A2, CYP6A13 |
| Ago-miR-1946a | Acetylcholinesterase, glutathione S-transferase omega 1 |
| Ago-miR-2886 | Acetylcholinesterase |
| Ago-miR-2899a | CYP18A1 |
| Ago-miR-3163 | Glutathione S-transferase sigma 1 |
| Ago-miR-3191-3p | Carboxylesterase |
| Ago-miR-3575 | CYP315A1, glutathione S-transferase sigma 2 |
| Ago-miR-4172-3p | CYP6J1, glutathione S-transferase sigma 1 |
| Ago-miR-4213-5p | Carboxylesterase, glutathione S-transferase sigma 2 |
| Ago-miR-466la-3p | Carboxylesterase |
| Ago-miR-4467a-1 | Sodium channel protein, calcium channel flower, CYP6A8, ryanodine receptor 44F, gamma-aminobutyric acid receptor, voltage-dependent calcium channel type A subunit alpha 1, nicotinic acetylcholine receptor alpha subunit |
| Ago-miR-4467a-2 | CYP6A2, CYP6A8, esterase FE4, ryanodine receptor 44F |
| Ago-miR-4783-5p | Glutamate-gated chloride channel, voltage-dependent T-type calcium channel subunit alpha 1 |
| Ago-miR-4973a-1-5p | Ryanodine receptor |
| Ago-miR-4973-5p-11 | Ryanodine receptor 44F, sodium channel protein para, voltage-dependent L-type calcium channel subunit beta 2, Nic acetylcholine receptor alpha 2, Glutamate-gated chloride channel, Voltage-dependent T-type calcium channel subunit alpha 1 |
| Ago-miR-4973 g-5p | Acetylcholinesterase, CYP6K1 |
| Ago-miR-6236 | Voltage-dependent calcium channel type A subunit alpha 1 |
| Ago-miR-6850-3p-1 | CYP6A8 |
| Ago-miR-7475a-5p | Venom carboxylesterase-6 |
| Ago-miR-8798a | CYP6A8 |
| Ago-miR-8798c | Voltage-dependent calcium channel type A subunit alpha 1 |
| Ago-novel-7 | Sodium channel protein 60E, CYP6A8, UDP-glucuronosyltransferase |
| Ago-novel-10a | Ryanodine receptor 44F |
| Ago-novel-16 | CYP18A1 |
| Ago-novel-19 | Acetylcholinesterase, CYP6A8, CYP6A2 |
| Ago-novel-24 | Acetylcholinesterase, CYP6A8 |
| Ago-novel-28 | Voltage-dependent calcium channel type A subunit alpha 1 |

from a range of organisms. Sattar et al. [36] identified 102 miRNAs in *A. gossypii* fed on susceptible and resistant (*Vat*⁻ and *Vat*⁺) melon plants. In the present study, we sequenced five small RNA libraries and identified 293 miRNAs from *A. gossypii* fed on various allelochemicals; our sequencing data greatly expands the scope of the resources available to study *A. gossypii* miRNAs. The size distribution pattern of the clean reads revealed that the five libraries were dominated by 22 nt sequences (Fig. 1), a result consistent with the known characteristics of miRNA [37]. The overall length distribution observed of the *A. gossypii* miRNA in this study was similar to distributions observed in several other insect species, including *Blattella germanica* [38] and *Apis mellifera* [39].

Our expression analysis based on read counts showed that 134 miRNAs were differentially expressed in the allelochemical-treated aphids, as compared to the control aphids, which clearly suggests that allelochemicals affect miRNA expression, thus implying a possible role for miRNAs in the regulation of the metabolism of allelochemicals in *A. gossypii*.

Our results showed that several miRNAs were up/down regulated in all four allelochemicals libraries, and these miRNAs may play important roles in the metabolism of plant allelochemicals in *A. gossypii*. For instance, miR-92b was up-regulated in all four allelochemical libraries, and suggests that miR-92b may involve in the response of *A. gossypii* to plant allelochemicals. In other

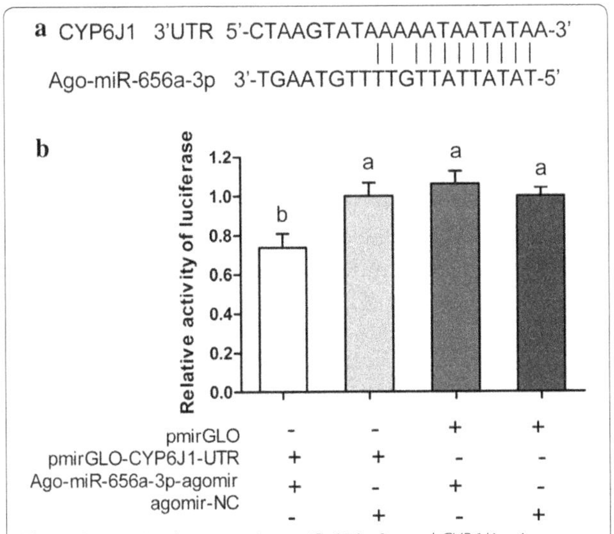

a CYP6J1 3'UTR 5'-CTAAGTATAAAAATAATATAA-3'
          | |  | | | | | | | | | |
   Ago-miR-656a-3p  3'-TGAATGTTTTGTTATTATAT-5'

**Fig. 5** Interaction between Ago-miR-656a-3p and *CYP6J1* using a dual fluorescent reporter system. **a** Predicted target sites of Ago-miR-656a-3p in the 3′ UTR of *CYP6J1*. **b** Luciferase reporter assays performed by co-transfection of Ago-miR-656a-3p agomir with a luciferase reporter gene linked to the 3′ UTR of *CYP6J1*. Firefly luciferase activity was normalized to Renilla luciferase activity and then normalized to the activity of the control group. The mathematical operators of "+" and "−" mean add and subtract. *Different letters* on the *bars* of the histogram indicate significant differences based on ANOVA followed by Tukey's HSD multiple comparison test ($P < 0.05$)

insect species, miR-92b has been reported to be involved in multiple biological processes. In *Drosophila*, miR-92b plays an important role in muscle development [40], and is essential for neuroblast self-renewal [41]. Meanwhile, miR-92b was classified as a stress responsive marker in *Eurosta solidaginis* [42].

In the present study, the expression levels of Ago-let-7-5p and Ago-miR-100-5p were down-regulated following allelochemical treatment. Let-7 and miR-100 are basic components of the let-7-complex (let-7-C), which is required for the development of normal morphology in *D. melanogaster* [43]. Given that allelochemicals can have great effects on insect development [26, 34, 35], the differential expression of let-7 and miR-100 might be attributable to the influence of plant allelochemicals. The similar phenomenon was also observed in the host adaption of *Myzus persicae* to nicotianae by the let-7 and miR-100 participating in regulating of the expression of *CYP6CY3* post-transcriptionally [44]. Further, Sattar et al. found that when apterous cotton aphids fed on *Vat⁻* melon with high susceptibility to aphids, the expression levels of let-7 and miR-100 were decreased in compare with the aphids that fed on resistant (*Vat⁺*) melon [36]. These results suggest that let-7 and miR-100 might be involved in the metabolism of xenobiotics of *A. gossypii*.

To further understand the possible roles of miRNAs in the metabolism of plant allelochemicals, the putative target genes of the newly identified miRNAs were predicted, and many of the predicted target genes were annotated to be involved in multiple biological processes. In addition, we found that several miRNAs were predicted to target genes from families known to be important in the metabolism of xenobiotics [45, 46]. Combining these results of the target prediction with our differential expression analysis, we found that some of the miRNAs predicted to target these xenobiotic metabolism-related genes were among the differentially express miRNAs. This suggested that miRNAs may be involved in insect metabolism of plant allelochemicals by regulating the expression of xenobitic metabolism genes.

## Conclusions

A total of 292 miRNAs was identified from *A. gossypii*, and the expression analysis results demonstrated that the transcript levels of these miRNAs were changed depend on the plant allelochemicals feeding by *A. gossypii*. The results of target prediction suggest that miRNAs may be involved in the metabolism of plant allelochemicals of *A. gossypii*, and these results represent an important new small RNA genomics resource for further studies on this topic.

## Methods

### Cotton aphid strain and cell culture

The strain of *A. gossypii* used in this study was collected in 1999 from cotton fields in the Xinjiang Uygur Autonomous Region, China, and has been maintained in our laboratory for more than 15 years. The aphids were reared on cotton seedings in controlled conditions of 20–23 °C, 60% relative humidity, and a photoperiod of 16:8 h (L:D), as described previously [47]. The mammalian HEK293T cell line was maintained at 37 °C under 5% $CO_2$ in DMEM high-glucose medium (Gibco) containing 10% fetal bovine serum (Gibco).

### Chemicals

Gossypol, 2-tridecanone, quercetin, and tannic acid were purchased from Sigma-Aldrich (St. Louis, MO, USA). Sucrose was purchased from Beijing Solarbio Science & Technology Co., Ltd (Beijing, China).

### *In vitro* feeding assay

Sterilized glass tubes open at both ends (3 cm in length, 2 cm diameter) were used for the in vitro feeding assays. One end of each tube was covered by two layers of parafilm, with the following solution sandwiched between the two parafilm layers (artificial diet): 200 μl of a 0.5 M sterile sucrose solution that contained 0.2% gossypol,

2-tridecanone, quercetin, and tannic acid). One hundred healthy aperous adults were gently placed into the tube with a brush, and the tube was sealed with a piece of Chinese art paper with solid glue. Aphids were allowed to feed on this artificial diet for 24 h. The same system and solution, minus the alleochemicals, was used as a control. Following the 24 h of feeding, the live aphids were collected for RNA extraction.

## RNA extraction and sequencing

Five samples of apterous *A. gossypii* adults collected from the feeding assays were used for the preparation of the small RNA libraries. Total RNA was isolated with TRIzol® reagent (Invitrogen, Carlsbad, CA, USA) following the manufacturer's instructions. The purity and the concentration of the RNA were assessed with a NAS-99 spectrophotometer (ACTGene, USA), and RNA integrity was evaluated with analysis on a 1% (w/v) agarose gel. About 10 μg of total RNA was isolated on a 15% denaturing polyacrylamide gel, and small RNA molecules ranging from 18 to 30 nt in length were purified and then ligated with 3′ and 5′ adapters. The ligated products were then reverse transcribed into cDNA with SuperScript II reverse transcriptase (Invitrogen, Carlsbad, CA, USA) following the manufacturer's protocol, and the resulting cDNA was amplified with PCR (15 cycles). Amplified cDNA products were purified with agarose gels and sent to the Beijing Genome Institute Inc. (China) for high-throughput sequencing with the Illumina Hiseq 2000 platform.

## Identification of miRNAs from sequencing data

*A. gossypii* miRNA candidates were identified using miRDeep2 software [48]. Raw sequencing reads from the five libraries were submitted as input into miRDeep2, and the data from each library were analyzed separately. The miRDeep2 analysis was performed using the default options and settings. After trimming the adaptor sequences and discarding rRNA, tRNA, snRNA, and snoRNA, as well as the sequences containing polyA tails from the raw reads, the small RNAs between 18 and 30 nt in length were selected for further analysis. The *Acyrthosiphon pisum* genome sequence was used as the reference genome, since the *A. gossypii* genome sequence was not available at the time of analysis.

## Expression profile of miRNAs in five libraries

In order to find the miRNAs that may play very important roles in the response of cotton aphids against plant alleochemicals, the differential expression of the identified miRNAs was analyzed. The read counts of the newly identified miRNA were analyzed using edgeR software; edgeR version 3.10.2 is available in Bioconductor version 3.1 (http://www.bioconductor.org/packages/release/bioc/html/edgeR.html) [49]. edgeR is a Bioconductor software package for analyzing the differential expression of digital gene expression data. Briefly, an overdispersed Poisson model is used to account for variability, and empirical Bayes methods are used to moderate the degree of overdispersion across the transcripts. The Benjamini-Hochberg method was used to adjust for multiple testing [50]. Only those miRNAs with a fold change $\geq 2$ and a false discovery rate (FDR) <0.05 were considered to be significant.

## Target gene prediction and analysis

Since the complete genome sequence of *A. gossypii* was not available at the time of analysis, the *A. gossypii* transcriptome database (unpublished) was used to predict the targets of the sequenced *A. gossypii* miRNAs. Two miRNA target prediction software programs were used, each with the default settings: miRanda (http://www.microrna.org/) [51] and RNAhybrid (http://bibiserv2.cebitec.uni-bielefeld.de/rnahybrid/) [52]. The miRNA target genes commonly predicted by both programs were selected for further analysis. The predicted target genes were then aligned using the BLASTX program from NCBI (http://blast.ncbi.nlm.nih.gov/Blast.cgi) (e value cut-off used was 1.0E−5), and these genes were mapped and annotated by BLAST2GO [53]. The GO terms from this analysis were used to define the functional categories of the predicted miRNA target genes.

## Quantitative real-time PCR

To validate the miRNA data obtained through our deep sequencing experiment, 4 conserved and 2 novel miRNA candidates were selected, and their expression was confirmed with quantitative real-time PCR (qRT-PCR) analysis. Aperous adults fed on an artificial diet (with 0.2% alleochemicals) for 24 h were used for total RNA isolation; total RNA was extracted using miRNeasy Mini Kit (Qiagen, Germany) following the manufacturer's protocol. First strand cDNA was synthesized from 2 μg of total RNA using miScript II RT kit (Qiagen) following the manufacturer's instructions. SYBR Green Master Mix (miScript SYBR Green PCR Kit, Qiagen) was used for miRNA expression assays, and qRT-PCR was performed as previously described [15]. Three biological replicates, with three technical replications for each, were evaluated for each sample. Analysis of the qRT-PCR data was carried out using the $2^{-\Delta\Delta Ct}$ method of relative quantification [54]. As an endogenous control, U6 snRNA was used to quantify the expression level of miRNA. The primers used for the qRT-PCR analysis are listed in Additional file 2: Table S1.

## Vector construction and dual luciferase reporter assay

The target prediction results were validated by selecting Ago-miR-656a-3p and *CYP6J1* that has a target site of Ago-miR-656a-3p in the 3′ UTR. The *CYP6J1* 3′ UTR sequence was synthesized by GenePharm Co. Ltd (Shanghai, China), and was inserted into the pmirGLO vector, generating the pmirGLO-CYP6J1-UTR target construct. The agomir (mimics) of Ago-miR-656a-3p was synthesized by GenePharm Co. Ltd (Shanghai, China). The HEK293T cells were cultured in a 24-well plate and were transfected with target plasmids and agomir of Ago-miR-656a-3p or agomir-NC using a Calcium Phosphate Cell Transfection Kit (Beyotime, Nanjing, China) according to the manufacturer's instruction. Each well contained 0.5 μg of the plasmid, with the final concentration of miRNA agomir reaching 100 nM/L. Luciferase assays were performed by using the Dual-Glo® Luciferase Assay System (Promega) at 24 h post-transfection. Normalized firefly luciferase activity (firefly luciferase activity/Renilla luciferase activity) was compared to that of the control pmirGLO Vector. The mean of the relative luciferase expression ratio (firefly luciferase/renilla luciferase) of the control was set to 1. For each transfection, luciferase activity was averaged from five replicates.

## Abbreviations
miRNAs: microRNAs; UTRs: untranslated regions; P450: cytochrome p450; PCR: polymerase chain reaction; qRT-PCR: quantitative real-time PCR; U: uracil; let-7-C: let-7-complex.

## Authors' contributions
KM and XG designed the research, analyzed the data and wrote the paper. KM, FL and PL performed the experiments. XC and YL participated in the data analysis. All authors read and approved the final manuscript.

## Acknowledgements
This work was supported by The National Natural Science Foundation of China (No. 31330064).

## Competing interests
The authors declare that they have no competing interests.

## Funding
The National Natural Science Foundation of China (No. 31330064).

## References
1. Asgari S. MicroRNA functions in insects. Insect Biochem Mol Biol. 2013;43:388–97.
2. Pritchard CC, Cheng HH, Tewari M. MicroRNA profiling: approaches and considerations. Nat Rev Genet. 2012;13:358–69.
3. Bartel DP. MicroRNAs: target recognition and regulatory functions. Cell. 2009;136:215–33.
4. Rigoutsos I. New tricks for animal microRNAs: targeting of amino acid coding regions at conserved and nonconserved sites. Cancer Res. 2009;69:3245–8.
5. Yokoi T, Nakajima M. microRNAs as mediators of drug toxicity. Annu Rev Pharmacol. 2013;53:377–400.
6. Asgari S. Role of microRNAs in insect host-microorganism interactions. Front Physiol. 2011;2:48.
7. Ibanez-Ventoso C, Vora M, Driscoll M. Sequence relationships among *C. elegans*, *D. melanogaster* and human microRNAs highlight the extensive conservation of microRNAs in biology. PLoS ONE. 2008;3:e2818.
8. Lee RC, Feinbaum RL, Ambros V. The *C. elegans* heterochronic gene lin-4 encodes small RNAs with antisense complementarity to lin-14. Cell. 1993;75:843–54.
9. Aravin AA, Lagos-Quintana M, Yalcin A, Zavolan M, Marks D, Snyder B, Gaasterland T, Meyer J, Tuschl T. The Small RNA profile during *Drosophila melanogaster* development. Dev Cell. 2003;5:337–50.
10. Yu XM, Zhou Q, Li SC, Luo QB, Cai YM, Lin WC, Chen H, Yang Y, Hu SN, Yu J. The Silkworm (*Bombyx mori*) microRNAs and their expressions in multiple developmental stages. PLoS ONE. 2008;3:e2997.
11. Cao J, Tong C, Wu X, Lv J, Yang Z, Jin Y. Identification of conserved microRNAs in *Bombyx mori* (silkworm) and regulation of fibroin L chain production by microRNAs in heterologous system. Insect Biochem Mol Biol. 2008;38:1066–71.
12. Zhang X, Zheng Y, Cao X, Ren R, Yu XQ, Jiang H. Identification and profiling of *Manduca sexta* microRNAs and their possible roles in regulating specific transcripts in fat body, hemocytes, and midgut. Insect Biochem Mol Biol. 2015;62:11–22.
13. Zhang X, Zheng Y, Jagadeeswaran G, Ren R, Sunkar R, Jiang H. Identification of conserved and novel microRNAs in *Manduca sexta* and their possible roles in the expression regulation of immunity-related genes. Insect Biochem Mol Biol. 2014;47:12–22.
14. Zhang X, Zheng Y, Jagadeeswaran G, Ren R, Sunkar R, Jiang H. Identification and developmental profiling of conserved and novel microRNAs in *Manduca sexta*. Insect Biochem Mol Biol. 2012;42:381–95.
15. Liang P, Feng B, Zhou XG, Gao XW. Identification and developmental profiling of microRNAs in diamondback moth, *Plutella xylostella* (L.). PLoS ONE. 2013;8:e78787.
16. Etebari K, Hussain M, Asgari S. Identification of microRNAs from *Plutella xylostella* larvae associated with parasitization by *Diadegma semiclausum*. Insect Biochem Mol Biol. 2013;43:309–18.
17. Agrawal N, Sachdev B, Rodrigues J, Sree KS, Bhatnagar RK. Development associated profiling of chitinase and microRNA of *Helicoverpa armigera* identified chitinase repressive microRNA. Sci Rep. 2013;3:2292.
18. Lomate PR, Mahajan NS, Kale SM, Gupta VS, Giri AP. Identification and expression profiling of *Helicoverpa armigera* microRNAs and their possible role in the regulation of digestive protease genes. Insect Biochem Mol Biol. 2014;54:129–37.
19. Lucas K, Raikhel AS. Insect microRNAs: biogenesis, expression profiling and biological functions. Insect Biochem Mol Biol. 2013;43:24–38.
20. Hussain M, Asgari S. MicroRNAs as mediators of insect host-pathogen interactions and immunity. J Insect Physiol. 2014;70:151–8.
21. Cao CW, Zhang J, Gao XW, Liang P, Guo HL. Overexpression of carboxylesterase gene associated with organophosphorous insecticide resistance in cotton aphids, *Aphis gossypii* (Glover). Pestic Biochem Physiol. 2008;90:175–80.
22. Blackman RL, Eastop VF. Aphids on the world's crops: an identification guide. New York: John Wiley and Sons; 1984.

23. Li ZQ, Zhang S, Luo JY, Wang CY, Lv LM, Dong SL, Cui JJ. Ecological adaption analysis of the cotton aphid (*Aphis gossypii*) in different phenotypes by transcriptome comparison. PLoS ONE. 2013;8:e83180.

24. Cao DP, Liu Y, Walker WB, Li JH, Wang GR. Molecular characterization of the *Aphis gossypii* olfactory receptor gene families. PLoS ONE. 2014;9:e101187.

25. Despres L, David JP, Gallet C. The evolutionary ecology of insect resistance to plant chemicals. Trends Ecol Evol. 2007;22:298–307.

26. Golawska S, Sprawka I, Lukasik I, Golawski A. Are naringenin and quercetin useful chemicals in pest-management strategies? J Pest Sci. 2014;87:173–80.

27. Lema C, Cunningham MJ. MicroRNAs and their implications in toxicological research. Toxicol Lett. 2010;198:100–5.

28. Choi YM, An S, Lee EM, Kim K, Choi SJ, Kim JS, Jang HH, An IS, Bae S. CYP1A1 is a target of miR-892a-mediated post-transcriptional repression. Int J Oncol. 2012;41:331–6.

29. Hong S, Guo Q, Wang W, Hu S, Fang F, Lv Y, Yu J, Zou F, Lei Z, Ma K, et al. Identification of differentially expressed microRNAs in *Culex pipiens* and their potential roles in pyrethroid resistance. Insect Biochem Mol Biol. 2014;55:39–50.

30. Lei Z, Lv Y, Wang W, Guo Q, Zou F, Hu S, Fang F, Tian M, Liu B, Liu X, et al. MiR-278-3p regulates pyrethroid resistance in *Culex pipiens pallens*. Parasitol Res. 2015;114:699–706.

31. Lau NC, Lim LP, Weinstein EG, Bartel DP. An abundant class of tiny RNAs with probable regulatory roles in *Caenorhabditis elegans*. Science. 2001;294:858–62.

32. Mi S, Cai T, Hu Y, Chen Y, Hodges E, Ni F, Wu L, Li S, Zhou H, Long C, et al. Sorting of small RNAs into *Arabidopsis* argonaute complexes is directed by the 5′ terminal nucleotide. Cell. 2008;133:116–27.

33. Mao YB, Cai WJ, Wang JW, Hong GJ, Tao XY, Wang LJ, Huang YP, Chen XY. Silencing a cotton bollworm P450 monooxygenase gene by plant-mediated RNAi impairs larval tolerance of gossypol. Nat Biotechnol. 2007;25:1307–13.

34. Celorio-Mancera MdlP, Ahn SJ, Vogel H, Heckel DG. Transcriptional responses underlying the hormetic and detrimental effects of the plant secondary metabolite gossypol on the generalist herbivore *Helicoverpa armigera*. BMC Genom. 2011;12:575.

35. Gao F, Zhu S, Sun Y, Du L, Parajulee M, Kang L, Ge F. Interactive effects of elevated $CO_2$ and cotton cultivar on tri-trophic interaction of *Gossypium hirsutum*, *Aphis gossyppii*, and *Propylaea japonica*. Environ Entomol. 2008;37:29–37.

36. Sattar S, Addo-Quaye C, Song Y, Anstead JA, Sunkar R, Thompson GA. Expression of small RNA in *Aphis gossypii* and its potential role in the resistance interaction with melon. PLoS ONE. 2012;7:e48579.

37. Bartel DP. MicroRNAs: genomics, biogenesis, mechanism, and function. Cell. 2004;116:281–97.

38. Cristino AS, Tanaka ED, Rubio M, Piulachs MD, Belles X. Deep sequencing of organ- and stage-specific microRNAs in the evolutionarily basal insect *Blattella germanica* (L.) (Dictyoptera, Blattellidae). PLoS ONE. 2011;6:e19350.

39. Liu F, Peng W, Li Z, Li W, Li L, Pan J, Zhang S, Miao Y, Chen S, Su S. Next-generation small RNA sequencing for microRNAs profiling in *Apis mellifera*: comparison between nurses and foragers. Insect Mol Biol. 2012;21:297–303.

40. Chen Z, Liang S, Zhao Y, Han Z. miR-92b regulates Mef2 levels through a negative-feedback circuit during *Drosophila* muscle development. Development. 2012;139:3543–52.

41. Yuva-Aydemir Y, Xu XL, Aydemir O, Gascon E, Sayin S, Zhou W, Hong Y, Gao FB. Downregulation of the host gene jigr1 by miR-92 Is essential for neuroblast self-renewal in *Drosophila*. PLoS Genet. 2015;11:e1005264.

42. Lyons PJ, Storey KB, Morin PJ. Expression of miRNAs in response to freezing and anoxia stresses in the freeze tolerant fly *Eurosta solidaginis*. Cryobiology. 2015;71:97–102.

43. Chawla G, Sokol NS. Hormonal activation of let-7-C microRNAs via EcR is required for adult *Drosophila melanogaster* morphology and function. Development. 2012;139:1788–97.

44. Peng TF, Pan YO, Gao XW, Xi JH, Zhang L, Ma KS, Wu YQ, Zhang JH, Shang QL. Reduced abundance of the *CYP6CY3*-targeting *let-7* and *miR-100* miRNAs accounts for host adaptation of *Myzus persicae* nicotianae. Insect Biochem Mol Biol. 2016;75:89–97.

45. Omiecinski CJ, Vanden Heuvel JP, Perdew GH, Peters JM. Xenobiotic metabolism, disposition, and regulation by receptors: from biochemical phenomenon to predictors of major toxicities. Toxicol Sci. 2011;120:49–75.

46. Li X, Schuler MA, Berenbaum MR. Molecular mechanisms of metabolic resistance to synthetic and natural xenobiotics. Annu Rev Entomol. 2007;52:231–53.

47. Ma KS, Li F, Liang PZ, Chen XW, Liu Y, Gao XW. Identification and validation of reference genes for the normalization of gene expression data in qRT-PCR analysis in *Aphis gossypii* (Hemiptera: Aphididae). J Insect Sci. 2016;16:1–9.

48. Friedlander MR, Mackowiak SD, Li N, Chen W, Rajewsky N. miRDeep2 accurately identifies known and hundreds of novel microRNA genes in seven animal clades. Nucleic Acids Res. 2012;40:37–52.

49. Robinson MD, McCarthy DJ, Smyth GK. edgeR: a bioconductor package for differential expression analysis of digital gene expression data. Bioinformatics. 2010;26:139–40.

50. Benjamini Y, Hochberg Y. Controlling the false discovery rate: a practical and powerful approach to multiple testing. J R Stat Soc B. 1995;57:289–300.

51. Betel D, Wilson M, Gabow A, Marks DS, Sander C. The microRNA.org resource: targets and expression. Nucleic Acids Res. 2008;36:149–53.

52. Rehmsmeier M, Steffen P, Hochsmann M, Giegerich R. Fast and effective prediction of microRNA/target duplexes. RNA. 2004;10:1507–17.

53. Conesa A, Gotz S, Garcia-Gomez JM, Terol J, Talon M, Robles M. Blast2GO: a universal tool for annotation, visualization and analysis in functional genomics research. Bioinformatics. 2005;21:3674–6.

54. Livak KJ, Schmittgen TD. Analysis of relative gene expression data using real-time quantitative PCR and the $2^{-\Delta\Delta CT}$ method. Methods. 2001;25:402–8.

# 6

# Immunoblot screening of CRISPR/Cas9-mediated gene knockouts without selection

Jason A. Estep, Erin L. Sternburg, Gissell A. Sanchez and Fedor V. Karginov[*]

## Abstract

**Background:** Targeted genomic editing using the CRISPR/Cas9 methodology has opened exciting new avenues in probing gene function in virtually any model system, including cultured mammalian cells. Depending on the desired mutation, several experimental options exist in the isolation of clonal lines, such as selection with introduced markers, or screening by PCR amplification of genomic DNA. However, streamlined approaches to establishing deletion and tagging mutants with minimal genomic perturbation are of interest in applying this methodology.

**Results:** We developed a procedure for rapid screening of clonal cell lines for the deletion of a protein of interest following CRISPR/Cas9 targeting in the absence of selective pressure based on dot immunoblots. To assess the technique, we probed clonal isolates of 293-TREx cells that were targeted with three separate sgRNAs against the HuR gene. Validation of knockout candidates by western blot indicated that the normalized protein abundances indicated by the dot blot serve as accurate predictors of deletion. In total, 32 independent biallelic deletion lines out of 248 screened clones were isolated, and recovery of null mutants ranged from 6 to 36 % for the individual sgRNAs. Genomic sequencing verified small deletions at the targeted locus.

**Conclusions:** Clonal screening for CRISPR/Cas9-mediated editing events using dot immunoblot is a straightforward and efficient approach that facilitates rapid generation of genomic mutants to study gene function.

**Keywords:** CRISPR/Cas9, Clonal selection, Screening, Dot blot

## Background

Manipulating protein levels and activities is a principal tool in understanding the functions and the relationships of these molecular components in cells. Transient perturbations involve chemical inhibition (and activation) with small molecules, overexpression from non-integrating vectors, or knockdown by RNA interference [1]. Often more experimentally desirable, stable genetic alteration can be achieved by integration of overexpression and shRNA constructs [2], or by genomic editing of the endogenous protein locus with transcription activator-like effector nucleases (TALENs) [3], zinc-finger nucleases [4, 5], and, more recently, by RNA-guided nucleases based on the clustered, regularly interspaced

short palindromic repeats (CRISPR)/Cas9 system [6–11]. Combinations of stable integration and modification with temporal control include the use of chemically inducible systems, such as tetracycline repression/activation [12, 13], tamoxifen control of Cre-ER recombination [14, 15], and more modular control of expression, localization and activity by small molecules [16–18].

The CRISPR/Cas module is an endogenous adaptive immunity system commonly used by bacteria and archaea to counteract phage infection and introduction of plasmid DNA [19–22]. Short (20–30 bp) sequence tags from the invaders are incorporated as spacers between direct repeats of a CRISPR locus. Its transcription and processing yields small crRNAs that associate with and guide CRISPR-associated (cas) protein(s) to complementary DNA targets for endonucleolytic cleavage. Due to its particularly simple makeup, the type II

*Correspondence: karginov@ucr.edu
Department of Cell Biology and Neuroscience, Institute for Integrative Genome Biology, University of California, Riverside, CA 92521, USA

CRISPR/Cas system of *Streptococcus pyogenes* has been adapted for genomic editing with great success: a single protein, Cas9, is required for crRNA binding and cleavage, and the RNA components have been engineered into a single guide RNA (sgRNA) [23]. Thus, targeting of nearly any genomic sequence is possible with the introduction of an sgRNA and Cas9 into the cells of interest [24, 25]. In its simplest form, creation of a knockout line depends on the sgRNA-guided dsDNA cleavage by the Cas endonuclease, followed by non-homologous end joining (NHEJ) repair of the site by the cell. A small, random deletion is often introduced at the repair site. Deletions that target the open reading frame and result in frame-shifts, particularly early in the mRNA, are very likely to yield a non-functional protein sequence and to target the mRNA for nonsense-mediated decay due to premature stop codons in the new frame (Fig. 1). In contrast, small in-frame deletions may not compromise the function of the protein. Aside from NHEJ-mediated deletions, the Cas cleavage can also stimulate homology-driven repair based on a supplied DNA template, and allow for larger deletions or insertion of tags and markers.

In the targeted modification strategies, the researcher has the option to isolate correct clones by introducing a selectable marker at the genomic locus. While this approach is beneficial if the sgRNA has low efficiency, or if the mutation confers a strong selective disadvantage to the cells, it requires the construction of repair template vectors containing the marker and homology arms specific to the locus [26]. Furthermore, it leaves a significantly-sized, transcribed insertion that may affect the expression of nearby genes, although it may be excised through an additional Cre-mediated step, if surrounded by loxP sites [27].

However, the relatively high efficiency of Cas9 targeting allows for clonal screening without selection. In a common approach, purification of genomic DNA from isolated clones permits PCR amplification followed by sequencing, deletion analysis or restriction digest [28]. In the present study, we describe a simple method for isolation of clonal deletion mutants based on CRISPR/Cas9 targeting combined with dot immunoblot analysis and validated by western blots and DNA sequencing. Screening directly for protein production ensures proper knockout, avoiding signal from in-frame deletions/mutations

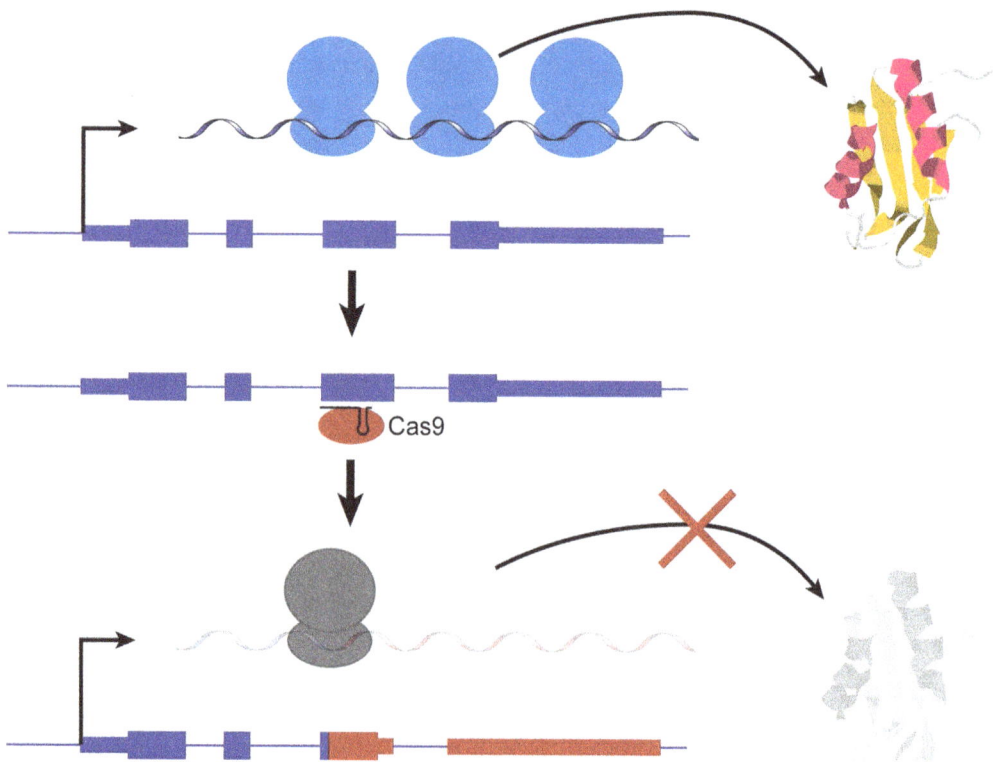

**Fig. 1** sgRNA/Cas9-mediated gene knockouts. sgRNA/Cas9 targeting to a gene exon induces a double-stranded DNA break that is repaired by non-homologous end joining, often introducing frameshift mutations (depicted in *red*) that lead to lack of functional protein production and likely cause nonsense-mediated decay of the mRNA

that may not eliminate its function. The approach can also be applied to screening for insertions of tags mediated by homologous recombination.

## Results and discussion

We wished to apply the CRISPR/Cas system to establish stable cell lines completely lacking specific protein function. As with other genetic mutant approaches, it is beneficial to isolate several independently derived knockout clones to control for secondary mutations. In the case of CRISPR/Cas, off-target cleavage events at sites of imperfect sgRNA complementarity are known to occur [29, 30], and the use of different sgRNAs can mitigate the possibility of spuriously observed phenotypes. We used pre-designed human genomic target sites for the *S. pyogenes* Cas9 effector nuclease, picking the reduced set of three sites closer to the 5′ end of the gene [31]. The targeted sites were introduced into sgRNA-producing context in a plasmid that also co-expresses the Cas9 protein, pSpCas9(BB) [24, 28].

We opted to develop a simple, selection-free strategy based on dot blots for the protein of interest to isolate

the knockout clones (Fig. 2a). First, a bulk population of cells is transfected with the sgRNA/Cas9 constructs. To ensure a high percentage of expressing cells, the transfection procedure may be sequentially repeated, and a parallel control transfection with a GFP marker is performed. Next, transfected cells are plated at limiting dilution concentrations into 96-well plates, and clonal populations are established. Setting aside a propagating aliquot for each, confluent clones are lysed in the plate with a passive lysis buffer. To identify knockout candidates, a small amount of lysate (1 μl) is blotted onto two nitrocellulose membranes, to be immunoblotted for the protein of interest, and a normalization control. Alternatively, a single membrane may be probed sequentially. Clones that exhibit little or no expression of the protein, while demonstrating a measurable amount of cell extract in the control blot, can be identified visually, or by quantifying the ratio of the corresponding dot intensities. In this fashion, screening directly for protein production ensures a functional knockout, as opposed to deletions/mutations that may not efficiently disrupt the protein. It should be noted that the extent of cross-reactivity of the antibody will affect

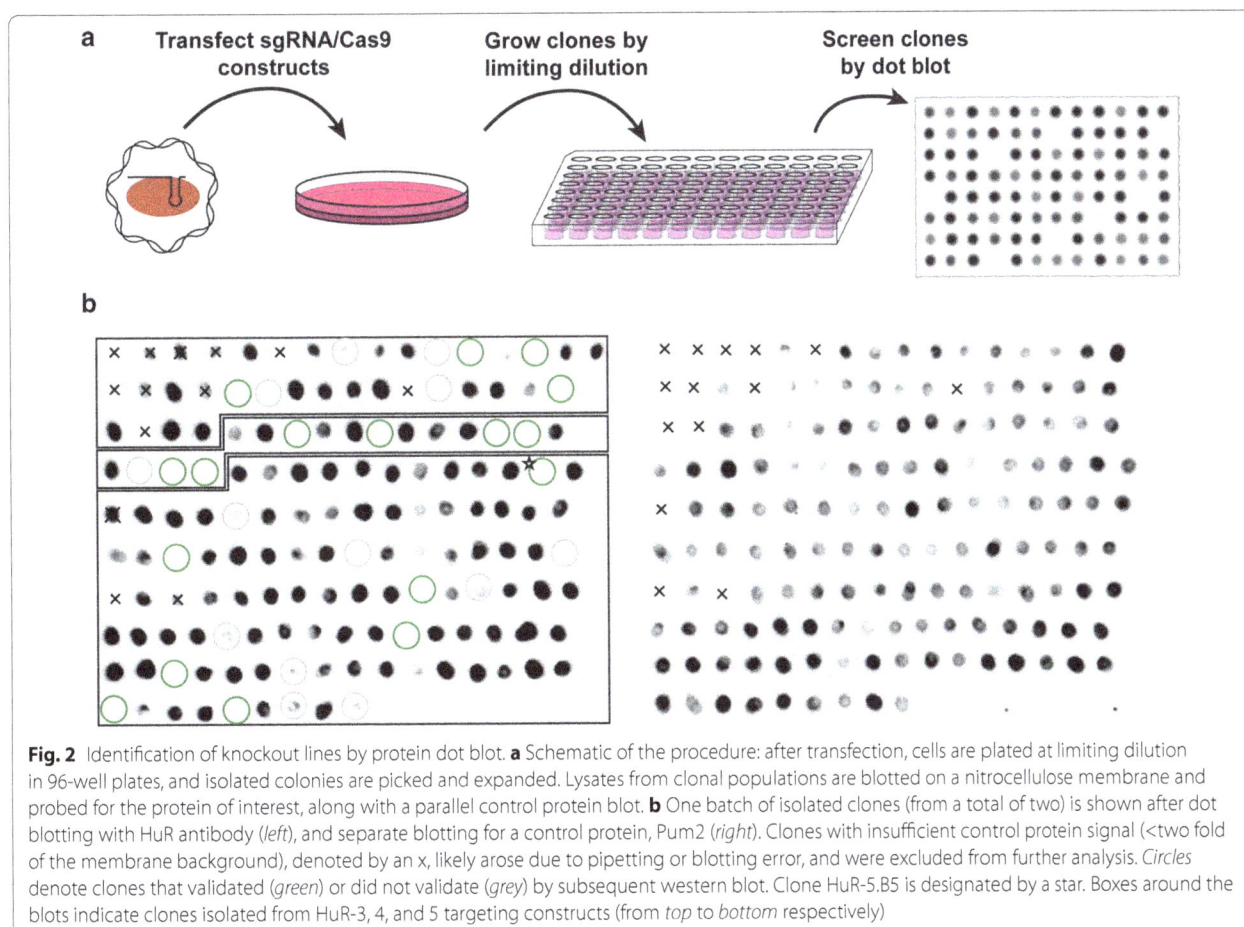

**Fig. 2** Identification of knockout lines by protein dot blot. **a** Schematic of the procedure: after transfection, cells are plated at limiting dilution in 96-well plates, and isolated colonies are picked and expanded. Lysates from clonal populations are blotted on a nitrocellulose membrane and probed for the protein of interest, along with a parallel control protein blot. **b** One batch of isolated clones (from a total of two) is shown after dot blotting with HuR antibody (*left*), and separate blotting for a control protein, Pum2 (*right*). Clones with insufficient control protein signal (<two fold of the membrane background), denoted by an x, likely arose due to pipetting or blotting error, and were excluded from further analysis. *Circles* denote clones that validated (*green*) or did not validate (*grey*) by subsequent western blot. Clone HuR-5.B5 is designated by a star. Boxes around the blots indicate clones isolated from HuR-3, 4, and 5 targeting constructs (from *top* to *bottom* respectively)

the accuracy of the method, since genuine knockout lines may still display a significant level of background signal from other proteins in the blot.

We applied the above strategy to generate a knockout of the HuR (ELAVL1) protein in 293-TREx cells, targeting exons 3, 4, and 5 with three distinct sgRNAs. Most of the expanded clones exhibited parental growth rates and morphology, while several lines grew substantially slower and/or with altered cell shape. However, these features did not correlate with the absence of HuR (see below), likely reflecting spurious mutations acquired due to CRISPR/Cas off-targeting or simply during the procedure. Thus, HuR is dispensable to normal growth rates under optimal conditions in 293-TREx cells; however, if the protein of interest is expected to affect the division or morphology of cells, it can form the basis of a phenotypic pre-screen during clonal expansion. Mutations that are substantially detrimental to growth rates can still be recovered, since the clonal isolation begins soon after the introduction of the sgRNA/Cas9.

Isolated clones were grown and tested for the presence of HuR with a dot blot (Fig. 2b, left), yielding several candidates with complete or near-complete lack of HuR (circled). In addition to visual inspection, HuR expression signal on the dot blots was quantified relative to a loading control (Fig. 2b, right). After eliminating clones with low control expression (less than twofold of the average background staining of the blot), the ratios displayed a range of >10⁴-fold in normalized HuR expression (Fig. 3a). A possible expectation was for the HuR levels to cluster into stepwise values representing wild-type, heterozygous, and homozygous mutant clones. Instead, the observed levels form a fairly smooth gradient, likely due to clonal variability in expression, as well as probable compensatory mechanisms affecting expression in heterozygotes. Of note, our estimates based on absolute protein

quantification in the literature [32] indicate that ~300 fmol (11 pg) of HuR were blotted, and we expect that substantially lower protein amounts should be reliably detected, provided suitable antibodies are available.

Forty four of the 248 identified candidates (examples shown in Fig. 2b, circles), including some with low but visible signal in the dot blot, were further tested for the absence of HuR by western blot (example in Fig. 3b). In an all-or-none fashion, 32 clones validated as complete knockouts, while 12 appeared to be false positives. The false positives may have resulted from pipetting error during blotting; from stochastically low, but non-zero HuR expression in some wild-type clones; and from selecting clones with comparatively fewer cells in the lysate (observed as low expression both for HuR and the control). Indeed, the validated KO lines had significantly lower normalized HuR signal (this remaining signal likely arising from antibody cross-reactivity) relative to the candidates that were not confirmed (Fig. 3a). This indicated that the HuR/control ratio is an accurate quantitative criterion in separating true knockouts from wild-type clones in our broad set of tested candidates, and can serve as a reliable selection filter. It should be noted that a modification to the procedure that would strip and re-probe the same blot for both proteins could eliminate some of the signal variability, at the expense of a slightly lengthier protocol. Finally, we verified the deletions in clone HuR-5.B5 by amplifying, cloning and sequencing the corresponding genomic region (Fig. 4). Four sequenced clones revealed two different short deletions, consistent with independent NHEJ events at two chromosomal locations. In both cases, the deletions occurred near the Cas9 cleavage site three nucleotides upstream of the protospacer-adjacent motif (NGG) [23, 33], as observed previously [24].

The effectiveness of generating homozygous knockouts varied substantially with the sgRNA used (Table 1),

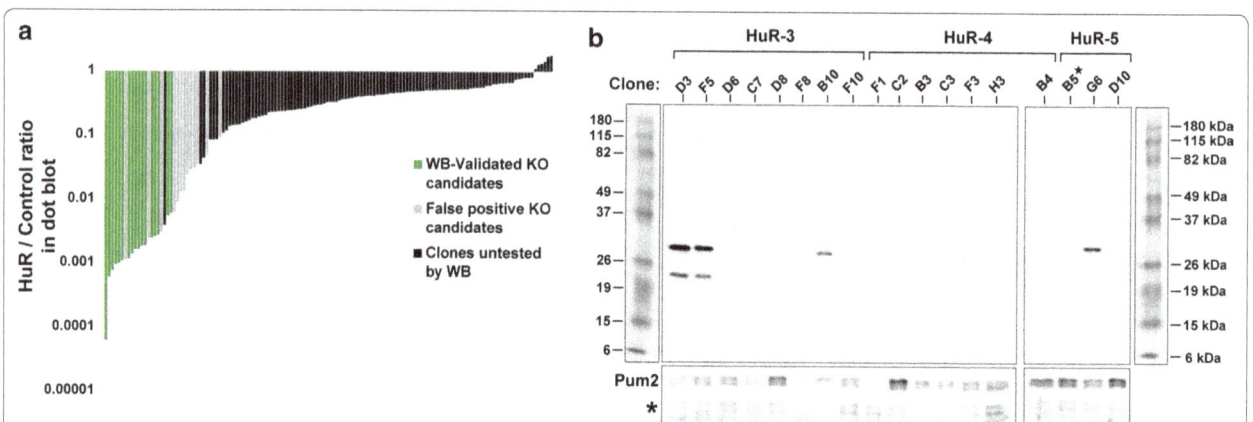

**Fig. 3** Quantification and validation of KO candidates. **a** Ratios of HuR to control protein signal in the dot blot shown in Fig. 2b, in increasing order. Clones that were confirmed or were not confirmed by western blot are shown in *green* and *grey*, respectively. **b** Western blot for HuR (*top panels*) of a subset of clones identified by dot blot. *Bottom panel* shows a control blot for Pum2 of the same membrane. Clone HuR-5.B5 is designated by a *star*

```
HuR-5 sgRNA target                                      CCGGATAAACGCAACCCCTC
293-TREx (parental)   CTTCTGCCTCCGACCGTTTGTCAAACCGGATAAACGCAACCCCTCTGGACAAACCTTTTCAACAAATCAAC
HuR-5.B5 sequence 1   CTTCTGCCTCCGACCGTTTGTCAAACCGGATAAACGCAACCC                GGAACAAATCAAC
HuR-5.B5 sequence 2   CTTCTGCCTCCGACCGTTTGTCAAACCGGATAAACGCAACCC                GGAACAAATCAAC
HuR-5.B5 sequence 3   CTTCTGCCTCCGACCGTTTGTCAAACCGGATAAAC             ACAAACCTTTTTAACAAATCAAC
HuR-5.B5 sequence 4   CTTCTGCCTCCGACCGTTTGTCAAACCGGATAAAC             ACAAACCTTTTTAACAAATCAAC
```

**Fig. 4** Sequence of the genomic deletions in clone HuR-5.B5. The genomic region surrounding the targeted site was PCR-amplified from wt 293-TREx cells and clone HuR-5.B5, and placed into a plasmid vector. Individual plasmid clones were sequenced. The Cas9 cleavage site is denoted by a *red triangle*

**Table 1 Number of analyzed and validated KO clonal lines for the sgRNAs used**

| SgRNA construct | # of clones tested by dot blot | # of validated KOs | KO percentage (%) |
|---|---|---|---|
| HuR-3 | 69 | 15 | 22 |
| HuR-4 | 22 | 8 | 36 |
| HuR-5 | 157 | 9 | 6 |

ranging from 6 to 36 % of the initially tested clones. Knockout production was reproducible, as an independent replicate of the experiment with guide RNAs 3 and 4 yielded 10 and 8 % of null mutations respectively (data not shown), with the variability likely arising from transfection efficiencies. Considering that these lines represent concurrent editing of both alleles, and most likely require a frameshift mutation, the observed KO rates compare favorably with the ~6–50 % rates of overall NHEJ-driven indels measured by sequencing or the SURVEYOR assay using similar delivery methods [34, 35]. The variability among the three tested sgRNAs also agrees with related observations for multiple guide RNAs at a given locus [24], presumably reflecting differences in sgRNA production, complex formation with Cas9, and targeting/cleavage efficiency [36]. However, CRISPR/Cas editing efficiencies may also show loci-specific variability, potentially explained by the dependence of Cas9 binding on chromatin accessibility, reflected in DNase hypersensitivity and CpG methylation [37, 38]. Nevertheless, screening of a reasonably small number of initial clones is likely to produce more than one independent deletion mutant using this simple procedure.

Aside from targeting individual genes for knockout, the CRISPR/Cas system allows many additional means of genome editing. We assessed whether our screening protocol can be used to achieve simultaneous knockout of tandemly arranged genes by inducing large chromosomal deletions between two sgRNA target sites. Small deletions have been previously achieved without selection [24], while larger biallelic and monoallelic deletions have been created after sorting for strong Cas9 expression [39], or in haploid embryonic stem cells [40]. In our hands, attempts to excise the ~250 Kbp human Ago4, 1, 3 locus by targeting Ago4 and Ago3 while blotting for Ago1 have not yielded any complete deletions (data not shown). Thus, elimination of large genomic fragments may require the use of homologous recombination templates spanning the deletion, enrichment for cells expressing Cas9 (without genomic insertion), and/or incorporation of selectable markers at the deletion site. Similarly, gene mutagenesis without deletion would require homologous templates involving selection or protein tagging. In these cases, dot blotting for the tag may still be used as a screening step. It should be pointed out that directly blotting for the protein under study is limited by antibody availability, and assaying of secreted proteins may necessitate the collection of culture media instead of cell pellets.

## Conclusions

The CRISPR/Cas system is transforming modern genetics and molecular biology by offering unprecedented ways of genomic manipulation. We have described a methodology to isolate simple deletion mutants of a gene of interest by CRISPR/Cas based on dot-blot screening for the resulting protein product, without the need for intervening selection steps. The advantages of the method lie in the ease of upfront preparation (only the sgRNA constructs need to be created, based on short synthetic oligonucleotides), the relative simplicity of the screening procedure, and the minimal genomic perturbations to the resulting clonal lines. For a typical knockout experiment, our estimates indicate a time saving of 4–5 days relative to direct western blotting, or on the order of weeks relative to design and construction of selection steps. The protocol can be modified to allow for editing with homologous recombination templates, by blot screening for protein tags introduced on the template. The presented techniques should expand the available toolkit for the application of biochemistry and molecular and cell biology approaches to the study of protein function in mammalian cell culture.

## Methods

### Preparation of Cas9-sgRNA plasmids

The pSpCas9(BB) plasmid was a gift from Feng Zhang (Addgene plasmid #42230). Single guide RNA sequences were constructed to target the pre-designed loci in the third, forth, and fifth exon of the HuR gene (HuR-3, TGTGAACTACGTGACCGCGA;HuR-4,CGGG CGAGCATACGACACCT;HuR-5,CCGGATAAACG CAACCCCTC) and cloned into the parental plasmid as previously described [28]. Clones were tested for correct inserts using Sanger sequencing.

### Cell Culture, transfections, and dilutions

293-TREx cells were grown in Dulbecco's Modification of Eagle's Medium (DMEM) with 4.5 g/L glucose, L-glutamine and sodium pyruvate (Corning Cellgro Ref:10-013-CV) that was supplemented with 10 % Fetal Bovine Serum (HyClone FBS Characterized Cat: SH30071.03) and 10 units/ml of Penicillin–Streptomycin (HyClone Cat: SV30010). Cultured plates were grown in a humidified incubator supplying 5 % $CO2$ at 37 °C.

Two independent transfection events were performed for each construct. First, cells were grown in a 6-well format to approximately 60 % confluency and transfected with 2.5 µg of the prepared Cas9-sgRNA plasmid using the calcium phosphate method. 24 h later cells were passaged into 10 cm plates. 24 h following passaging, cells were again transfected with 12.5 µg plasmid using the TransIT-LT1 reagent (Mirus, Cat: MIR 2300). Cells were allowed to reach 100 % confluency before seeding out into individual clones. After the transfections, efficiency was estimated to be 70–80 % by GFP expression in a parallel transfection with a modified version of the pMSCV-PIG vector.

To isolate individual clones, cells were seeded at a density of 3–5 cells per 96-well. Colony growth in less that 60 % of the wells was observed over 3–4 weeks, and media was changed weekly to prevent drying of wells. Clone colonies that were visible to the eye were picked into 96-well plates by aspirating the media, dispensing 2 µl of fresh media onto the colony, scraping across the colony with the pipette tip, and drawing the media back. 24–48 h later, isolated clones that were growing as a clump were reseeded by pipette mixing in the well to encourage monolayer cell growth.

### Dot blot and western blot

Individual clones were grown until 100 % confluency and collected for dot blot analysis, dislodging the monolayer of cells by pipeting within the well. 90 µl of the 100 µl total volume was then removed, spun down, and the cell pellet lysed in 10 µl of 1× Passive Lysis Buffer (Promega's Dual-Luciferase Reporter Assay System, Cat: E1910). The remainder of the cells were propagated. 1 µl of cell lysate was pipetted onto dry nitrocellulose membrane (Bio-Rad, Cat: 162-0115) to form a dot. Each sample was blotted twice on two separate membranes, creating two identical patterns of samples, blocked in 5 % milk in TBST for 1 h at room temperature, and blotted for HuR (Santa Cruz Biotechnology, clone 3A2, Cat: sc-5261, 1:1000–1:5000 dilution) and Pum2 (Bethyl Labs, Cat: A300-202A, 1:1000 dilution) in 5 % milk in TBST for 1 h, with 3 × 5 min TBST washes in between incubation steps. Anti-Mouse-HRP-conjugated secondary antibody was from Cell Signaling (anti-mouse, Cat: 7076). Anti-Rabbit-HRP-Conjugated Secondary Antibody was from Cell Signaling (anti-rabbit, Cat: 7074). Blotted membranes were imaged on a Biorad ChemiDoc MP imager and quantified by Image Lab: identical circles were centered on each dot, and the signal volume was computed by the software. Local background values for each dot were computed and subtracted by Image Lab. Western blotting was performed as described above.

### Sequencing of genomic DNA

To amplify a genomic fragment flanking the HuR-5 target region, primers were designed containing BbsI restriction sites with distinct overhangs (HuR-5 For, CGACTTGAAG ACCTCACCTGTGATGGGCTCAGAGGAACC; HuR-5 Rev, CGACTTGAAGACCTAAACAGACTTCTAGCCTG GCCCAC). Genomic DNA of the parental 293-TREx line and the HuR-5.B5 mutant was isolated by standard SDS/Proteinase K lysis followed by phenol/chloroform extraction and isopropanol precipitation. The target region was PCR amplified using Q5 High Fidelity DNA Polymerase (New England Biolabs, Cat:M0491S), and cloned by Golden Gate cloning into a modified pRL-TK plasmid bearing two consecutive BbsI sites (ThermoFisher Scientific Cat# FERFD1014). Isolated bacterial colonies were mini-prepped and Sanger sequenced.

**Abbreviations**
CRISPR: clustered, regularly interspaced short palindromic repeats; sgRNA: single guide RNA; shRNA: short hairpin RNA; TALEN: transcription activator-like effector nuclease; crRNA: CRISPR RNA; NHEJ: non-homologous end joining.

**Authors' contributions**
FVK designed the project; ELS and GAS designed and cloned the sgRNA/Cas9 constructs; JAE performed the transfection, clone expansion, dot blotting, western blot, sequencing validation and analyzed data; FVK and JAE wrote and edited the manuscript. All authors read and approved the final manuscript.

**Acknowledgements**
The authors would like to thank Megan Lee and Alyssa Rodriguez for help with the experiments, the IIGB sequencing core for technical assistance, and Feng Zhang for sharing reagents.

## Competing interests
The authors declare that they have no competing interests.

## References
1. Carthew RW, Sontheimer EJ. Origins and mechanisms of miRNAs and siRNAs. Cell. 2009;136(4):642–55.
2. Fellmann C, Lowe SW. Stable RNA interference rules for silencing. Nat Cell Biol. 2014;16(1):10–8.
3. Sun N, Zhao H. Transcription activator-like effector nucleases (TALENs): a highly efficient and versatile tool for genome editing. Biotechnol Bioeng. 2013;110(7):1811–21.
4. Bibikova M, Golic M, Golic KG, Carroll D. Targeted chromosomal cleavage and mutagenesis in *Drosophila* using zinc-finger nucleases. Genetics. 2002;161(3):1169–75.
5. Wood AJ, Lo TW, Zeitler B, Pickle CS, Ralston EJ, Lee AH, Amora R, Miller JC, Leung E, Meng X, et al. Targeted genome editing across species using ZFNs and TALENs. Science. 2011;333(6040):307.
6. Jiang W, Marraffini LA. CRISPR-Cas: new tools for genetic manipulations from bacterial immunity systems. Annu Rev Microbiol. 2015;69:209–28.
7. Sontheimer EJ, Barrangou R. The Bacterial origins of the CRISPR genome-editing revolution. Hum Gene Ther. 2015;26(7):413–24.
8. Sternberg SH, Doudna JA. Expanding the biologist's toolkit with CRISPR-Cas9. Mol Cell. 2015;58(4):568–74.
9. Barrangou R, Marraffini LA. CRISPR-Cas systems: prokaryotes upgrade to adaptive immunity. Mol Cell. 2014;54(2):234–44.
10. Wiedenheft B, Sternberg SH, Doudna JA. RNA-guided genetic silencing systems in bacteria and archaea. Nature. 2012;482(7385):331–8.
11. Karginov FV, Hannon GJ. The CRISPR system: small RNA-guided defense in bacteria and archaea. Mol Cell. 2010;37(1):7–19.
12. Gossen M, Bujard H. Tight control of gene expression in mammalian cells by tetracycline-responsive promoters. Proc Natl Acad Sci USA. 1992;89(12):5547–51.
13. Schonig K, Bujard H, Gossen M. The power of reversibility regulating gene activities via tetracycline-controlled transcription. Methods Enzymol. 2010;477:429–53.
14. Feil R, Brocard J, Mascrez B, LeMeur M, Metzger D, Chambon P. Ligand-activated site-specific recombination in mice. Proc Natl Acad Sci USA. 1996;93(20):10887–90.
15. Feil R, Wagner J, Metzger D, Chambon P. Regulation of Cre recombinase activity by mutated estrogen receptor ligand-binding domains. Biochem Biophys Res Commun. 1997;237(3):752–7.
16. Rivera VM, Clackson T, Natesan S, Pollock R, Amara JF, Keenan T, Magari SR, Phillips T, Courage NL, Cerasoli F Jr, et al. A humanized system for pharmacologic control of gene expression. Nat Med. 1996;2(9):1028–32.
17. Ho SN, Biggar SR, Spencer DM, Schreiber SL, Crabtree GR. Dimeric ligands define a role for transcriptional activation domains in reinitiation. Nature. 1996;382(6594):822–6.
18. Liang FS, Ho WQ, Crabtree GR. Engineering the ABA plant stress pathway for regulation of induced proximity. Sci Signal. 2011;4(164):2.
19. Bolotin A, Quinquis B, Sorokin A, Ehrlich SD. Clustered regularly inter-spaced short palindrome repeats (CRISPRs) have spacers of extrachromo-somal origin. Microbiology. 2005;151(Pt 8):2551–61.
20. Mojica FJ, Diez-Villasenor C, Garcia-Martinez J, Soria E. Intervening sequences of regularly spaced prokaryotic repeats derive from foreign genetic elements. J Mol Evol. 2005;60(2):174–82.
21. Pourcel C, Salvignol G, Vergnaud G. CRISPR elements in *Yersinia pestis* acquire new repeats by preferential uptake of bacteriophage DNA, and provide additional tools for evolutionary studies. Microbiology. 2005;151(Pt 3):653–63.
22. Barrangou R, Fremaux C, Deveau H, Richards M, Boyaval P, Moineau S, Romero DA, Horvath P. CRISPR provides acquired resistance against viruses in prokaryotes. Science. 2007;315(5819):1709–12.
23. Jinek M, Chylinski K, Fonfara I, Hauer M, Doudna JA, Charpentier E. A pro-grammable dual-RNA-guided DNA endonuclease in adaptive bacterial immunity. Science. 2012;337(6096):816–21.
24. Cong L, Ran FA, Cox D, Lin S, Barretto R, Habib N, Hsu PD, Wu X, Jiang W, Marraffini LA, et al. Multiplex genome engineering using CRISPR/Cas systems. Science. 2013;339(6121):819–23.
25. Mali P, Yang L, Esvelt KM, Aach J, Guell M, DiCarlo JE, Norville JE, Church GM. RNA-guided human genome engineering via Cas9. Science. 2013;339(6121):823–6.
26. Kim H, Ishidate T, Ghanta KS, Seth M, Conte D Jr, Shirayama M, Mello CC. A co-CRISPR strategy for efficient genome editing in *Caenorhabditis elegans*. Genetics. 2014;197(4):1069–80.
27. Dickinson DJ, Ward JD, Reiner DJ, Goldstein B. Engineering the *Caeno-rhabditis elegans* genome using Cas9-triggered homologous recombina-tion. Nat Methods. 2013;10(10):1028–34.
28. Ran FA, Hsu PD, Wright J, Agarwala V, Scott DA, Zhang F. Genome engi-neering using the CRISPR-Cas9 system. Nat Protoc. 2013;8(11):2281–308.
29. Wu X, Kriz AJ, Sharp PA. Target specificity of the CRISPR-Cas9 system. Quant Biol. 2014;2(2):59–70.
30. Fu Y, Foden JA, Khayter C, Maeder ML, Reyon D, Joung JK, Sander JD. High-frequency off-target mutagenesis induced by CRISPR-Cas nucleases in human cells. Nat Biotechnol. 2013;31(9):822–6.
31. Aach J, Mali P, Church GM. CasFinder: Flexible algorithm for identifying specific Cas9 targets in genomes. bioRxiv. 2014: 005074.
32. Wisniewski JR, Hein MY, Cox J, Mann M. A "proteomic ruler" for protein copy number and concentration estimation without spike-in standards. Mol Cell Proteom. 2014;13(12):3497–506.
33. Garneau JE, Dupuis ME, Villion M, Romero DA, Barrangou R, Boyaval P, Fremaux C, Horvath P, Magadan AH, Moineau S. The CRISPR/Cas bacte-rial immune system cleaves bacteriophage and plasmid DNA. Nature. 2010;468(7320):67–71.
34. Jinek M, East A, Cheng A, Lin S, Ma E, Doudna J. RNA-programmed genome editing in human cells. Elife. 2013;2:e00471.
35. Hsu PD, Scott DA, Weinstein JA, Ran FA, Konermann S, Agarwala V, Li Y, Fine EJ, Wu X, Shalem O, et al. DNA targeting specificity of RNA-guided Cas9 nucleases. Nat Biotechnol. 2013;31(9):827–32.
36. Wang T, Wei JJ, Sabatini DM, Lander ES. Genetic screens in human cells using the CRISPR-Cas9 system. Science. 2014;343(6166):80–4.
37. Wu X, Scott DA, Kriz AJ, Chiu AC, Hsu PD, Dadon DB, Cheng AW, Trevino AE, Konermann S, Chen S, et al. Genome-wide binding of the CRISPR endonuclease Cas9 in mammalian cells. Nat Biotechnol. 2014;32(7):670–6.
38. Kuscu C, Arslan S, Singh R, Thorpe J, Adli M. Genome-wide analysis reveals characteristics of off-target sites bound by the Cas9 endonuclease. Nat Biotechnol. 2014;32(7):677–83.
39. Canver MC, Bauer DE, Dass A, Yien YY, Chung J, Masuda T, Maeda T, Paw BH, Orkin SH. Characterization of genomic deletion efficiency mediated by clustered regularly interspaced palindromic repeats (CRISPR)/Cas9 nuclease system in mammalian cells. J Biol Chem. 2014;289(31):21312–24.
40. Horii T, Morita S, Kimura M, Kobayashi R, Tamura D, Takahashi RU, Kimura H, Suetake I, Ohata H, Okamoto K, et al. Genome engineering of mam-malian haploid embryonic stem cells using the Cas9/RNA system. PeerJ. 2013;1:e230.

# 7

# RUNX1 induces DNA replication independent active DNA demethylation at *SPI1* regulatory regions

Shubham Goyal, Takahiro Suzuki, Jing-Ru Li, Shiori Maeda, Mami Kishima, Hajime Nishimura, Yuri Shimizu and Harukazu Suzuki*

## Abstract

**Background:** SPI1 is an essential transcription factor (TF) for the hematopoietic lineage, in which its expression is tightly controlled through a −17-kb upstream regulatory region and a promoter region. Both regulatory regions are demethylated during hematopoietic development, although how the change of DNA methylation status is performed is still unknown.

**Results:** We found that the ectopic overexpression of RUNX1 (another key TF in hematopoiesis) in HEK-293T cells induces almost complete DNA demethylation at the −17-kb upstream regulatory region and partial but significant DNA demethylation at the proximal promoter region. This DNA demethylation occurred in mitomycin-C-treated nonproliferating cells at both regulatory regions, suggesting active DNA demethylation. Furthermore, ectopic RUNX1 expression induced significant endogenous *SPI1* expression, although its expression level was much lower than that of natively *SPI1*-expressing monocyte cells.

**Conclusions:** These results suggest the novel role of RUNX1 as an inducer of DNA demethylation at the *SPI1* regulatory regions, although the mechanism of RUNX1-induced DNA demethylation remains to be explored.

**Keywords:** *SPI1*, RUNX1, DNA demethylation, Endogenous expression

## Background

SPI1 is a hematopoietic lineage-specific TF belonging to the ETS family [1]. It plays an important role in the development of myeloid and lymphoid cells and is highly expressed in monocytes [2, 3]. The expression of *SPI1* is regulated by the combined activity of a proximal promoter and an upstream regulatory element (URE) located −17-kb upstream of the transcription start site (also known as the distal promoter region) in human [3, 4]. Deletion of the URE region contributes to the exacerbation of acute myeloid leukemia (AML) or erythroleukemia [5, 6]. The *SPI1* regulatory regions are differentially methylated in *SPI1*-expressing and nonexpressing cell lines [7, 8]. This differential pattern is also maintained

during hematopoietic differentiation, in which ES cells are hypermethylated while hematopoietic stem cells become hypomethylated [7–9]. Meanwhile, abnormal hypermethylation of the *SPI1* regulatory regions is frequently observed in myeloma cell lines with downregulated *SPI1* [10]. Thus, this methylation change seems to be important for *SPI1* expression. However, no molecular mechanism behind the change in DNA methylation status at *SPI1* regulatory regions has been reported yet.

RUNX1 is another TF that is essential for the regulation and maintenance of mammalian hematopoiesis [11]. A previous study reported that RUNX1 regulates *SPI1* at both transcriptional and epigenetic levels [9]. RUNX1 binding at the conserved sites in the URE of *SPI1* is critical for the onset of *SPI1* expression during hematopoietic stem cell formation, making *SPI1* the direct downstream target of RUNX1 [12]. At the pre-hematopoietic or hemangioblast stage, the inception of RUNX1 expression

*Correspondence: harukazu.suzuki@riken.jp
Division of Genomic Technologies, RIKEN Center for Life Science Technologies, 1-7-22 Suehiro-cho, Tsurumi-ku, Yokohama, Kanagawa 230-0045, Japan

induces chromatin remodeling at the regulatory regions of *SPI1*, in which the binding of RUNX1 to the *SPI1* regulatory region is essential [9]. Moreover, several studies have revealed that chromatin remodeling is coupled with DNA demethylation during embryonic development [13]. Therefore, we hypothesize that RUNX1 may also be involved in recruiting the DNA methylation status change at the *SPI1* regulatory regions.

Here, we describe that the ectopic expression of RUNX1 in HEK-293T cells induces DNA demethylation at the two functionally active regulatory regions of *SPI1* in a replication-independent manner.

## Results

### RUNX1 overexpression induced DNA demethylation at *SPI1* −17-kb URE

*SPI1* contains two regulatory regions; one is the −17-kb URE that lies upstream of the transcription start site (also known as the distal promoter region) and the other is the proximal promoter region (Fig. 1). While the −17-kb URE contains three RUNX1 binding sites (Fig. 2a), the proximal promoter region does not contain any such sites [14]. To analyze how RUNX1 expression affects the DNA methylation status of *SPI1* regulatory regions, we first focused on the −17-kb URE. We transduced RUNX1-overexpressing lentivirus into HEK-293T cells, a cell line that does not express RUNX1, followed by DNA methylation analysis by bisulfite sequencing. We observed drastic DNA demethylation in RUNX1-overexpressing HEK-293T cells (Fig. 2b). This DNA demethylation appeared to be RUNX1-specific because we did not see any of these changes in either MCS-transduced (multiple cloning sites: MCS) or wild-type HEK-293T cells. The region as a whole was significantly demethylated (p < 0.001)

(Fig. 2c). The DNA demethylation of all individual CpG sites investigated at the −17-kb URE was also found to be significant (p < 0.005) (Fig. 2d).

### RUNX1 overexpression induced partial DNA demethylation at *SPI1* proximal promoter

The proximal promoter region of *SPI1* does not contain any binding site for RUNX1 [14] (Fig. 3a); therefore, we wondered whether RUNX1 can still induce any change in DNA methylation in this region. Our results showed that RUNX1 can induce partial demethylation in the proximal promoter region, with about 30% of CpGs being demethylated (Fig. 3b). Although the demethylation was incomplete, the region as a whole was significantly demethylated (p < 0.0001; Fig. 3c) and the change at most of the CpG sites was also found to be significant (p < 0.005; Fig. 3d), in comparison to that of MCS-transduced cells.

### Molecular mechanism of RUNX1-induced DNA demethylation at *SPI1* regulatory regions

DNA demethylation can be active or passive in nature, where active demethylation occurs by the enzymatic activity of TET enzymes in the absence of replication, while passive demethylation occurs slowly during several rounds of replication [15]. To identify the type of RUNX1-induced demethylation at both *SPI1* regulatory regions, we used mitomycin-C to arrest cell growth at the $G_1$ phase. Different concentrations of mitomycin-C were individually tested to identify the concentration when the ratio of Edu positive cells was zero (no DNA synthesis, see "Methods") (Fig. 4a). We observed drastic DNA demethylation in RUNX1-transduced cells at the *SPI1* −17-kb URE region, even upon mitomycin-C treatment, which would prevent passive DNA

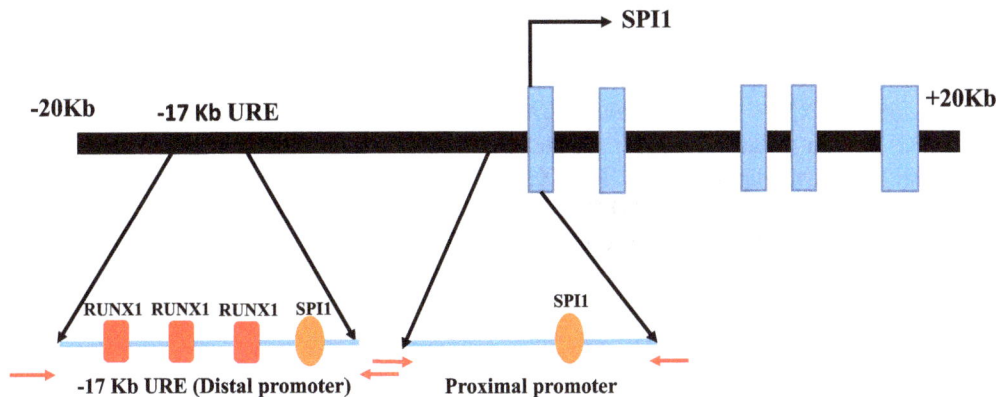

**Fig. 1** Schematic representation of *SPI1* gene structure: Simplified scheme of the human *SPI1* locus, indicating the two regulatory regions of *SPI1*: distal promoter (−17-kb URE) and proximal promoter. Binding sites for RUNX1 and SPI1 are shown by *red* and *orange boxes*, respectively. The *red horizontal arrows* in the URE and the proximal promoter indicate primers designed for methylation study

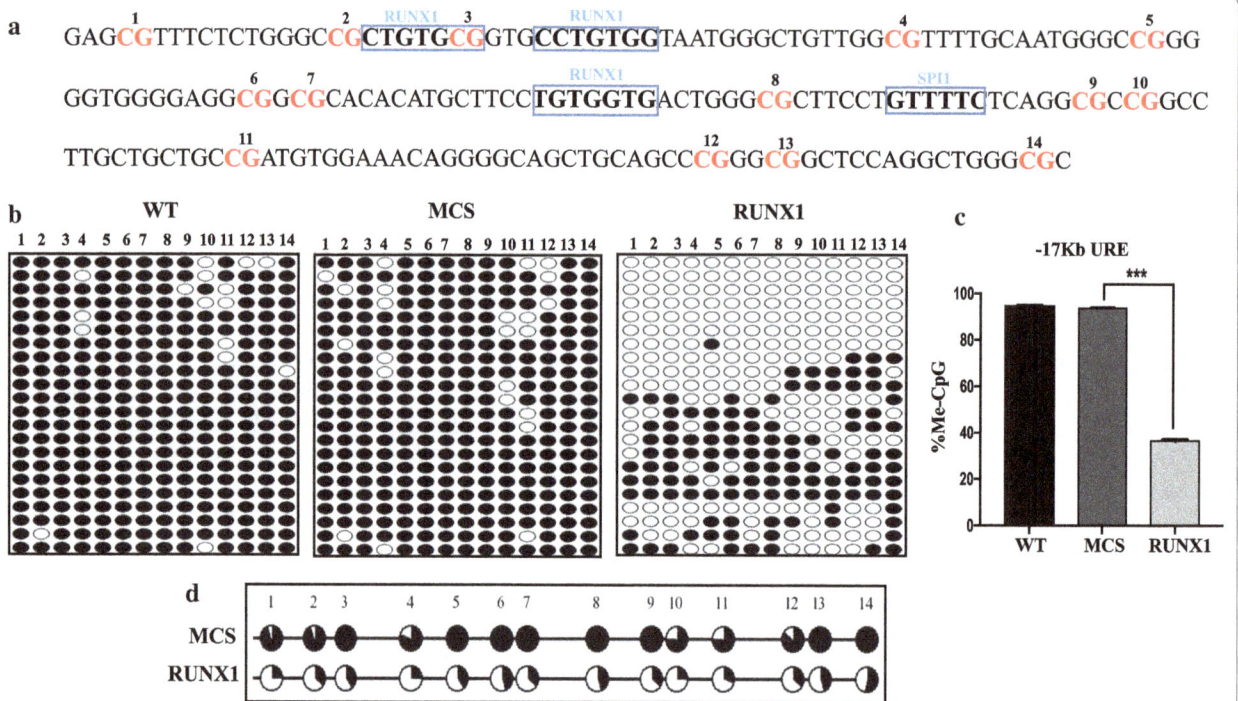

**Fig. 2** DNA methylation status analysis of *SPI1* −17-kb URE in RUNX1-overexpressing HEK-293T cells: **a** Target sequence of human *SPI1* −17-kb URE. The *number* represents the CpG sites and RUNX1 and SPI1 binding sites are shown in *bold letters*. **b** Methylation pattern diagrams of wild-type, MCS, and RUNX1-overexpressing HEK-293T cells at the *SPI1* −17-kb URE region. *Each row of circles* represents the result of a single amplicon and *each column* represents those of a single CpG site. *Black circles* denote methylated CpGs and *open circles* represent unmethylated ones. The CpG numbers shown correspond to the target nucleotide sequence, shown in **a**. **c** Quantification of percent CpG methylation at the −17-kb URE in wild-type, MCS, and RUNX1-overexpressing HEK-293T cells. The p-value calculated from the comparison between MCS and RUNX1-overexpressing HEK-293T cells was significant (***p < 0.001). **d** Comparison of methylation status graph at each individual CpG site between MCS and RUNX1-overexpressing HEK-293T cells. All of the CpGs were significantly demethylated (p < 0.005)

demethylation (Fig. 4b). The statistical analysis revealed that the region as a whole was significantly demethylated (p < 0.0001) (Fig. 4c). Significant DNA demethylation was found at each CpG site investigated (Fig. 4d), the same as in the RUNX1-overexpressing cells not treated with mitomycin-C, as shown in Fig. 2d. Interestingly, we also observed RUNX1-induced active DNA demethylation at the proximal promoter region in mitomycin-C-treated cells (Fig. 4e–g). These results demonstrate the ability of RUNX1 to induce active DNA demethylation at both *SPI1* regulatory regions. Recent studies in zebrafish have highlighted that RUNX1 induces the expression of dnmt3bb.1, a DNMT3 paralog in this animal model [16]. Therefore, we next examined whether RUNX1 overexpression enhanced the expression of genes encoding enzymes that are known to be involved in both DNA methylation and demethylation. Specifically, we compared the expression levels of *DNMT1*, *DNMT3A* and *3B*, *TET1-3*, and *IDH1* and *IDH2* between MCS and RUNX1-transduced cells. However, no difference in the expression of these genes was observed (Fig. 4h).

### RUNX1 induces significant *SPI1* expression, but much less than in monocytes

DNA methylation at gene regulatory regions generally suppresses gene expression, possibly by blocking transcription factor binding at those regions. To examine whether RUNX1-induced DNA demethylation at *SPI1* regulatory regions can affect *SPI1* endogenous expression, we measured *SPI1* expression in RUNX1-overexpressing HEK-293T cells by qRT-PCR. *SPI1* endogenous expression was significantly (p < 0.005) induced by the RUNX1 ectopic expression, which was estimated to involve a 12-fold upregulation, from the differences in ΔCt values (Fig. 5). However, the *SPI1* expression level for RUNX1-overexpressing HEK-293T cells was still far less than that of monocytes; only 3% of *SPI1* mRNA expression was estimated.

### Discussion

In this study, we found that the ectopic expression of RUNX1 in wild-type HEK-293T cells converts the methylation status of both *SPI1* regulatory regions from hypermethylated to hypomethylated; the results also

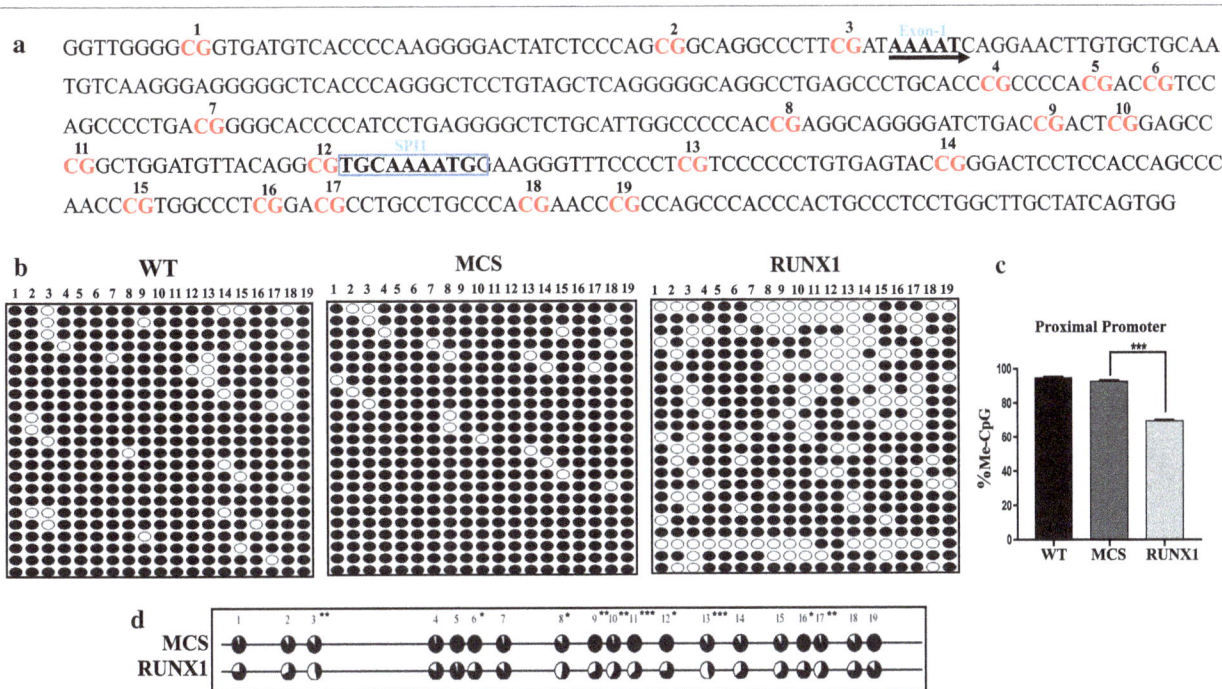

**Fig. 3** DNA methylation status analysis of *SPI1* proximal promoter in RUNX1-overexpressing HEK-293T cells: **a** Target sequence of the human *SPI1* proximal promoter region. The *number* represents the CpG sites and the SPI1 binding site is also shown. **b** Methylation pattern diagrams of wild-type, MCS, and RUNX1-overexpressing HEK-293T cells at the *SPI1* proximal promoter region. *Each row of circles* represents the result of a single amplicon and *each column* represents a single CpG site. *Black circles* denote methylated CpGs and *open circles* represent unmethylated ones. The CpG numbers shown correspond to the target nucleotide sequence, shown in **a**. **c** Quantification of percent CpG methylation at the proximal promoter region in wild-type, MCS, and RUNX1-overexpressing HEK-293T cells. The p value calculated from the comparison between MCS and RUNX1-overexpressing HEK-293T cells was significant (***p < 0.0001). **d** Comparison of methylation status graph at each individual CpG site between MCS and RUNX1-overexpressing HEK-293T cells. The significantly demethylated CpG sites are marked with *asterisks*, ***p < 0.005, **p < 0.01, *p < 0.05

showed that the induced demethylation was replication-independent active DNA demethylation. Further, RUNX1 overexpression did not change gene expression of enzymes involved in both DNA methylation and demethylation. It has been reported that chromatin remodeling of the *SPI1* URE region in hemangioblasts is induced by the binding of RUNX1, which also accompanies the DNA demethylation of this region [9]. This suggests that RUNX1 binding directly recruits the DNA demethylating machinery. Actually, several TFs have recently shown to be involved in DNA demethylation by recruiting DNA demethylation machinery [17]. On the other hand, the proximal promoter region revealed partial but significant DNA demethylation, although it does not contain any binding sites for RUNX1 and neither any mechanism of its binding is known [12, 14]. Thus, it was interesting to ponder how this replication-independent active DNA demethylation occurs at the proximal promoter region of *SPI1* by the overexpression of RUNX1. Previous chromatin immunoprecipitation (ChIP) data have shed light on the binding of various transcription factors in the URE and proximal promoter regions [8]. Therefore,

we speculate that other RUNX1-regulated transcription factor(s) may also be able to induce DNA demethylation at both regulatory regions. Thus, we can conclude that RUNX1 could be directly or indirectly responsible for inducing DNA demethylation at the *SPI1* regulatory regions, although the actual process remains to be confirmed.

Our study also showed that RUNX1 overexpression is not sufficient for inducing the higher level of endogenous *SPI1* expression in HEK-293T cells. The partial demethylation at the proximal promoter region may be responsible for the low level of *SPI1* expression. In fact, the proximal promoter region of *SPI1* is completely hypomethylated in monocytes, in which *SPI1* is expressed at a higher level [18, 19]. Previous reports also suggest the positive correlation between DNA demethylation at gene regulatory regions and gene expression, revealing that active genes are generally hypomethylated while inactive genes are generally hypermethylated at their promoter regions [7, 20]. Furthermore, the cells with high *SPI1* expression regulate this expression by forming an autoregulatory loop between its URE and the proximal promoter region [4,

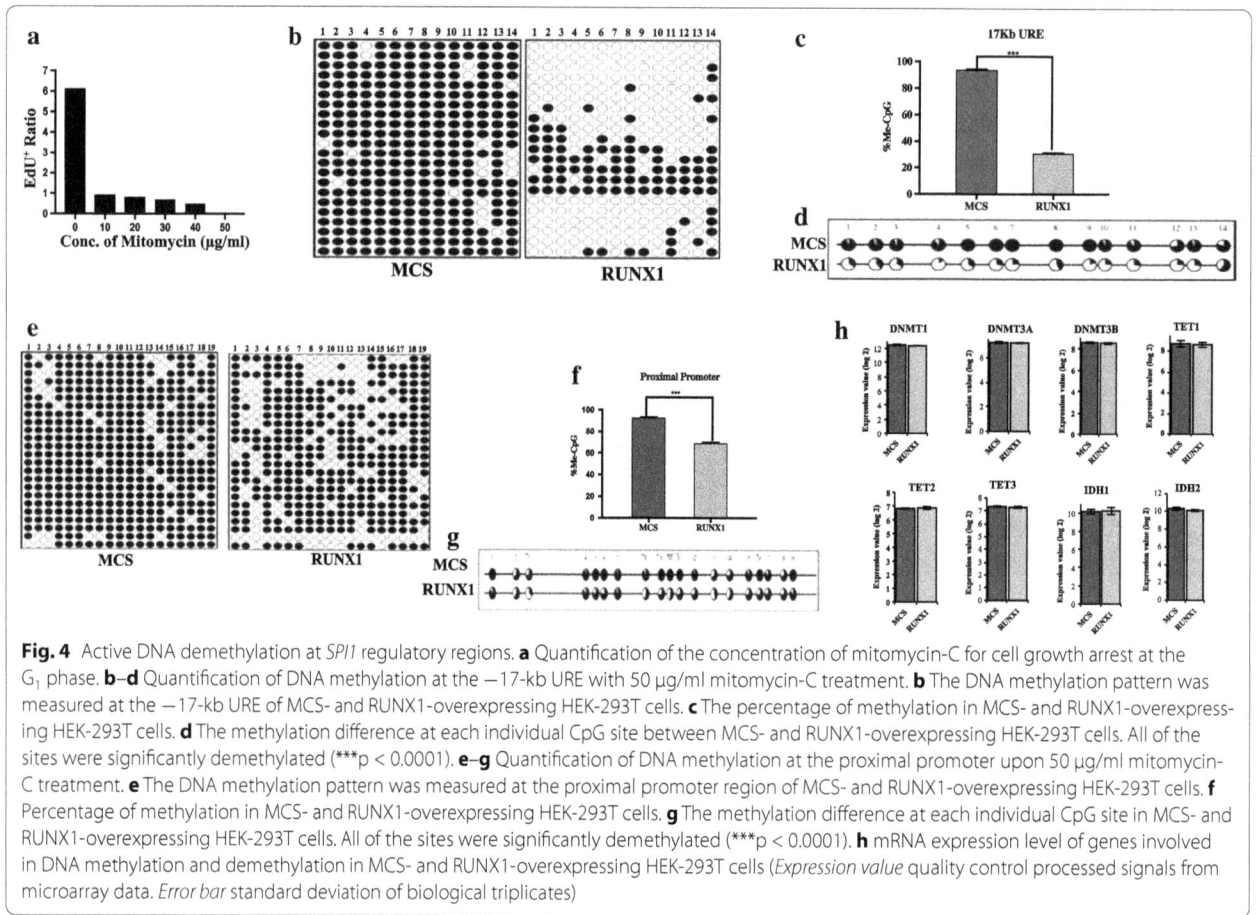

**Fig. 4** Active DNA demethylation at *SPI1* regulatory regions. **a** Quantification of the concentration of mitomycin-C for cell growth arrest at the G₁ phase. **b–d** Quantification of DNA methylation at the −17-kb URE with 50 µg/ml mitomycin-C treatment. **b** The DNA methylation pattern was measured at the −17-kb URE of MCS- and RUNX1-overexpressing HEK-293T cells. **c** The percentage of methylation in MCS- and RUNX1-overexpressing HEK-293T cells. **d** The methylation difference at each individual CpG site between MCS- and RUNX1-overexpressing HEK-293T cells. All of the sites were significantly demethylated (***p < 0.0001). **e–g** Quantification of DNA methylation at the proximal promoter upon 50 µg/ml mitomycin-C treatment. **e** The DNA methylation pattern was measured at the proximal promoter region of MCS- and RUNX1-overexpressing HEK-293T cells. **f** Percentage of methylation in MCS- and RUNX1-overexpressing HEK-293T cells. **g** The methylation difference at each individual CpG site in MCS- and RUNX1-overexpressing HEK-293T cells. All of the sites were significantly demethylated (***p < 0.0001). **h** mRNA expression level of genes involved in DNA methylation and demethylation in MCS- and RUNX1-overexpressing HEK-293T cells (*Expression value* quality control processed signals from microarray data. *Error bar* standard deviation of biological triplicates)

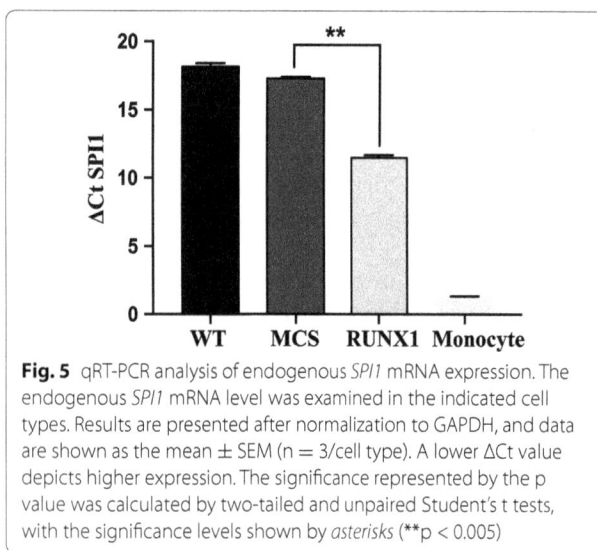

**Fig. 5** qRT-PCR analysis of endogenous *SPI1* mRNA expression. The endogenous *SPI1* mRNA level was examined in the indicated cell types. Results are presented after normalization to GAPDH, and data are shown as the mean ± SEM (n = 3/cell type). A lower ΔCt value depicts higher expression. The significance represented by the p value was calculated by two-tailed and unpaired Student's t tests, with the significance levels shown by *asterisks* (**p < 0.005)

21], which are in close proximity to each other [22]. Thus, our data suggest that the autoregulatory loop may not form well due to incomplete demethylation at the proximal promoter. The presence of other transcription factors

that are expressed in SPI1-expressing hematopoietic cells could also be necessary to induce higher endogenous expression of *SPI1*.

We used HEK-293T cells in this study because they are from a cell line that does not express RUNX1. The propensity of these cells to undergo transfection, their availability, and the possibility of avoiding passage biasness during analysis also make them easy to use. Since our results are derived from nonhematopoietic cells by using an artificial overexpression system, it would be preferable for our results to be evaluated further using hematopoietic cell lines in which RUNX1 and SPI1 are endogenously expressed. However, DNA demethylation at *SPI1* regulatory regions already occurs at the hematopoietic stem cell stage at which RUNX1 is expressed. As the next step, it may be necessary to perform such evaluation in hematopoietic cell lines under specific conditions or with gene manipulation.

## Conclusions

To our knowledge, this is the first study evaluating the potential role of RUNX1 as a demethylating inducer. Our results provide a hint about how demethylation occurs at *SPI1* regulatory regions.

## Methods

### Cell culture and RUNX1 over expression

HEK-293T cells, a human embryonic cell line provided by Riken Cell Bank (RCB), were cultured in High-Glucose Dulbecco's Modified Eagle Medium (DMEM, Wako, Japan) supplemented with 10% fetal bovine serum (Lonza, Basel, Switzerland) and 2 mM penicillin–streptomycin (Sigma-Aldrich, St. Louis, MO, USA) at 37 °C in 5% $CO_2$.

RUNX1 overexpression in HEK-293T cells was carried out by using the lentivirus transduction method. The lentivirus for RUNX1 overexpression and its control MCS were prepared in a mixture of packaging constructs, pCMV-VSV-G-RSV-Rev and pCAG-HIVgp (provided by RIKEN BRC) in accordance with a previously reported protocol [19]. HEK-293T cells ($1 \times 10^6$ cells per well) were seeded in a poly-D-lysine-coated 12-well plate, followed by transduction on the second day by adding 8 μg/ml polybrene and 10 multiplicity of infection (MOI) of lentivirus in each well. One day after the transduction, the medium was replaced with fresh DMEM medium containing 1 μg/ml puromycin (Thermo Fisher Scientific, Waltham, MA, USA) for the selection of transduced cells. Cells were harvested 7 days after the transduction. Wild-type HEK-293T cells were also kept as a control.

### Bisulfite sequencing

Genomic DNA was isolated from the harvested cells using an All Prep DNA/RNA Mini Kit (Qiagen, Hilden, Germany), in accordance with the manufacturer's protocol. The DNA quality was assessed by using a Nanodrop 1000 (Thermo Fisher Scientific, Waltham, MA, USA). DNA bisulfite conversion and purification were performed in accordance with the protocol of the EZ DNA Methylation Gold Kit (Zymo Research, Irvine, CA, USA). For amplification of the −17-kb URE (326 bp) and the proximal promoter region (448 bp), previously reported PCR primers [10] were used (Table 1). The amplified PCR products were checked on a 1.5% agarose gel, purified using the PCR purification kit (Qiagen, Hilden, Germany), and then cloned using the Target Clone plus kit (Toyobo, Osaka, Japan). Plasmid DNA was isolated using the QIAprep 96 Turbo miniprep kit (Qiagen, Hilden, Germany) and sequenced in an ABI 3730xl DNA Analyzer. The sequence analysis, including the calculation of methylation percentage, methylation pattern change, and comparison at each CpG site, was performed using the online software QUMA (http://quma.cdb.riken.jp/).

### Cell cycle arrest

HEK-293T cells ($2 \times 10^6$ cells per well) were plated into a six-well plate, treated with different concentrations of mitomycin-C (Sigma-Aldrich, St. Louis, MO), and

**Table 1  List of primers used for bisulfite PCR and qRT-PCR**

| Bisulfite primers | |
|---|---|
| SPI1 proximal promoter (Fwd) | GAGATTTTTTGTATGTAGTGTAAGA |
| SPI1 proximal promoter (Rv) | TAACTTCCCACTAATAACAAACCA |
| SPI1 −17-kb URE (Fwd) | GTTGGATATTTTTTTGAGGTTTGG |
| SPI1 −17-kb URE (Rv) | TAAAACCTAAAACTACTAAACCCA |
| Real-time primers | |
| GAPDH (Fwd) | GAAATCCCATCACCATCTTCCAGG |
| GAPDH (Rv) | GAGCCCCAGCCTTCTCCATG |
| SPI1 (Fwd) | TCCTGAGGGGCTCTGCATT |
| SPI1 (Rv) | TGTCATAGGGCACCAGGTCTT |

incubated for 4 h at 37 °C in 5% $CO_2$. To check cell cycle arrest (at the $G_1$ phase), the proliferation assay was performed by following the Click-iT® EdU Cytometry Cell Proliferation Assay protocol (Thermo Fisher Scientific, Waltham, MA, USA); for assessment, FACS was used. After the confirmation of cell cycle arrest at 50 μg/ml mitomycin-C, the cells were plated and transduced with RUNX1 lentivirus on the second day by following the above-mentioned protocol. Wild-type 293T cells were also treated with 50 μg/ml mitomycin-C and transduced with MCS, which were kept as a control.

### DNA microarray

Total RNA was isolated by using the Nucleospin® RNA kit protocol (Macherey–Nagel, Germany). Five hundred nanograms of total RNA was amplified using the Ambion total RNA amplification kit (Ambion, Carlsbad, CA), followed by hybridization of the synthesized cRNA with the Human HT-12 v4 Expression BeadChip kit (Illumina, San Diego, CA, USA), in accordance with the manufacturer's protocol. Scanning of the chip was performed using Illumina BeadScan and BeadStudio (version 3.1). Data were processed with a package from Bioconductor (lumi) [23, 24] using the free software environment R (http://www.r-project.org/). The microarray data has been registered in gene expression omnibus (https://www.ncbi.nlm.nih.gov/geo/) in NCBI (GSE95308).

### qRT-PCR

Total RNA was isolated using the Nucleospin® RNA kit protocol (Macherey–Nagel, Germany). Reverse transcription of total RNA was performed using the Prime Script RT kit (Takara, Takara Bio, Japan); qRT-PCR was performed with the ABI PRISM® 7500 sequence detection system (Applied Biosystems, USA) using SYBR Premix Ex Taq™ II (Tli RNaseH plus, Takara Bio, Japan) and gene-specific primers for SPI1 and GAPDH (Table 1). PCR cycling conditions consisted of initial denaturation at 95 °C for 10 s, followed by 40 cycles at 95 °C for 5 s and 60 °C for 30 s. Data was analyzed using the $2^{-\Delta\Delta Ct}$

method [25]. Human peripheral blood CD14$^+$ monocytes (Lonza, Basel, Switzerland) were kept as a positive control for the comparison of *SPI1* expression.

## Statistical analysis

All of the statistical analyses for DNA methylation status and calculation of methylation percentage were performed using the online quantification tool QUMA. The significance of differences between the transduced and untransduced cells was determined by using unpaired, two-tailed, and Student's tests (t test), where $p < 0.05$ was considered significant.

## Abbreviations

TF: transcription factor; URE: upstream regulatory element; AML: acute myeloid leukemia; MCS: multiple cloning sites; ChIP: chromatin immunoprecipitation; MOI: multiplicity of infection.

## Authors' contributions

Conceived and designed the experiments: SG, TS, and HS. Performed the experiments: SG, HN, MK, SM, YS. Analyzed the data: SG, TS, HS, and JRL. Contributed reagents/materials/analysis tools: SG, TS, JRL, HN, MK, SM, YS. Wrote the paper: SG, HS. All authors read and approved the final manuscript.

## Acknowledgements

The authors wish to thank RIKEN-CLST for all the facilities.

## Competing interests

The authors declare that they have no competing interests.

## Funding

This work was supported by a research grant from MEXT for RIKEN Center for Life Science Technologies. The funding agency has no role in the design of study, data collection, data interpretation, decision to publish, or preparation of the manuscript.

## References

1. Iwasaki H, Somoza C, Shigematsu H, Duprez E, Iwasaki-Arai J, Mizuno S-I, Arinobu Y, Geary K, Zhang P, Dayaram T, et al. Distinctive and indispensable roles of PU. 1 in maintenance of hematopoietic stem cells and their differentiation. Blood. 2005;106:1590–600.
2. Miyamoto T, Iwasaki H, Reizis B, Ye M, Graf T, Weissman IL, Akashi K. Myeloid or lymphoid promiscuity as a critical step in hematopoietic lineage commitment. Dev Cell. 2002;3:137–47.
3. Chen HM, Zhang P, Voso MT, Hohaus S, Gonzalez DA, Glass CK, Zhang DE, Tenen DG. Neutrophils and monocytes express high levels of PU.1 (Spi-1) but not Spi-B. Blood. 1995;85:2918–28.
4. Okuno Y, Huang G, Rosenbauer F, Evans EK, Radomska HS, Iwasaki H, Akashi K, Moreau-gachelin F, Li Y, Tenen DG. Potential autoregulation of transcription factor PU.1 by an upstream regulatory element. Mol Cell Biol. 2005;25:2832–45.

5. Durual S, Rideau A, Ruault-Jungblut S, Cossali D, Beris P, Piguet V, Matthes T. Lentiviral PU.1 overexpression restores differentiation in myeloid leukemic blasts. Leukemia. 2007;21:1050–9.
6. Curik N, Burda P, Vargova K, Pospisil V, Belickova M, Vlckova P, Savvulidi F, Necas E, Hajkova H, Haskovec C, et al. 5-azacitidine in aggressive myelodysplastic syndromes regulates chromatin structure at PU.1 gene and cell differentiation capacity. Leukemia. 2012;26:1804–11.
7. Amaravadi L, Klemsz MJ. DNA methylation and chromatin structure regulate PU.1 expression. DNA Cell Bio. 1999;18:875–84.
8. Hoogenkamp M, Krysinska H, Ingram R, Huang G, Barlow R, Clarke D, Ebralidze A, Zhang P, Tagoh H, Cockerill PN, et al. The Pu.1 locus is differentially regulated at the level of chromatin structure and noncoding transcription by alternate mechanisms at distinct developmental stages of hematopoiesis. Mol Cell Biol. 2007;27:7425–38.
9. Hoogenkamp M, Lichtinger M, Krysinska H, Lancrin C, Clarke D, Williamson A, Mazzarella L, Ingram R, Jorgensen H, Fisher A, et al. Early chromatin unfolding by RUNX1: a molecular explanation for differential requirements during specification versus maintenance of the hematopoietic gene expression program. Blood. 2009;114:299–309.
10. Tatetsu H, Ueno S, Hata H, Yamada Y, Takeya M, Mitsuya H, Tenen DG, Okuno Y. Down-regulation of PU.1 by methylation of distal regulatory elements and the promoter is required for myeloma cell growth. Cancer Res. 2007;67:5328–36.
11. Ichikawa M, Goyama S, Asai T, Kawazu M, Nakagawa M, Takeshita M, Chiba S, Ogawa S, Kurokawa M. AML1/RUNX1 negatively regulates quiescent hematopoietic stem cells in adult hematopoiesis. J Immunol. 2008;180:4402–8.
12. Huang G, Zhang P, Hirai H, Elf S, Yan X, Chen Z, Koschmieder S, Okuno Y, Dayaram T, Growney JD, et al. PU.1 is a major downstream target of AML1 (RUNX1) in adult mouse hematopoiesis. Nat Genet. 2008;40:51–60.
13. Sakurai K, Hoang M, Kim Y, Mathiyakom N, Kim Y. DNA methylation and chromatin dynamics in embryonic stem cell regulation. OA Stem Cells. 2014;2:1–8.
14. Lichtinger M, Ingram R, Hannah R, Müller D, Clarke D, Assi SA, Lie-A-Ling M, Noailles L, Vijayabaskar MS, Wu M, et al. RUNX1 reshapes the epigenetic landscape at the onset of haematopoiesis. EMBO J. 2012;31:4318–33.
15. Hackett JA, Surani MA. DNA methylation dynamics during the mammalian life cycle. Philos Trans R Soc B. 2013;368:20110328.
16. Gore AV, Athans B, Iben JR, Johnson K, Russanova V, Castranova D, Pham VN, Butler MG, et al. Epigenetic regulation of hematopoiesis by DNA methylation. eLife. 2016;5:e11813.
17. Marchal C, Miotto B. Emerging concept in DNA methylation: role of transcription factors in shaping DNA methylation patterns. J Cell Physiol. 2015;230:743–51.
18. Rica DL, Rodríguez-Ubreva J, García M, Islam AB, Urquiza JM, Hernando H, Christensen J, Helin K, Gómez-Vaquero C, Ballestar E. PU.1 target genes undergo Tet2-coupled demethylation and DNMT3b-mediated methylation in monocyte-to-osteoclast differentiation. Genome Biol. 2013;14:R99.
19. Suzuki T, Nakano-Ikegaya M, Yabukami-Okuda H, de Hoon M, Severin J, Saga-Hatano S, Shin JW, Kubosaki A, Simon C, Hasegawa Y, et al. Reconstruction of monocyte transcriptional regulatory network accompanies monocytic functions in human fibroblasts. PLoS ONE. 2012;7:e33474.
20. Wagner JR, Busche S, Ge B, Kwan T, Pastinen T, Blanchette M. The relationship between DNA methylation, genetic and expression inter-individual variation in untransformed human fibroblasts. Genome Biol. 2014;15:R37.
21. Staber PB, Zhang P, Ye M, Welner RS, Levantini E, Di Ruscio A, Ebralidze AK, Bach C, Zhang H, Zhang J, et al. The Runx-PU.1 pathway preserves normal and AML/ETO9a leukemic stem cells. Blood. 2014;124:2391–9.
22. Ebralidze AK, Guibal FC, Steidl U, Ebralidze AK, Guibal FC, Steidl U, Zhang P, Lee S, Bartholdy B, Jorda MA, et al. *PU.1* expression is modulated by the balance of functional sense and antisense RNAs regulated by a shared cis -regulatory element by the balance of functional sense and antisense RNAs regulated by a shared cis -regulatory element. Genes Dev. 2008;22:2085–92.
23. Lin SM, Du P, Huber W, Kibbe WA. Model-based variance-stabilizing transformation for Illumina microarray data. Nucleic Acids Res. 2008;36:1–9.
24. Du P, Kibbe WA, Lin SM. Lumi: a pipeline for processing Illumina microarray. Bioinformatics. 2008;24:1547–8.
25. Livak KJ, Schmittgen TD. Analysis of relative gene expression data using real-time quantitative PCR and the $2^{-\Delta\Delta CT}$ method. Methods. 2001;25:402–8.

8

# Heterogeneous pattern of DNA methylation in developmentally important genes correlates with its chromatin conformation

Puja Sinha[1], Kiran Singh[2] and Manisha Sachan[1*]

## Abstract

**Background:** DNA methylation is a major epigenetic modification, playing a crucial role in the development and differentiation of higher organisms. DNA methylation is also known to regulate transcription by gene repression. Various developmental genes such as c-mos, HoxB5, Sox11, and Sry show tissue-specific gene expression that was shown to be regulated by promoter DNA methylation. The aim of the present study is to investigate the establishment of chromatin marks (active or repressive) in relation to heterogeneous methylation in the promoter regions of these developmentally important genes.

**Results:** Chromatin-immunoprecipitation (ChIP) assays were performed to immuno-precipitate chromatin by antibodies against both active (H3K4me3) and repressive (H3K9me3) chromatin regions. The analysis of ChIP results showed that both the percentage input and fold enrichment of activated chromatin was higher in tissues expressing the respective genes as compared to the tissues not expressing the same set of genes. This was true for all the genes selected for the study (c-mos, HoxB5, Sox11, and Sry). These findings illustrate that inconsistent DNA methylation patterns (sporadic, mosaic and heterogeneous) may also influence gene regulation, thereby resulting in the modulation of chromatin conformation.

**Conclusions:** These findings illustrate that various patterns of DNA methylation (asynchronous, mosaic and heterogeneous) correlates with chromatin modification, resulting in the gene regulation.

**Keywords:** DNA methylation, Chromatin conformation, Chromatin-immunoprecipitation (ChIP), Sox11, C-mos, HoxB5, Tissue-specific gene expression

## Background

DNA methylation is a major component of epigenetic mechanism which is stably inherited and playing important roles during transcriptional regulation [1, 2]. This event implicates the cytosine residues to be directly modified which is immediately followed by guanine residues (CpGs). These CpG dinucleotides are extremely underrepresented in mammalian genomes and are usually present in small stretches known as CpG islands. These

CpG islands are found in approximately 70% of annotated gene promoters and are normally unmethylated in normal somatic tissues. Methylation of these CpG targets leads to loss of gene expression during embryonic development [3]. DNA methylation patterns are well regulated, non-random and tissue-specific which is consistent with its functional importance [4]. DNA methylation is also involved in other fundamental processes like X chromosome inactivation, genomic imprinting, suppression of retrotransposon elements, and is essential for normal development [5]. The principal importance of DNA methylation has been studied in myriad of biological context and the mechanism involves the influence

*Correspondence: manishas@mnnit.ac.in; manishas77@rediffmail.com
[1] Department of Biotechnology, Motilal Nehru National Institute of Technology, Allahabad 211004, India
Full list of author information is available at the end of the article

of overall chromatin structure. DNA methylation causes the attraction of repressive complexes and the recruitment of methyl CpG binding proteins (including MeCP2) thereby making the chromatin structure inaccessible for binding of transcription factors leading to the formation of closed chromatin structure [6]. Lucifero et al. [7, 8] reported differences in DNA methylation between maternal and paternal alleles of many imprinted genes (Snrpn, Mest and Peg3). Acquisition of methylation was asynchoronous and heterogeneous at these different genes however the varied patterns of methylation on this subset of imprinted genes have not been analyzed for their role in regulating gene expression. Snrpn methylation data showed unique mosaic pattern of specific cytosine methylation during post natal oocyte development. E-cadherin, a tumor suppressor gene, showed de novo methylation of upstream and downstream regions in neoplastic tissues as predominantly methylated/unmethylated CpG islands [9].

Alternatively DNA methylation plays a pivotal role in embryonic development and differentiation. Developmental genes belonging to different gene families such as Hox and Sox family elicits distinct developmental programs to regulate developmental plasticity during evolution. Various developmentally important genes have been selected for the present study (HoxB5, c-mos, Sox11 and Sry) showing varied temporal and spatial tissue-specific gene expression profile and this variation has been shown to be influenced by DNA methylation patterns directly or indirectly during ontogenesis [11–14].

Therefore the present study focuses to understand the correlation of inconsistency and heterogeneity in DNA methylation patterns during developmental time with altered chromatin conformation and thereby gene regulation. However no reports were found to support this evidence enlightening its relevance in regulation of gene expression and chromatin modifications. Distribution of active (open) or inactive (closed) chromatin marks are studied using ChIP (chromatin-immunoprecipitation) assay for gaining insight towards a gross correlation between the heterogeneous methylation pattern and the distribution of the two chromatin states leading to transcriptional regulation.

## Methods
### Genomic DNA extraction
The Parkes strain of mouse was used in this study. Work was approved by the Institutional Ethics Committee (IEC, Ref No. 176/R&C/13-14) of Motilal Nehru National Institute of Technology Allahabad and the Project Ref. No. is BT/PR13317/GBD/27/252/2009. Mice were randomly bred and maintained under laboratory conditions.

Animals were killed by cervical dislocation and different somatic and germinal tissues of both fetal and adult were excised for DNA isolation. Mouse genomic DNA was isolated in three independent sets from adult tissues while 6–7 fetuses were pooled in fetal/neonatal stage especially whole mesonephros gonadal complex from 12.5dpc embryos. DNA was isolated with the help of standard protocol using Proteinase K digestion (50 µg/ml) at 37 °C for 12–14 h. Phenol: chloroform: isoamylalcohol/chloroform: isoamyl alcohol extraction was done at 25 °C. Finally, DNA precipitation was done by adding 1/30th volume of 3 M sodium acetate (pH 5.0) along with two volumes of absolute chilled ethanol.

### Sodium bisulfite treatment
Sodium bisulfite treatment converts cytosine residues to uracil except the methylated cytosine residues which imparts unique distinguishing pattern between methylated and non-methylated cytosines in DNA. The bisulfite conversion was done using the BisulFlash™ DNA Modification Kit (Cat. P-1026, USA) according to manufacturer's instructions. 400–500 ng of input DNA was used as starting template for bisulfite conversion reaction. The bisulfite converted DNA approx. 1.0–2.0 µl was used for each PCR reaction and was amplified in duplicates which was then pooled together during purification. The primers (forward and reverse each with 20 µM) for c-mos gene were as follows: Region (1970–2189 bp): FP 5'-TGTTTTATGT GATTGTTTTATTTG-3', RP 5'-CACAAAAACACCAT AATAAATAAC-3'. Region (2242 bp-2537 bp): FP 5'-ATT TATTATGGTGTTTTTGTGGTTA-3', RP 5'-ATTCACT AACTTCAAATCCAAATAC-3'. PCR conditions include denaturation @ 94 °C, annealing @ 58 °C and 54 °C for region I and II respectively, extension @ 72 °C repeated for 30 cycles.

### Cloning and sequencing of positive clones
Amplified PCR products were gel excised and purified from low melting agarose using a DNA purification kit (nucleopore) as pcr instructions manual. The ligation of purified PCR product was done in T-vector using an InsTAclone™ PCR cloning kit (Thermo Scientific USA) and the ligated product was finally transformed and cloned in E.coli DH5α. Positive transformed clones were selected on the basis of blue-white screening and plasmid isolation was done using an alkaline lysis method (Thermo Scientific Plasmid Miniprep Kit). Sequencing of 8-10 independent clones were done by automated DNA sequencer (ABI 3130 genetic analyzer). Plasmid DNA 50–100 ng was used for sequencing with BigDye® Terminator v3.1 Cycle Sequencing Kit as per manufacturer's instructions.

## RNA isolation and quantitative real time PCR

Total RNA was isolated from mouse tissues using TRI reagent (sigma) according to the manufacturer's protocol. Concentration of RNA was determined spectrophotometrically at OD 260 nm both before and after DNAse treatment. RNA was reverse transcribed in equal amount using oligo-dT(18) primers and 200 U of MMLV Reverse transcriptase (NEB). cDNA 0.5 μl was used in PCR amplification using SsoFast™ EvaGreen Supermix with Low ROX (2X) along with 5 μM each of forward and reverse primers in real time step one plus PCR machine (ABI). PCR reactions were carried out in triplicates for both *GAPDH* and *c-mos*. Expression of mRNA level was estimated and normalized relative to the mRNA level of *GAPDH*. The average of the cycle threshold (Ct) values and standard deviation were determined. Fold change in gene expression was calculated according to $2^{-\Delta\Delta Ct}$ method [10]. The primer sequences of *GAPDH* are FP: 5′GGAGCCAAACGGGTCATCATCTC3′ and RP-5′GAGGGGCCATCCACAGTCTTCT 3′; *c-mos gene* FP 5′-TACGCCACGACAACATAGTTCG-3′ RP 5′-CTTGCTCACTGATCAAAATGTTGG-3′.

## Chromatin-immunoprecipitation (ChIP)

ChIP assay was performed according to the instructions manual (Diagenode ChIP kit Cat. No. kch-orgHIS-012). Chromatin was isolated from different somatic (brain, spleen and kidney) and germinal tissues (testis) of adult, fetal and neonatal stages of mouse. The excised tissues were homogenized and subjected to collagenase treatment (50–200 U/ml) followed by incubation for 2–3 h at 37 °C. Single cell suspension was made by pipetting during the incubation time and cell counting was performed using haemocytometer. The minimum number of cells required to perform ChIP experiments is $1 \times 10^6$ cells. Cell cross-linking was done by adding 37% formaldehyde (w/v, final concentration 1%) kept for 10 min at 25 °C on a rotating wheel followed by quenching with 1.25 M glycine (final concentration 125 mM) for 5 min at 25 °C, centrifuged at 4 °C for 5–8 min. Supernatant was discarded and the cell pellet was resuspended in lysis buffer (containing protease inhibitors). The cell suspension was subjected to sonication using a sonicator (SKN-IIDN) at the rate of 3 s ON/1 s OFF for 3–4 cycles for obtaining the desired chromatin range from 200–800 bp. The sheared chromatin was then processed for pre-clearing by adding an IP-incubation mix and pre-blocked beads. Antibodies specific for capturing the desired protein and interacting DNA were used (H3K4me3, Diagenode MAb-152-050 and H3K9me3, Diagenode, MAb-146-050, concentration 1 μg/μl). Negative control IgG antibody (Diagenode, C15400001 (C15200001) was used which binds with non-specific target and the associated DNA fragments were immuno-precipitated. The addition of specific antibodies was followed by incubation on a rotating wheel at 4 °C for overnight. Bead washing with wash buffer—1, 2 and 3 removes non-associated DNA fragments and Protein/DNA complexes were found to get eluted from pre-blocked beads by the addition of elution buffer. The eluted complex was reversibly cross-linked and purified using phenol: chloroform: iso-amyl alcohol/chloroform: iso-amyl alcohol. DNA fragments were precipitated by adding DNA precipitant, DNA co-precipitant and absolute chilled ethanol. The DNA pellet was resuspended in 30 μl of milliQ water and the relative amount of specifically immunoprecipitated DNA was analyzed through PCR amplification using quantitative real-time PCR (ABI step one plus) with 1.0 μl of DNA, SsoFast™ EvaGreen Supermix (2X) with Low ROX (Biorad) and gene specific primers forward and reverse 5 μM each. Control primers *GAPDH* (c17021045, Diagenode used as positive control against activated chromatin regions) and *TSH2β* (c17021042, Diagenode used as positive control against repressed chromatin regions) were used. The percentage input and fold enrichment was calculated which represents the enrichment of certain histone modifications on specific region using the ChIP reactions performed in triplicate.

The primers used for various ChIP reactions in different developmental genes were shown in Table 1.

## Results

We have selected three more developmentally important genes whose methylation pattern has been already studied in different tissues include kidney, brain, spleen, testis and mesonephros gonadal cells (MGCs) [11–13]. In HoxB5 methylation analysis was performed in fetal and adult stages of mouse kidney and spleen tissues whereas the same analysis was carried out in fetal and adult stages

**Table 1 Shows the primer sequence of different genes used for ChIP-qPCR reactions**

| GENE | Forward primer | Reverse primer |
| --- | --- | --- |
| c-mos | 5′-TGGGGAAGTGCCTCAAGTAT-3′ | 5′-AGGTCCAAGTGCAAAATGCT-3′ |
| HoxB5 | 5′-AGTCCCTGCCCTGCACTAA-3′ | 5′-GCCTCGTCTATTTCGGTGAA-3′ |
| Sox11 | 5′-TCCAGGTCCTTAT CCACCAG-3′ | 5′-GACGACCTCATGTTCG ACCT-3′ |
| Sry | 5′-GTCAAGCGCCCCATGAAT-3′ | 5′-CAGCTGCTTGCTGATCTCTG-3′ |

of mouse brain and kidney tissues in Sox11. Promoter methylation pattern of Sry was determined in mouse 12.5 dpc mesonephros gonadal complex (MGCs) and in adult testis. Comparing the methylation status of HoxB5 and Sox11, higher methylation was found in adult as compared to fetus. In Sry, less methylation was present in fetal gonads as compared to fetal liver. The work done in the above cited references showed a direct inverse correlation between tissue-specific developmental methylation pattern and gene expression.

### Methylation and expression pattern of c-mos gene

c-mos gene was analyzed for its DNA methylation patterns in both somatic and germinal tissues through sodium bisulfite genomic DNA sequencing with the aim at distinguishing methylated cytosines from unmethylated cytosine residues. Methylation was mapped in the regulatory/coding region of c-mos gene targeting a total of 25 CpG sites. Two pairs of primers covering 12 and 13 CpG sites respectively (from 1970–2189 and 2242–2537 bp) were designed with the PCR amplicon sizes of 220 and 296 bp respectively (Fig. 1a). In case of germinal tissues, adult testis shows no methylation as shown in (Fig. 1c) while adult ovary has a very few sites methylated (Fig. 1d). The methylation pattern in somatic tissues (adult kidney) is about 55% and seems to be heterogeneous and asynchronous (Fig. 1e) while germinal tissues possess only about 7% methylation. The methylation data for adult kidney shows that sites #8 and #12 were fully methylated while sites #1, #2, #3, #11, #14, #21, #23, #24 and #25 were more than 50% methylated and sites #5 and #10 showed 50% methylation. The remaining sites (#4, #6, #7, #9, #13, #15, #16, #17, #18, #19, #20 and #22) were less than 50% methylated. The real time quantitative gene expression data for c-mos showed an increased expression in adult testis and ovary whereas adult kidney showed repressed expression (Fig. 1b).

### ChIP (Chromatin-Immunoprecipitation) results

The results of Chromatin-Immunoprecipitation demonstrate the occupancy of the fractionated DNA fragments precipitated with a particular antibody against activated (H3K4me3) or repressed (H3K9me3) chromatin domains for specific gene in adult, fetal and neonatal stages of various somatic and germinal tissues of mice. The data were represented as percentage input and in terms of fold enrichment (FE).

#### c-mos

The chromatin interaction results of c-mos gene in adult testis illustrates that the percentage input of activated chromatin (H3K4me3) was higher than adult kidney (Fig. 1f, h). In a similar way fold enrichment was also

higher (19 fold) in activated chromatin regions of adult testis as compared to adult kidney (Fig. 1g, i).

#### HoxB5

The previous studies on HoxB5 promoter methylation by Sachan et al. (2006) demonstrated that fetal tissues of kidney and spleen showed expression of HoxB5 whereas no expression was detected in adult tissues of kidney and spleen. Higher methylation in adult kidney (18.6%) and adult spleen (74.55%) was observed in contrast to fetal kidney (2%) and fetal spleen (30.45%). We have performed ChIP for these tissues of adult and fetal stages. The ChIP results of HoxB5 gene in fetal kidney and spleen illustrate that the percentage input of activated chromatin (H3K4me3) was higher as compared to adult kidney and spleen (Fig. 2a, c, e, g). Fold enrichment data measured lower fold change in adult kidney when it is compared to fetal kidney. In contrast fetal kidney possesses 18 fold higher enrichment of activated chromatin in contrast to repressed (Fig. 2b, d). Similarly fold enrichment values showed higher fold change (3 times) for activated chromatin regions in the fetal spleen and (4 times) for repressed chromatin (H3K9me3) in adult spleen (Fig. 2f, h).

#### Sox11

ChIP data for Sox11 shows that the higher percentage input of activated chromatin (H3K4me3) fraction was measured in the 0dpp brain as compared to adult brain and the same is true for 0dpp kidney and adult kidney respectively (Fig. 3a, c, e, g). Fold enrichment values were 2.2 fold higher in the 0dpp brain for activated chromatin regions, while adult brain has 1.7 times higher enrichment for repressed chromatin (Fig. 3d, b). Higher fold enrichment was found (2.0 fold) in activated chromatin regions for 0dpp kidney while adult kidney shows 5 times higher fold enrichment in repressed chromatin regions (Fig. 3h, f).

#### Sry

The data suggest that the higher percentage input was observed in the activated chromatin region (H3K4me3) in 12.5dpc testis (MGCs) while higher repressed chromatin marks (H3K9me3) were found in adult testis (Fig. 4c, a). Fold enrichment was approximately 12 times higher for activated chromatin in 12.5dpc testis while sixfolds higher enrichment for the repressed region in adult testis (Fig. 4d, b).

### Discussion

The heterogeneous methylation pattern present in the somatic tissues (adult kidney shows the average percentage of all CpGs around 55) and almost complete absence

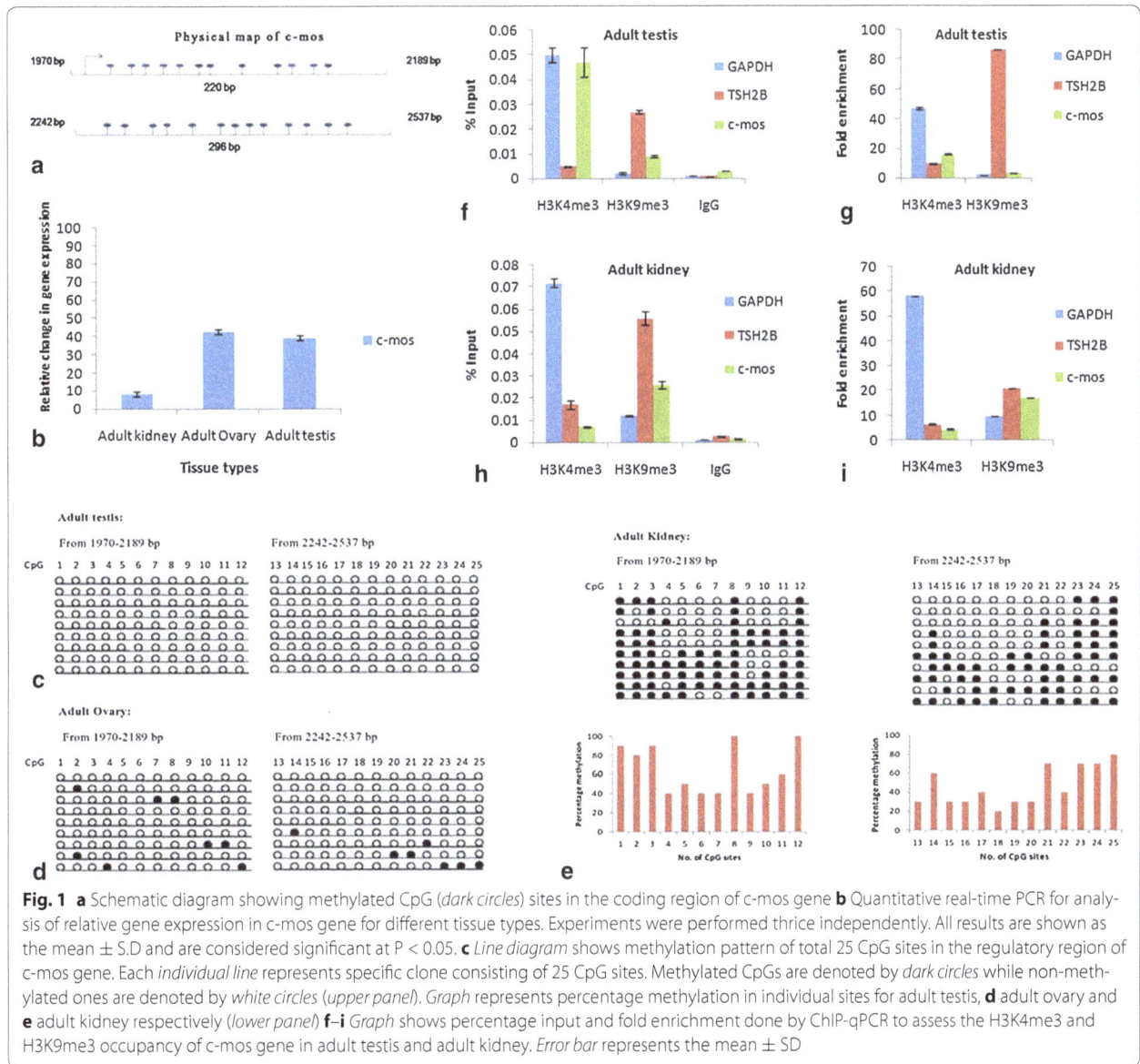

**Fig. 1** **a** Schematic diagram showing methylated CpG (*dark circles*) sites in the coding region of c-mos gene **b** Quantitative real-time PCR for analysis of relative gene expression in c-mos gene for different tissue types. Experiments were performed thrice independently. All results are shown as the mean ± S.D and are considered significant at P < 0.05. **c** *Line diagram* shows methylation pattern of total 25 CpG sites in the regulatory region of c-mos gene. Each *individual line* represents specific clone consisting of 25 CpG sites. Methylated CpGs are denoted by *dark circles* while non-methylated ones are denoted by *white circles* (*upper panel*). *Graph* represents percentage methylation in individual sites for adult testis, **d** adult ovary and **e** adult kidney respectively (*lower panel*) **f–i** *Graph* shows percentage input and fold enrichment done by ChIP-qPCR to assess the H3K4me3 and H3K9me3 occupancy of c-mos gene in adult testis and adult kidney. *Error bar* represents the mean ± SD

of methylation in the germinal tissues (testis and ovary) of c-mos gene correlates inversely with its tissue specific expression. c-mos displays higher expression in adult testis and ovary and almost no expression in somatic tissues. It further strengthens the crucial role of c-mos gene in meiotic maturation of germinal tissues [14, 15] and in spermatogenesis [16] and oogenesis [17]. These finding correlates with the chromatin–immunoprecipitation results that displays the fractionation of chromatin fragments having the higher percentage input of activated chromatin regions (H3K4me3) in the expressing tissue (germinal tissues) in contrast to repressed chromatin regions (H3K9me3) in non-expressing tissues (somatic). Similarly HoxB5, a member of Hox family genes playing important roles in antero-posterior axial body patterning

and the RNA transcripts were found in a wide range of fetal tissues such as lung, gut, kidney, liver and spleen [18–20]. Examination of the methylation pattern of HoxB5 shows higher methylation density in adult as compared to fetal stages of mouse tissues (Kidney and spleen) [11]. ChIP data also suggests higher input fraction of activated chromatin during fetal/neonatal stages of tissues in comparison to their adult counter parts.

Another developmental gene, Sox11, whose expression was prominent in developing central nervous system of mouse embryos, suggests its important role in neuronal growth, maturation and survival [21–23]. Promoter methylation mapping of Sox11 reveals higher methylation in both adult brain and kidney (average percentage of all CpGs around 48 and 68% in brain and kidney

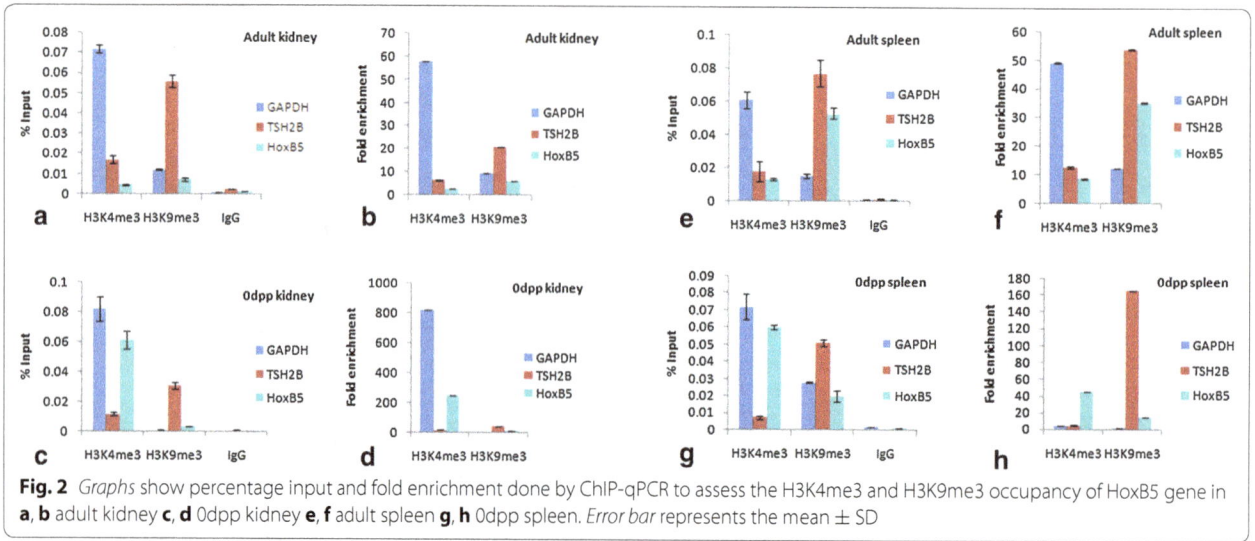

**Fig. 2** *Graphs* show percentage input and fold enrichment done by ChIP-qPCR to assess the H3K4me3 and H3K9me3 occupancy of HoxB5 gene in **a**, **b** adult kidney **c**, **d** 0dpp kidney **e**, **f** adult spleen **g**, **h** 0dpp spleen. *Error bar* represents the mean ± SD

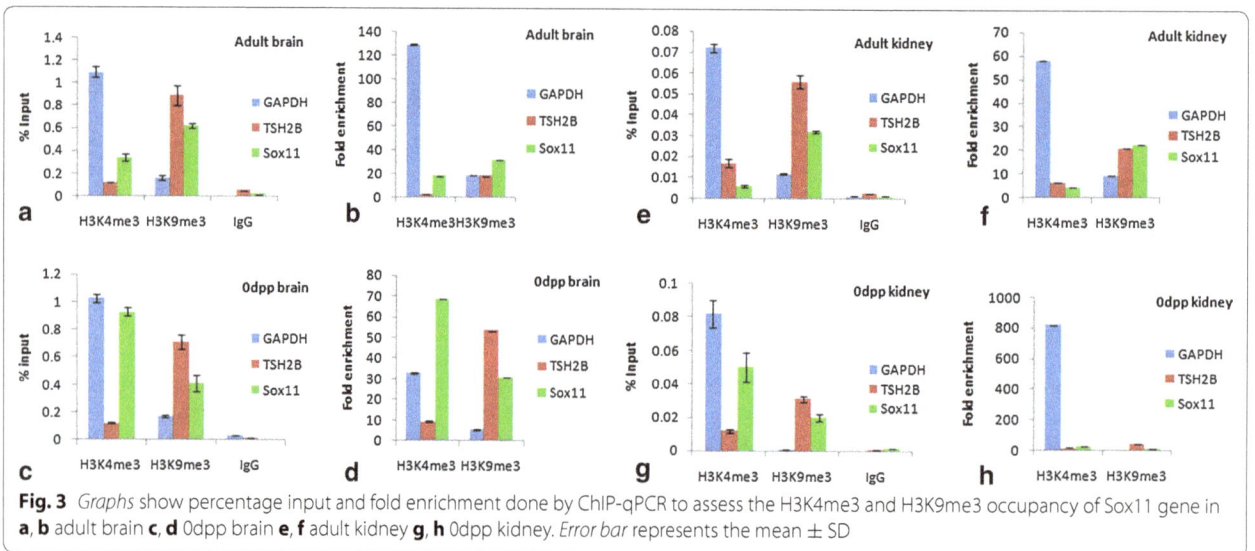

**Fig. 3** *Graphs* show percentage input and fold enrichment done by ChIP-qPCR to assess the H3K4me3 and H3K9me3 occupancy of Sox11 gene in **a**, **b** adult brain **c**, **d** 0dpp brain **e**, **f** adult kidney **g**, **h** 0dpp kidney. *Error bar* represents the mean ± SD

respectively) while the fetal and neonatal stages of the same tissues showed lower methylation density (14 and 31% in fetal brain and kidney respectively) [12]. The ChIP results of Sox11 in the present study are highly consistent with its tissue specific gene expression which correlates with changes in chromatin conformation.

A study by Nishino et al. [13] demonstrate that the mechanistic role of Sry gene regulation occurs by DNA methylation during mouse gonadal development and the Sry transcripts were expressed only in pre Sertoli cells of developing male gonad during 10.5–12.5dpc (average percentage of all CpGs around 62% and <3% in 11.5dpc liver and gonads respectively) [24, 25]. Hypermethylated status of the two regions analyzed was maintained in the tissues that did not express Sry however the methylation pattern was not purely consistent. Our ChIP results

coincides with this finding as higher percentage input and fold enrichment values of activated chromatin fragments in fetal mesonephros gonads are in order with tissue specific gene expression [13] and inversely correlates with the promoter methylation of Sry gene.

Curradi et al. [26] analyzed the inhibitory effects of promoter DNA methylation through in vitro experiments and the results showed that a definite number of methylated CpG residues are required to organize a stable, diffusible chromatin structure. Histone deacetylation significantly imparts to transcriptional repression only when the number of modified cytosine residues is adequate to suppress the gene over a long range. These data suggest the importance of methylated cytosine residues for its ability to create a repressive effect and this repression can propagate in both directions for several hundred

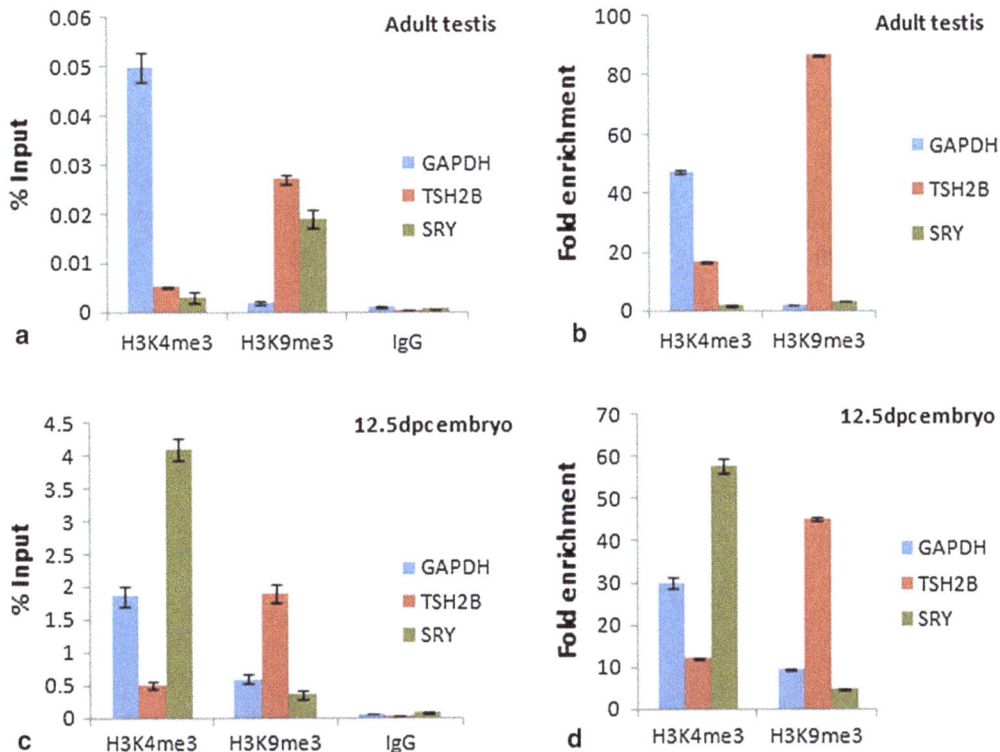

**Fig. 4** *Graph* show percentage input and fold enrichment done by ChIP-qPCR to assess the H3K4me3 and H3K9me3 occupancy of Sry gene in **a, b** adult testis **c, d** 12.5dpc embryo. *Error bar* represents the mean ± SD

base pairs of the altered DNA only when enough number of methylated CpGs are present [26].

The potential role of DNA methylation is conserved from yeast to mammals. Some recent reports exhibited tissue and life-stage specific gene expression and its regulation by DNA methylation in species other than mammals suggesting functional similarity between transcriptional regulation [27, 28]. However, other studies also show impact of heterogeneous DNA methylation on gene expression including Oct-4 whose expression is confined to the germ line and pluripotent stem cells, which is critical for normal mouse development. The epigenetic regulation of Oct-4 show heterogeneous DNA methylation in its regulatory region among the population of adult somatic cells [29]. Similarly the heterogeneity in DNA methylation patterns across the distinct promoter and 5′UTR regions of differentiating stem cells and during reprogramming events suggest new insight into the methylation dynamics pertaining to embryonic stem cells [30, 31]. It has also been reported that the methylation patterns present in several imprinted genes and repetitive elements of embryonic germ (EG) cells may exhibit DNA methylation changes occurring in the germ cell lineage during the differentiation process [32]. Induction of mosaic DNA methylation patterns also leads

to tumorigenesis by participating in important cancer related pathways and thus causing loss of co-expression connectivity in colorectal cancer [33]. The presence of epigenetic heterogeneity of developmentally important genes has important implications for assisted reproduction outcomes [34]. Another significant role was observed in the epigenetic regulation of stress reactivity in humans [35] and the formation and maintenance of memory in plasticity genes [36].

These findings elucidate that the tissue and stage specific gene regulation of several developmentally important genes of mouse is mediated through promoter/regulatory regions DNA methylation. The ChIP results are also consistently indicating a direct correlation between the percentage of input chromatin fraction and the level of gene expression.

## Conclusions

Transcriptional regulation mediated by the establishment of tissue specific methylation patterns in the promoter or regulatory regions of genes examined in this study might speculate to be a crucial mechanism which may be different from non-developmental genes. This might reflect the specialized functions of various developmental genes during the process of embryonic development. A significant level of heterogeneous/asynchronous methylation exists as the development progresses from embryo to adult and in most of the cases this heterogeneity is associated with transcriptional gene regulation and chromatin modifications. Further, higher density of methylation present in the promoter regions might render the DNA inaccessible (either fully or partially) for binding of transcription factors, leading to altered chromatin assembly and gene expression.

## Abbreviations
dpc: days post coitum; dpp: days post partum; ChIP: chromatin immunoprecipitation; MeCP2: methyl cytosine binding protein 2.

## Authors' contributions
PS carried out the experimental work and drafted the manuscript. MS and KS supervised the study. All authors read and approved the final manuscript.

## Author details
[1] Department of Biotechnology, Motilal Nehru National Institute of Technology, Allahabad 211004, India. [2] Department of Molecular and Human Genetics, Banaras Hindu University, Varanasi, India.

## Acknowledgements
This work was supported by Department of Biotechnology (DBT), Government of India. We thank Cytogenetic Lab, Department of Zoology, Banaras Hindu University (BHU) for providing animal house facility.

## Competing interests
The authors declare that they have no competing interests.

## References
1. Kulis M, Esteller M. DNA methylation and cancer. Adv Genet. 2010;70:27–56.
2. Suzuki MM, Bird A. DNA methylation landscapes: provocative insights from epigenomics. Nature Rev Genet. 2008;9(6):465–76.
3. Deaton AM, Bird A. CpG islands and the regulation of transcription. Genes Dev. 2011;25:1010–22.
4. Illingworth R, Kerr A, Desousa D, Jørgensen H, Ellis P, Stalker J, Jackson D, Clee C, Plumb R, Rogers J, Humphray S, Cox T, Langford C, Bird A. A novel CpG island set identifies tissue-specific methylation at developmental gene loci. PLoS Biol. 2008;6(1):37–51.
5. Chen ZX, Riggs AD. DNA methylation and demethylation in mammals. J Biol Chem. 2011;286(21):18347–53.
6. Singal R, Ginder GD. DNA methylation. Blood. 1999;93(12):4059–70.
7. Lucifero D, Mertineit C, Clarke HJ, Bestor TH, Trasler JM. Methylation dynamics of imprinted genes in mouse germ cells. Genomics. 2002;79(4):530–8.
8. Lucifero D, Mann MR, Bartolomei MS, Trasler JM. Gene-specific timing and epigenetic memory in oocyte imprinting. Hum Mol Genet. 2004;13(8):839–49.
9. Graff JR, Herman JG, Lapidus RG, Chopra H, Xu R, Jarrard DF, Isaacs WB, Pitha PM, Davidson NE, Baylin SB. E-cadherin expression is silenced by DNA hypermethylation in human breast and prostate carcinomas. Cancer Res. 1995;55(22):5195–9.
10. Livak KJ, Schmittgen TD. Analysis of relative gene expression data using real-time quantitative PCR and the 2(-Delta Delta C(T)) Method. Methods. 2001;25:402–8.
11. Sachan M, Raman R. Developmental methylation of the regulatory region of HoxB5 gene in mouse correlates with its tissue-specific expression. Gene. 2006;380(2):151–8.
12. Pamnani M, Sinha P, Nara S, Sachan M. Study of promoter DNA methylation of Sox11 and its correlation with tissue-specific expression in the laboratory mouse. Gene. 2014;552(1):133–9.
13. Nishino K, Hattori N, Tanaka S, Shiota K. DNA methylation-mediated control of sry gene expression in mouse gonadal development. J Biol Chem. 2004;279(21):22306–13.
14. Kanduri C, Raman R. Tissue-specific characterization of DNA methylation in the gonad-specific proto-oncogene, c-mos, in the male laboratory mouse. Int J Dev Biol. 1999;43:91–4.
15. Mutter GL, Grills GS, Wolgemuth DJ. Evidence for the involvement of the proto-oncogene c-mos in mammalian meiotic maturation and possibly very early embryogenesis. EMBO J. 1988;7(3):683–9.
16. Cao SF, Li D, Yuan Q, Guan X, Xu C. Spatial and temporal expression of c-mos in mouse testis during postnatal development. Asian J Androl. 2008;10(2):277–85.
17. Jałocha I, Gabryś MS, Bal J. The crucial role of the proto-oncogene c-mos in regulation of oocyte maturation. Postepy Hig Med Dosw. 2010;64:636–41.
18. Hershko AY, Kafri T, Fainsod A, Razin A. Methylation of HoxA5 and HoxB5 and its relevance to expression during mouse development. Gene. 2003;302(1–2):65–72.
19. Krumlauf RO, Holland PW, McVEY JH, Hogan BL. Developmental and spatial patterns of expression of the mouse homeobox gene, Hox 2.1. Development. 1987;99(4):603–17.
20. Fu M, Lui VC, Sham MH, Cheung AN, Tam PK. HOXB5 expression is spatially and temporarily regulated in human embryonic gut during neural crest cell colonization and differentiation of enteric neuroblasts. Dev Dyn. 2003;228(1):1–10.
21. Hargrave M, Wright E, Kun J, Emery J, Cooper L, Koopman P. Expression of the Sox11 gene in mouse embryos suggests roles in neuronal maturation and epithelio-mesenchymal induction. Dev Dyn. 1997;210(2):79–86.
22. Jankowski MP, Cornuet PK, McIlwrath S, Koerber HR, Albers KM. SRY-box containing gene 11 (Sox11) transcription factor is required for neuron survival and neurite growth. Neuroscience. 2006;143(2):501–14.
23. Wang Z, Reynolds A, Kirry A, Nienhaus C, Blackmore MG. Overexpression of Sox11 promotes corticospinal tract regeneration after spinal injury while interfering with functional recovery. J Neurosci. 2015;35(7):3139–45.
24. Rossi P, Dolci S, Albanesi C, Grimaldi P, Geremia R. Direct evidence that the mouse sex-determining gene Sry is expressed in the somatic cells of male fetal gonads and in the germ cell line in the adult testis. Mol Reprod Dev. 1993;34(4):369–73.
25. Hacker A, Capel B, Goodfellow P, Lovell-Badge R. Expression of Sry, the mouse sex determining gene. Development. 1995;121:1603–14.
26. Curradi M, Izzo A, Badaracco G, Landsberger N. Molecular mechanisms of gene silencing mediated by DNA methylation. Mol Cell Biol. 2002;22(9):3157–73.
27. Li S, Zhu Y, Zhi L, Han X, Shen J, Liu Y, Yao J, Yang X. DNA methylation variation trends during the embryonic development of chicken. PLoS ONE. 2016;11(7):e0159230.
28. Mayasich SA, Bemis LT, Clarke BL. DNA methylation in a sea lamprey vasotocin receptor gene promoter correlates with tissue-and life-stage-specific mRNA expression. Comp Biochem Physiol. 2016;202:56–66.
29. Marikawa Y, Fujita TC, Alarcón VB. Heterogeneous DNA methylation status of the regulatory element of the mouse Oct4 gene in adult somatic cell population. Cloning Stem Cells. 2005;7(1):8–16.
30. Shao X, Zhang C, Sun MA, Lu X, Xie H. Deciphering the heterogeneity in DNA methylation patterns during stem cell differentiation and reprogramming. BMC Genom. 2014;15:978–88.
31. Singer ZS, Yong J, Tischler J, Hackett JA, Altinok A, Surani MA, Cai L, Elowitz MB. Dynamic heterogeneity and DNA methylation in embryonic stem cells. Mol Cell. 2014;55(2):319–31.
32. Shovlin TC, Durcova-Hills G, Surani A, McLaren A. Heterogeneity in imprinted methylation patterns of pluripotent embryonic

germ cells derived from pre-migratory mouse germ cells. Dev Biol. 2008;313(2):674–81.

33. Wang Q, Jia P, Cheng F, Zhao Z. Heterogeneous DNA methylation contributes to tumorigenesis through inducing the loss of coexpression connectivity in colorectal cancer. Genes Chromosomes Cancer. 2015;54(2):110–21.

34. Kuhtz J, Schneider E, Hajj NE, Zimmermann L, Fust O, Linek B, Seufert R, Hahn T, Schorsch M, Haaf T. Epigenetic heterogeneity of developmentally important genes in human sperm: implications for assisted reproduction outcome. Epigenetics. 2014;9(12):1648–58.

35. Houtepen LC, Vinkers CH, Carrillo-Roa T, Hiemstra M, van Lier PA, Meeus W, Branje S, Heim CM, Nemeroff CB, Mill J, Schalkwyk LC, Creyghton MP, Kahn RS, Joëls M, Binder EB, Boks MP. Genome-wide DNA methylation levels and altered cortisol stress reactivity following childhood trauma in humans. Nat Commun. 2016;7:10967.

36. Halder R, Hennion M, Vidal RO, Shomroni O, Rahman RU, Rajput A, Centeno TP, Bebber FV, Capece V, Vizcaino JCG, Schuetz AL, Burkhardt S, Benito E, Sala MN, Javan SB, Haass C, Schmid B, Fischer A, Bonn S. DNA methylation changes in plasticity genes accompany the formation and maintenance of memory. Nature Neurosci. 2016;19:102–17.

# 9

# Mitochondrial RNA processing in absence of tRNA punctuations in octocorals

Gaurav G. Shimpi[1,4*], Sergio Vargas[1] , Angelo Poliseno[1] and Gert Wörheide[1,2,3*]

## Abstract

**Background:** Mitogenome diversity is staggering among early branching animals with respect to size, gene density, content and order, and number of tRNA genes, especially in cnidarians. This last point is of special interest as tRNA cleavage drives the maturation of mitochondrial mRNAs and is a primary mechanism for mt-RNA processing in animals. Mitochondrial RNA processing in non-bilaterian metazoans, some of which possess a single tRNA gene in their mitogenomes, is essentially unstudied despite its importance in understanding the evolution of mitochondrial transcription in animals.

**Results:** We characterized the mature mitochondrial mRNA transcripts in a species of the octocoral genus *Sinularia* (Alcyoniidae: Octocorallia), and defined precise boundaries of transcription units using different molecular methods. Most mt-mRNAs were polycistronic units containing two or three genes and 5′ and/or 3′ untranslated regions of varied length. The octocoral specific, mtDNA-encoded mismatch repair gene, the *mtMutS*, was found to undergo alternative polyadenylation, and exhibited differential expression of alternate transcripts suggesting a unique regulatory mechanism for this gene. In addition, a long noncoding RNA complementary to the *ATP6* gene (lnc*ATP6*) potentially involved in antisense regulation was detected.

**Conclusions:** Mt-mRNA processing in octocorals possessing a single mt-tRNA is complex. Considering the variety of mitogenome arrangements known in cnidarians, and in general among non-bilaterian metazoans, our findings provide a first glimpse into the complex mtDNA transcription, mt-mRNA processing, and regulation among early branching animals and represent a first step towards understanding its functional and evolutionary implications.

**Keywords:** Mitochondrial RNA, Polycistronic transcripts, Alternative polyadenylation (APA), Long noncoding RNA (lncRNA), Mitogenome, Cnidaria, Octocorals, tRNA punctuation, Non-bilaterian

## Background

Two major evolutionary events occurred early in the animal history forging the majority of animals, as we know them today: the origin of their multicellularity, and the origin of bilateral symmetry. Multiple genomic changes accompanied these morphological transitions, and different genome sequencing projects give us a glimpse into these changes [1, 2]. Undoubtedly, these transitions also correlate with multiple changes in mitochondrial genome (mitogenome) architecture and organization [3].

The metazoan mitochondrial genome underwent reductive evolution, transferring most of its genome content to the nucleus [4, 5]. The majority of these alterations in mitogenome content include the loss of ribosomal proteins and some tRNA genes, changes in the genetic code, disappearance of introns, and further compaction of mitochondrial DNA (mtDNA). As an aftermath, a quintessential animal mitochondrial genome harbors only 13 genes encoding essential energy pathway proteins, 2 ribosomal RNA genes and 22 transfer RNA genes. This composition is nearly invariable among bilaterians in terms of gene content [6]. However, alterations in mitogenome content, size and organization are more prominent and peculiar among non-bilaterian animals. The mitogenomes of non-bilaterian metazoan phyla

*Correspondence: nature.gaurav@gmail.com; woerheide@lmu.de
[1] Department of Earth and Environmental Sciences, Palaeontology & Geobiology, Ludwig-Maximilians-Universität München, Richard-Wagner Str. 10, 80333 Munich, Germany
Full list of author information is available at the end of the article

comprise several novelties compared to the canonical animal mitogenomes [7]. These include, the presence of group I and group II introns in sponges and scleractinians [8–12], additional protein coding genes and/ or unknown ORFs and gene duplications in anthozoans [13–16], linear mitogenomes in calcisponges and medusozoans [17, 18], among other. Therefore, as a hotspot of mitochondrial genome diversity, early branching animals present a unique opportunity to understand the evolution of mitochondrial genome architectures as well as the fundamental processes governing its functionality and maintenance.

The significantly reduced but extremely crucial repertoire of genes present in animal mitogenomes is fundamental to its molecular and cellular functions, and to gain a deeper understanding of these processes it is essential to understand the expression and processing of mitochondrial gene transcripts. The majority of information available so far on the mitogenome transcription originates from bilaterian members of the animal kingdom [19–22]. In this regard, the canonical vertebrate mitogenome is known to transcribe symmetrically as polycistronic precursors spanning the entire heavy (H-) and light (L-) strands [23]. The 22 tRNAs interspersed throughout the mitogenome serve as punctuation marks that are recognized and cleaved at 5′ and 3′ ends by the mitochondrial RNase P and RNase Z, respectively [24, 25]. The genes within these precursors are simultaneously liberated for maturation following this mt-tRNA processing step. Consequently, in most bilaterian metazoans, most mature mitochondrial mRNAs are monocistronic, with ATP8–ATP6 and ND4L–ND4, which are known to exist as bicistronic elements, as the only exceptions. Finally, all messengers end with the post-transcriptional addition of 40–45 adenosines for maturation, which also completes the stop codon at 3′ end of mRNA in most cases [26]. Bilaterian mt-mRNAs are either essentially devoid of the untranslated regions (UTRs) or tend to have very short UTRs consisting of 1–2 nucleotides flanking the mature mRNAs; a few notable exceptions such as COI, COII, ND5 and ND6 genes, which possess slightly longer 3′ UTRs complementary to the genes on opposite strand [23, 27]. This transcription model is, however, based on the study of a small number of bilaterians [19–22]. Recent studies on medusozoan members possessing linear mitogenomes do provide some insights into the mt-transcription in non-bilaterians [28, 29]. However, a detailed exploration of mitochondrial RNA processing and characterization of UTRs is still lacking for most non-bilaterian animals, including anthozoans with circular mitogenomes more similar to the stereotypical animal mitochondrial genomes.

Among non-bilaterian metazoans, octocorals (Octocorallia: Anthozoa) are unique due to their atypical mitochondrial genomes. As many as five different gene arrangements have been reported among the octocorals studied so far [30–33], all with an exceptionally reduced complement of transfer RNAs (i.e. a single tRNA$^{Met}$ gene) and the presence of an additional gene, a mismatch repair gene (mtMutS) [13, 34], closely related to the non-eukaryotic MutS7 lineage from epsilon-proteobacteria or DNA viruses [35]. This gene has been predicted to have a self-contained mismatch DNA repair function [35], and it has been speculated to play a role in the slow rate of mtDNA evolution observed among octocorals [36, 37], and to be responsible for various genome rearrangements through intramolecular recombination [31]. Considering the evolutionary trend towards a reduced mitogenome in Metazoa [38], the occurrence of such a large gene, such as the mtMutS, occupying nearly 16% of the octocoral mitogenome is somewhat surprising. The presence of a mtMutS mRNA transcript suggested its availability for translation [35]. However, 20 years after its discovery [13] and despite of being extensively used for phylogenetic studies of octocorals [39], a thorough understanding of its transcriptional processing and maturation, and in general of its biology is lacking.

Octocoral mitogenomes exhibit five different gene arrangements, all containing a single gene for tRNA$^{Met}$. Cleavage of this tRNA from the precursor polycistronic RNA would only result in linearization of precursors. The way in which the individual mitochondrial gene mRNAs are released for maturation from the long polycistronic precursor remains to be determined. In absence of knowledge on the precise boundaries of the mitochondrial mRNA in octocorals, despite all the novelties they confer, the understanding of the biology and evolution of animal mitochondria remain incomplete. Here we characterize the mitogenome transcription of an early branching non-bilaterian metazoan, the octocoral *Sinularia* cf. *cruciata*. (Alcyoniidae: Octocorallia). We describe the 5′ and 3′ boundaries and UTRs of mature mitochondrial mRNAs and characterize the transcription of the *mtMutS* gene. Our results provide the first glimpse to the unique features and complexity of the mitochondrial transcriptome in non-bilaterians.

## Results

### The mitogenome of *Sinularia* cf. *cruciata*

The complete mitogenome of *Sinularia* cf. *cruciata* was 18,730 bp in length and included, similar to other octocorals, 14 protein-coding genes (PCGs) (*ATP6, ATP8, COI–III, CytB, ND1–6, ND4L* and *mtMutS*), two ribosomal RNAs (12S and 16S) and a single transfer RNA (tRNA$^{Met}$). Most PCGs and the two rRNA genes were encoded on the H-strand. *ATP6, ATP8, COII, COIII* and tRNA$^{Met}$ were encoded on the L-strand. Gene order was consistent with that of other octocorals with mitogenome arrangement 'A' [31]. Base composition of the mitogenome was A, 30.2%; C, 16.5%; G, 19.3%; T, 33.9% and G+C, 35.8%. Among the 14 PCGs, the *mtMutS* (2982 bp) was the longest and *ATP8* (216 bp) the shortest. All PCGs had ATG as the start codon, while the stop codons TAA and TAG were predominant among PCGs; *COI* was an exception, having an incomplete termination codon (T) (see Additional file 1: Table S1). Except for *ND2* and *ND4* (13 bp overlap), the remaining genes were separated by intergenic regions (IGRs) of different lengths (see Additional file 1: Table S1, Figure S1A). The longest IGR (112 bp) was found between *COII* and *COI*, while the shortest (4 bp) was located between *12S* and *ND1*. The mitogenome of *Sinularia* cf. *cruciata* was 12 bp shorter than that of *Sinularia peculiaris* and the two species had same base composition and GC content. Sequence variability within the two *Sinularia* species was 2.31%, excluding nucleotide ambiguities (0.05%) and gaps (0.16%). The most variable genes were *mtMutS, ND2, ND5, ND4* and *COII* (Additional file 1: Figure S1B).

## Recovering mitogenomic transcripts from RNA-Seq data of *S.* cf. *cruciata* and other octocorals

In order to understand mtDNA expression and processing in octocorals, an RNA-Seq library of *Sinularia* cf. *cruciata* was screened for mitochondria-mapping reads (hereafter mt-reads). A total of 4153 reads out of 26 M pairs were mapped to the sequenced *S.* cf. *cruciata* mitogenome. This resulted in a partial mitogenome assembly covering 62.8% of its length and leaving 37.2% of the genome uncovered. With the exception of *12S* and *16S* rRNAs, which exhibited very high coverage, none of the PCGs present in the mitogenome were completely covered (Fig. 1). Despite low coverage, reads spanning IGRs were detected pointing towards the collinear expression of *CytB–ND6, ND4L–mtMutS, ND5–ND4, ATP8–ATP6,* and *COII–ATP8*.

Additionally, RNA-Seq data from two published octocoral transcriptome studies were screened for mitochondrial reads. These included *Gorgonia ventalina* (Genome arrangement A; SRR935078–87) and *Corallium rubrum* (Genome arrangement C; SRX675792) [31, 40–42]. Hence, they provided us with the opportunity to explore mt-transcription in octocorals with different mitogenome arrangements. In case of *G. ventalina*, 55,783 out of 302 M reads were mapped to the *Pseudopterogorgia*

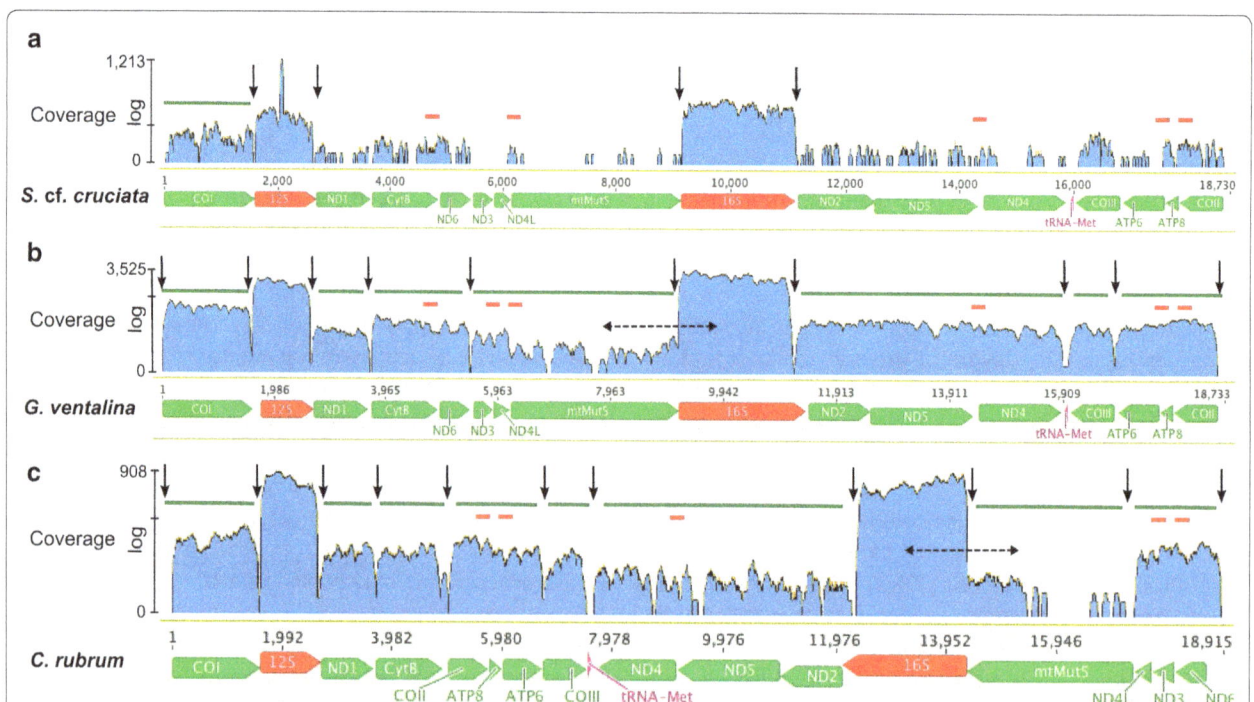

**Fig. 1** RNA-Seq (*log scale*) coverage of the mitochondrial genome of octocorals **a** *Sinularia* cf. *cruciata*, **b** *Gorgonia ventalina*, and **c** *Corallium rubrum*. The *X*-axis represents positions of the mitochondrial genes. *Green* and *red blocks* below coverage represent PCGs and rRNAs, respectively. *Y*-axis shows the log scale RNA-Seq coverage: log (coverage + 1)/log (maximum coverage + 1), which has an indicator at the mean coverage level. *Black vertical arrows* indicate abrupt drop in coverage. *Green bars* above coverage represent putative transcriptional units of PCGs, whereas *small red bars* below show presence of reads for IGRs. *Dashed horizontal bidirectional arrows* denote the areas indicating presence of unprocessed partial *mtMutS*-16S RNA transcripts

*bipinnata* mitogenome (NC_008157), a species closely related to *G. ventalina*. Almost the entire mitogenome was covered except for 2.1% uncovered sequence data, which mainly included IGRs between putative transcription units and a region of the *mtMutS* gene. Based on the observation of read-pairs spanning IGRs, the entire L-strand genes, namely *COII, ATP6, ATP8, COIII*, were detected as a collinear unit, whereas, for the H- strand the *COI–12S–ND1*, *CytB–ND6*, *ND3–ND4–mtMutS*, *ND2–ND5–ND4* were detected as collinear transcriptional units. However, as judged by sudden drop in coverage at IGRs between genes and the difference in expression levels of *COI* and *12S* as well as *12S* and *ND1* it is likely that these genes are actually monocistronic units and that the detected collinearity results from sequencing of low abundance premature RNA or unprocessed intermediates of these genes.

In the case of *C. rubrum*, 17,126 out of 241M reads were mapped to its mitogenome (AB700136). 8.8% (1661 nt) data was missing. *COI* was present as a single transcriptional unit whereas other PCGs were observed to occur as a collinear unit as follows: *ND1–CytB, COII–ATP8–ATP6–COIII, ND6–ND3–ND4L*, and *ND2–ND5–ND4* (with low coverage). We were unable to assign *mtMutS* to any of the transcription units due to small number of reads mapped to it. (see Fig. 2 for a proposed scheme of mitogenome expression for arrangement 'A' and 'C').

Independently of the sequencing depth, in all three analyzed transcriptomes *12S* and *16S* were the most abundant transcripts (Additional file 2). Moreover, we observed that none of the three analyzed transcriptomes contained reads to cover the complete *mtMutS* gene transcripts. No RNA-Seq reads could be mapped to the IGR region between *COII-COI* and *ND6-COI* in *S.* cf. *cruciata*/*G. ventalina*, and *C. rubrum*, respectively. These regions fold into a stable stem-loop structure with a 33 bp conserved motif in all octocorals studied here (Additional file 3), and represent an inversion of polarities, thus they could function as control regions (CR)/origin of replication (OriH) in octocorals with 'A' and 'C' type mitogenomes, respectively.

## RT-PCR corroborates the presence of mature polycistronic mRNA transcripts in the mitochondrial transcriptome of *S.* cf. *cruciata*

Depending on the sequencing depth of the RNA-Seq libraries, the RNA-Seq data may contain immature/unprocessed precursor RNA that could lead to the detection of false polycistronic mRNAs. In order to verify the presence of genuine mature polycistronic mRNA transcripts, we conducted RT-PCR experiment using primers binding substantially up/downstream from the start/stop codons of putative consecutive genes and amplifying

their IGRs. Using DNA as a template, amplification was observed using all primer pairs screened (Table 1). Using cDNA synthesized with an anchored oligo(dT) primer as a PCR template, bicistronic and tricistronic transcripts corresponding to the *CytB–ND6, ND2–ND5*, and *ND3–ND4L–mtMutS, COII–ATP8–ATP6*, respectively, were detected corroborating the result obtained from the analysis of RNA-Seq data suggesting that these regions are polycistronic transcription units.

## UTR mapping of the mitochondrial protein coding genes in *S.* cf. *cruciata*

The UTRs of several mitochondrial mature transcriptional units were mapped using 5′/3′ RACE and circularized RT-PCR. The *COI* mRNA has a 4 bp 5′ UTR upstream to the start codon. The 3′ end of this gene, which lacks proper stop codon could not be deduced with enough certainty, but is likely that the partial stop codon is completed by polyadenylation yielding a monocistronic unit, in agreement with the transcriptome and RT-PCR results, which suggested a monocistronic nature of *COI*.

For *CytB* three different 5′ ends were detected. One of these 5′ ends initiated exactly two codons (6 bp) downstream (position 3683) from the annotated start, without any 5′ UTR. The other two were downstream from this 5′ end at positions 3926 and 3970. The first two messengers were detected using both RACE and cRT-PCR, whereas the third one was observed only with cRT-PCR. No 3′ end could be detected for *CytB* mRNA further supporting its co-expression with *ND6* in a bicistronic unit. Moreover, a polyadenylated *ND6* mRNA 3′ end was detected with an 8 bp 3′ UTR.

Solely based on cRT-PCR, the mature mRNA ends were detected for the *ND2–ND5–ND4* tricistronic unit. The 5′ consists of a single base before the start codon at position 11,173 (not 11,146 as it is annotated in GenBank). A 44 bp long 3′ UTR was found after the stop codon at position 15,868.

The analysis of transcriptomic data indicated the presence of a *COII–ATP8–ATP6* tricistronic mRNA. RT-PCR results and end mapping corroborated this observation. *COII* mRNA was found to be flanked by a 3 bp 5′ UTR, whereas an 83 bp long 3′ UTR was detected after the stop codon of *ATP6*. The precise ends of the protein coding genes *ND1* and *COIII*, which based on transcriptome and RT-PCR results likely are monocistronic messages, could not be determined. For more details on the UTRs of the mature mt-mRNAs of *S.* cf. *cruciata* see Fig. 3.

## The *mtMutS* is alternatively transcribed and its transcripts are differentially expressed

For the *mtMutS*, 5′-RACE mapped a 31 bp 5′ UTR upstream of the *ND3* start codon further corroborating

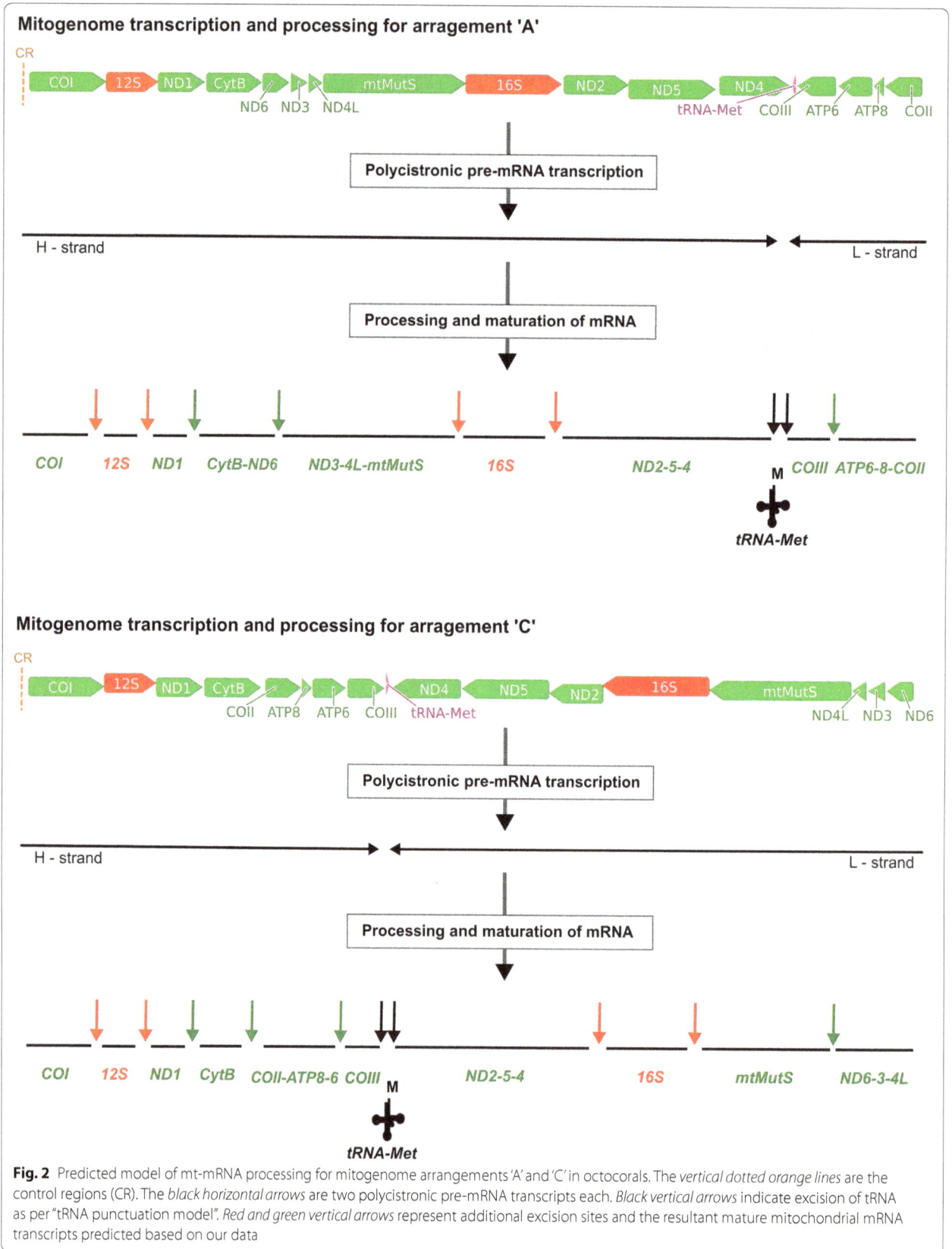

**Fig. 2** Predicted model of mt-mRNA processing for mitogenome arrangements 'A' and 'C' in octocorals. The *vertical dotted orange lines* are the control regions (CR). The *black horizontal arrows* are two polycistronic pre-mRNA transcripts each. *Black vertical arrows* indicate excision of tRNA as per "tRNA punctuation model". *Red and green vertical arrows* represent additional excision sites and the resultant mature mitochondrial mRNA transcripts predicted based on our data

**Table 1** RT-PCR screening for mature mRNA transcripts spanning two or more genes

| Sr. no. | Primer pair forward/reverse | Regions | Length (bp) | igr+Gene[a] | RT-PCR | |
|---|---|---|---|---|---|---|
| | | | | | DNA | cDNA |
| 1 | C2-18519/C1-839 | *COII–COI* | 1051 | 951 | ☑ | ☒ |
| 2 | C1-743/12S-2240 | *COI–12S* | 1498 | 658 | ☑ | ☒ |
| 3 | N1-3534/CB-4128 | *ND1–CytB* | 595 | 479 | ☑ | ☒ |
| 4 | CB-4539/N6-4995 | *CytB–ND6* | 447 | 142 | ☑ | ☑ |
| 5 | CB-4539/N3-5625 | *CytB–ND6–ND3* | 1064 | ND6+171 | ☑ | ☒ |
| 6 | N6-4974/N3-5625 | *ND6–ND3* | 641 | 171 | ☑ | ☑ |
| 7 | N6-4974/N3-5820 | *ND6–ND3* | 859 | 389 | ☑ | ☒ |
| 8 | N6-4974/N4L-6120 | *ND6–ND3–ND4L* | 1159 | ND3+291 | ☑ | ☒ |
| 9 | N3-5499/M-6780 | *ND3–ND4L–mtMutS* | 1282 | ND4L+638 | ☑ | ☑ |
| 10 | N3-5499/M-7199 | *ND3–ND4L–mtMutS* | 1701 | ND4L+1057 | ☑ | ☑ |
| 11 | N3-5499/M-8286 | *ND3–ND4L–mtMutS* | 2788 | ND4L+2456 | ☑ | ☑ |
| 12 | N3-5499/M-8826 | *ND3–ND4L–mtMutS* | 3328 | ND4L+2684 | ☑ | ☑ |
| 13 | M-8714/16S-9396 | *mtMutS–16S* | 683 | 259 | ☑ | ☑ |
| 14 | 16S-10954/N2-11753 | *16S–ND2* | 800 | 639 | ☑ | ☒ |
| 15 | N2-12177/N5-12612 | *ND2–ND5* | 448 | 106 | ☑ | ☑ |
| 16 | N5-12961/N4-15314 | *ND5–ND4* | 2354 | 990 | ☑ | ☑ |
| 17 | N4-14807/C3-16468 | *ND4–COIII* | 1662 | 598 | ☑ | ☒ |
| 18 | N4-15714/A6-17048 | *ND4–COIII–ATP6* | 1335 | COIII+225 | ☑ | ☒ |

☑, positive amplification; ☒, no amplification

Primer codes = first part indicates gene and number indicates 5′ end of primer corresponding to *S.* cf. *cruciata* mitogenome

[a] Indicates bps into the gene downstream to the igr

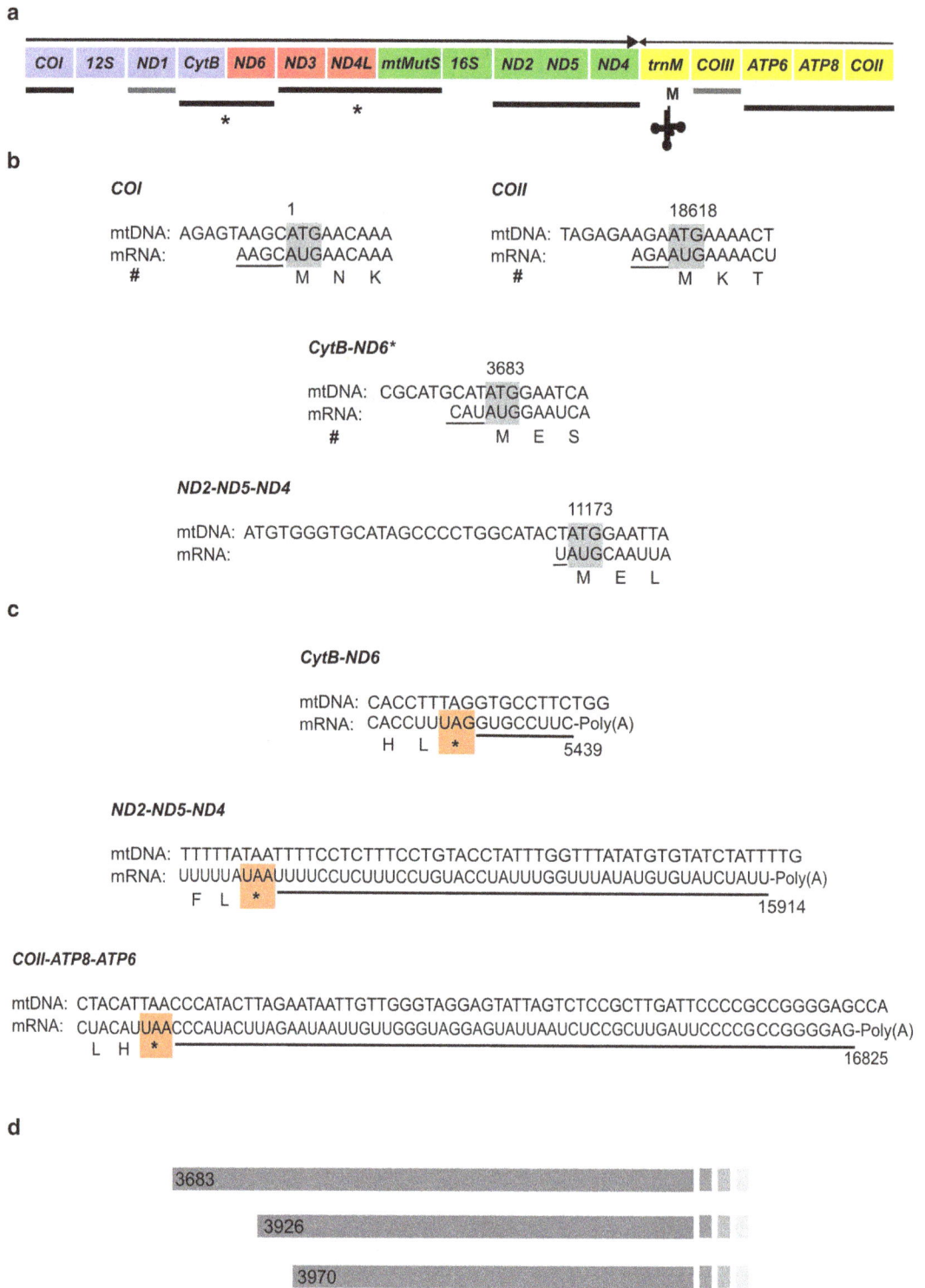

**Fig. 3** Mapped 5′ and 3′ ends of mature mitochondrial mRNAs. **a** Schematic representation of *Sinularia* mito-transcriptome arrangement. *Arrows above* show transcription orientation. *Lines below* denote the transcription units (mono- and polycistronic transcripts). *Black lines* below indicate transcription units for which one or both the ends are known, whereas *grey lines* (below *ND1* and *COIII*) indicate units for which ends remained unknown. *Asterisk* shows the transcription units for which alternate ends were detected. **b** Summary of 5′ end mapping for mt-mtRNAs. The 5′ UTR regions are *underlined*. *Shaded boxes* depict start codons. *Hash* indicates detection using both, RACE and cRT-PCR methods. Nucleotide positions of the first base of the start codons are indicated. **c** Summary of 3′ end mapping for mt-mtRNAs. The 3′ UTR regions are underlined. *Colored boxes* high-light stop codons. Nucleotide positions of the last base of stop codons are indicated. **d** Alternative starting positions (5′ ends) of *CytB–ND6* mRNA

that this gene is transcribed as a tricistronic unit together with *ND3* and *ND4L* (Fig. 4a, b). Using several *mtMutS*-specific primers we were unable to detect any alternate 5′ end using RACE PCR. However, using circularized RT-PCR an internal *mtMutS* 5′ end was detected 387 bp downstream (position 6542) of the annotated *mtMutS* start codon. The 3′ end for this particular transcript was at a position 8950, which is 188 bp upstream to the annotated stop codon. This transcript ended with UAA and is clearly followed by a polyA-tail. However, this stop codon is not in frame.

Somewhat surprisingly, at least six different 3′ ends were detected for the *mtMutS* gene mRNA using RACE and cRT-PCR. These were found to end at position 6746, 6771, 6911, 8761, 8950 (as described above), and 8977 besides the annotated in-frame stop codon at position-9135 (Fig. 4c). These messages are polyadenylated and end with GAA or UAA. Notably, none of these end-codons are in-frame. RT-PCR screening confirmed the presence of all possible transcript variants described above (see Additional file 4: Table S2 for primer details).

Further confirmation of alternative polyadenylation (APA) of *mtMutS* messengers in the normal mt-mRNA pool was attained using RHAPA (RNase H alternative polyadenylation assay).

The *mtMutS* transcripts were differentially expressed. The central region, which includes a partial domain III of *mtMutS,* was 6.35 ± 0.3-fold more abundant than the 3′ end of the gene whereas the expression of the transcript containing the 5′ part of the *mtMutS* mRNA was 1.8 ± 0.15-fold higher than that of the extreme 3′end of the transcript (Fig. 5). This observation suggests the existence of different variants of the same gene under normal conditions in the mature mt-mRNA pool of *Sinularia* cf. *cruciata* (For details on the primers used for this assay see Additional file 4: Table S2 and Additional file 5 for results).

### Antisense *ATP6* mRNA

Using 5′ RACE we detected antisense mRNA transcripts complementary to the *ATP6.* Five different starting points were determined for these transcripts (i.e. positions

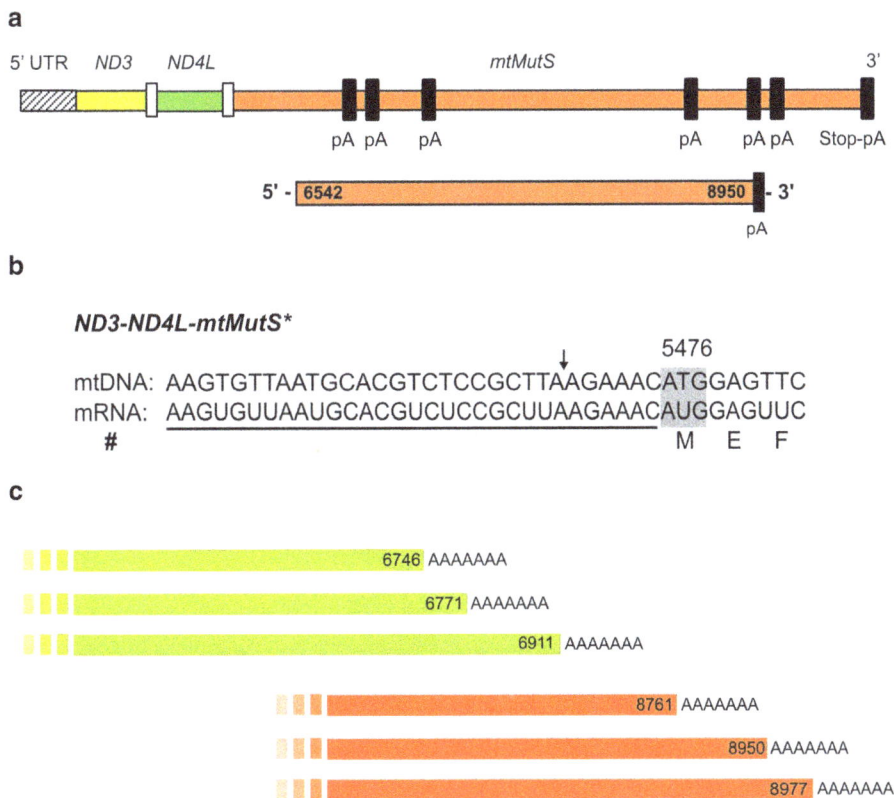

**Fig. 4** The *mtMutS* mRNA transcripts. **a** Schematic of the *mtMutS* gene as a tricistronic transcription unit with different poly(A) tail positions (not to scale) shown as *dark blocks*. pA = poly(A)-tail. Below is the internal *mtMutS* transcript. **b** The 5′ end of *ND3–ND4L–mtMutS* tricistronic transcript. The 5′ UTR region is *underlined*. *Shaded box* indicate the start codon. The *arrow above* indicates 11 bp deletion in *S.* cf. *cruciata* compared to *S. piculiaris*. *Hash* indicates detection using both, RACE and cRT-PCR methods. **c** Alternatively polyadenylated *mtMutS* mRNAs; the position of the poly(A) start is indicated

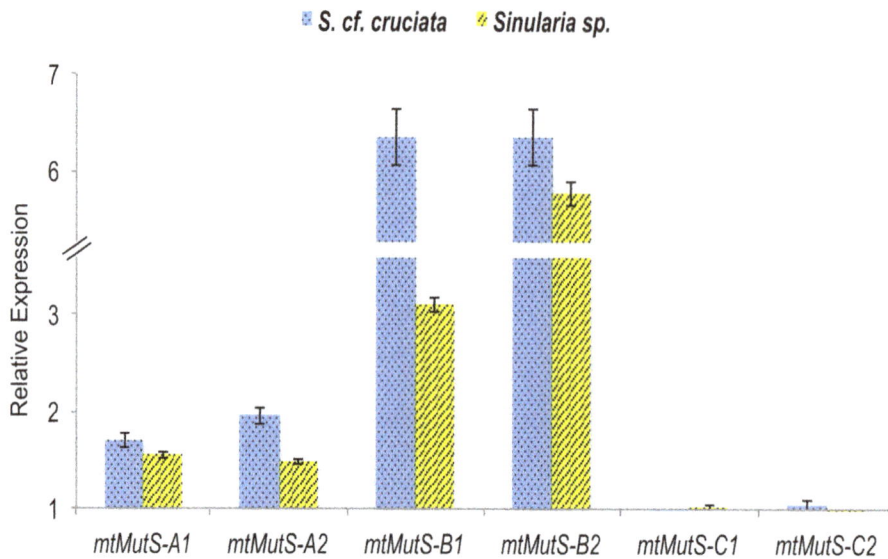

**Fig. 5** Relative quantification of alternate *mtMutS* mRNA transcript abundance. Quantification of alternatively polyadenylated *mtMutS* transcripts. The *ACTB* gene served as reference. The comparison was performed against the 3′ region cleaved-off after RNase H digestion (*mtMutS*-C1, C2) (see Additional file 4: Table S2 for primer details). Data shows relative expression ±SD of technical triplicates for two *Sinularia* species

16,837, 16,866, 16,870, 16,891, and 16,920), which were longer than 200 bp (reverse primer used binds the antisense messenger at position 17,048), polyadenylated (i.e. reverse transcribed using anchored oligo-dT), and lacked open reading frames (ORFs). Therefore, this RNA species can be categorized as long noncoding RNA (lncRNA) [43]. Antisense strand-specific RT-PCR of an internal region of these lnc*ATP6* transcripts further corroborated their presence (see Additional file 6).

## Discussion

Using different experimental approaches we describe the mitogenome expression patterns of an early branching animal. The precise ends of most mature mt-mRNAs were assessed for the first time in octocorals. Most mature protein-coding mRNAs detected were bicistronic or tricistronic units, with the exception of *COI*, *ND1* and *COIII*. The occurrence of polycistronic mature mRNAs potentially stems from the paucity of tRNA punctuation marks in the octocoral mitogenome. The majority of mature transcription units were found to possess 5′ and 3′ UTRs, contrary to what is known for bilaterians in general. Moreover, the occurrence of APA sites of the *mtMutS* mRNA transcript and long non-coding antisense *ATP6* RNA (lnc*ATP6*) provide a glimpse into a unique and potentially complex mitochondrial transcription mechanism in octocorals and likely in other early branching metazoans with non-canonical mitogenomes.

The evolution of specific mechanisms for the expression of the mitogenome was necessitated by the reduced gene

repertoire and compact nature of this crucial cell organelle with its own genome evolved from an α-proteobacterium-like ancestor [44, 45]. In cnidarians, which contrary to a typical animal mitogenome containing 37 (13 PCGs, 2 rRNA and 22 tRNA) genes in total, harbor only 16 ± 2 (13 or 14 PCGs, two rRNAs and one or two tRNA) genes, reduction in gene content, but not genome size, is remarkable to some extent. Our understanding of the mitogenome transcription/expression patterns and regulation is currently limited to a handful of bilaterian members of the animal kingdom [19–21]. However, such attempts are lacking for non-bilaterians. Information on the mitochondrial transcriptomes of sea anemone [46], hydrozoan [28] and jellyfish [29] have shown the potential complexity of mitochondrial transcription and highlighted the need for a better understanding of the evolutionary processes leading to different strategies of mitochondrial transcription and regulation.

The processing of the mt-tRNAs interspersed in the mitogenomes of most animals provide a mechanism to liberate monocistronic protein-coding mRNAs from polycistronic precursors, leading to their maturation and availability for translation [24]. Most studies in animals so far point to the generality of the tRNA punctuation model of mitochondrial mRNA processing with only the occurrence of two bicistronic transcription units (i.e. *ATP8–ATP6* and *ND4L–ND4*) apparently due to the overlap of their ORFs [20, 21, 23]. Cnidarians, however, possess only one or two tRNA genes in their mitogenomes. Our observation of four polycistronic units comprising 11 (out of 14) different PCGs is staggering, and provides evidence for a potentially unique mechanism of mitochondrial mRNA processing, expression and regulation in octocorals. These findings may also apply to other animals exhibiting a paucity of mt-tRNA genes in their mitogenome, for example, chaetognaths [47], some demosponges [48] and other cnidarians [49].

The use of RNA-Seq data provides a unique opportunity to explore mitogenome expression in non-model organisms such as non-bilaterian metazoans. Nevertheless, we observed that very few reads mapped to the published/sequenced mitochondrial genomes (<0.018%). This has been observed before using different NGS methods in sea anemones (0.053% reads) [46], and hydrozoans (<0.003% reads) [28], where very few reads could be identified as mitochondrial. We assembled almost the entire mitogenome of *G. ventalina* with the exception of <3% of its nucleotides mostly belonging to IGRs flanking mature transcription units and to the *mtMutS* gene. In the case of *Corallium* and *Sinularia*, and despite the difference in the number of reads available for mapping in these two species, mapping did not result in a sufficient number of reads to produce reliable contigs. Yet, the presence of read-pairs spanning multiple genes allowed us to postulate the presence of polycistronic mt-mRNAs in these species. Emblem et al. [46] suggested that low mitogenome copy number per haploid nuclear genome results in low-level expression of mitochondrial mRNAs and is responsible for the depleted number of mitochondrial reads observed in three different sea anemones. This may explain our results as well. The poor coverage observed for the *mtMutS* gene remains puzzling. The absence of a full *mtMutS* transcript can be attributed to low coverage in case of *S.* cf. *cruciata*. However, a full *mtMutS* is also absent from the other transcriptomes analyzed that have higher sequencing depths. It should also be noted that there are marked differences in the *mtMutS* coverage despite of similar sequencing depths in these two transcriptomes (relative well-covered with higher expression in *G. ventalina* but not in *C. rubrum*) (see Fig. 2). Large size, the occurrence of APA, and lack of a persistent requirement of the gene product under normal physiological conditions (relative to the genes involved in oxidative phosphorylation) potentially explain the low expression of the *mtMutS* gene under such conditions.

Previous studies on octocorals have proposed the *COII–COI* IGR as potential CR/oriH in the octocorals with mitogenome arrangement 'A' [30–32]. The observed absence of RNA-Seq reads in *G. ventalina* supports this proposition. The presence of very similar stem-loop structure in the *ND6–COI* IGR in *C. rubrum*, along with absence of RNA-Seq reads in this region of this species, suggests that this IGR is the potential CR in octocorals

with 'C' type mitogenome arrangement and corroborates earlier predictions in this regard [32, 42] (Additional file 3).

Untranslated regions (UTRs) flanking the mature mRNA transcripts play a crucial role in the post-transcriptional regulation of gene expression [50]. In mitochondria, mature mRNA transcripts are generally devoid of UTRs, or these are only few (≤3) nucleotides long [23]. The presence of 5′ UTRs for transcriptional units such as *COI*, *ND3–ND4L–mtMutS*, *ND2–ND5–ND4*, and *COII–ATP8–ATP6* mRNA, and of 3′ UTRs for *CytB–ND6*, *ND4*, and *COII–ATP8–ATP6* suggests a putative role of these elements in the regulation of these genes and represent, to our knowledge, the first report of the presence of long UTRs in the mature mt-mRNAs of non-bilaterians.

Different studies have detected at least five mitogenome arrangements in octocorals so far, all of which appear to preserve four conserved gene blocks. The inversion or translocation of one of these blocks at a time is proposed to have led to five different mitogenome arrangements [31, 33]. It has been suggested that the occurrence of the genes in conserved clusters is selectively advantageous, for instance, as the genes can be co-transcribed and processed in a similar way [31, 51]. However, the evidence on selection favoring a particular mitochondria gene order is sparse in cnidarians, as they exhibit high diversity of mitogenome arrangements with no sharing of gene boundaries, particularly in the subclass Hexacorallia [31]. The transcriptional units we detected encompass genes from two distinct adjacent gene blocks (e.g. the polycistronic transcription units *CytB–ND6* and *ND3–ND4L–mtMutS*), contradicting the hypothesis of co-transcription as a selective force in keeping these genes together in conserved blocks in the mitochondria of octocorals. Evidence for selection favoring a particular mitochondrial gene order in cnidarians and other metazoan taxa is scarce as well [31, 52]. Hence, maintenance of synteny within the four conserved gene blocks detected so far appears to result from the lack of recombination-hotspots that promote genome rearrangements within, and/or their prevalence between these conserved gene blocks. Expression studies in future would be required to corroborate this initial observation. However, our results indicate that different mitogenome rearrangements detected in octocorals have different mature mt-mRNA transcript structures and transcriptional patterns (see *Corallium* vs. *Sinularia–Gorgonia* in Fig. 2). This transcriptional diversity highlights the potential complexity of mitochondrial transcription among non-bilaterians.

Assuming the tRNA punctuation model holds true, in three out of five mitogenome arrangements observed in octocorals (i.e. A, C and D), the tRNA[Met] lies at the end of either H- or L-strand and its processing would liberate both coding and non-coding parts of the polycistronic precursor RNA. In *S.* cf. *cruciata*, the simultaneous or sequential processing of rRNAs could provide a mechanism for the liberation of *COI* and *ND2–ND5–ND4*. However, the excision of other transcription units from the polycistronic precursor remains to be explained. Secondary structures such as stem-loops are likely involved in maturation of pre-mRNA in octocorals, as is the case in hydrozoans and other animals [19, 28]. In *S.* cf. *cruciata*, IGRs where excision is required in order to liberate detected transcriptional units (i.e. *ND1–CytB*, *ND6–ND3*, *COIII–ATP6*) form one or more stem-loop structures (Additional file 7A). We pose that the enzymes involved in mRNA maturation recognize the conserved 11 bp motifs (Additional file 7B) present in the IGRs flanking transcription units and cleave them from the precursor to be available for maturation. This motif is absent in the IGRs between individual genes within the above-mentioned transcription units.

The *mtMutS* gene present in octocorals is thought to underpin several peculiar processes not present in other animal mitochondria. The presence of a large gene not involved in energy production within a streamlined organelle genome dedicated to this task is mysterious as well as interesting. The second largest gene in *Sinularia* cf. *cruciata* mitogenome, *ND5* (1818 bp long) is known to be the most tightly regulated protein-coding gene in other animals [53]. Thus we hypothesize that the transcription of *mtMutS* is tightly regulated as well. In favor of this hypothesis, we observed distinct *mtMutS* variants resulting from the use of different internal polyadenylation sites within the *mtMutS* gene. In contrast to its function in plant mitochondria and bacteria, which polyadenylate RNA to promote their degradation [54], polyadenylation is used in mammals to provide stability to the mature mRNA and create the stop codon, if it is not complete [26]. Polyadenylated truncated transcripts destined to degradation have also been detected in mammals [27], but, their abundance is low and they are generally difficult to detect using standard methods. All the *mtMutS* variants reported here were readily detectable indicating their potential functional role.

Interestingly, the alternate *mtMutS* transcripts were differentially expressed with a transcript variant encompassing Domain III and V of *mtMutS* (position 6542–8950) being more abundant under normal conditions than either the 5′ or 3′ end regions of the gene. These domains are either structurally important (e.g. Domain III) or have important biochemical functionality (e.g. ATPase; Domain V) [35, 55]. APA plays a crucial role in regulating gene expression [56]. Hence, APA of the *mtMutS* gene may have a regulatory function allowing a tight control of the expression of this gene in octocoral mitochondria.

The use of the RHAPA method avoided a limitation of Northern blotting that has a tendency to detect unprocessed transcripts and/or degradation intermediates. Specifically, transcripts having partial *mtMutS* continued into the 16S were readily detected with cRT-PCR, and also visible in the *G. ventalina* and *C. rubrum* transcriptome data (Fig. 1). In addition, RHAPA also proved helpful in determining the relative abundances of APA transcripts when coupled with qPCR. However, this resulted in lack of information on sizes of each of the alternatively polyadenylated transcripts. Attempts to obtain size information on each type of alternatively polyadenylated *mtMutS* transcripts using a common 5′ primer and oligo(dT) led to the detection of false alternate transcripts of different sizes with apparent artificial deletions in the sequence. This appears to be caused by the presence of direct repeats in this fairly long gene, coupled with the template-switching property of the reverse transcriptase (RT) used for cDNA synthesis, a phenomenon reported earlier by Cocquet et al. [57]. Additional efforts involving target RNA capture methods [58] would be needed to determine the sizes of individual alternate *mtMutS* transcripts in the future. The precise start and end points of each *mtMutS* mRNA variant deserve to be determined to better understand how the start–stop codons are chosen during translation. Additionally, protein studies need to be conducted in order to corroborate the localization and functionality of these transcripts and their products.

Long noncoding RNAs (lncRNAs) have been recently described in the mitochondria of mammals, primarily for *ND5*, *ND6* and *CytB*, and have been shown to interact with their mRNA complements stabilizing them and/or blocking the access of mitochondrial ribosomes, thereby inhibiting translation [59]. The presence of an lncRNA transcript for *ATP6* (lnc*ATP6*) in *Sinularia* mitochondria is striking, and indicates that the regulation of mitochondrial expression using lncRNAs evolved early in Metazoa and is ancient.

More than 99% of the mitochondrial proteome is encoded by the nuclear genome [60]. The loss of mt-tRNAs in cnidarians is suggested to have occurred in association with the loss of nuclear-encoded mt-aminoacyl-tRNA synthetases [61] and indicates a greater nuclear dependency in cnidarians relative to other animals. The retention of a single mt-tRNA for formyl-methionine is interesting and likely reflects its very specific mitochondrial function or the necessity of an excision starting point that triggers the mt-mRNA maturation cascade. Our findings, together with the general paucity of tRNAs and the varied mitogenome rearrangements observed in octocorals indicate a highly complex and perhaps a unique system for mRNA processing in the mitochondria of these organisms.

## Conclusions

Recent studies on the human mitochondrial transcriptome revealed an unexpected complexity in expression, processing, and regulation of mt-mRNAs [19, 62]. Our results shed first light on the potentially more complex nature of these processes in the mitochondria of early branching animals by virtue of their "special" and diverse mitogenomes. Overall, due to the lack of tRNA punctuation marks, mitochondrial mRNA processing in octocorals appears to be drastically different. The presence of polycistronic mature mRNAs for the majority of genes provides evidence for the complexity of the transcription process in these animals. The occurrence of alternately polyadenylated transcripts for the *mtMutS* gene and their differential expression, the existence of 5′ and 3′ UTRs, and the presence of lnc*ATP6* transcripts are additional features highlighting the diverse set of post-transcriptional modifications and regulatory mechanisms used among octocorals. More research will contribute to better understand the mitochondrial biology of early branching animals from a functional perspective. This will certainly increase our knowledge on the innovations that shaped the evolution of these organisms.

## Methods

### Specimens

Coral colonies were obtained from a commercial source and maintained in a closed circuit seawater aquarium at Molecular Geo- and Palaeobiology lab, LMU, Munich. Two species of the genus *Sinularia* were used in this study. *Sinularia* cf. *cruciata* (Lab Voucher Code: GW1725) was utilized for the majority of the experiments, whereas additionally, a second *Sinularia* sp. (Lab Voucher Code: GW2911) was used for RT-PCR screening, RHAPA and antisense mRNA detection (see below). All the references to the nucleotide positions refer to the full mitochondrial genome of *Sinularia* cf. *cruciata* (GenBank Accession: KY462727), which was sequenced completely (details are provided in Additional file 1).

### Total RNA extraction and cDNA synthesis

TRIzol reagent (Invitrogen, USA) was utilized for the extraction of total RNA as per the manufacture's instructions. RNA was dissolved in 100 μl DEPC treated water and contaminating DNA was eliminated from RNA extracts by performing a DNase (RQ1 RNase-free DNase, Promega, USA) treatment at 37 °C for 30 min. Treated RNA was purified after inactivation of the DNase and its purity was determined using a Nanodrop ND-1000 spectrophotometer (Thermo Fisher Scientific, USA). RNA samples with absorbance at OD260/280 and

OD260/230 ratios ~2.0 were used for further analysis. RNA integrity was also verified by 1% agarose gel electrophoresis as well as using a Bioanalyzer (Agilent Inc.). RNA extracts with a RIN value ≥7.5 were used for cDNA synthesis (data not shown); these extracts were stored at −80 °C until use.

### RNA-Seq and read mapping

RNA-Seq reads for *Gorgonia ventalina* (SRR935078–SRR935087) and *Corallium rubrum* (SRR1552943–SRR1552945 and SRR1553369) were downloaded from NCBI's Short Read Archive, imported in Geneious and mapped against the mitochondrial genomes of *Pseudopterogorgia bipinnata* (NC_008157), in the case of *G. ventalina*, or *Corallium rubrum* (AB700136). In the case of *Sinularia* cf. *cruciata* ca. $26 \times 10^6$ 50 bp pairs of reads were sequenced, imported in Geneious® 8.1.8 (Biomatters) [63] and mapped to the mitochondrial genome we sequenced for this species. The mapping was done using a low sensitivity strategy that avoids remapping reads to previously build contigs based on the previous mapping rounds. The mapping results were screened in Geneious to find gaps in coverage flanking putative transcription units. Additionally, we screened the mapped reads to assess whether read pairs spanning adjacent genes could be found among the sequences. These reads, i.e. read-pairs linking adjacent genes and spanning intergenic regions, were taken as evidence of collinearity.

### Reverse-transcription PCR (RT-PCR)

RNA extracts were PCR controlled in order to detect amplifiable levels of small DNA fragments. Only RNA extracts devoid of any amplification were used in RT-PCR experiments. For each sample, ~1 μg of total RNA was reverse transcribed in 20 μl reactions using the ProtoScript® II First Strand cDNA Synthesis Kit (New England Biolabs, USA) with an anchored oligo-(dT) primer and following the manufacture's instructions.

RT-PCR sequencing primers were designed using the sequenced mitochondrial genomic sequence of *Sinularia* cf. *cruciata* (GenBank Accession: KY462727). Screening for the presence of the polycistronic mRNAs was done using the primers enlisted in Table 1.

### Analysis of 5′ and 3′ ends

Two different approaches were used to determine the transcript ends of mature mitochondrial mRNA species.

#### Circularized RT-PCR (cRT-PCR)

Isolation of mRNA from total RNA was performed using Dynabeads® mRNA magnetic beads (Invitrogen). 100 ng of polyA-selected mRNA as well as total RNA were circularized using T4 RNA ligase I (New England Biolabs) following the manufacture's protocol. The circularized RNA was purified and used for cRT-PCR and 5'/3' end screening using the method described earlier [64]. Briefly, cDNA synthesis was performed as described above using gene-specific reverse primers binding near the 5' end of linear RNA resulting in production of first strand that contains 5' end, the ligation site, and the 3' end of the molecule. These first strands were subjected to PCR amplification using specific primer pairs (see Table 2 for primer details).

### Rapid amplification of cDNA ends (5' and 3' RACE)

First strand synthesis was performed to obtain the template for 5' and 3' RACE PCRs using SMART™ RACE cDNA Amplification Kit (CLONETECH Inc.) following the supplier's protocol. Approximately ~1 μg of total RNA was used to obtain two separate pools of 5'-RACE-Ready cDNA and 3'-RACE-Ready cDNA. RACE PCR reactions were performed using different gene-specific primers paired with adaptors primers as per the supplier's instructions (see Table 2).

### Cloning, sequencing and sequence analysis of amplified products

Amplified products were either extracted from 1% agarose gels or purified using the NucleoSpin Gel and PCR Purification Kit (MACHEREY–NAGEL, Germany) and cloned using a TOPO TA Cloning Kit (Invitrogen, USA). The clones obtained were PCR amplified, precipitated and sequenced using ABI BigDye v3.1 sequencing chemistry on an ABI 3730 DNA Analyzer Sequencing instrument. Sequences obtained were analyzed and aligned to the mitogenomes of *S. piculiaris* (NC_018379) and the sequenced *S.* cf. *cruciata* (Accession No. KY462727) using Geneious® 6.1.6 software (Biomatters) [63].

### Detection and quantification of alternative polyadenylation (APA)

RNase H alternative polyadenylation assay or RHAPA [65] was employed to determine and quantify alternative polyadenylation (APA) of the *mtMutS* mRNA transcripts. For the first time, we coupled this assay with quantitative real-time PCR technique (qPCR),

### Table 2  Primers used in this study

| Gene | 5' RACE | Sequence | 3' RACE | Sequence |
|---|---|---|---|---|
| A. Primers used in 5' and 3' RACE amplification | | | | |
| COI | C1-839[a] | ATCATAGCATAGACCATACC | | |
|  | C1-1144 | CATAGTGGAAGTGAGCTACTAC | | |
| CytB–ND6 | CB-4128[a] | GCTCCCCAAAAGGACATTTGTC | | |
| ND3–ND4–mtMutS | N3-5602 | CACATTCATAGACCGACACTT | N4L-6071 | GCCATTATGGTTAACTATTAC |
|  | M-6402 | ACGAAGCAACTTGTTCAATGG | M-6363 | ATTGAACAAGTTGCYTCGTTACTTG |
|  | M-6719 | CCGGGTTACTTTGTCCCTGTCCG | M-8714 | GCCCTCTCAATATGGCATTG |
|  | | | M-6655 | CAGCCATGAATGGGCATAG |
|  | | | M-8067 | GCATTAAGCGGGGCTATTGCGG |
| COII | C2-18519 | CCATAACAGGACTAGCAGCATC | | |

| Gene | 3' Facing | Sequence | 5' Facing | Sequence |
|---|---|---|---|---|
| B. Primers used in cRT-PCR | | | | |
| COI | C1-1324 | TACTCGGATTTCCCTGATGC | C1-251 | AACCAATTTCCGAATCCTCCG |
| CytB–ND6 | N6-4962 | TTTGGTTAGTTATTGCCTTT | CB-4128[a] | GCTCCCCAAAAGGACATTTGTC |
| ND3–ND4–mtMutS | M-8624 | TGATTCGCCAGTTCGGTGCT | M-6780[a] | TTAAACCTACCCCCGAGTCC |
| ND2–ND5 | N5-14119 | GCTCAGTTTGGAAGTTTGGC | N2-11753[a] | ACATCGGGAGCCCACATA |
| ND4 | N4-15714 | TTTTGGGCAACTTTCTCC | N4-14532 | CAGAGACCACTCTAACGCTTGTTG |
| COII–ATP8–ATP6 | C2-18161 | GGTTGAAGGTCACTCGTAGGTATC | A6-17048 | GGGTTCGCAATGATTAGTAATGGAATGT |

| Gene | | | Primer | Sequence |
|---|---|---|---|---|
| C. Gene-specific primers used for RT for cRT-PCR analysis | | | | |
| ATP6 | | | A6-17092 | TTAGCAGCCAATCGAACACC |
| ND4 | | | N4-15295 | GCGTCTACCTGTCTGCAAGT |
| mtMutS | | | M-8826 | CATTTCGGGATGGTAGCTCC |

[a] Indicates the other primers used for RT during cRT-PCR

which allows for the accurate estimation of the abundance of the alternative transcripts. The oligonucleotide 5′-CATTTCGGGATGGTAGCTCC-3′ was used to remove the 3′ end of the polyadenylated complete *mtMutS* messenger. This primer hybridizes to the *mtMutS* mRNA between positions 8807 and 8826. After hybridization, the DNA–RNA hybrids were digested with RNase H and the resulting mRNA was purified using RNA Clean & Concentrator™ kit (Zymo Research) and reverse transcribed using oligo(dT) as described above. Only alternatively polyadenylated *mtMutS* forms should be present in the resulting cDNA after RNase H digestion. A control RT-PCR using primers binding to the adjacent regions of the RNase H digested site ensured the successful digestion of the 3′ including and poly(A) tail of the mature, full *mtMutS* mRNA species. Afterwards, a quantitative real-time PCR (qPCR) assay was performed to determine the abundance-levels of the transcripts using primers binding upstream and downstream of digested mRNA region. For primer details see Additional file 4: Table 2.

### Strand-specific RT-PCR
Strand-specific RT-PCR was performed as described previously [66] for the detection of antisense RNA transcripts of the *ATP6* gene in both *Sinularia* species. In earlier, RACE experiment, we observed *ATP6* clones containing a RACE adaptor ligated at 3′end of 5′ RACE-ready cDNA. This suggested the presence of an antisense transcript for this gene, as it has also been noted previously in porcine brain [67]. Two antisense strand-specific primers were used for cDNA synthesis (AAR1. 5′-TTACTCCTACTGCCCATATTG-3′ and AAR2. 5′-TGTAGTTCGGATAATTGGGGG-3′), whereas a sense strand-specific primer (SAF. 5′-TTAGCAGC-CAATCGAACACC-3′) and an anchored oligo(dT) were employed separately for first strand synthesis. For RT-PCR, the primer AAR1-SAF pair was used.

### Quantitative Real-time RT-PCR (qPCR) and data analysis
The Rotor-Gene Q 2plex system (QIAGEN) was utilized for qPCR experiment. The KAPA SYBR FAST universal mastermix (Peqlab) was used in 15 µl reactions containing 1 µl diluted cDNA, 7.5 µl 2X mastermix, and 250–400 nM of each primer. A two-step qPCR including an initial denaturation step of 3 min at 95 °C followed by 40 cycles of 95 °C for 10 s and 60 °C for 20 s was performed. A non-template control was always included in each assay and melting curve analysis was performed at the end of each qPCR to confirm amplification specificity. In addition, amplification products were also checked by agarose gel electrophoresis after each assay.

Fluorescence data obtained after qPCR was analyzed using LinRegPCR, which determines Cq values and PCR efficiencies [68]. These values were further used to analyze mitochondrial expression; statistical tests were performed using REST 2009 (QIAGEN) as described previously [69].

### Additional files

**Additional file 1.** Mitochondrial genome of *Sinularia* cf. *cruciata*.

**Additional file 2.** Mitochondrial gene expression determined by RNA-Seq for *Sinularia* cf. *cruciata*, *Gorgonia ventalina*, and *Corallium rubrum*.

**Additional file 3.** Stem-loop structures and conserved motifs of putative control regions from octocorals mitogenomes studied.

**Additional file 4.** Additional tables (1, 2).

**Additional file 5.** RHAPA analysis.

**Additional file 6.** Antisense strand-specific RT-PCR.

**Additional file 7.** Stem-loop structures and conserved motif of IGRs between detected transcriptional units from *S*. cf. *cruciata*.

### Abbreviations
*COI*: cytochrome oxidase, subunit I; *COII*: cytochrome oxidase, subunit II; *ATP6*: ATP synthase, subunit 6; *ATP8*: ATP synthase, subunit 8; *ND5*: NADH dehydrogenase, subunit 5; *ND4*: NADH dehydrogenase, subunit 4; *ND4L*: NADH dehydrogenase, subunit 4L; *ND6*: NADH dehydrogenase, subunit 6; *CytB*: cytochrome b; *ND1*: NADH dehydrogenase, subunit 1; *ND2*: NADH dehydrogenase, subunit 2; *ND3*: NADH dehydrogenase, subunit 3; *COIII*: cytochrome oxidase, subunit III; *12S*: small subunit of mitochondrial ribosomal RNA gene; *16S*: large subunit of mitochondrial ribosomal RNA gene.

### Authors' contributions
GGS, SV, and GW conceived and designed the experiments; GGS performed the experiments and analyzed the data; AP helped in bridging the mitogenome sequence gaps and analyzing the genome; GGS drafted the manuscript; and SV and GW helped with its revision. All authors read and approved the final manuscript.

### Author details
[1] Department of Earth and Environmental Sciences, Palaeontology & Geobiology, Ludwig-Maximilians-Universität München, Richard-Wagner Str. 10, 80333 Munich, Germany. [2] GeoBio-Center, Ludwig-Maximilians-Universität München (LMU), Richard-Wagner Str. 10, Munich 80333, Germany. [3] SNSB–Bavarian State Collections of Palaeontology and Geology, Richard-Wagner Str. 10, 80333 Munich, Germany. [4] Present Address: Marine Biotechnology and Ecology (MBE) Division, CSIR-Central Salt and Marine Chemicals Research Institute (CSMCRI), Bhavnagar, Gujarat 364002, India.

### Acknowledgements
We are thankful to the editor, the reviewers of the journal, and two anonymous reviewers of Axios Review for insightful feedback on the manuscript. We would like to acknowledge Gabi Büttner and Simone Schätzle for their assistance in the laboratory, Dr. Dirk Erpenbeck and Dr. Oliver Voigt for their constructive discussions, and Dr. Peter Naumann's for his assistance in the aquarium. SV is indebted to M. Vargas Villalobos, S. Vargas Villalobos, S. Vargas Villalobos and N. Villalobos Trigueros for their constant support.

### Competing interests
The authors declare that they have no competing interests.

**Funding**
The research was funded by DAAD (German Academic Exchange Service) Ph.D. scholarship awarded to GGS. (91540344).

**References**
1. GIGA Community of Scientists. The global invertebrate genomics alliance (GIGA): developing community resources to study diverse invertebrate genomes. J Hered. 2014;105(1):1–18.
2. Bhattacharya D, Agrawal S, Aranda M, Baumgarten S, Belcaid M, Drake JL, Erwin D, Foret S, Gates RD, Gruber DF, et al. Comparative genomics explains the evolutionary success of reef-forming corals. Elife. 2016;5:e13288.
3. Lavrov DV. Key transitions in animal evolution: a mitochondrial DNA perspective. Integr Comp Biol. 2007;47:734–43.
4. Berg OG, Kurland CG. Why mitochondrial genes are most often found in nuclei. Mol Biol Evol. 2000;17:951–61.
5. Adams KL, Palmer JD. Evolution of mitochondrial gene content: gene loss and transfer to the nucleus. Mol Phylogenet Evol. 2003;29(3):380–95.
6. Boore JL. Animal mitochondrial genomes. Nucleic Acids Res. 1999;27(8):1767–80.
7. Lavrov DV, Pett W. Animal mitochondrial DNA as we don't know it: mt-genome organization and evolution in non-bilaterian lineages. Genome Biol Evol. 2016;8(9):2896–913.
8. Rot C, Goldfarb I, Ilan M, Huchon D. Putative cross-kingdom horizontal gene transfer in sponge (Porifera) mitochondria. BMC Evol Biol. 2006;6:71.
9. Erpenbeck D, Aryasari R, Hooper JN, Wörheide G. A mitochondrial intron in a verongid sponge. J Mol Evol. 2015;80(1):13–7.
10. Szitenberg A, Rot C, Ilan M, Huchon D. Diversity of sponge mitochondrial introns revealed by cox 1 sequences of Tetillidae. BMC Evol Biol. 2010;10:288.
11. van Oppen MJ, Catmull J, McDonald BJ, Hislop NR, Hagerman PJ, Miller DJ. The mitochondrial genome of Acropora tenuis (Cnidaria; Scleractinia) contains a large group I intron and a candidate control region. J Mol Evol. 2002;55(1):1–13.
12. Huchon D, Szitenberg A, Shefer S, Ilan M, Feldstein T. Mitochondrial group I and group II introns in the sponge orders Agelasida and Axinellida. BMC Evol Biol. 2015;15:278.
13. Pont-Kingdon GA, Okada NA, Macfarlane JL, Beagley CT, Wolstenholme DR, Cavalier-Smith T, Clark-Walker GD. A coral mitochondrial mutS gene. Nature. 1995;375:109–11.
14. Flot JF, Tillier S. The mitochondrial genome of Pocillopora (Cnidaria: Scleractinia) contains two variable regions: the putative D-loop and a novel ORF of unknown function. Gene. 2007;401(1–2):80–7.
15. Park E, Song JI, Won YJ. The complete mitochondrial genome of Calicogorgia granulosa (Anthozoa: Octocorallia): potential gene novelty in unidentified ORFs formed by repeat expansion and segmental duplication. Gene. 2011;486(1–2):81–7.
16. Lin MF, Kitahara MV, Luo H, Tracey D, Geller J, Fukami H, Miller DJ, Chen CA. Mitochondrial genome rearrangements in the Scleractinia/Corallimorpharia complex: implications for coral phylogeny. Genome Biol Evol. 2014;6(5):1086–95.
17. Lavrov DV, Pett W, Voigt O, Wörheide G, Forget L, Lang BF, Kayal E. Mitochondrial DNA of Clathrina clathrus (Calcarea, Calcinea): six linear chromosomes, fragmented rRNAs, tRNA editing, and a novel genetic code. Mol Biol Evol. 2013;30(4):865–80.
18. Voigt O, Erpenbeck D, Wörheide G. A fragmented metazoan organellar genome: the two mitochondrial chromosomes of Hydra magnipapillata. BMC genom. 2008;9:350.
19. Mercer TR, Neph S, Dinger ME, Crawford J, Smith MA, Shearwood AM, Haugen E, Bracken CP, Rackham O, Stamatoyannopoulos JA, et al. The human mitochondrial transcriptome. Cell. 2011;146(4):645–58.
20. Coucheron DH, Nymark M, Breines R, Karlsen BO, Andreassen M, Jørgensen TE, Moum T, Johansen SD. Characterization of mitochondrial mRNAs in codfish reveals unique features compared to mammals. Curr Genet. 2011;57:213–22.
21. Stewart JB, Beckenbach AT. Characterization of mature mitochondrial transcripts in Drosophila, and the implications for the tRNA punctuation model in arthropods. Gene. 2009;445(1–2):49–57.
22. Markova S, Filipi K, Searle JB, Kotlik P. Mapping 3' transcript ends in the bank vole (Clethrionomys glareolus) mitochondrial genome with RNA-Seq. BMC Genom. 2015;16:870.
23. Temperley RJ, Wydro M, Lightowlers RN, Chrzanowska-Lightowlers ZM. Human mitochondrial mRNAs-like members of all families, similar but different. Biochim Biophys Acta. 2010;1797(6–7):1081–5.
24. Ojala D, Montoya J, Attardi G. tRNA punctuation model of RNA processing in human mitochondria. Nature. 1981;290(5806):470–4.
25. Rossmanith W. Of P and Z: mitochondrial tRNA processing enzymes. Biochim Biophys Acta. 2012;1819(9–10):1017–26.
26. Nagaike T, Suzuki T, Ueda T. Polyadenylation in mammalian mitochondria: insights from recent studies. Biochim Biophys Acta. 2008;1779(4):266–9.
27. Slomovic S, Laufer D, Geiger D, Schuster G. Polyadenylation and degradation of human mitochondrial RNA: the prokaryotic past leaves its mark. Mol Cell Biol. 2005;25(15):6427–35.
28. Kayal E, Bentlage B, Cartwright P, Yanagihara AA, Lindsay DJ, Hopcroft RR, Collins AG. Phylogenetic analysis of higher-level relationships within Hydroidolina (Cnidaria: Hydrozoa) using mitochondrial genome data and insight into their mitochondrial transcription. PeerJ. 2015;3:e1403.
29. Kayal E, Bentlage B, Collins AG. Insights into the transcriptional and translational mechanisms of linear organellar chromosomes in the box jellyfish Alatina alata (Cnidaria: Medusozoa: Cubozoa). RNA Biol. 2016;13:1–11.
30. Brugler MR, France SC. The mitochondrial genome of a deep-sea bamboo coral (Cnidaria, Anthozoa, Octocorallia, Isididae): genome structure and putative origins of replication are not conserved among octocorals. J Mol Evol. 2008;67(2):125–36.
31. Brockman SA, McFadden CS. The mitochondrial genome of Paraminabea aldersladei (Cnidaria: Anthozoa: Octocorallia) supports intramolecular recombination as the primary mechanism of gene rearrangement in octocoral mitochondrial genomes. Genome biol evol. 2012;4:994–1006.
32. Uda K, Komeda Y, Koyama H, Koga K, Fujita T, Iwasaki N, Suzuki T. Complete mitochondrial genomes of two Japanese precious corals, Paracorallium japonicum and Corallium konojoi (Cnidaria, Octocorallia, Coralliidae): notable differences in gene arrangement. Gene. 2011;476:27–37.
33. Figueroa DF, Baco AR. Octocoral mitochondrial genomes provide insights into the phylogenetic history of gene order rearrangements, order reversals, and cnidarian phylogenetics. Genome Biol Evol. 2015;7(1):391–409.
34. Pont-Kingdon G, Okada NA, Macfarlane JL, Beagley CT, Watkins-Sims CD, Cavalier-Smith T, Clark-Walker GD, Wolstenholme DR. Mitochondrial DNA of the coral Sarcophyton glaucum contains a gene for a homologue of bacterial MutS: a possible case of gene transfer from the nucleus to the mitochondrion. J Mol Evol. 1998;46:419–31.
35. Bilewitch JP, Degnan SM. A unique horizontal gene transfer event has provided the octocoral mitochondrial genome with an active mismatch repair gene that has potential for an unusual self-contained function. BMC Evol Biol. 2011;11:228.
36. Shearer TL, Van Oppen MJH, Romano SL, Wörheide G. Slow mitochondrial DNA sequence evolution in the Anthozoa (Cnidaria). Mol Ecol. 2002;11:2475–87.
37. Hellberg ME. No variation and low synonymous substitution rates in coral mtDNA despite high nuclear variation. BMC Evol Biol. 2006;6:24.
38. Lynch M, Koskella B, Schaack S. Mutation pressure and the evolution of organelle genomic architecture. Science (New York). 2006;311:1727–30.
39. McFadden CS, Benayahu Y, Pante E, Thoma JN, Nevarez PA, France SC. Limitations of mitochondrial gene barcoding in Octocorallia. Mol ecol resour. 2011;11:19–31.
40. Burge CA, Mouchka ME, Harvell CD, Roberts S. Immune response of the Caribbean sea fan, Gorgonia ventalina, exposed to an Aplanochytrium parasite as revealed by transcriptome sequencing. Front Physiol. 2013;4:180.
41. Pratlong M, Haguenauer A, Chabrol O, Klopp C, Pontarotti P, Aurelle D. The red coral (Corallium rubrum) transcriptome: a new resource for population genetics and local adaptation studies. Mol Ecol Resour. 2015;15(5):1205–15.
42. Uda K, Komeda Y, Fujita T, Iwasaki N, Bavestrello G, Giovine M, Cattaneo-Vietti R, Suzuki T. Complete mitochondrial genomes of the Japanese pink coral (Corallium elatius) and the Mediterranean red coral (Corallium rubrum): a reevaluation of the phylogeny of the family Coralliidae based on molecular data. Comp Biochem Physiol Part D Genomics Proteom. 2013;8(3):209–19.
43. Rinn JL, Chang HY. Genome regulation by long noncoding RNAs. Annu Rev Biochem. 2012;81:145–66.

44. Gray MW. Origin and evolution of organelle genomes. Curr Opin Genet Dev. 1993;3(6):884–90.

45. Dyall SD, Brown MT, Johnson PJ. Ancient invasions: from endosymbionts to organelles. Science. 2004;304(5668):253–7.

46. Emblem A, Okkenhaug S, Weiss ES, Denver DR, Karlsen BO, Moum T, Johansen SD. Sea anemones possess dynamic mitogenome structures. Mol Phylogenet Evol. 2014;75:184–93.

47. Helfenbein KG, Fourcade HM, Vanjani RG, Boore JL. The mitochondrial genome of *Paraspadella gotoi* is highly reduced and reveals that chaetognaths are a sister group to protostomes. Proc Natl Acad Sci USA. 2004;101(29):10639–43.

48. Wang X, Lavrov DV. Seventeen new complete mtDNA sequences reveal extensive mitochondrial genome evolution within the Demospongiae. PLoS ONE. 2008;3:e2723.

49. Beagley CT, Okimoto R, Wolstenholme DR. The mitochondrial genome of the sea anemone *Metridium senile* (Cnidaria): introns, a paucity of tRNA genes, and a near-standard genetic code. Genetics. 1998;148(3):1091–108.

50. Mignone F, Gissi C, Liuni S, Pesole G. Untranslated regions of mRNAs. Genome Biol. 2002;3(3):REVIEWS0004.

51. Dowton M, Campbell NJ. Intramitochondrial recombination—is it why some mitochondrial genes sleep around? Trends Ecol Evol. 2001;16(6):269–71.

52. Dowton M, Cameron SL, Dowavic JI, Austin AD, Whiting MF. Characterization of 67 mitochondrial tRNA gene rearrangements in the hymenoptera Suggests that mitochondrial tRNA gene position is selectively neutral. Mol Biol Evol. 2009;26(7):1607–17.

53. Bai Y, Shakeley RM, Attardi G. Tight control of respiration by NADH dehydrogenase ND5 subunit gene expression in mouse mitochondria. Mol Cell Biol. 2000;20(3):805–15.

54. Mohanty BK, Kushner SR. Bacterial/archaeal/organellar polyadenylation. Wiley Interdiscip Rev RNA. 2011;2(2):256–76.

55. Obmolova G, Ban C, Hsieh P, Yang W. Crystal structures of mismatch repair protein MutS and its complex with a substrate DNA. Nature. 2000;407(6805):703–10.

56. Lutz CS, Moreira A. Alternative mRNA polyadenylation in eukaryotes: an effective regulator of gene expression. Wiley Interdiscip Rev RNA. 2011;2(1):22–31.

57. Cocquet J, Chong A, Zhang G, Veitia RA. Reverse transcriptase template switching and false alternative transcripts. Genomics. 2006;88:127–31.

58. Mercer TR, Gerhardt DJ, Dinger ME, Crawford J, Trapnell C, Jeddeloh JA, Mattick JS, Rinn JL. Targeted RNA sequencing reveals the deep complexity of the human transcriptome. Nat Biotechnol. 2012;30(1):U99–147.

59. Rackham O, Shearwood AM, Mercer TR, Davies SM, Mattick JS, Filipovska A. Long noncoding RNAs are generated from the mitochondrial genome and regulated by nuclear-encoded proteins. RNA. 2011;17(12):2085–93.

60. Salinas-Giege T, Giege R, Giege P. tRNA biology in mitochondria. Int J Mol Sci. 2015;16(3):4518–59.

61. Haen KM, Pett W, Lavrov DV. Parallel loss of nuclear-encoded mitochondrial aminoacyl-tRNA synthetases and mtDNA-encoded tRNAs in Cnidaria. Mol Biol Evol. 2010;27(10):2216–9.

62. Rackham O, Mercer TR, Filipovska A. The human mitochondrial transcriptome and the RNA-binding proteins that regulate its expression. Wiley Interdiscip Rev RNA. 2012;3(5):675–95.

63. Kearse M, Moir R, Wilson A, Stones-Havas S, Cheung M, Sturrock S, Buxton S, Cooper A, Markowitz S, Duran C, et al. Geneious basic: an integrated and extendable desktop software platform for the organization and analysis of sequence data. Bioinformatics. 2012;28(12):1647–9.

64. Slomovic S, Schuster G. Circularized RT-PCR (cRT-PCR): analysis of the 5′ ends, 3′ ends, and poly(A) tails of RNA. Methods Enzymol. 2013;530:227–51.

65. Cornett AL, Lutz CS. RHAPA: a new method to quantify alternative polyadenylation. Methods Mol Biol. 2014;1125:157–67.

66. Ho EC, Donaldson ME, Saville BJ. Detection of antisense RNA transcripts by strand-specific RT-PCR. Methods Mol Biol. 2010;630:125–38.

67. Michel U, Stringaris AK, Nau R, Rieckmann P. Differential expression of sense and antisense transcripts of the mitochondrial DNA region coding for ATPase 6 in fetal and adult porcine brain: identification of novel unusually assembled mitochondrial RNAs. Biochem Biophys Res Commun. 2000;271(1):170–80.

68. Ruijter JM, Ramakers C, Hoogaars WMH, Karlen Y, Bakker O, van den Hoff MJB, Moorman AFM. Amplification efficiency: linking baseline and bias in the analysis of quantitative PCR data. Nucleic Acids Res. 2009;37:e45.

69. Pfaffl MW, Horgan GW, Dempfle L. Relative expression software tool (REST) for group-wise comparison and statistical analysis of relative expression results in real-time PCR. Nucleic Acids Res. 2002;30(9):e36.

# 10

# Myelodysplasia-associated mutations in serine/arginine-rich splicing factor SRSF2 lead to alternative splicing of CDC25C

Lindsey Skrdlant[1], Jeremy M. Stark[2] and Ren-Jang Lin[1]* ⓘ

## Abstract

**Background:** Serine–arginine rich splicing factor 2 (SRSF2) is a protein known for its role in RNA splicing and genome stability. It has been recently discovered that SRSF2, along with other splicing regulators, is frequently mutated in patients with myelodysplastic syndrome (MDS). The most common MDS mutations in SRSF2 occur at proline 95; the mutant proteins are shown to have different RNA binding preferences, which may contribute to splicing changes detected in mutant cells. However, the influence of these SRSF2 MDS-associated mutations on specific splicing events remains poorly understood.

**Results:** A tetracycline-inducible TF-1 erythroleukemia cell line was transduced with retroviruses to create cell lines expressing HA-tagged wildtype SRSF2, SRSF2 with proline 95 point mutations found in MDS, or SRSF2 with a deletion of one of the four major domains of the protein. Effects of these mutants on apoptosis and specific alternative splicing events were evaluated. Cells were also treated with DNA damaging drugs for comparison. MDS-related P95 point mutants of SRSF2 were expressed and phosphorylated at similar levels as wildtype SRSF2. However, cells expressing mutant SRSF2 exhibited higher levels of apoptosis than cells expressing wildtype SRSF2. Regarding alternative splicing events, in nearly all examined cases, SRSF2 P95 mutants acted in a similar fashion as the wildtype SRSF2. However, cells expressing SRSF2 P95 mutants had a percent increase in the C5 spliced isoform of cell division cycle 25C (CDC25C). The same alternative splicing of CDC25C was detected by treating cells with DNA damaging drugs, such as cisplatin, camptothecin, and trichostatin A at appropriate dosage. However, unlike DNA damaging drugs, SRSF2 P95 mutants did not activate the Ataxia telangiectasia mutated (ATM) pathway.

**Conclusion:** SRSF2 P95 mutants lead to alternative splicing of CDC25C in a manner that is not dependent on the DNA damage response.

**Keywords:** SRSF2, Myelodysplastic syndromes, RNA splicing, CDC25C, DNA damage response

## Background

Serine–arginine rich splicing factor 2 (SRSF2), previously named SC35, is a member of the SR protein family of splicing regulators. The primary role of SR proteins is to regulate splice site selection for both constitutive and alternative splicing. In addition to their role in splicing, SR proteins are also involved in the maintenance of genome stability through the prevention of R-loop structure formation during transcription [1–4]. Structurally, SRSF2 can be divided into specific protein domains (Fig. 1a). The RNA recognition motif (RRM) is involved in binding to exonic or intronic splicing enhancers in the nascent pre-mRNA transcript. The eponymous serine–arginine rich domain (RS) is primarily involved in protein–protein interactions with other spliceosomal components and with the 7SK complex involved in transcriptional elongation [3, 5, 6]. In addition, serine phosphorylation at the RS domain significantly affects the subnuclear localization and functions of SRSF2 [7–11].

*Correspondence: RLin@coh.org
[1] Department of Molecular and Cellular Biology, Irell & Manella Graduate School of Biological Sciences, Beckman Research Institute of the City of Hope, Duarte, CA 91010, USA
Full list of author information is available at the end of the article

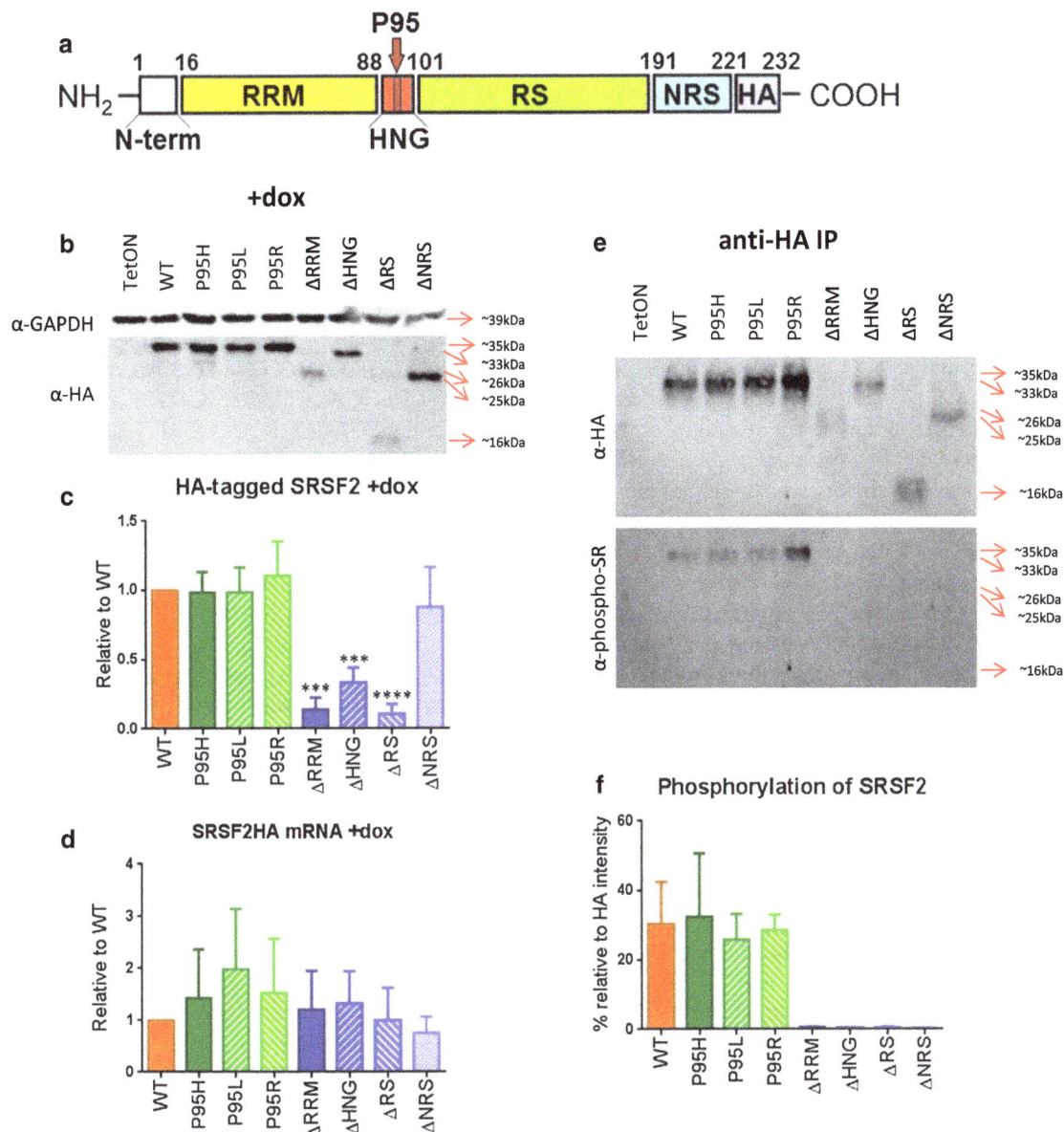

**Fig. 1** Expression and phosphorylation of SRSF2 P95 and deletion mutants. **a** Schematic of SRSF2HA ORF showing the primary protein domains removed to create SRSF2 deletion mutants. *RRM* RNA recognition motif; *HNG* hinge region; *RS* arginine/serine-rich domain; *NRS* nuclear retention signal; *HA* hemagglutinin tag. Numbers denote amino acids defining the domain boundaries within the full length protein. Location of the P95 amino acid within the hinge region is denoted with a *large red arrow*. **b** Western blot of protein isolated from cell lines after 48 h doxycycline induction. **c** Average HA-tagged protein expression in induced cell lines, normalized to GAPDH expression. Statistics are based on one-way analysis of variance (ANOVA) with comparison to SRSF2[WT] cell line (n = 5). **d** qRT-PCR analysis of HA-tagged SRSF2 mRNA expression, normalized to SRSF2[WT]. No statistical significance using one-way ANOVA with comparison to SRSF2[WT] (n = 4). **e** Western blot using anti-HA and anti- phospho-SR antibodies after immunoprecipitation of HA-tagged SRSF2 protein. **f** Analysis of phospho-SR signal normalized to HA-tagged protein signal. Statistics use one-way ANOVA with comparison to SRSF2[WT] (n = 3). *p* values: *≤0.05; **≤0.01; ***≤0.001; ****≤0.0001

The short amino acid sequence between the RRM and RS domains is the hinge region (HNG). The exact function of this region is largely unknown, though studies with SRSF1 (ASF/SF2) suggest that it may have an important role in kinase docking [12]. The last major domain of SRSF2 is the nuclear retention signal (NRS) at the C-terminal end of the protein, which is unique to SRSF2. Deletion of this NRS allows SRSF2 to shuttle between the nucleus and cytoplasm, similar to the shuttling function of other SR proteins that aid in mRNA export [9].

It was recently discovered that SRSF2 is mutated in 10–15 % of patients with Myelodysplastic syndrome (MDS) and 25–30 % of patients with chronic myelomonocytic leukemia (CMML) [13–19]. Both of these diseases are aging-associated hematopoietic disorders that occur primarily in individuals over the age of 60 [20]. The only effective long term treatment for either disease is a bone marrow transplant, which is often not possible to perform due to both the age of the patients and a high relapse rate in patients with advanced disease [21–24]. While the cause of these disorders is still unknown, alternative splicing in genes related to hematopoiesis and cell cycle regulation, such as CDC25C and RUNX1, have been found in patients with MDS or AML [25, 26]. For both of these diseases, patients with SRSF2 mutation have a miscoding of the proline at position 95 (P95) to a histidine, arginine, or leucine during the early stages of the disease. These mutations persist throughout the disease [18, 27]. Recent research has also shown that P95 mutations of SRSF2 affect the ability of SRSF2 to bind its canonical splicing enhancer sequences in RNA [28, 29]. In addition, the P95H mutation of SRSF2 can increase death of hematopoietic cells and cause changes in hematopoiesis [28, 30]. Results from these studies support the notion that these mutations of SRSF2 are involved in MDS pathogenesis. However, the mechanism for how these mutations lead to disease development is still unknown.

We have constructed stable cell lines expressing from a Tet-inducible promoter HA-tagged wildtype SRSF2 (SRSF2$^{WT}$), HA-tagged SRSF2 with point mutations found in patients with MDS (SRSF2$^{P95H}$, SRSF2$^{P95L}$, and SRSF2$^{P95R}$), and HA-tagged SRSF2 with in-frame deletions of each of the four major domains of the protein (SRSF2$^{\Delta RRM}$ = deletion of the RNA recognition motif, SRSF2$^{\Delta HNG}$ = deletion of the hinge region, SRSF2$^{\Delta RS}$ = deletion of the arginine/serine-rich domain, and SRSF2$^{\Delta NRS}$ = deletion of the nuclear retention signal) in TF-1 erythroleukemia cells (Fig. 1a). Our data showed that while the SRSF2$^{P95R/L/H}$ mutations did not affect cellular localization of the protein, they did increase early apoptosis and affect the alternative splicing of CDC25C towards a shorter isoform (CDC25C-C5) that has previously been shown to be upregulated when DNA is damaged in breast cancer cells exposing to sub-lethal levels of doxorubicin and cisplatin [31]. Interestingly, we found SRSF2 mutant induced alternative splicing of CDC25C does not require activation of the DNA damage response pathway that is activated with cisplatin treatment.

## Methods
### Plasmid construction
The pRevTRE-SC35HA (SRSF2HA) tet-inducible plasmid was a gift from Xiang-Dong Fu's lab at UCSD

[2]. Mutations of P95 were produced using site-directed mutagenesis with the QuikChange Kit (Agilent) with the following primers: SRSF2-P95X sense (5′-CTACGGCCGCCDCCCGGACTCAC-3′) and SRSF2-P95X antisense (5′-GTGAGTCCGGGHGGCGGCCGTAG-3′), where D is A, T, and G and H is A, T, and C [32]. SRSF2 deletion mutants were produced using the InFusion cloning system (Clontech) and primers that overlapped the deletion sites: ΔRRM F1 (ATGTGGAGGGTATGACCTCCATGGCGCGCTACGGC), ΔRRM R1 (GCCGTAGCGCGCCATGGAGGTCATACCCTCCACAT), ΔHNG F1 (CGAGCTGCGGGTGCAAAGCCGCCGGGGACCA)ΔHNGR(GGTCCCCGGCGGCTTTGCACCCGCAGCTCG), ΔRS F1 (CCCGGACTCACACCACCCTCCCCCAGTGTCCA), ΔRS R1 (TGGACACTGGGGGAGGGTGGTGTGAGTCCGGG),ΔNRS F1 (GGTCTCGGTCCAGGAGTCTCGAGTACCCATACGACG), ΔNRS R1 (CGTCGTATGGGTACTCGAGACTCCTGGACCGAGACC). SRSF2HA gene sequence was confirmed using Sanger sequencing. The lentiviral plasmid for expression of the reverse Tet transactivator protein (pHIV7-rtTA-V15), the plasmids for lentiviral packaging (pC-GP, pCMV-rev, pCMV-G), and the plasmids for retrovirus packaging (pCMV-GP and pCMV-G) were gifts from the lab of Jiing-Kwan Yee at City of Hope.

### Production and transduction of lentivirus and retrovirus
For the rtTA lentivirus, 293T cells were transfected with pHIV7-rtTA, pC-GP-2, pCMV-rev2, and pCMV-G using calcium phosphate [32]. After 48 h, cell supernatant was collected, filtered through a 0.45 μm filter, and precipitated in a 10 % PEG solution overnight at 4 °C. Virus was then concentrated by centrifuging the PEG solution 30 min at $2000 \times g$. Supernatant was decanted and virus pellet was resuspended in the remaining solution. Virus was tittered on HT1080 cells by G418 resistance. TF-1 cells were transduced with rtTA lentivirus at an MOI of 0.003 for 48 h and then diluted to single cells and plated in 96-well plates for G418 selection. Single clones were collected and assayed for doxycyline induction potential. Clone #6 was chosen as the TF-1 TetON parental cell line.

For the SRSF2HA retroviruses, 293T cells were transfected with the respective pRevTRE-SRSF2HA plasmid, pCMV-GP, and pCMV-G sing calcium phosphate transfection. After 48 h, virus was collected as above. Virus was tittered on HT1080 cells by hygromycin B selection. TF-1 TetON cells were transduced with resulting retroviruses at an MOI of 0.1 for 48 h. Each group was then selected for 2 weeks using hygromycin B to produce the TF-1 TetON SRSF2HA cell lines.

## cDNA synthesis and RT-PCR

$1 \times 10^6$ cells were pelleted and resuspended in 100 µl Trizol. After 5 min incubation, 100 µl 100 % EtOH was added and RNA was isolated using Direct-zol kit (Zymo Research). 1 µg RNA was used for cDNA synthesis using either SuperScript III RT (Invitrogen) or PrimeScript RT (Clontech). Each RT-PCR reaction contained 10 µl 2× Taq (BioPioneer), 0.5 µM each primer, and 1 µl cDNA. Primer list and reaction conditions can be found in Additional file 1: Table S1.

## Western Blot Analysis

$1 \times 10^6$ cells were washed 1× with PBS and then resuspended in 50 µl dH20. 50 µl 2× Laemmli buffer, prewarmed to 85 °C, was added to each sample. Samples were incubated 10 min at 95 °C. Samples were then aliquoted and flash frozen in liquid nitrogen to be stored at −80 °C. For SDS-PAGE, samples were diluted 3.5-fold using 1× Laemmli buffer, and 7 µl was loaded per lane. Following SDS-PAGE, proteins were transferred to PVDF membrane. The membranes were blocked for 1 h at RT with either 5 % (w/v) nonfat dry milk in TBST (50 mM Tris–Cl, pH 7.5. 150 mM NaCl, 0.1 % Triton X-100) or 3 % (w/v) BSA in TBST. Membranes were washed 3× with TBST. Primary antibodies [rabbit anti-HA tag (Abcam; ab9110; 1:10,000), mouse anti-GAPDH (Invitrogen; ZG003; 1:5000), mouse anti-SR proteins [1H4] (Santa Cruz Biotech; sc-13,509; 1:200), rabbit anti-CDC25C [E303] (GeneTex; GTX61135; 1:500), rabbit anti-phospho CHK2 (Cell Signaling Technology; 2661; 1:2000), and mouse anti-phospho p53 [16G8] (Cell Signaling Technology; 9286; 1:500)] were diluted as stated in TBST and incubated overnight at 4 °C. Membranes were washed 3× with TBST. Secondary antibodies [goat anti-rabbit IgG-IRDye 680LT (Li-Cor; 926-68021; 1:20,000), goat anti-mouse IgG-IRDye 800CW (Li-Cor; 926-32,210; 1:20,000), goat anti-rabbit-HRP (BioRad; 170–6515; 1:3000), and goat anti-mouse-HRP (BioRad; 170–6516; 1:3000)] were diluted as stated in TBST and incubated with the membranes for 4 h at RT. Membranes were washed 3×, 5 min each wash, and subsequently scanned with an Odyssey machine (for Li-Cor secondaries) or exposed to X-ray film (for HRP secondaries).

## Immunoprecipitation

$1 \times 10^7$ cells were pelleted for 5 min at 400×g. Supernatant was removed and cells were resuspended in a buffer containing 50 mM HEPES (pH 7.8), 3 mM MgCl$_2$, 300 mM NaCl, 1 mM DTT, 0.1 mM PMSF, 1X Halt™ Protease Inhibitor Cocktail (Thermoscientific), 5 mM NaF, 2 mM NaVO$_3$, 0.5U/µl DNase I, and 5U/µl benzonase. Samples were incubated at 4 °C for 1 h with constant rotation, after which samples were centrifuged for 3 min

at 4 °C, 10,000×g. Supernatant was added to 50 µl anti-HA agarose beads (Sigma) and total volume was brought to 200 µl with TBS. Samples were incubated overnight at 4 °C with constant end-over-end rotation. Agarose beads were pelleted for 10 s at 16,000g, and supernatant was removed as the unbound fraction. Pellet was washed 5× with TBST. Samples were eluted by incubating agarose beads with 50 µl 1 mg/ml HA peptide (ThermoFisher Scientific) at 30 °C for 15 min. Eluate was assayed by Western blot.

## Immunofluoresecence

Glass coverslips (Fisher Scientific; 22 × 22–1.5) were incubated with 0.01 % poly-L-lysine (Sigma) for 10 min. Poly-L-lysine solution was removed and coverslips were air-dried. $2 \times 10^5$ cells were incubated on coverslips 1 h at 37 °C. Coverslips were washed 2× with PBS. Cells were fixed by incubating coverslips with 4 % (w/v) paraformaldehyde in PBS for 20 min at RT. Coverslips were then washed 2× with PBS and incubated 15 min at RT with 0.1 % (v/v) Triton X-100 in PBS. Cells were blocked by incubating coverslips overnight at 4 °C in 10 % (v/v) FBS in PBS. Coverslips were incubated with primary antibodies [rabbit anti-HA tag (Abcam; ab9110), mouse anti-SC35 (Abcam; ab11826), and mouse anti-γH2AX (ab11174)] each diluted 1:100 in the blocking solution for 1 h at RT. Coverslips were washed 5× with PBS and then incubated 1 h at RT with secondary antibodies [goat anti-rabbit IgG-AF647 (Jackson ImmunoResearch; 111-605-003) and goat anti-mouse IgG-AF488 (Jackson ImmunoResearch; 115-545-003)] diluted 1:100 in the blocking solution. Coverslips were then washed 5× with PBS and incubated 10 min at RT with 1× CellMask Orange (LifeTechnologies). Coverslips were then washed 3× with PBS and mounted on glass slides using VectaShield Hardset Mounting media with DAPI (Vector Laboratories). Slides were incubated overnight at 4 °C before imaged using a Zeiss LSM 700 confocal microscope. Analysis for HA-tagged protein localization was done using Image-Pro Plus (Media Cybernetics). The whole cell area was outlined in the Cell Mask Orange channel image and copied into the HA protein channel image. The nucleus was outlined in the DAPI channel image and also copied into the HA protein channel image. The integrated optical density (IOD) was analyzed independently for the whole cell and nucleus. The IOD for the cytoplasm alone was calculated by subtracting the IOD for the nucleus from the IOD of the whole cell.

## Apoptosis assay

$1 \times 10^5$ cells were washed 1× with PBS. Cells were stained for 15 min using FITC-Annexin V and PI from FITC Annexin V Apoptosis Kit I (BD Pharmigen).

Samples were then analyzed on a Cyan FACS machine. Cells were considered to be apoptotic based upon positive FITC detection. The stage of apoptosis was determined by the level of PI staining of the FITC-positive cells. Early apoptosis lacked any PI staining. Middle apoptosis had positive, albeit low levels of, PI staining. Late apoptosis was defined by high levels of PI staining.

### Cell proliferation assay

$5 \times 10^6$ cells were stained on day 0 in 10 µM CFSE (carboxyfluorescein succinimidyl ester, Thermo Fisher Scientific) for 15 min at 37 °C. Each day including day 0, half of the cell culture was removed from the flask, washed 2× with PBS, fixed for 15 min in 4 % (w/v) paraformaldehyde in water, washed 2× with PBS, and stored at 4 °C in the dark. After all time points for a single biological replicate were collected, samples were analyzed on a Cyan FACS machine. Data are plotted by using CFSE log mean.

### Cell cycle analysis

$1 \times 10^6$ cells were washed 2× with PBS. Cells were resuspended in 500 µl ice-cold 80 % EtOH and incubated overnight at 4 °C. Cells were washed 2× with PBS, and resuspended in 0.25 % Triton X-100, 20 µg/ml PI (propidium iodide), and 0.1 ng/ml RNase A. Cells were incubated overnight at 4 °C. Samples were analyzed on a Cyan FACS machine. Data were analyzed using the cell cycle tool in FlowJo v6.

## Results
### Characterization of SRSF2-HA-expressing cell lines

We started our study by engineering a TF-1 TetON cell line. The TF-1 cell line was chosen due to its classification as an erythroblast, allowing for inducible expression of an HA-tagged SRSF2 in a cell type similar to those commonly affected in MDS patients in order to examine phenotypic changes that may occur as a result of SRSF2 mutations. Furthermore, by PCR and Sanger sequencing, none of the common MDS-related mutations in *U2AF1(S34)*, *SF3B1(K700)*, or *SRSF2(P95)* were detected in the parental TF-1 cell line (data not shown).

We used site directed mutagenesis to produce eight different SRSF2 constructs: wildtype (SRSF2$^{WT}$), each of the three MDS- and CMML-related point mutations (SRSF2$^{P95H}$, SRSF2$^{P95L}$, and SRSF2$^{P95R}$), and four deletions for each of the primary domains of SRSF2 for comparison (SRSF2$^{\Delta RRM}$, SRSF2$^{\Delta HNG}$, SRSF2$^{\Delta RS}$, and SRSF2$^{\Delta NRS}$) (Fig. 1a).

In order to assay for transgene expression, each of the SRSF2 cell lines (WT, P95H, P95L, P95R, ΔRRM, ΔHNG, ΔRS, and ΔNRS) and the parental TF-1 TetON cell line were cultured for 48 h, both with and without induction using 2 µg/ml doxycycline. After 48 h, protein and RNA were isolated from each line. Western blotting using anti-HA antibodies detected the expression of HA-tagged proteins with the expected size for each construct upon doxycycline induction (Fig. 1b). A small amount of the HA-tagged protein was expressed from the minimal CMV promoter without doxycycline induction (Additional file 2: Figure S1). The expression levels for SRSF2$^{\Delta RRM}$, SRSF2$^{\Delta HNG}$, and SRSF2$^{\Delta RS}$ were noticeably lower than others (Fig. 1b, c). qRT-PCR of the ectopically expressed SRSF2-HA mRNAs showed no significant differences in the RNA level between the WT and mutants (Fig. 1d). Therefore, it appears that a deletion of the RRM, HNG, or RS domain negatively affects protein stability and/or efficient translation.

Since the hinge region of SRSF1 is required for kinase docking [12], we wanted to determine whether the disease-related P95 mutations of SRSF2, which occur in the hinge region, would affect the phosphorylation level of SRSF2. To this end, we immunoprecipitated the HA-tagged SRSF2 using an antibody against the HA tag. Western blots of the immunoprecipitated samples showed that SRSF2$^{P95H}$, SRSF2$^{P95L}$, and SRSF2$^{P95R}$ were phosphorylated to the same level as SRSF2$^{WT}$, as measured by anti-phospho-SR antibody (1H4 monoclonal antibody) while none of the deletion mutants showed detectable levels of phosphorylation (Fig. 1e, f).

As stated previously, unlike other SR proteins that are shuttled between the nucleus and the cytoplasm, SRSF2 is confined to the nucleus [9]. To examine whether any of the mutations would affect the cellular localization of SRSF2, we stained the tet-induced cells with anti-HA antibodies. Immunofluorescence results from the anti-HA staining showed that SRSF2$^{WT}$, SRSF2$^{P95H}$, SRSF2$^{P95L}$, SRSF2$^{P95R}$, SRSF2$^{\Delta RRM}$, and SRSF2$^{\Delta HNG}$ were all confined to the nucleus (Fig. 2a; Additional file 3: Figure S2a). SRSF2$^{\Delta NRS}$ was located in both the nucleus and cytoplasm in agreement with previous studies showing that deletion of the NRS allows SRSF2 to shuttle between the nucleus and cytoplasm [9]. Unexpectedly, SRSF2$^{\Delta RS}$ was also present in both the nucleus and cytoplasm when it was expected to be located only in the cytoplasm due to loss of the primary nuclear localization signal [9, 33–35]. It is possible that the remaining presence of the NRS allows for partial nuclear localization. We also stained each of the cell lines with an anti-SC35 antibody that specifically stains the nuclear phosphorylated SC35 speckles [33, 36]. Previous research has shown that the phosphorylation state determines whether SRSF2/SC35 is localized within nuclear speckles [10]; therefore the pan-nuclear staining of HA-tagged SRSF2 indicated that a substantial portion of the HA-tagged protein is not in a phosphorylation state reserved for speckle localization. We also observed nuclear SC35-speckles with

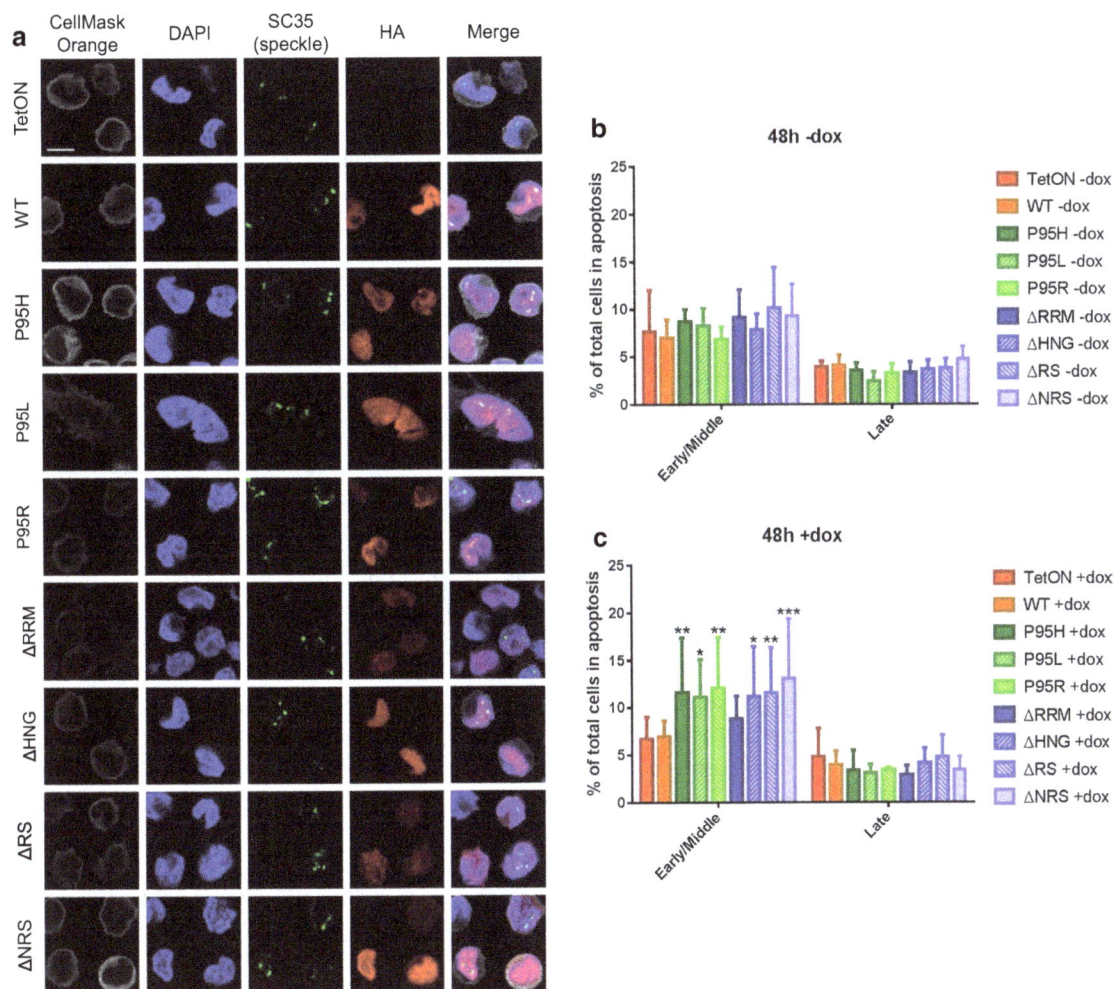

**Fig. 2** Apoptosis and subcellular localization of P95 and deletion SRSF2 mutants. **a** Representative immunofluorescence images of SRSF2HA cell lines (n = 3). CellMask Orange stained cytoplasm, DAPI stained DNA in the nucleus, anti-SC35 speckle marker stained SC35/SRSF2 in the nuclear speckles, anti-HA stained HA-tagged SRSF2 protein. **b** Average percentage of cells in early or late apoptosis at 48 h in the absence of doxycycline induction. **c** Average percentage of cells in early or late apoptosis at 48 h post-doxycycline induction. Statistics use two-way ANOVA with comparison to TF-1 TetON (n = 4). *Asterisks* are *p* values as in Fig. 1

similar sizes and numbers among the cell lines, indicating the expression of the SRSF2-HA proteins did not significantly alter the formation or the dynamics of nuclear speckles (Fig. 2a).

Factors that can initiate an apoptotic signaling cascade, such as Fas, Fas-L, tumor necrosis factor-a, and tumor necrosis factor-related apoptosis initiating ligand, are upregulated in bone marrow cells of MDS patients [37, 38], and it is thought that this increase in apoptosis in the bone marrow may contribute to the cytopenias that are a hallmark of MDS [39]. To assay our cell lines for changes in apoptosis, FITC-Annexin V and propidium iodide (PI) staining were used to determine apoptosis levels in each of the TF-1 cell lines (Additional file 3: Figure S2b).

Without doxycycline induction, there was no significant difference in apoptosis levels between any of the cell lines (Fig. 2b). However, with the doxycycline-induced expression of any of the three P95 mutant SRSF2 transgenes, cells showed a significant increase in early/middle apoptosis as compared to the SRSF$^{WT}$ and parental TetON cell lines (Fig. 2c). This result suggests that SRSF2 P95 mutations may be pro-apoptotic. We also noticed that SRSF2$^{\Delta HNG}$, SRSF2$^{\Delta RS}$, and SRSF2$^{\Delta NRS}$ expressing cells also showed an increase in early/middle apoptosis as compared to the control cell lines, whereas SRSF2$^{\Delta RRM}$ showed no effect on apoptosis. We have also measured cell proliferation using CFSE staining. Upon staining, CFSE covalently binds intracellular molecules such as proteins, and as the cells divide the existing CFSE-bound proteins are redistributed to the daughter cells resulted in decreasing CFSE staining per cell. Thus, cell proliferation can be accurately measured by monitoring CFSE staining in cell culture [40]. By using CFSE staining, we did not observe significant change of cell proliferation with any of the SRSF2 mutants (Additional file 3: Figure S2c). Thus, increased apoptosis in SRSF2$^{P95H/L/R}$ is consistent with increased apoptosis of MDS cells.

### Alternative splicing changes

SRSF2 is known to regulate alternative splicing of a number of pre-mRNAs including its own [41]. There are six splice variants of SRSF2 described in the database, and four of them were detected by RT-PCR in TF-1 cells using the primers depicted (Fig. 3a, b). Among these, splice isoforms v3, v4, and v6 are substrates of nonsense-mediated decay (NMD) because they all have at least one intron that is more than 50 nucleotides downstream from the stop codon (Fig. 3a) [42]. The splice variant 5 (v5) has an intron 7 nucleotides downstream from the stop codon, which does not activate NMD, and is the predominant RNA species (Fig. 3b). As previously shown [41], SRSF2$^{WT}$ overexpression promoted inclusion of alternative exons and resulted in an increase in isoform 4 expression as detected by RT-PCR (Fig. 3b, c). Similarly, expression of a P95 point mutant (SRSF2$^{P95H}$, SRSF2$^{P95L}$, or SRSF2$^{P95R}$) also led to an increase of v4, indicating that these P95 mutations of SRSF2 do not affect its ability to auto-regulate its exon inclusion. On the contrary, expression of the RRM, RS, or HNG deletion mutants did not cause a significant increase of v4, while expression of SRSF2$^{\Delta NRS}$ did (Fig. 3c). This supports previous findings that deletion of the NRS does not affect the alternative splicing function of SRSF2, whereas both the RRM and RS domains are required for this function [9, 33, 43–45].

We next investigated whether any of the SRSF2 mutants would alter splicing of other mRNAs that have been previously shown to be affected by SRSF2. To begin with, E2F1-mediated overexpression of SRSF2 has been shown to promote alternative splicing of a set of apoptosis-related genes to their pro-apoptotic form [46]. We assayed alternative splicing of BCL-X, caspase-8, and caspase-9 by RT-PCR. At 48 h post-induction, we did not observe significant changes in alternative splicing in any of the SRSF2-expressing TF-1 cell lines (Additional file 4: Figure S3). Next, we assayed splice variants of BAP1 and TRA2A, which are found to change in mouse fibroblasts upon SRSF2 knock out [47]. We observed no significant splicing changes in these genes with doxycycline-induced overexpression of SRSF2$^{WT}$ or any of the SRF2 mutants in TF-1 cells (Additional file 5: Figure S4).

We then analyzed the alternative splicing of CDC25C, a dual-specific phosphatase critical for the G2/M checkpoint pathway of the cell cycle. CDC25C has previously been found to be alternatively spliced in the bone marrow of patients with MDS [25]. In addition, CDC25C is part of the common deleted region in 5q- syndrome, the only subtype of MDS that has a defined cytogenetic cause and effective chemotherapy treatment [48–51]. Furthermore, haploinsufficiency of CDC25C is required for effective lenalidomide treatment of patients with 5q- syndrome [52]. RT-PCR analysis of CDC25C revealed a distinctive change in the levels of the C5 isoform, which lacks exons 3, 5, and 6, relative to C1 among the SRSF2 mutants. Specifically, expression of the MDS-associated point mutants (SRSF2$^{P95H}$, SRSF2$^{P95L}$, and SRSF2$^{P95R}$) each caused an increase in the CDC25C C5/C1 ratio (Fig. 4b, c). In contrast, expression of SRSF2$^{WT}$ or the deletion mutants did not affect this ratio.

We then considered that this alternative splicing change may cause expression of a truncated CDC25C protein. Western blotting using an anti-CDC25C antibody, specific for the C-terminus of the protein, detected the ~60 kDa CDC25C C1 protein and an increased expression of a smaller protein isoform that corresponds to the expected ~45 kDa CDC25C C5 protein (Fig. 4d, e). This result indicated that expression of SRSF2$^{P95H}$, SRSF2$^{P95R}$, or SRSF2$^{P95L}$ causes a splicing change of CDC25C mRNA, whereas expression of the SRSF2 deletion mutants or the wild-type protein does not produce the alternative splicing change.

### CDC25C alternative splicing and the DNA damage response

The elevated ratio of CDC25C C5/C1 detected in our TF-1 SRSF2-P95 mutants, has also been reported to be induced in breast cancer cells by treatment with sublethal doses of DNA-damaging agents, such as cisplatin (CIS) and doxorubicin [31]. To test whether DNA damaging agents would also alter alternative splicing of CDC25C in hematopoietic cells like TF-1, we treated TF-1 cells with various concentrations of CIS. As shown in Fig. 5a, b the ratio of C5/C1 indeed increased when cells were treated with 40–50 μM of CIS for 12 h. CIS forms DNA crosslinks that can cause double-stranded breaks that activate the DNA damage response via the ATM pathway [53–55]. We also tested other drugs that induce the ATM DNA damage response pathway: camptothecin (CPT), a topoisomerase I poison that causes chromosomal breaks [56–58], and trichostatin A (TSA), a histone deacetylase inhibitor [59, 60]. Similar to the effects observed with CIS treatment, CPT at 20–200 nM for 6 h or TSA at 5–20 μM for 12 h also caused an increase of the C5 splice variant of

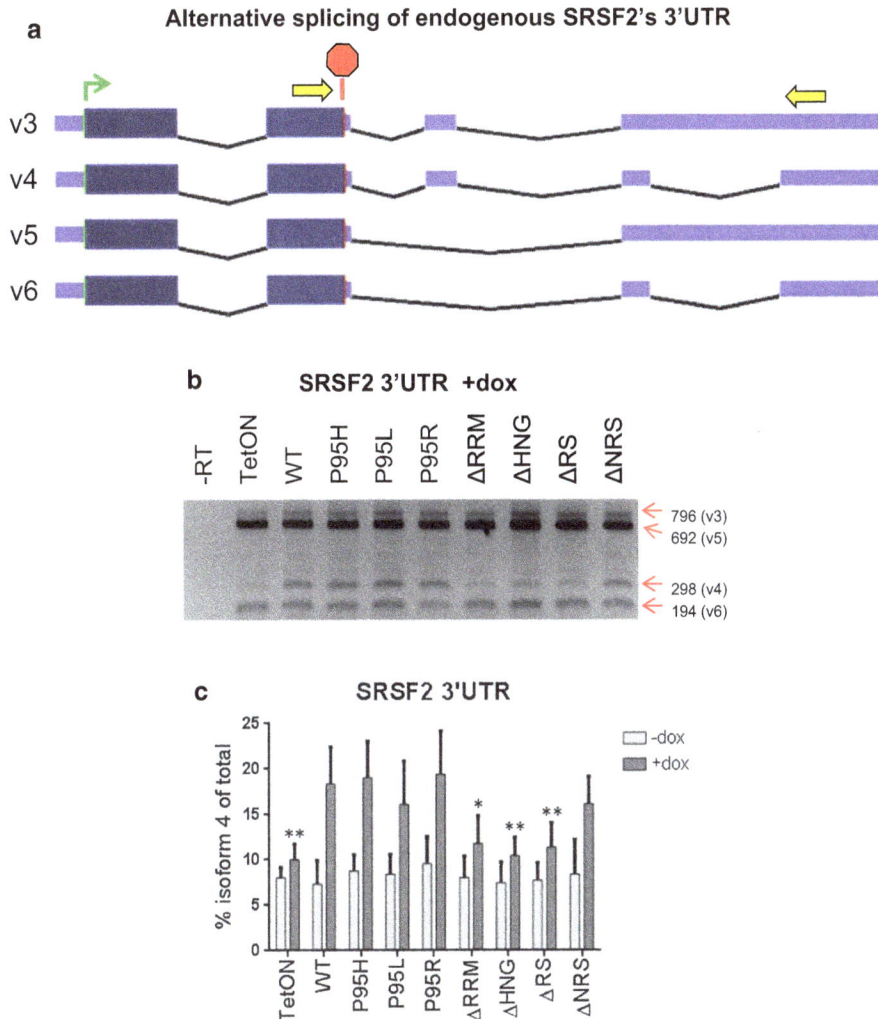

**Fig. 3** Alternative splicing at the 3'UTR of endogenous SRSF2 transcripts. **a** Schematic of previously described SRSF2 isoforms (splice variants 3, 4, 5, and 6) that result from alternative splicing of the 3'UTR. *Yellow arrows* depict general location of primers used in RT-PCR. **b** Representative gel from RT-PCR of endogenous SRSF2's 3'UTR after 48 h treatment with 2 μg/ml doxycycline. The sizes in base pairs of the PCR products corresponding to the splice variants are labeled on the right. **c** *Bar graphs* depicting changes in % isoform 4 (v4) of total RNA. Statistics are based on two-way ANOVA with comparison to SRSF2$^{WT}$ (n = 4). *Asterisks* are *p*values as in Fig. 1; note that the data were compared to the WT and not to the tetON

CDC25C (Fig. 5c, d; Additional file 6: Figure S5a, b). We noticed that some RT-PCR signals were weak or barely detectable—this was evident in RNA samples from cells treated with high drug doses or extended time—likely resulted from low RNA yields under those conditions. Similar weak RT-PCR signals were also observed and used to quantify the ratio of C5 to C1 in heavily treated cells by Albert et al. [31]. We also analyzed the protein levels of the CDC25C isoforms by immune-blotting and found evidence consistent with a percent increase of the C5 protein isoform in cells treated with CIS, CPT, or TSA at a dosage that caused CDC25C alternative splicing (Additional file 6: Figure S5c, d).

Since agents that activate the ATM-mediated DNA damage response cause a similar splicing change (elevated CDC25C C5/C1) as the MDS-associated SRSF2 mutants, we wondered whether these point mutants may also activated the ATM-mediated DNA damage response. To test this, we examined whether the phosphorylation of ATM substrates (p53, CHK2, and H2A.X [61–68]) increases in the drug-treated cells and/or in the SRSF2 mutant expressing cells. Cell lysates were analyzed using Western blotting for phospho-p53 and phospho-pCHK2. Treatment of the TF-1 TetON cell line with CIS, CPT, or TSA led to an increase in both phospho-p53 and phospho-CHK2 compared to untreated TF-TetON cells,

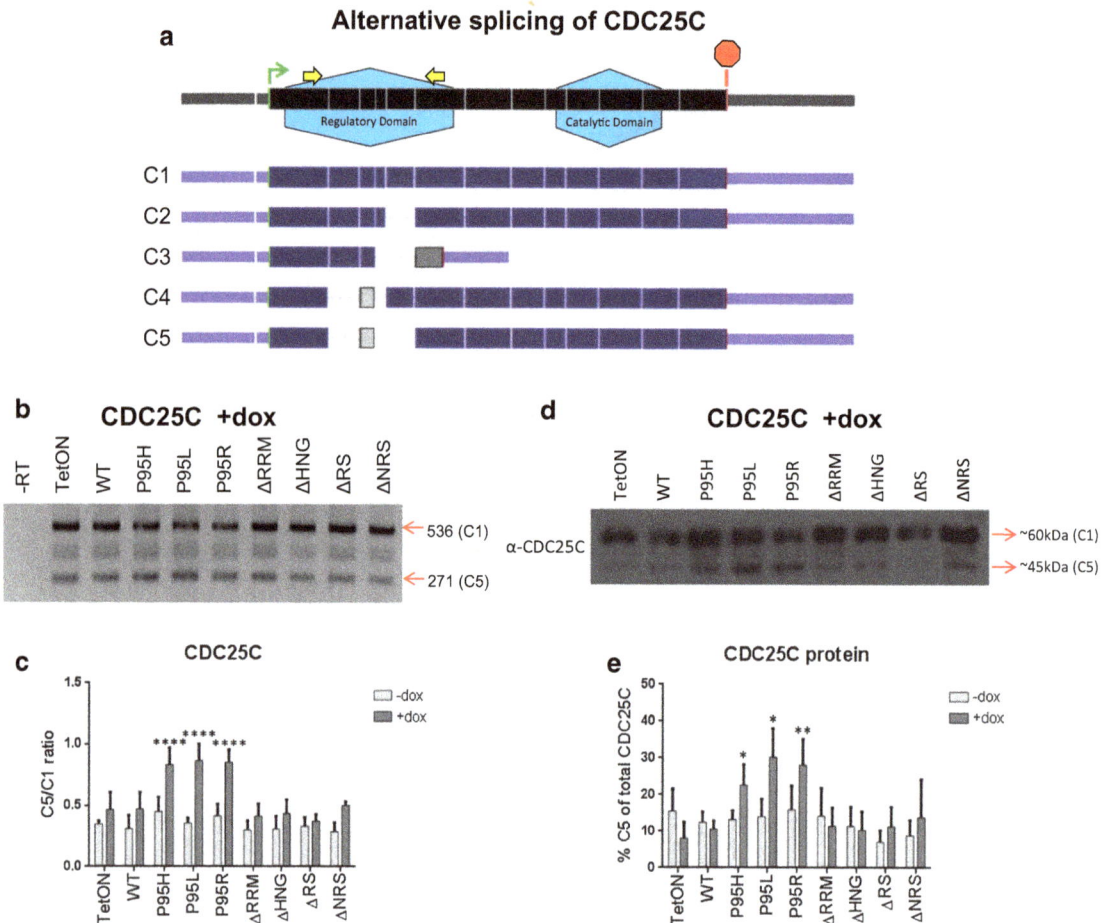

**Fig. 4** Alternative splicing of CDC25C in SRSF2 mutant cell lines. **a** Schematic of previously described five CDC25C isoforms. *Black* is the CDC25C RNA with all 14 exons (the exon boundaries are marked). The general locations for the N-terminal regulatory domain and the C-terminal catalytic domain, as well as the RT-PCR primers, are annotated. *Dark blue* is the known CDC25C alternative splicing transcripts. *Gray* depicts areas of frame shift that occurs with alternative splicing change in C3, C4, and C5. **b** Representative gel from RT-PCR of CDC25C after 48 h treatment with 2 μg/ml doxycycline. The sizes in base pairs of the PCR products corresponding to the splice variants are marked on the right. **c** *Bar graphs* depicting ratio of C5/C1 isoforms. Statistics are based on two-way ANOVA with comparison to SRSF2$^{WT}$ (n = 5). **d** Western blot depicting CDC25C protein expression in the presence of doxycyline induction. **e** Graph depicting the Western results shown in D for CDC25C protein expression at 48 h post-induction. Statistics are based on two-way ANOVA with comparison to SRSF2WT (n = 5). *Asterisks* are *p* values as in Fig. 1

as expected (Fig. 5e). However, ectopically expressing SRSF2 in TF-1 cells did not show a significant increase in phospho-p53 or phospho-CHK2 compared to doxycyline-treated TF-1 TetON, regardless of whether the SRSF2 was WT or a P95 mutant. We also tested levels of γH2A.X phosphorylation, a general marker for genome instability [69], in each of these cell lines. Again, CIS, CPT, and TSA treatment each increased γH2A.X levels while none of the SRSF2 cell lines had an increase in γH2A.X (Fig. 5e; Additional file 7: Figure S6). In addition, by propidium iodide staining, none of the MDS-related point mutants induced G2/M arrest (Additional file 8: Figure S7a). To test whether apoptosis played a role in CDC25C alternative splicing, we added caspase

inhibitor zVAD to block apoptosis in cells expressing SRSF2-P95 mutants. The ratio between splice variant C5 and C1 increased when SRSF2-P95 mutant expression was induced as expected; however, the addition of zVAD had no observable effect on CDC25C alternative splicing regardless whether SRSF2-P95 mutant expression was induced or not (Additional file 8: Figure S7B, C). This result indicated that apoptosis, although occurred in SRSF2-P95 mutant cells, was not responsible for the CDC25C alternative splicing. Taken together, it appears that the induction of CDC25C alternative splicing by MDS-related SRSF2 point mutations does not require activation of classical ATM DNA damage response pathway.

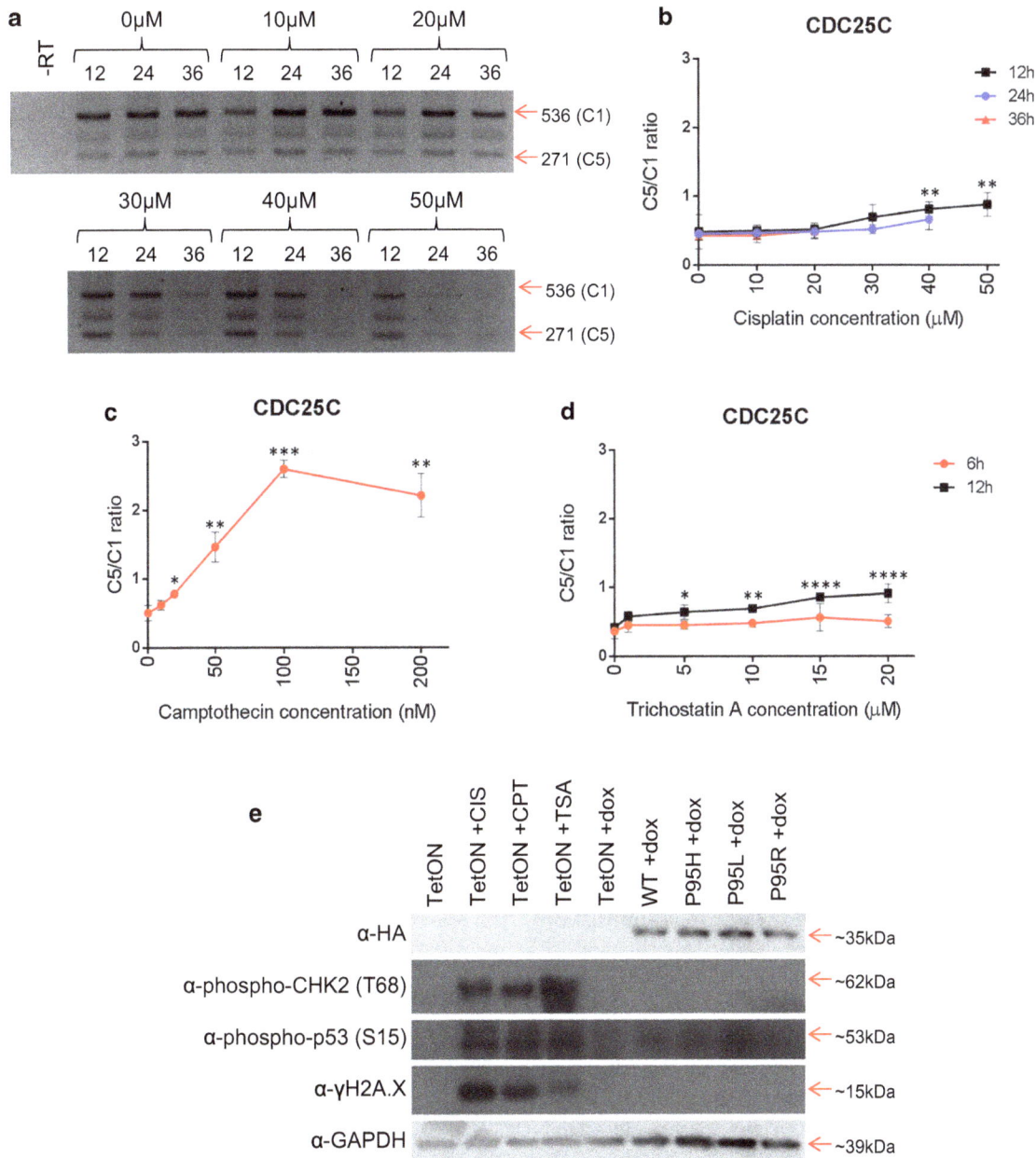

**Fig. 5** Drug-induced CDC25C alternative splicing. **a** Representative RT-PCR gel of alternative splicing of CDC25C in response to cisplatin treatment. (**b** to **d**) *Line graph* depicting average C5/C1 ratios for CDC25C in TF-1 TetON cells treated with **b** cisplatin, **c** CPT, and **d** TSA treated TF-1 TetON cells. Statistics are determined by two-way ANOVA with comparison to 0 μM (n = 4). **e** Western blot of HA-tagged protein, phosphorylated CHK2, phosphorylated p53, and GAPDH after no treatment (*lane 1*), 12 h 50 μM CIS (*lane 2*), 6 h 200 nM CPT (*lane 3*), 12 h 20 μM TSA (*lane 4*), or 48 h doxycycline treatment (*lanes 5–9*). *Asterisks* are *p* values as in Fig. 1

## Discussion

Known functions of SRSF2 include regulation of constitutive and alternative splicing and maintenance of genome stability through the prevention of transcription-related R-loops [2, 3, 70]. Nearly half of all patients with MDS have mutation in a gene that regulates splicing. One of the most common mutations is the point mutation of proline 95 of SRSF2 to histidine, leucine, or arginine, which occurs in 15–20 % of all MDS patients. Recent research has shown that this mutation affects the RNA recognition specificity of SRSF2 and that the P95H mutation in a mouse model correlates with a phenotype similar to MDS [28–30]. However, the effect of SRSF2 mutation on distinct alternative splicing events remains poorly understood. In this study, we show that MDS-related SRSF2 mutants act similar to wildtype SRSF2 in many ways, including phosphorylation and splicing of a select group of alternative splicing events. However, there is an alternative splicing change in CDC25C that is unique to SRSF2$^{P95}$ mutants. Chemotherapeutics that induce the classical ATM DNA damage response, including cisplatin, camptothecin, and trichostatin A, also induce an identical alternative splicing change in CDC25C.

Recent research has shown that various types of DNA damage can lead to post-translational modification of RNA-binding proteins, many of which have roles in the regulation of RNA splicing [71–75]. In fact, cisplatin treatment has been shown to cause increased translocation of SR protein kinases SRPK1 and SRPK2 to the nucleus leading to hyperphosphorylation of SRSF2 and reducing acetylation of SRSF2 [11]. In addition, activation of ATM- or ATR-mediated DNA damage response pathway has been shown to result in multiple changes in specific alternative splicing events, such as alternative splicing of apoptotic genes (BCL-X and caspase-8) [11, 76], DNA damage response genes (CHK1 and CHK2) [77, 78], and regulators of RNA splicing itself (TRA2, U2AF1, SRSF2 and SRSF1) [79–82]. However, the full changes in alternative splicing and their consequences for the DNA damage response still remain largely unknown.

Although it is known that CDC25C alternative splicing occurs after treatment with genotoxic drugs, the affects of this alternative splicing on the DNA damage response (DDR) and cell fate is still undetermined. CDC25C protein undergoes various post-translation modifications during the DDR in order to either induce the G2/M checkpoint or apoptosis. One of the major post-translational modifications is phosphorylation at Ser216 by CHK2, which leads to interaction of CDC25C with the 14-3-3 complex and subsequent translocation of the CDC25C protein to the cytoplasm [83]. Without CDC25C present in the nucleus to activate CDK1 by dephosphorylation at Tyr15, the cell enters G2/M arrest. The C5 isoform of CDC25C is upregulated during S phase, a cell cycle phase during which DNA repair frequently occurs due to DNA replication; however, the role of the C5 isoform in the S phase is not clear [84]. The alternative splicing of CDC25C, and its related protein CDC25A, is notable because it occurs primarily in the region of each transcript that encodes for the regulatory domain of the protein while leaving the catalytic domain intact (Fig. 4a) [84]. It is possible that alternative splicing of CDC25C, in conjunction with post-translational modification, plays an important regulatory role in cell fate after DNA damage. Interestingly, our results showed that the SRSF2 P95 point mutants cause the same alternative

CDC25C splicing change without activating conventional DDR.

CDC25C is part of the commonly deleted region in 5q-syndrome, and effective treatment of 5q- syndrome with lenalidomide requires haploinsufficiency of CDC25C [48–52]. Furthermore, alternative splicing of CDC25C has already been observed in bone marrow samples from patients with MDS [25]. In addition, mutations in DNA damage response genes, such as p53 (TP53) and ATM, are found in MDS patients without 5-q deletion [85], and MDS-related U2AF1 mutations lead to alternative splicing in the DNA damage response genes ATR and FANCA [86]. Therefore, it appears that CDC25C may be recurrently misregulated in MDS through gene deletion, aberrant DNA damage response, or aberrant splicing factors like SRSF2-P95 mutants and U2AF1 mutants.

## Conclusion

In conclusion, our data support a model where MDS-related SRSF2 mutants lead to alternative splicing of CDC25C that increases the percentage of the C5 isoform among all the CDC25C variants. This alternative splicing event can also be induced by cisplatin, camptothecin, or trichostatin A treatment. However, the SRSF2 point mutants do not lead to phosphorylation of CHK2 or p53, which is readily detected in the drug-treated cells. Therefore, SRSF2P95H/L/R-mediated CDC25C alternative splicing changes do not require activation of the classical ATM DDR. The mechanism of CDC25C alternative splicing and its downstream effects may be pertinent to the role of SRSF2's point mutation in the development of myelodysplastic syndrome.

## Additional files

**Additional file 1.** Supplementary information: supplemental figure legends, supplemental references, and Table S1.

**Additional file 2: Figure S1.** HA-tagged SRSF2 protein expression in uninduced TF-1 cell lines.

**Additional file 3: Figure S2.** Apoptosis, subcellular localization, and cell proliferation of SRSF2 mutants.

**Additional file 4: Figure S3.** Alternative splicing of apoptosis genes.

**Additional file 5: Figure S4.** Alternative splicing of genes from a previous study using SRSF2 depletion.

**Additional file 6: Figure S5.** Alternative splicing of CDC25C in TF-1 cells treated with CIS, CPT, or TSA.

**Additional file 7: Figure S6.** DNA damage in SRSF2 mutant cell lines.

**Additional file 8: Figure S7.** Cell cycle analysis of SRSF2 mutant cells and the effect of an apoptosis inhibitor on CDC25C alternative splicing.

## Abbreviations
SRSF2: serine–arginine rich splicing factor 2; MDS: myelodysplastic syndrome; CDC25C: cell division cycle 25C; ATM: ataxia telangiectasia mutated; RRM: RNA

recognition motif; RS: serine–arginine rich domain; HNG: hinge region; NRS: nuclear retention signal; CMML: chronic myelomonocytic leukemia.

**Authors' contributions**
LS and RJL conceived the project; LS, JMS, and RJL designed experiments; LS performed experiments; LS, JMS, and RJL analyzed the data; and LS, JMS, and RJL wrote the paper. All authors read and approved the final manuscript.

**Author details**
[1] Department of Molecular and Cellular Biology, Irell & Manella Graduate School of Biological Sciences, Beckman Research Institute of the City of Hope, Duarte, CA 91010, USA. [2] Department of Cancer Genetics and Epigenetics, Irell & Manella Graduate School of Biological Sciences, Beckman Research Institute of the City of Hope, Duarte, CA 91010, USA.

**Acknowledgements**
We thank Xiang-Dong Fu of UCSD for plasmid pRevTRE-SC35HA (SRSF2HA) and Jiing-Kuan Yee of City of Hope for plasmids pHIV7-rtTA, pC-GP-2, pCMV-rev2, pCMV-G, and pCMV-GP.

**Competing interests**
The authors declared that they have no competing interests.

**Funding**
LS was supported in part by the H.N. & Frances Berger Foundation Fellowship and the Norman and Melinda Payson Fellowship. This work was supported by grants from the Beckman Research Institute to RJL, National Institutes of Health Grant R01-CA120954 to JMS, and by NIH Grant P30-CA033572 for shared research core facilities at City of Hope.

**References**
1. Li X, Manley JL. Inactivation of the SR protein splicing factor ASF/SF2 results in genomic instability. Cell. 2005;122(3):365–78.
2. Xiao R, Sun Y, Ding JH, Lin S, Rose DW, Rosenfeld MG, Fu XD, Li X. Splicing regulator SC35 is essential for genomic stability and cell proliferation during mammalian organogenesis. Mol Cell Biol. 2007;27(15):5393–402.
3. Lin S, Coutinho-Mansfield G, Wang D, Pandit S, Fu XD. The splicing factor SC35 has an active role in transcriptional elongation. Nat Struct Mol Biol. 2008;15(8):819–26.
4. Sollier J, Stork CT, Garcia-Rubio ML, Paulsen RD, Aguilera A, Cimprich KA. Transcription-coupled nucleotide excision repair factors promote R-loop-induced genome instability. Mol Cell. 2014;56(6):777–85.
5. Ji X, Zhou Y, Pandit S, Huang J, Li H, Lin CY, Xiao R, Burge CB, Fu XD. SR proteins collaborate with 7SK and promoter-associated nascent RNA to release paused polymerase. Cell. 2013;153(4):855–68.
6. Fu XD, Maniatis T. The 35-kDa mammalian splicing factor SC35 mediates specific interactions between U1 and U2 small nuclear ribonucleoprotein particles at the 3' splice site. Proc Natl Acad Sci USA. 1992;89(5):1725–9.
7. Caceres JF, Screaton GR, Krainer AR. A specific subset of SR proteins shuttles continuously between the nucleus and the cytoplasm. Genes Dev. 1998;12(1):55–66.
8. Wang C, Chua K, Seghezzi W, Lees E, Gozani O, Reed R. Phosphorylation of spliceosomal protein SAP 155 coupled with splicing catalysis. Genes Dev. 1998;12(10):1409–14.
9. Cazalla D, Zhu J, Manche L, Huber E, Krainer AR, Caceres JF. Nuclear export and retention signals in the RS domain of SR proteins. Mol Cell Biol. 2002;22(19):6871–82.
10. Hall LL, Smith KP, Byron M, Lawrence JB. Molecular anatomy of a speckle. Anat Rec A Discov Mol Cell Evol Biol. 2006;288(7):664–75.
11. Edmond V, Moysan E, Khochbin S, Matthias P, Brambilla C, Brambilla E, Gazzeri S, Eymin B. Acetylation and phosphorylation of SRSF2 control cell fate decision in response to cisplatin. EMBO J. 2011;30(3):510–23.
12. Ngo JCK, Chakrabarti S, Ding J-H, Velazquez-Dones A, Nolen B, Aubol BE, Adams JA, Fu X-D, Ghosh G. Interplay between SRPK and Clk/Sty kinases in phosphorylation of the splicing factor ASF/SF2 is regulated by a docking motif in ASF/SF2. Mol Cell. 2005;20(1):77–89.
13. Patnaik MM, Lasho TL, Finke CM, Hanson CA, Hodnefield JM, Knudson RA, Ketterling RP, Pardanani A, Tefferi A. Spliceosome mutations involving SRSF2, SF3B1, and U2AF35 in chronic myelomonocytic leukemia: prevalence, clinical correlates, and prognostic relevance. Am J Hematol. 2013;88(3):201–6.
14. Itzykson R, Solary E. An evolutionary perspective on chronic myelomonocytic leukemia. Leukemia. 2013;27(7):1441–50.
15. Yoshida K, Sanada M, Shiraishi Y, Nowak D, Nagata Y, Yamamoto R, Sato Y, Sato-Otsubo A, Kon A, Nagasaki M, et al. Frequent pathway mutations of splicing machinery in myelodysplasia. Nature. 2011;478(7367):64–9.
16. Damm F, Kosmider O, Gelsi-Boyer V, Renneville A, Carbuccia N, Hidalgo-Curtis C, Della Valle V, Couronne L, Scourzic L, Chesnais V, et al. Mutations affecting mRNA splicing define distinct clinical phenotypes and correlate with patient outcome in myelodysplastic syndromes. Blood. 2012;119(14):3211–8.
17. Thol F, Kade S, Schlarmann C, Loffeld P, Morgan M, Krauter J, Wlodarski MW, Kolking B, Wichmann M, Gorlich K, et al. Frequency and prognostic impact of mutations in SRSF2, U2AF1, and ZRSR2 in patients with myelodysplastic syndromes. Blood. 2012;119(15):3578–84.
18. Wu SJ, Kuo YY, Hou HA, Li LY, Tseng MH, Huang CF, Lee FY, Liu MC, Liu CW, Lin CT, et al. The clinical implication of SRSF2 mutation in patients with myelodysplastic syndrome and its stability during disease evolution. Blood. 2012;120(15):3106–11.
19. Federmann B, Abele M, Rosero Cuesta DS, Vogel W, Boiocchi L, Kanz L, Quintanilla-Martinez L, Orazi A, Bonzheim I, Fend F. The detection of SRSF2 mutations in routinely processed bone marrow biopsies is useful in the diagnosis of chronic myelomonocytic leukemia. Hum Pathol. 2014;45(12):2471–9.
20. Ma X, Does M, Raza A, Mayne ST. Myelodysplastic syndromes: incidence and survival in the United States. Cancer. 2007;109(8):1536–42.
21. Koreth J, Pidala J, Perez WS, Deeg HJ, Garcia-Manero G, Malcovati L, Cazzola M, Park S, Itzykson R, Ades L, et al. Role of reduced-intensity conditioning allogeneic hematopoietic stem-cell transplantation in older patients with de novo myelodysplastic syndromes: an international collaborative decision analysis. J Clin Oncol. 2013;31(21):2662–70.
22. Negrin RS, Haeuber DH, Nagler A, Kobayashi Y, Sklar J, Donlon T, Vincent M, Greenberg PL. Maintenance treatment of patients with myelodysplastic syndromes using recombinant human granulocyte colony-stimulating factor. Blood. 1990;76(1):36–43.
23. Legare RD, Gilliland DG. Myelodysplastic syndrome. Curr Opin Hematol. 1995;2(4):283–92.
24. Appelbaum FR, Anderson J. Allogeneic bone marrow transplantation for myelodysplastic syndrome: outcomes analysis according to IPSS score. Leukemia. 1998;12(Suppl 1):S25–9.
25. Caudill JS, Porcher JC, Steensma DP. Aberrant pre-mRNA splicing of a highly conserved cell cycle regulator, CDC25C, in myelodysplastic syndromes. Leuk Lymphoma. 2008;49(5):989–93.
26. Makishima H, Visconte V, Sakaguchi H, Jankowska AM, Abu Kar S, Jerez A, Przychodzen B, Bupathi M, Guinta K, Afable MG, et al. Mutations in the spliceosome machinery, a novel and ubiquitous pathway in leukemogenesis. Blood. 2012;119(14):3203–10.
27. Matsuda K, Ishida F, Ito T, Nakazawa H, Miura S, Taira C, Sueki A, Kobayashi Y, Honda T. Spliceosome-related gene mutations in myelodysplastic syndrome can be used as stable markers for monitoring minimal residual disease during follow-up. Leuk Res. 2012;36(11):1393–7.
28. Kim E, Ilagan JO, Liang Y, Daubner GM, Lee SC, Ramakrishnan A, Li Y, Chung YR, Micol JB, Murphy ME, et al. SRSF2 mutations contribute to myelodysplasia by mutant-specific effects on exon recognition. Cancer Cell. 2015;27(5):617–30.
29. Zhang J, Lieu YK, Ali AM, Penson A, Reggio KS, Rabadan R, Raza A, Mukherjee S, Manley JL. Disease-associated mutation in SRSF2 misregulates splicing by altering RNA-binding affinities. Proc Natl Acad Sci USA. 2015;112(34):E4726–34.
30. Komeno Y, Huang YJ, Qiu J, Lin L, Xu Y, Zhou Y, Chen L, Monterroza DD, Li H, DeKelver RC, et al. SRSF2 is essential for hematopoiesis, and its myelodysplastic syndrome-related mutations dysregulate alternative pre-mRNA splicing. Mol Cell Biol. 2015;35(17):3071–82.

31. Albert H, Battaglia E, Monteiro C, Bagrel D. Genotoxic stress modulates CDC25C phosphatase alternative splicing in human breast cancer cell lines. Mol Oncol. 2012;6(5):542–52.

32. Li MJ, Rossi JJ. Lentivirus transduction of hematopoietic cells. Cold Spring Harb Protoc. 2007;2007(5):pdb.prot4755.

33. Caceres JF, Misteli T, Screaton GR, Spector DL, Krainer AR. Role of the modular domains of SR proteins in subnuclear localization and alternative splicing specificity. J Cell Biol. 1997;138(2):225–38.

34. Li H, Bingham PM. Arginine/serine-rich domains of the su(wa) and tra RNA processing regulators target proteins to a subnuclear compartment implicated in splicing. Cell. 1991;67(2):335–42.

35. Hedley ML, Amrein H, Maniatis T. An amino acid sequence motif sufficient for subnuclear localization of an arginine/serine-rich splicing factor. Proc Natl Acad Sci USA. 1995;92(25):11524–8.

36. Misteli T, Caceres JF, Spector DL. The dynamics of a pre-mRNA splicing factor in living cells. Nature. 1997;387(6632):523–7.

37. Zang DY, Goodwin RG, Loken MR, Bryant E, Deeg HJ. Expression of tumor necrosis factor-related apoptosis-inducing ligand, Apo2L, and its receptors in myelodysplastic syndrome: effects on in vitro hemopoiesis. Blood. 2001;98(10):3058–65.

38. Gersuk GM, Beckham C, Loken MR, Kiener P, Anderson JE, Farrand A, Troutt AB, Ledbetter JA, Deeg HJ. A role for tumour necrosis factor-alpha, Fas and Fas-Ligand in marrow failure associated with myelodysplastic syndrome. Br J Haematol. 1998;103(1):176–88.

39. Parker JE, Fishlock KL, Mijovic A, Czepulkowski B, Pagliuca A, Mufti GJ. 'Low-risk' myelodysplastic syndrome is associated with excessive apoptosis and an increased ratio of pro-versus anti-apoptotic bcl-2-related proteins. Br J Haematol. 1998;103(4):1075–82.

40. Lyons AB, Blake SJ, Doherty KV. Flow cytometric analysis of cell division by dilution of CFSE and related dyes. Curr Protoc Cytom. 2013;9:9–11.

41. Sureau A, Gattoni R, Dooghe Y, Stevenin J, Soret J. SC35 autoregulates its expression by promoting splicing events that destabilize its mRNAs. EMBO J. 2001;20(7):1785–96.

42. Nagy E, Maquat LE. A rule for termination-codon position within intron-containing genes: when nonsense affects RNA abundance. Trends Biochem Sci. 1998;23(6):198–9.

43. Mayeda A, Screaton GR, Chandler SD, Fu XD, Krainer AR. Substrate specificities of SR proteins in constitutive splicing are determined by their RNA recognition motifs and composite pre-mRNA exonic elements. Mol Cell Biol. 1999;19(3):1853–63.

44. Kongruttanachok N, Phuangphairoj C, Thongnak A, Ponyeam W, Rattanatanyong P, Pornthanakasem W, Mutirangura A. Replication independent DNA double-strand break retention may prevent genomic instability. Mol Cancer. 2010;9:70.

45. Kumar A, Rai PS, Upadhya R, Vishwanatha Prasada KS, Rao BS, Satyamoorthy K. Gamma-radiation induces cellular sensitivity and aberrant methylation in human tumor cell lines. Int J Radiat Biol. 2011;87(11):1086–96.

46. Merdzhanova G, Edmond V, De Seranno S, Van den Broeck A, Corcos L, Brambilla C, Brambilla E, Gazzeri S, Eymin B. E2F1 controls alternative splicing pattern of genes involved in apoptosis through upregulation of the splicing factor SC35. Cell Death Differ. 2008;15(12):1815–23.

47. Pandit S, Zhou Y, Shiue L, Coutinho-Mansfield G, Li H, Qiu J, Huang J, Yeo GW, Ares M Jr, Fu XD. Genome-wide analysis reveals SR protein cooperation and competition in regulated splicing. Mol Cell. 2013;50(2):223–35.

48. Bunn HF. 5q- and disordered haematopoiesis. Clin Haematol. 1986;15(4):1023–35.

49. Pedersen B, Kerndrup G. Specific minor chromosome deletions consistently occurring in myelodysplastic syndromes. Cancer Genet Cytogenet. 1986;23(1):61–75.

50. Ebert BL. Deletion 5q in myelodysplastic syndrome: a paradigm for the study of hemizygous deletions in cancer. Leukemia. 2009;23(7):1252–6.

51. Jadersten M, Karsan A. Clonal evolution in myelodysplastic syndromes with isolated del (5q): the importance of genetic monitoring. Haematologica. 2011;96(2):177–80.

52. Wei S, Chen X, Rocha K, Epling-Burnette PK, Djeu JY, Liu Q, Byrd J, Sokol L, Lawrence N, Pireddu R, et al. A critical role for phosphatase haplodeficiency in the selective suppression of deletion 5q MDS by lenalidomide. Proc Natl Acad Sci USA. 2009;106(31):12974–9.

53. Jordan P, Carmo-Fonseca M. Molecular mechanisms involved in cisplatin cytotoxicity. Cell Mol Life Sci. 2000;57(8–9):1229–35.

54. Colton SL, Xu XS, Wang YA, Wang G. The involvement of ataxia-telangiectasia mutated protein activation in nucleotide excision repair-facilitated cell survival with cisplatin treatment. J Biol Chem. 2006;281(37):27117–25.

55. Mu XF, Jin XL, Farnham MM, Li Y, O'Neill C. DNA damage-sensing kinases mediate the mouse 2-cell embryo's response to genotoxic stress. Biol Reprod. 2011;85(3):524–35.

56. Ryan AJ, Squires S, Strutt HL, Johnson RT. Camptothecin cytotoxicity in mammalian cells is associated with the induction of persistent double strand breaks in replicating DNA. Nucleic Acids Res. 1991;19(12):3295–300.

57. Sordet O, Redon CE, Guirouilh-Barbat J, Smith S, Solier S, Douarre C, Conti C, Nakamura AJ, Das BB, Nicolas E, et al. Ataxia telangiectasia mutated activation by transcription- and topoisomerase I-induced DNA double-strand breaks. EMBO Rep. 2009;10(8):887–93.

58. Marinello J, Chillemi G, Bueno S, Manzo SG, Capranico G. Antisense transcripts enhanced by camptothecin at divergent CpG-island promoters associated with bursts of topoisomerase I-DNA cleavage complex and R-loop formation. Nucleic Acids Res. 2013;41(22):10110–23.

59. Lee JS. Activation of ATM-dependent DNA damage signal pathway by a histone deacetylase inhibitor, trichostatin A. Cancer Res Treat. 2007;39(3):125–30.

60. Ju R, Muller MT. Histone deacetylase inhibitors activate p21(WAF1) expression via ATM. Cancer Res. 2003;63(11):2891–7.

61. Banin S, Moyal L, Shieh S, Taya Y, Anderson CW, Chessa L, Smorodinsky NI, Prives C, Reiss Y, Shiloh Y, et al. Enhanced phosphorylation of p53 by ATM in response to DNA damage. Science. 1998;281(5383):1674–7.

62. Canman CE, Lim DS, Cimprich KA, Taya Y, Tamai K, Sakaguchi K, Appella E, Kastan MB, Siliciano JD. Activation of the ATM kinase by ionizing radiation and phosphorylation of p53. Science. 1998;281(5383):1677–9.

63. Chaturvedi P, Eng WK, Zhu Y, Mattern MR, Mishra R, Hurle MR, Zhang X, Annan RS, Lu Q, Faucette LF, et al. Mammalian Chk2 is a downstream effector of the ATM-dependent DNA damage checkpoint pathway. Oncogene. 1999;18(28):4047–54.

64. Matsuoka S, Huang M, Elledge SJ. Linkage of ATM to cell cycle regulation by the Chk2 protein kinase. Science. 1998;282(5395):1893–7.

65. Matsuoka S, Rotman G, Ogawa A, Shiloh Y, Tamai K, Elledge SJ. Ataxia telangiectasia-mutated phosphorylates Chk2 in vivo and in vitro. Proc Natl Acad Sci USA. 2000;97(19):10389–94.

66. Burma S, Chen BP, Murphy M, Kurimasa A, Chen DJ. ATM phosphorylates histone H2AX in response to DNA double-strand breaks. J Biol Chem. 2001;276(45):42462–7.

67. Fernandez-Capetillo O, Chen HT, Celeste A, Ward I, Romanienko PJ, Morales JC, Naka K, Xia Z, Camerini-Otero RD, Motoyama N, et al. DNA damage-induced G2-M checkpoint activation by histone H2AX and 53BP1. Nat Cell Biol. 2002;4(12):993–7.

68. Stiff T, O'Driscoll M, Rief N, Iwabuchi K, Lobrich M, Jeggo PA. ATM and DNA-PK function redundantly to phosphorylate H2AX after exposure to ionizing radiation. Cancer Res. 2004;64(7):2390–6.

69. Dickey JS, Redon CE, Nakamura AJ, Baird BJ, Sedelnikova OA, Bonner WM. H2AX: functional roles and potential applications. Chromosoma. 2009;118(6):683–92.

70. Paulsen RD, Soni DV, Wollman R, Hahn AT, Yee MC, Guan A, Hesley JA, Miller SC, Cromwell EF, Solow-Cordero DE, et al. A genome-wide siRNA screen reveals diverse cellular processes and pathways that mediate genome stability. Mol Cell. 2009;35(2):228–39.

71. Beli P, Lukashchuk N, Wagner SA, Weinert BT, Olsen JV, Baskcomb L, Mann M, Jackson SP, Choudhary C. Proteomic investigations reveal a role for RNA processing factor THRAP3 in the DNA damage response. Mol Cell. 2012;46(2):212–25.

72. Bennetzen MV, Larsen DH, Bunkenborg J, Bartek J, Lukas J, Andersen JS. Site-specific phosphorylation dynamics of the nuclear proteome during the DNA damage response. Mol Cell Proteomics. 2010;9(6):1314–23.

73. Blasius M, Forment JV, Thakkar N, Wagner SA, Choudhary C, Jackson SP. A phospho-proteomic screen identifies substrates of the checkpoint kinase Chk1. Genome Biol. 2011;12(8):R78.

74. Adamson B, Smogorzewska A, Sigoillot FD, King RW, Elledge SJ. A genome-wide homologous recombination screen identifies the RNA-binding protein RBMX as a component of the DNA-damage response. Nat Cell Biol. 2012;14(3):318–28.

75. Mastrocola AS, Kim SH, Trinh AT, Rodenkirch LA, Tibbetts RS. The RNA-binding protein fused in sarcoma (FUS) functions downstream of poly(ADP-ribose) polymerase (PARP) in response to DNA damage. J Biol Chem. 2013;288(34):24731–41.

76. Shkreta L, Michelle L, Toutant J, Tremblay ML, Chabot B. The DNA damage response pathway regulates the alternative splicing of the apoptotic mediator Bcl-x. J Biol Chem. 2011;286(1):331–40.

77. Best A, James K, Dalgliesh C, Hong E, Kheirolahi-Kouhestani M, Curk T, Xu Y, Danilenko M, Hussain R, Keavney B, et al. Human Tra2 proteins jointly control a CHEK1 splicing switch among alternative and constitutive target exons. Nat Commun. 2014;5:4760.

78. Paronetto MP, Minana B, Valcarcel J. The Ewing sarcoma protein regulates DNA damage-induced alternative splicing. Mol Cell. 2011;43(3):353–68.

79. Katzenberger RJ, Marengo MS, Wassarman DA. Control of alternative splicing by signal-dependent degradation of splicing-regulatory proteins. J Biol Chem. 2009;284(16):10737–46.

80. Filippov V, Filippova M, Duerksen-Hughes PJ. The early response to DNA damage can lead to activation of alternative splicing activity resulting in CD44 splice pattern changes. Cancer Res. 2007;67(16):7621–30.

81. Solier S, Barb J, Zeeberg BR, Varma S, Ryan MC, Kohn KW, Weinstein JN, Munson PJ, Pommier Y. Genome-wide analysis of novel splice variants induced by topoisomerase I poisoning shows preferential occurrence in genes encoding splicing factors. Cancer Res. 2010;70(20):8055–65.

82. Colla S, Ong DS, Ogoti Y, Marchesini M, Mistry NA, Clise-Dwyer K, Ang SA, Storti P, Viale A, Giuliani N, et al. Telomere dysfunction drives aberrant hematopoietic differentiation and myelodysplastic syndrome. Cancer Cell. 2015;27(5):644–57.

83. Bulavin DV, Higashimoto Y, Popoff IJ, Gaarde WA, Basrur V, Potapova O, Appella E, Fornace AJ Jr. Initiation of a G2/M checkpoint after ultraviolet radiation requires p38 kinase. Nature. 2001;411(6833):102–7.

84. Wegener S, Hampe W, Herrmann D, Schaller HC. Alternative splicing in the regulatory region of the human phosphatases CDC25A and CDC25C. Eur J Cell Biol. 2000;79(11):810–5.

85. Haferlach T, Nagata Y, Grossmann V, Okuno Y, Bacher U, Nagae G, Schnittger S, Sanada M, Kon A, Alpermann T, et al. Landscape of genetic lesions in 944 patients with myelodysplastic syndromes. Leukemia. 2014;28(2):241–7.

86. Ilagan JO, Ramakrishnan A, Hayes B, Murphy ME, Zebari AS, Bradley P, Bradley RK. U2AF1 mutations alter splice site recognition in hematological malignancies. Genome Res. 2015;25(1):14–26.

# 11

# TLK1B mediated phosphorylation of Rad9 regulates its nuclear/cytoplasmic localization and cell cycle checkpoint

Sanket Awate and Arrigo De Benedetti*

## Abstract

**Background:** The Tousled like kinase 1B (TLK1B) is critical for DNA repair and survival of cells. Upon DNA damage, Chk1 phosphorylates TLK1B at S457 leading to its transient inhibition. Once TLK1B regains its kinase activity it phosphorylates Rad9 at S328. In this work we investigated the significance of this mechanism by overexpressing mutant TLK1B in which the inhibitory phosphorylation site was eliminated.

**Results and discussion:** These cells expressing TLK1B resistant to DNA damage showed constitutive phosphorylation of Rad9 S328 that occurred even in the presence of hydroxyurea (HU), and this resulted in a delayed checkpoint recovery. One possible explanation was that premature phosphorylation of Rad9 caused its dissociation from 9-1-1 at stalled replication forks, resulting in their collapse and prolonged activation of the S-phase checkpoint. We found that phosphorylation of Rad9 at S328 results in its dissociation from chromatin and redistribution to the cytoplasm. This results in double stranded breaks formation with concomitant activation of ATM and phosphorylation of H2AX. Furthermore, a Rad9 (S328D) phosphomimic mutant was exclusively localized to the cytoplasm and not the chromatin. Another Rad9 phosphomimic mutant (T355D), which is also a site phosphorylated by TLK1, localized normally. In cells expressing the mutant TLK1B treated with HU, Rad9 association with Hus1 and WRN was greatly reduced, suggesting again that its phosphorylation causes its premature release from stalled forks.

**Conclusions:** We propose that normally, the inactivation of TLK1B following replication arrest and genotoxic stress functions to allow the retention of 9-1-1 at the sites of damage or stalled forks. Following reactivation of TLK1B, whose synthesis is concomitantly induced by genotoxins, Rad9 is hyperphosphorylated at S328, resulting in its dissociation and inactivation of the checkpoint that occurs once repair is complete.

**Keywords:** DNA damage response, Replication stress, TLK1, TLK1B, Rad9, pRad9 S328, WRN, γH2AX, 9-1-1 complex

## Background

The Tousled (*Tsl*) gene was first identified in the plant *Arabidopsis thaliana*. Recessive *Tsl* mutants show defects in leaf and flower development [1]. This was proposed to be linked to a replicative defect during organogenesis, but it may also result from failure to protect the genome from DNA damage [2–4], resulting in developmental aberrations [5, 6]. Animal homologs of Tousled, known as Tousled like kinases (TLKs), are found from *Caenorhabditis*

*elegans* to mammals. They are generally considered as genes of metazoans and are not found in yeast, although they are present in unicellular trypanosomes [7]. In mammals their activity is cell cycle regulated with maximal activity found in the S-phase. After many years of study, only a few direct "interacting" substrates of TLKs have been identified, namely the histone chaperone Asf1 [8], histone H3 [9], Rad9 [10], and Aurora B kinase [5]. As evident from their substrates, TLKs play a major role in chromatin assembly [10, 11], transcription [4, 12], DNA repair [3, 10, 13], and condensation of chromosomes at mitosis [5, 6]. In humans two structurally similar TLK genes (TLK1 and TLK2) with several splice variants have

*Correspondence: adeben@lsuhsc.edu
Department of Biochemistry and Molecular Biology, Louisiana State University Health Sciences Center, 1501 Kings Highway, Shreveport, LA 71130, USA

been identified. A splice variant of TLK1, TLK1B that lacks the first 237 amino acids was identified in our lab. TLK1 and TLK1B interact with similar substrates, are believed to have similar enzymatic functions and are often referred to as TLK1/1B. Our previous studies have shown that translation of TLK1B is induced by DNA damage through the activation of the mTOR-eIF4E pathway. We have shown that elevated expression of TLK1B promotes cell survival after irradiation (IR) or doxorubicin [13] and UV [3] by facilitating DNA repair and promoting chromatin assembly after repair. Expression of a dominant-negative mutant of TLK1B renders mammalian cells sensitive to IR [6]. Thus, the human homolog, TLK1B, has invoked interest because of its established role in cell survival after DNA damage [3, 9, 13]. Identification of Rad9 as a substrate for TLK1/1B attributes a direct role of TLK1/1B in DNA repair [14]. Our previous work suggests that TLK1/1B's chaperone activity, independent of its kinase activity, helps in the recruitment of Rad9 at the break site. We had previously shown some evidence that TLK1/1B kinase activity is important for the dissociation of Rad9-Rad1-Hus1 (9-1-1) complex from a double stranded break (DSB) [14].

Rad9 plays a major role in DNA repair, cell cycle checkpoint and apoptosis. Aberrant Rad9 expression has been linked to breast, lung, thyroid, skin and prostate tumorigenesis [15]. Rad9 is a part of 9-1-1 heterotrimeric complex which is required for activation of ATR. Rad9, Rad1 or Hus1 KO mice are embryonic lethal [16, 17]. Loss of Rad9 produces a defect in ATR signaling and increases the sensitivity of the cells towards genotoxic stress [18]. In response to replication stress RPA directs the clamp loader RAD17–replication factor C (RFC) to load the 9-1-1 complex at the 5′ end of the double strand-single strand DNA junctions [19, 20]. Chromatin-bound 9-1-1 complex acts as a scaffold for the recruitment of various DNA repair proteins and polymerases at the DNA damage break site. It ensures filling of gaps and efficient repair of DNA [21, 22]. Recently it has been shown that 9-1-1 complex is required for the recruitment of WRN protein at stalled replication forks and this interaction is important for the fork recovery [23]. WRN belongs to the RecQ family of DNA helicases. Loss of WRN gives rise to a genetic disease known as Werner syndrome (WS). It is characterized by pre-mature ageing and pre-disposition to cancer [24, 25]. Cells derived from the WS patients show a prolonged S-phase, a reduced life-span, and an increase in genomic instability [26, 27]. It has been shown that WRN stabilizes the stalled replication forks and the loss of WRN leads to the fork collapse and increase in DSBs that are repaired through recombination [28]. WRN interacts with the 9-1-1 complex to maintain genomic stability by preventing accumulation of DSBs at the damaged forks [23].

Activation of the DNA damage induced checkpoint mediates rapid and transient inhibition of TLK activity. This transient inhibition in response to DNA damage requires ATM and Chk1 function. Chk1 directly phosphorylates TLK1 at S695 which is equivalent to S457 of TLK1B [29]. Once TLK1/1B regains its kinase activity it phosphorylates Rad9 at S328 [10] and T355 [30]. Rad9 S328 phosphorylation follows the pattern of TLK1/1B activity wherein it is inhibited immediately after DNA damage and gets phosphorylated when TLK1/1B regains its activity [14, 31]. The reason for this transient inactivation of TLK1/1B still remains a question as there is a lack of direct evidence for the role of this inhibitory phosphorylation with regards to its effect on Rad9 and ATR mediated cell signaling. In order to answer this question we have made an S457A mutant (Mut) of TLK1B that lacks the inhibitory phosphorylation site. In this study we show that Mut TLK1B remains active in the presence of DNA damage, and cells overexpressing it display an altered cell cycle checkpoint and delay in cell cycle progression upon recovery from hydroxyurea (HU). Mut TLK1B overexpressing cells show an increase in phosphorylation of Rad9 at S328. In response to HU mediated replication arrest, Mut TLK1B overexpressing cells show a massive reduction in the formation of Rad9 foci and reduced association of Rad9 with the chromatin. Mut TLK1B expressing cells show reduced association of Rad9 with Hus1 and WRN suggesting an early dissociation of 9-1-1 complex. In response to HU, these cells show an increased accumulation of DSBs which are marked by an increase in association of p-ATM (S1981) and γ-H2AX with the chromatin. Cellular fractionation data of Mut TLK1B overexpressing cells showed an increase in the accumulation of total Rad9 and p-Rad9 (S328) in the cytoplasm. Furthermore, studies with Rad9 S328D phosphomimetic mutant suggest that the phosphorylation of Rad9 at S328 alone is sufficient to prevent Rad9 localization into the nucleus. Our results indicate that in response to replication arrest, transient inhibition of TLK1/1B is crucial to maintain localization of Rad9 into the nucleus at damage sites.

## Results
### Mut TLK1B increases phosphorylation of Rad9 at S328 both in presence and absence of DNA damage
We generated a S457A TLK1B mutant which lacks the Chk1 inhibitory phosphorylation site and cloned it into the pIRES2-EGFP vector. We then generated HEK293 stable cell lines overexpressing mutant (Mut) and wild-type (Wt) TLK1B, after selection for high GFP positive cells (methods). Figure 1a confirms the overexpression of

**Fig. 1** **a** Overexpression of Wt TLK1B and Mut TLK1B in stably transfected HEK293 cells. **b** TLK1B Wt and Mut cells or empty vector controls (EV) were treated for 2 h with doxo to promote the Chk1-dpendent phosphorylation of TLK1B (S457). TLK1B S457 phospho specific antibody recognizes only the overexpressed Wt TLK1B and not the Mut TLK1B. Note that the p-TLK1B *band* seen in the Mut lane corresponds to the endogenous TLK1B and not the transfected TLK1B Mut. **c** The basal phosphorylation of Rad9 (S328) is enhanced in cells expressing Mut TLK1B with respect to EV or cells expressing Wt TLK1B. **d** The phosphorylation of Rad9 (S328) persists in damage resistant Mut TLK1B expressing cells treated with doxo. **e** Pattern of Rad9 (S328) phosphorylation after recovery from HU in the cells expressing Mut TLK1B or Wt TLK1B. Note that TLK1B overexpression is unaffected by HU treatment in these cells

Wt and Mut TLK1B in HEK293 cells. Further, after treatment with doxorubicin (doxo, topoisomerase II inhibitor that causes accumulation of DSBs upon which Rad9 is loaded) Wt TLK1B shows an increase in phosphorylation at S457 whereas the phospho-specific antibody doesn't recognize the overexpressed Mut TLK1B (Fig. 1b). The mutant and empty vector (EV) show a slight band of p-TLK1B S457 attributable to the endogenous TLK1B that is translationally induced after doxo treatment [9, 32]. In comparison with the Wt and EV, the basal level of Rad9 phosphorylation is elevated in the Mut TLK1B expressing cells (Fig. 1c). In the Mut TLK1B expressing cells the phosphorylation of Rad9 is not suppressed after DNA damage, either with doxo (Fig. 1d) or HU (Fig. 1e), which suggests that the Mut kinase is active even in the presence of DNA damage. In Wt cells, Rad9

phosphorylation at S328 is initially very low and progressively increases by 6–8 h of recovery from HU (Fig. 1e). In Mut TLK1B cells, the phosphorylation of Rad9 is elevated after release from HU but decreases after 8 h of recovery, possibly due to the activation of a phosphatase. HU depletes the deoxyribonucleotide pools that results in replication fork stalling, leading to recruitment of Rad9.

## Cells expressing Mut TLK1B show defects in formation of Rad9 foci

When cells are exposed to replication stress Rad9 protein is redistributed to form discrete nuclear foci at the sites of DNA damage [33]. The recruitment of the 9-1-1 complex at the sites of DNA damage can be visualized by looking at the formation of Rad9 foci [34]. We showed that in response to DNA damage Rad9 S328 phosphorylation persists in the cells overexpressing Mut TLK1B. We wanted to examine whether TLK1B mutant cells show any defects in formation of Rad9 foci. In order to look at the Rad9 foci associated with chromatin, soluble proteins were removed before fixation (as described in "Methods" section). In comparison with the Wt, cells expressing Mut TLK1B showed a drastic reduction in the formation of Rad9 foci. The Wt cells had much brighter and greater number of Rad9 foci (Fig. 2a, b) in comparison with the Mut TLK1B expressing cells. When the soluble proteins were not removed before fixation we saw an increased accumulation of Rad9 in the cytoplasm in the Mut TLK1B expressing cells (Additional file 1: Figure S1).

## Cells expressing Mut TLK1B show reduction in association of Rad9 with the chromatin

It has been shown that replication stress stimulates the association of Rad9 with the chromatin [35]. We next wanted to see if this phosphorylation may result in early dissociation of Rad9 from the chromatin in these cells. To study association and dissociation kinetics of Rad9 in the cells expressing Wt and Mut TLK1B we performed chromatin fractionation assay after treating cells with HU and allowing them to recover. In the Wt TLK1B expressing cells Rad9 was largely associated with the chromatin fraction at recovery time t = 0 (Fig. 3a). Rad9 was found in the chromatin bound fraction until 4 h of recovery period. However at around 6 to 8 h of recovery time the Rad9 levels decreased in the chromatin bound fraction. A very different pattern was seen in the Mut TLK1B expressing cells (Fig. 3a). These cells had very low levels of Rad9 bound to the chromatin fraction until 4 h of recovery period (significantly lower than Wt expressing cells). However after 6 h of recovery time the Rad9 levels increased in the chromatin bound fraction. We believe that this is due the accumulation of DNA damage (including DSBs) at this time, and that this may trigger a mechanism for the re-import of Rad9 into the nucleus that is independent on S328 phosphorylation by TLK1B.

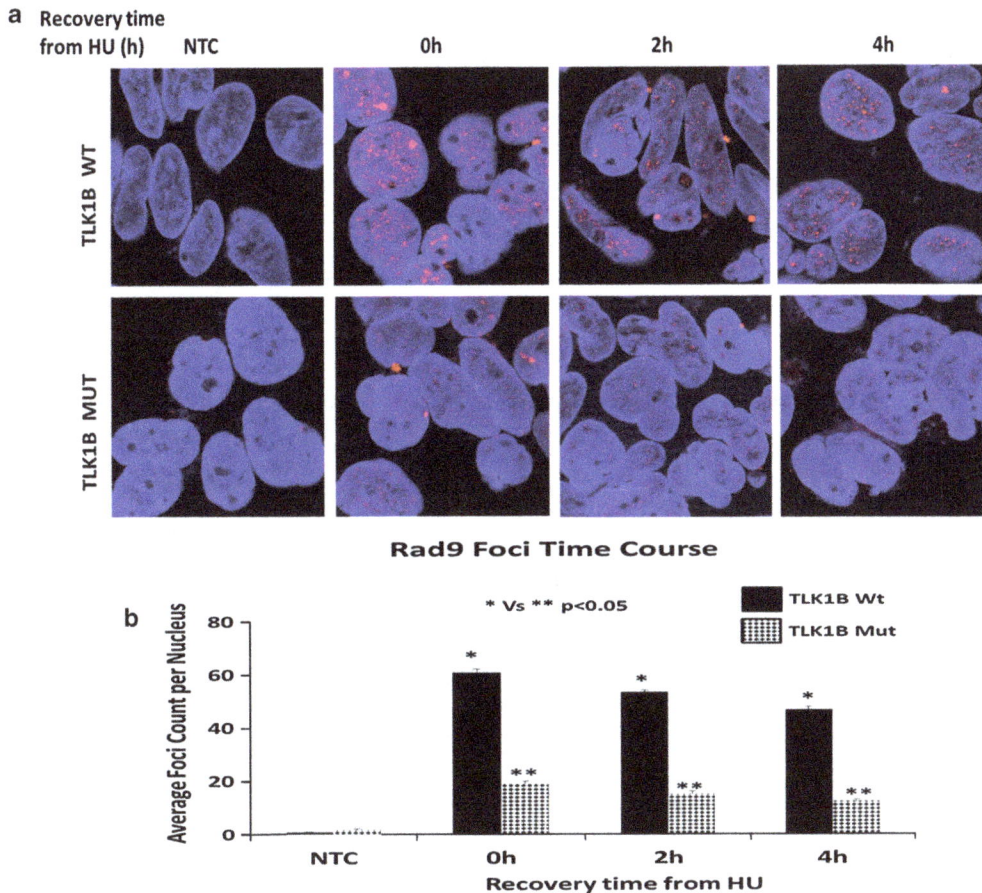

Fig. 2 a Rad9 foci in the TLK1B Wt and Mut overexpressing cells were examined after removing the soluble proteins and fixing chromatin bound proteins (methods). b Quantitative analysis of the Rad9 foci per cell averaged over 100 cells per data point. All the experiments were performed independently at least three times. Values are mean ± SE

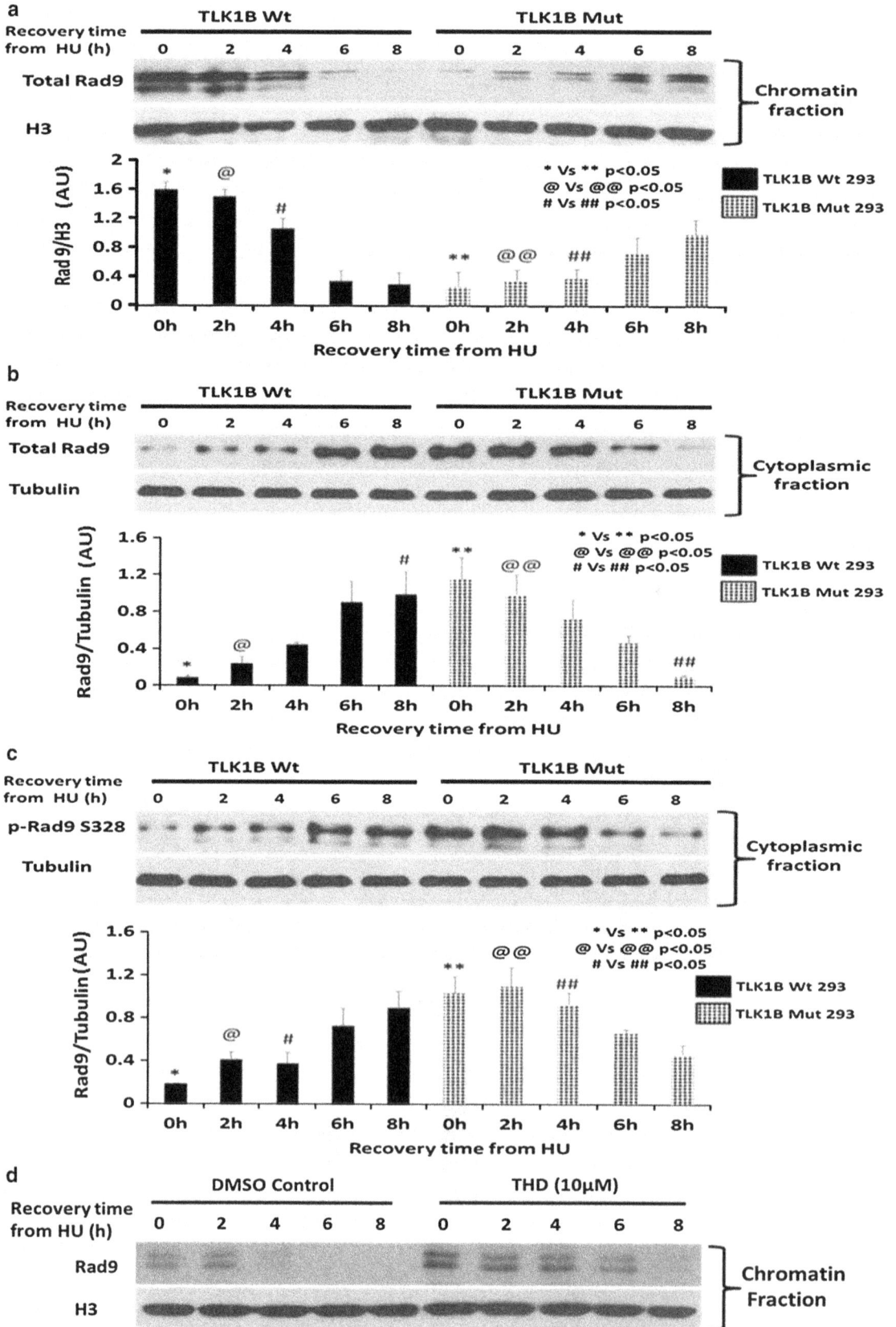

**Fig. 3 a** Distribution of Rad9 in the chromatin-bound fraction in Wt and Mut TLK1B expressing cells in response to treatment and recovery from HU. Values are mean ± standard error for n = 3. **b** Distribution of Rad9 in the cytoplasmic fraction in Wt and Mut TLK1B expressing cells. Note that the Rad9 in the cytoplasm appears as a single band, but in the chromatin fraction appears as multiple bands. Rad9 gets multiply phosphorylated only once it gets loaded onto the chromatin. Values are mean ± SE for n = 3. **c** Distribution of p-Rad9 (S328) in the cytoplasmic fraction in Wt vs Mut TLK1B expressing cells. Note that p-Rad9 (S328) is predominantly cytoplasmic and is elevated in Mut TLK1B expressing cells during replication stress. Values are mean ± standard error for n = 3. **d** Distribution of Rad9 in the chromatin-bound fraction in HEK293 cells in response to treatment and recovery from HU in the presence or absence of TLK inhibitor (THD)

## Cells expressing Mut TLK1B show an increased accumulation of Rad9 and p-Rad9 S328 in the cytoplasmic fraction

Through fractionation we obtained cytoplasmic, soluble nuclear and chromatin bound proteins. Correct distribution was confirmed by immunoblots using tubulin and ORC2 antibodies (Fig. 4a). We expected that the phosphorylation of Rad9 at S328 would cause the release of Rad9 from chromatin fraction into the soluble fraction. When we analyzed the cytoplasmic fraction we saw a reciprocal distribution where the decrease of Rad9 from the chromatin fraction (Fig. 3a) corresponded with its increase in the cytoplasmic fraction (Fig. 3b) and vice versa. It should be noted that Rad9 in the cytoplasm appears as a single band, but gets phosphorylated on multiple residues once it gets loaded onto the chromatin, resulting in the appearance of multiple bands [36]. In the Wt TLK1B expressing cells until 2–4 h of recovery time, low levels of Rad9 were present in the cytoplasmic fraction as most of it was bound to the chromatin. However, around 6 h of recovery in these cells Rad9 levels decreased in the chromatin bound fraction and correspondingly increased in the cytoplasmic fraction. In the Mut TLK1B expressing cells, most of Rad9 was found in the cytoplasm until 4 h of recovery; however, around 6–8 h of recovery time Rad9 levels in the cytoplasm gradually decreased and correspondingly increased in the chromatin bound fraction. During replication stress Mut TLK1B expressing cells had increased levels of p-Rad9 S328 (Fig. 3c) in the cytoplasm. The pattern of Rad9 S328 phosphorylation in the cytoplasmic fraction exactly matched the pattern of total Rad9 (Fig. 3b) suggesting that the phosphorylation of Rad9 at S328 may be responsible for its accumulation in the cytoplasm. Previously using an adeno-HO-mediated cleavage system in MM3MG cells, we had shown that overexpression of kinase dead TLK delays the release of Rad9 and Rad17 upon repair of the DSB [14]. In our previous study we identified specific TLK inhibitors that prevent the TLK-mediated phosphorylation of Rad9 at S328 and cause defects in checkpoint recovery [31]. Mut TLK1B overexpressing cells, showing an increased TLK1B activity, showed an early dissociation of the Rad9 from the chromatin. We next wanted to examine if inhibition of

TLKs with specific inhibitor Thioridazine hydrochloride (THD) leads to an increase in association of Rad9 with the chromatin. As expected, we saw an increased association of Rad9 onto the chromatin in the presence of THD (Fig. 3d). Also Rad9 remained bound onto the chromatin for longer time in the presence of the THD (Fig. 3d).

## Phosphomimetic Rad9 S328D mutant is sufficient to accumulate Rad9 in the cytoplasm

We next wanted to examine if phosphorylation of Rad9 at S328 is sufficient to alter the nuclear localization of Rad9. In order to do so we generated a phosphomimetic flag-tagged Rad9 S328D mutant (as described in "Methods" section). Fractionation was performed on the cells overexpressing the flag-tagged Rad9 S328D mutant and flag-tagged Wt Rad9. These cells were incubated with or without HU. After fractionating the cytoplasmic, soluble nuclear and chromatin bound proteins the distribution of Rad9 was examined by immunoblotting with either anti-flag antibody to specifically detect overexpressed flag-tagged Wt or Mut Rad9 (Fig. 4b) or anti-Rad9 antibody to detect both the endogenous and overexpressed Rad9 (Fig. 4c). Immunoblotting with anti-flag antibody showed that the overexpressed Rad9 S328D mutant was localized exclusively in the cytoplasm while the overexpressed Wt Rad9 was present in all the three fractions (Fig. 4b). These results confirm that the phosphorylation of Rad9 at S328 is sufficient to accumulate Rad9 in the cytoplasm. As previously mentioned multiple Rad9 bands appear in the chromatin fraction due to phosphorylation of Rad9 at multiple sites. Cellular fractionation results were confirmed by immunolocalization (as described in "Methods" section), which showed that the flag-tagged Rad9 (S328D) localized to the cytoplasm whereas the Wt Rad9 was nuclear (Fig. 4d).

## Phosphomimic Rad9-T355D does not affect its nuclear/cytoplasmic localization

Scott Davey`s group has recently found Rad9 T355 as another phosphorylation site of TLK1 [30]. Interestingly T355 lies next to nuclear localization sequence (NLS), which starts at the residue 356 and ends at 364. Bioinformatics analysis with cNLS mapper predicted that Rad9 is localized exclusively in the nucleus, and T355D mutation

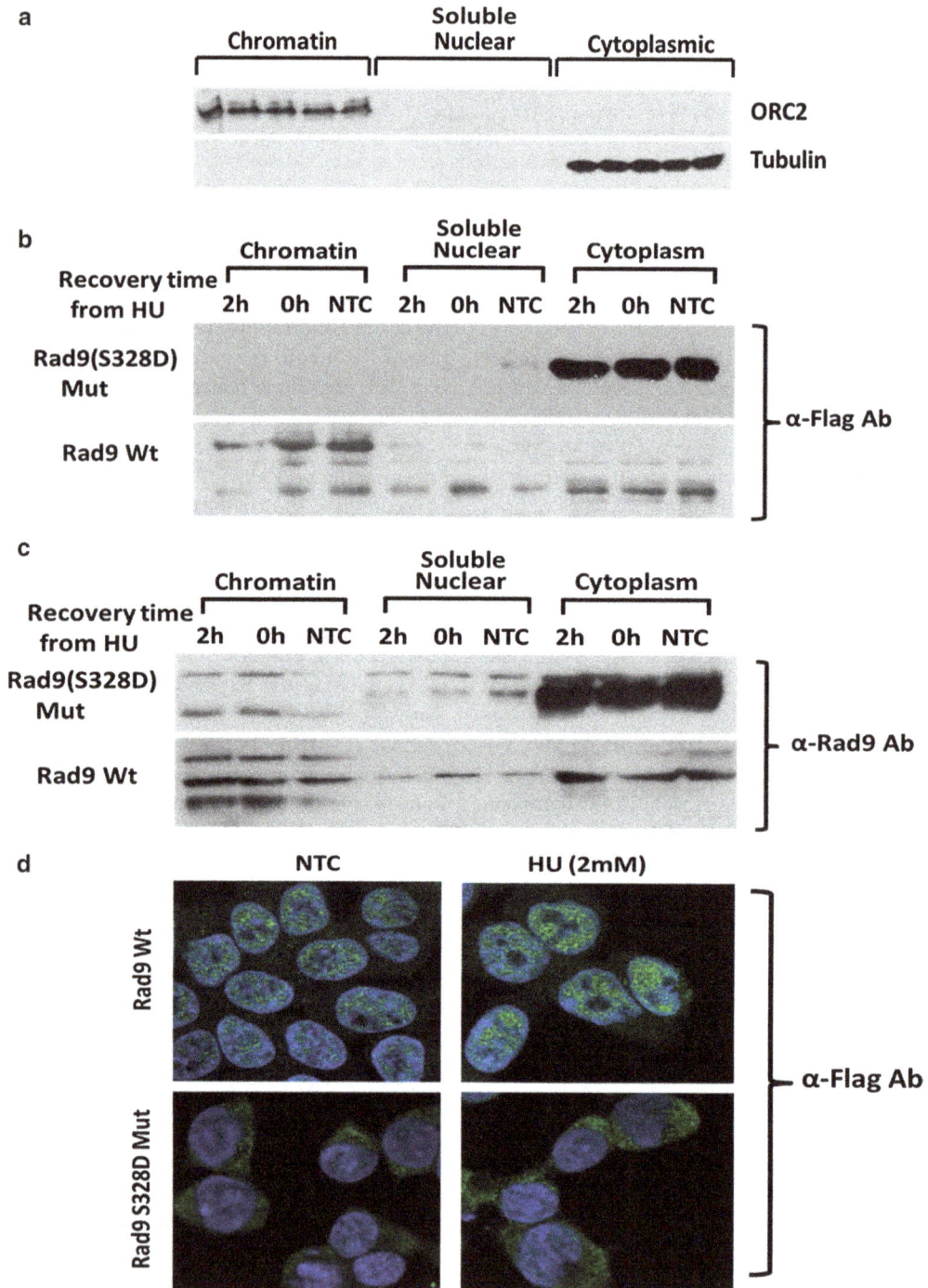

**Fig. 4 a** Verification of fractionation procedure by probing for ORC2 which is chromatin-bound and tubulin which is cytoplasmic. **b** α-Flag Ab detects only the overexpressed flag tagged Wt and Rad9 (S328D) mutant. **c** α-Rad9 Ab detects both the endogenous and the overexpressed Wt and Rad9 (S328D) mutant. **d** Immunolocalization shows that the Rad9 (S328D) mutant localizes primarily in the cytoplasm in contrast to the WT protein which is mostly nuclear

in Rad9 would cause a partial cytoplasmic accumulation. Our cellular fractionation data of Mut TLK1B overexpressing cells showed an increase in the accumulation of

total p-Rad9 (T355) in the cytoplasm but it was unclear if this simply reflected the total Rad9 level, i.e., phosphorylated also at S328 (Additional file 2: Figure S2A). We next wanted to examine if phosphorylation of Rad9 at T355 can alter the nuclear localization of Rad9. In order to do so we generated a phosphomimetic flag-tagged T355D mutant. The Rad9 T355D mutant was also found in the chromatin and nucleoplasmic fractions, although the chromatin to cytoplasmic ratio for the T355D mutant was not identical to the wt Rad9 (Additional file 2: Figure S2B).

## Mut TLK1B expressing cells show reduced association of Rad9 with Hus1 and WRN

Stable 9-1-1 complex is required for the recruitment of WRN at the stalled replication forks. Replication stress stimulates the interaction of Rad9 with Hus1 [37] and WRN [23]. Inactivation resistant Mut TLK1B accumulates Rad9 in the cytoplasm which would affect recruitment of WRN at stalled replication forks. In order to examine the stability of the 9-1-1 complex and WRN recruitment we examined the interaction of Rad9 with Hus1 and WRN in the Wt and Mut TLK1B expressing cells. Cell lysates were immunoprecipitated with an anti-Rad9 antibody and analyzed for the association with Hus1 and WRN. In the untreated Wt TLK1B expressing cells Rad9 was associated with Hus1 (Fig. 5b) and WRN (Fig. 5c) and this was stimulated upon treatment with HU. However, Mut TLK1B expressing cells showed reduced association of Rad9 with Hus1 (Fig. 5b) and WRN (Fig. 5c), suggesting that the Mut TLK1B reduces the stability of 9-1-1 and its interaction with WRN.

We next wanted to examine if the overexpressed Rad9 S328D mutant (flag tagged) can interact with WRN and Hus1 since this association should occur largely on chromatin. Immunoprecipitation of the overexpressed Rad9 S328D mutant using flag antibody showed that the interaction of mutant Rad9 with WRN and Hus1 was greatly impaired, while the overexpressed wt Rad9 was able to bind normally (Fig. 5e, f). In this case, the addition of HU did not increase the association of wt Rad9 with WRN and Hus1 probably due to the overexpression of the protein.

## Mut TLK1B expressing cells show increased amount of DNA damage

It has been shown that the interaction between WRN and the 9-1-1 complex prevent DSB formation at the stalled replication forks [38]. The fraction of WRN bound to chromatin (Fig. 6b) after treatment and release from HU is significantly reduced in the Mut TLK1B expressing cells, particularly at recovery period of 0 h ($p < 0.05$), when maximum number of fork are stalled. We next wanted to examine if the Mut TLK1B expressing cells show an increase in DSBs. In response to DSBs, ATM is autophosphorylated at S1981. This autophosphorylation

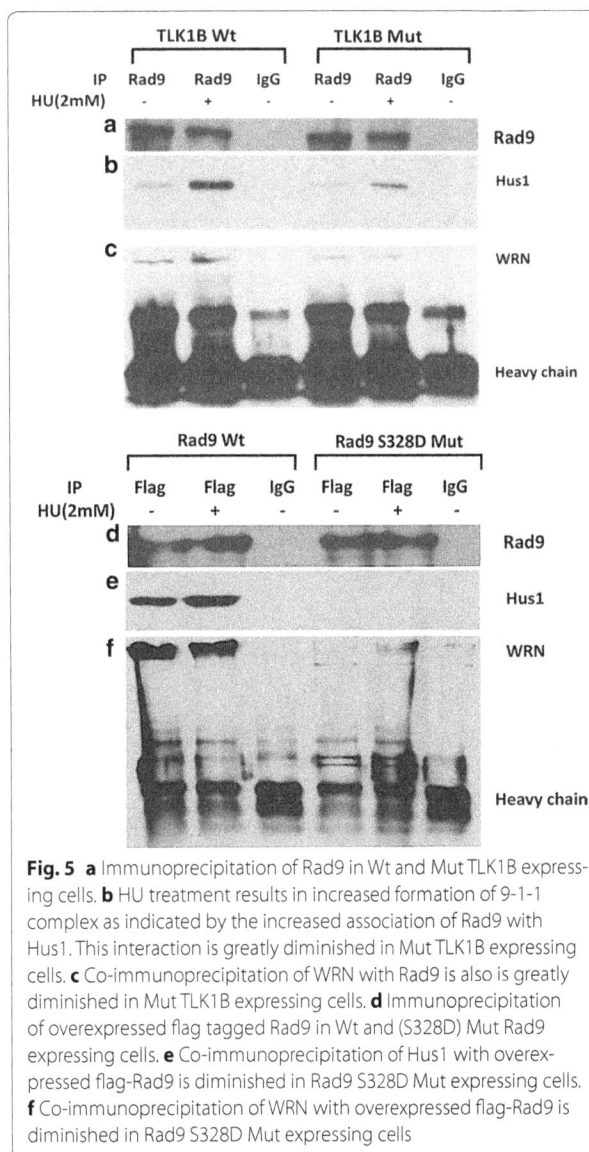

**Fig. 5** **a** Immunoprecipitation of Rad9 in Wt and Mut TLK1B expressing cells. **b** HU treatment results in increased formation of 9-1-1 complex as indicated by the increased association of Rad9 with Hus1. This interaction is greatly diminished in Mut TLK1B expressing cells. **c** Co-immunoprecipitation of WRN with Rad9 is also is greatly diminished in Mut TLK1B expressing cells. **d** Immunoprecipitation of overexpressed flag tagged Rad9 in Wt and (S328D) Mut Rad9 expressing cells. **e** Co-immunoprecipitation of Hus1 with overexpressed flag-Rad9 is diminished in Rad9 S328D Mut expressing cells. **f** Co-immunoprecipitation of WRN with overexpressed flag-Rad9 is diminished in Rad9 S328D Mut expressing cells

stabilizes ATM at the DSBs and is a marker of ATM activation. Prolonged treatment with HU generates DSBs. In the chromatin bound fraction of Wt TLK1B expressing cells at the recovery period of 0 h we saw increased levels of p-ATM S1981 (Fig. 6c). However after 4 h of recovery there was a decrease in p-ATM S1981 suggesting that most of the DSBs have been repaired. In comparison, Mut TLK1B expressing cells showed increased levels of p-ATM (Fig. 6c) until 8 h of recovery time (statistically significant). Activated ATM phosphorylates H2AX at S139, which is known as γ-H2AX. In the Wt TLK1B expressing cells γ-H2AX was present only until 4 h of recovery period; however, the Mut TLK1B expressing cells showed increased levels of γ-H2AX till 8 h of recovery time (Fig. 6d). These results suggest that in response

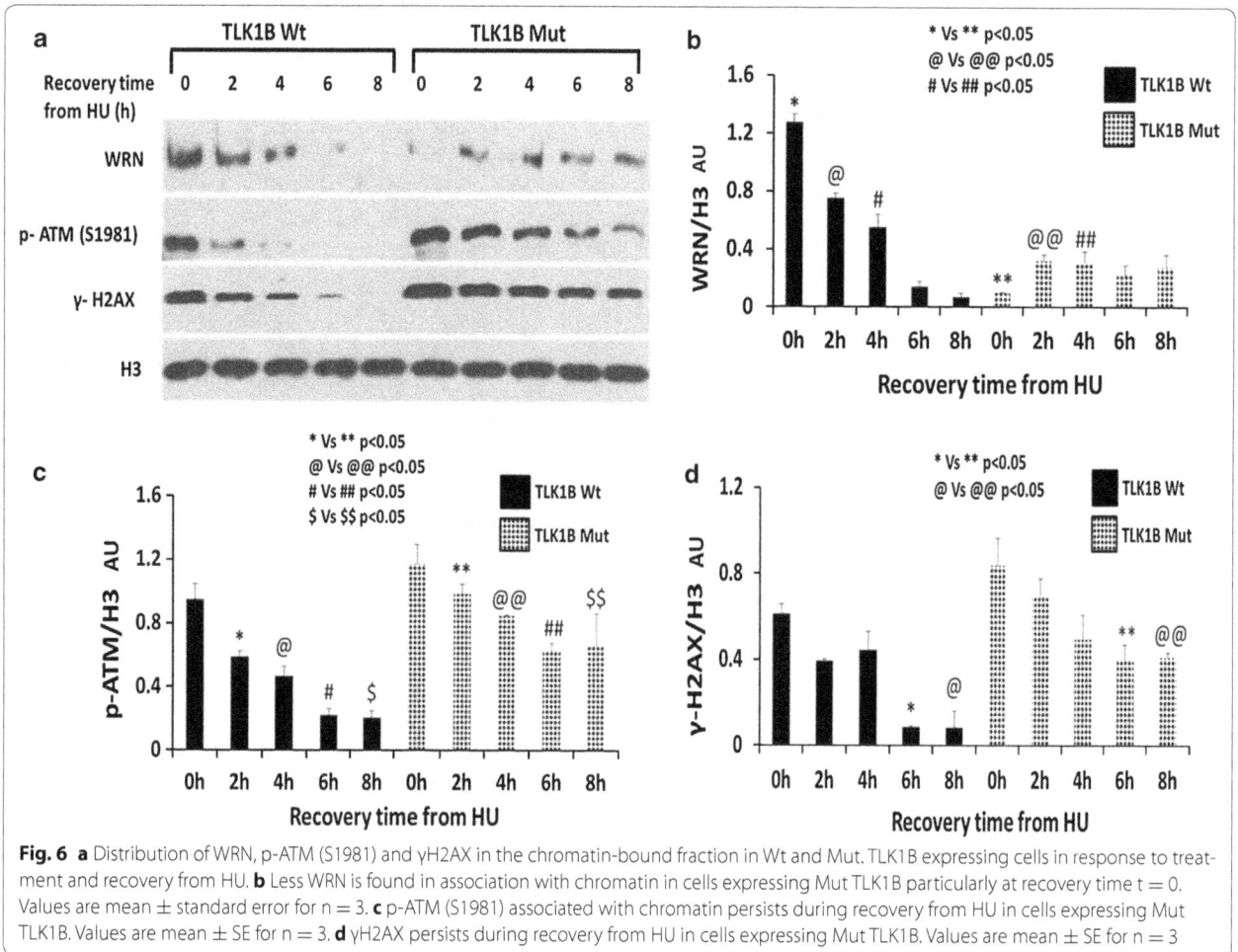

**Fig. 6 a** Distribution of WRN, p-ATM (S1981) and γH2AX in the chromatin-bound fraction in Wt and Mut. TLK1B expressing cells in response to treatment and recovery from HU. **b** Less WRN is found in association with chromatin in cells expressing Mut TLK1B particularly at recovery time t = 0. Values are mean ± standard error for n = 3. **c** p-ATM (S1981) associated with chromatin persists during recovery from HU in cells expressing Mut TLK1B. Values are mean ± SE for n = 3. **d** γH2AX persists during recovery from HU in cells expressing Mut TLK1B. Values are mean ± SE for n = 3

to replication stress Mut TLK1B expressing cells have an increase in DSBs that persist beyond 8 h of recovery.

Activation of ATM is essential for formation of γ-H2AX [39]. Studies from several groups have shown that ATM activation and H2AX phosphorylation can occur even in the absence of DSBs. [40–42]. Since we observed both activation and persistence of ATM and γ-H2AX in Mut TLK1B expressing cells, we probed for presence of DSBs in these cells by comet assays. We observed comets with large tail length and tail moment in the Mut TLK1B expressing cells (Fig. 7) after treatment and recovery from HU, whereas comets were not present at these times in control cells.

## HU treatment of Mut TLK1B expressing cells show a delay in cell cycle progression and recovery

We wanted to study the effect of this damage resistant kinase on cell cycle progression and recovery from HU induced replication arrest. Treatment with HU causes the cells to accumulate at the G1/S boundary. Washing away the HU allows completion of S-phase and then

synchronous entry into G2/M-phase, and then re-entry into the G1-phase of the next cycle. Cells overexpressing Wt TLK1B (Fig. 8a) when released from HU complete S-phase at around 8 h of recovery, synchronously enter into G2/M-phase at around 10 h and then re-enter into the G1 after 12 h. At around 14 h all the Wt cells have recovered. In contrast, Mut TLK1B overexpressing cells (Fig. 8b) remain arrested in the S-phase for 10 h. By 12 h only 7 % of Mut TLK1B expressing cells re-enter back into the G1-phase, in comparison to 35 % of cells expressing Wt TLK1B. By 14 h Mut TLK1B expressing cells still remain in the S- and G2-phase in comparison with the cells expressing Wt TLK1B which mostly re-enter into G-1 phase by that time. Thus, in response to replication stress cells expressing inactivation resistant Mut TLK1B arrest in the S- and G2-phase and show delay in re-entry into the cell cycle. We should stress that in the absence of HU the cell cycle profile is undistinguishable from Wt TLK1B or parental cells (not shown here). Comet assays also showed no presence of DSBs in the absence of HU treatment.

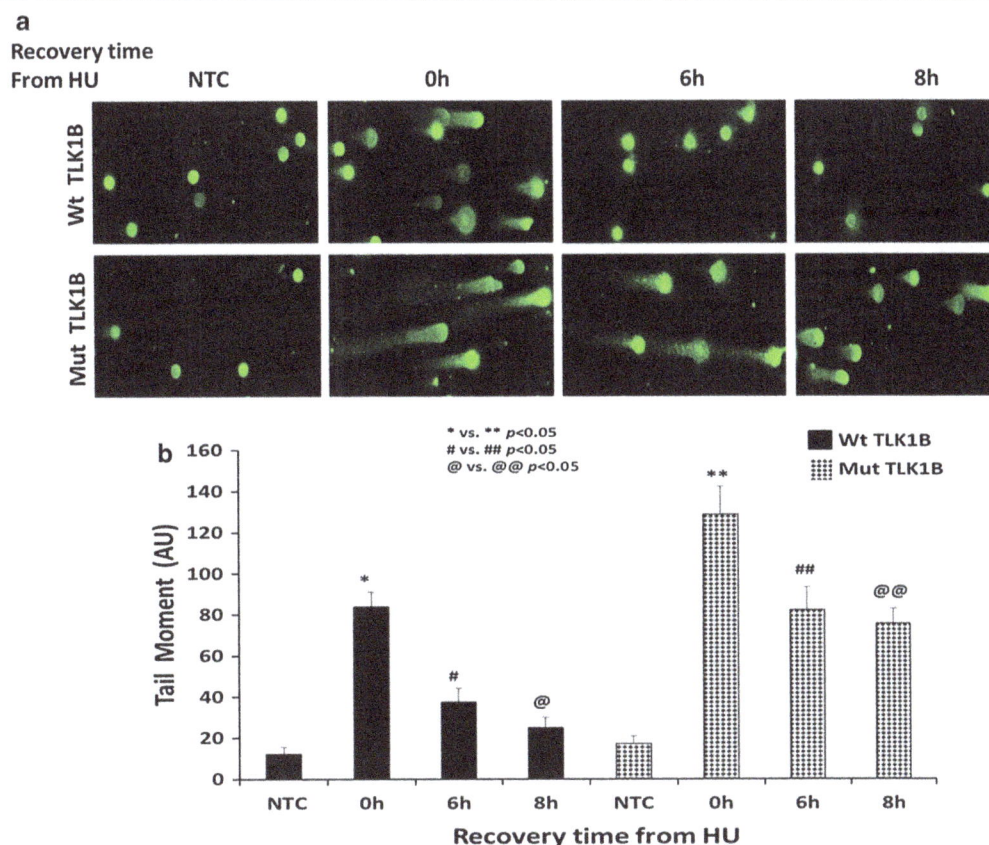

**Fig. 7** Comet assays was performed to measure the amount of unrepaired DNA damage in untreated cells or cells treated with 2 mM HU and recovery for indicated time points. **a** Representative images of the different time points. **b** Tail moments (tail DNA % × length of tail) as quantified for each cell using ImageJ OpenComet plugin. 50 comet images were measured for each treatment

## Mut TLK1B expressing cells display an altered checkpoint control

ATR is the primary kinase that phosphorylates RPA2 at T21 after HU treatment [43, 44]. Phosphorylation of RPA by ATR stimulates DNA synthesis and prevents ssDNA accumulation during replication stress [45]. Since less Rad9 was associated with the chromatin in the Mut TLK1B expressing cells, we expected to see a reduced phosphorylation of the ATR substrates in these cells. As expected, we observed a reduced phosphorylation of RPA2 T21 in the Mut TLK1B expressing cells (Fig. 9a). However, we didn't observe a reduction in Chk1 phosphorylation (Fig. 9a). In Mut TLK1B expressing cells we saw an increased accumulation of DSBs and ATM activation. Since we observed Chk1 phosphorylation in the Mut TLK1B expressing cells, we wanted to examine if this was mediated via ATM. It has been shown that ATM and ATR pathways exhibit a high degree of crosstalk [46–49]. ATM has been shown to phosphorylate Chk1 in vitro and increase in DSB accumulation has been shown to induce ATM mediated phosphorylation of Chk1 [50, 51].

Treatment with a specific ATM inhibitor (KU-55933) showed a slight reduction in Chk1 phosphorylation in the Mut cell line (Fig. 9b), suggesting that the increased Chk1 phosphorylation in these cells could be due to an increase in ATM activity, as shown in Fig. 6. Since we know that ATR is inhibited in these cells due to the lack of chromatin bound Rad9, there is the possibility that part of the Chk1 phosphorylation is instead due to the activity of DNA-PKcs in Mut TLK1B expressing cells [52–54]—this needs to be investigated further.

## Discussion

Tousled Like kinases (TLK) are serine/threonine kinases that play an important role in DNA repair. TLK overexpression is observed in multiple cancers and often corresponds to reduced sensitivity towards radiotherapy or chemotherapy due to the efficient repair in those tumors [32, 55]. Our lab identified TLK1B, which is a splice variant of TLK1 gene from a library of mRNAs that are translationally upregulated by overexpression of translation initiation factor 4E. TLK1B is known to protect cells

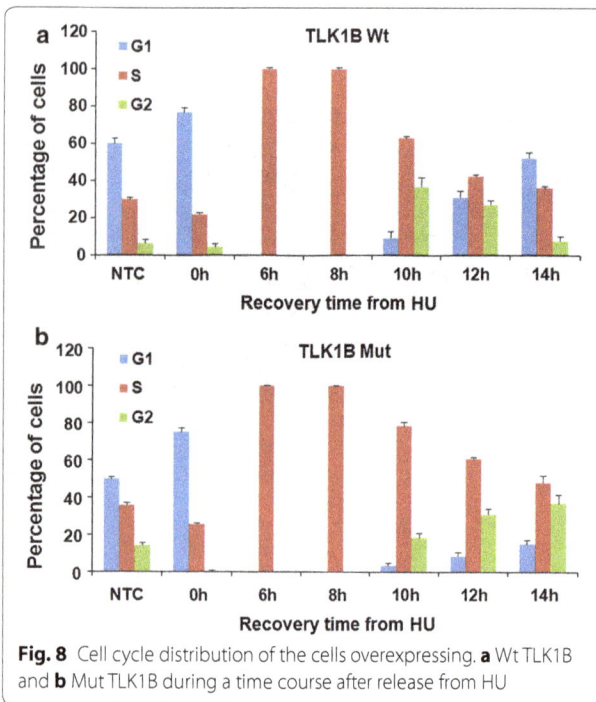

**Fig. 8** Cell cycle distribution of the cells overexpressing. **a** Wt TLK1B and **b** Mut TLK1B during a time course after release from HU

from genotoxic stress and is translationally upregulated in response to stress and DNA damage via mTOR-eIF4E pathway [13, 56]. At the same time, genotoxic stress leads to Chk1 mediated transient inactivation of TLK [11]. TLK1B promotes repair of damaged DNA in cooperation with Rad9 by facilitating the assembly of repair proteins to the sites of DNA damage. TLKs are the only kinases that phosphorylate Rad9 at S328. Rad9 is aberrantly expressed in prostate, breast, thyroid, skin, lung, and gastric cancers [57, 58]. It plays a major role in cell cycle checkpoint and DNA damage repair. It is essential for genomic stability as frequent chromosomal breakage is observed in cells in which both the Rad9 alleles are inactivated [59, 60]. Rad9, a member of PCNA-like 9-1-1 complex, contains 110 amino acid long C-terminal region that does not share homology with PCNA. This C-terminal region of Rad9 is extensively modified by phosphorylation. Some residues are constitutively phosphorylated while some are transiently phosphorylated in response to DNA damage and cell cycle position [36, 61]. In response to DNA damage, once TLK1B regains its kinase activity it transiently phosphorylates Rad9 at S328. In this study we wanted to elucidate the significance of this phosphorylation.

It has been shown that the stable 9-1-1 complex is required for WRN localization at the stalled replication forks [23], and is necessary for the ATR-dependent phosphorylation of WRN following replication fork stalling [62]. The possible role of the TLK1B mediated

phosphorylation of Rad9 on its interaction with WRN was not known. In this work we investigated the significance of the inhibition of TLK1B activity following replication stress and its consequence for Rad9 phosphorylation in relation to the DNA damage response (DDR) and checkpoint recovery. In previous studies using a kinase dead TLK1B we have found that the release of Rad9 from a DSB induced with HO nuclease was delayed well beyond the time required to repair the break [14]. Conversely, in this work we report that constitutive phosphorylation of Rad9 by the mutant TLK1B after release from HU results in its dissociation from chromatin and translocation to the cytoplasm (Fig. 10). Consistent with this model, the Rad9 (S328D) mutant was found exclusively in the cytoplasm and didn't associate with the chromatin. We also studied the consequences of this effect of inactivation resistant TLK1B mutant on DDR and cell cycle checkpoint. We found that the cells delayed in cell cycle recovery following release from HU and remained in the S- and G2 phase for several more hours compared to the cell expressing WT TLK1B. Our explanation for these results is that premature phosphorylation of Rad9 mediated by the inactivation resistant TLK1B mutant leads to the dissociation of the 9-1-1 complex from sites of damage and stalled forks (Fig. 10). This dissociation leads to fork collapse and generation of DSBs marked by increase in γH2AX and activated ATM. ATM activates Chk1, marked by its phosphorylation at S317 and S345, which mediates cell cycle arrest. The premature dissociation of the 9-1-1 complex from stalled forks in Mut TLK1B expressing cells results in lesser amount of Rad9 associated with Hus1 and WRN. In fact, there was less WRN in association with chromatin after release from HU in Mut TLK1B expressing cells, which would be expected to result in increased fork collapse, as WRN promotes replication fork recovery [62]. Thus, in this study we have identified the significance of Rad9 S328 phosphorylation and have shown that the presence of the damage resistant active TLK1B targets Rad9 to the cytoplasm. We should emphasize that loss of Rad9 from the nucleus can increase the instances of chromosomal breakage leading to genomic instability. Scott Davey's group reported that TLK1 phosphorylates primarily Rad9(T335) and that plays a key role in cell cycle progression and G2/M checkpoint exit [30]. In our Mut TLK1B overexpressing cells, the phosphorylation of Rad9 (T355) was very weak, but more importantly, we could not see a difference in cytoplasmic/nuclear redistribution when a T355D mutant was used. The accumulation of Rad9 in the cytoplasm is consistent only with the S328 phosphorylation data and the S328D substitution, whereas the T335D phosphomimetic mutation was not sufficient to accumulate Rad9 in the cytoplasm.

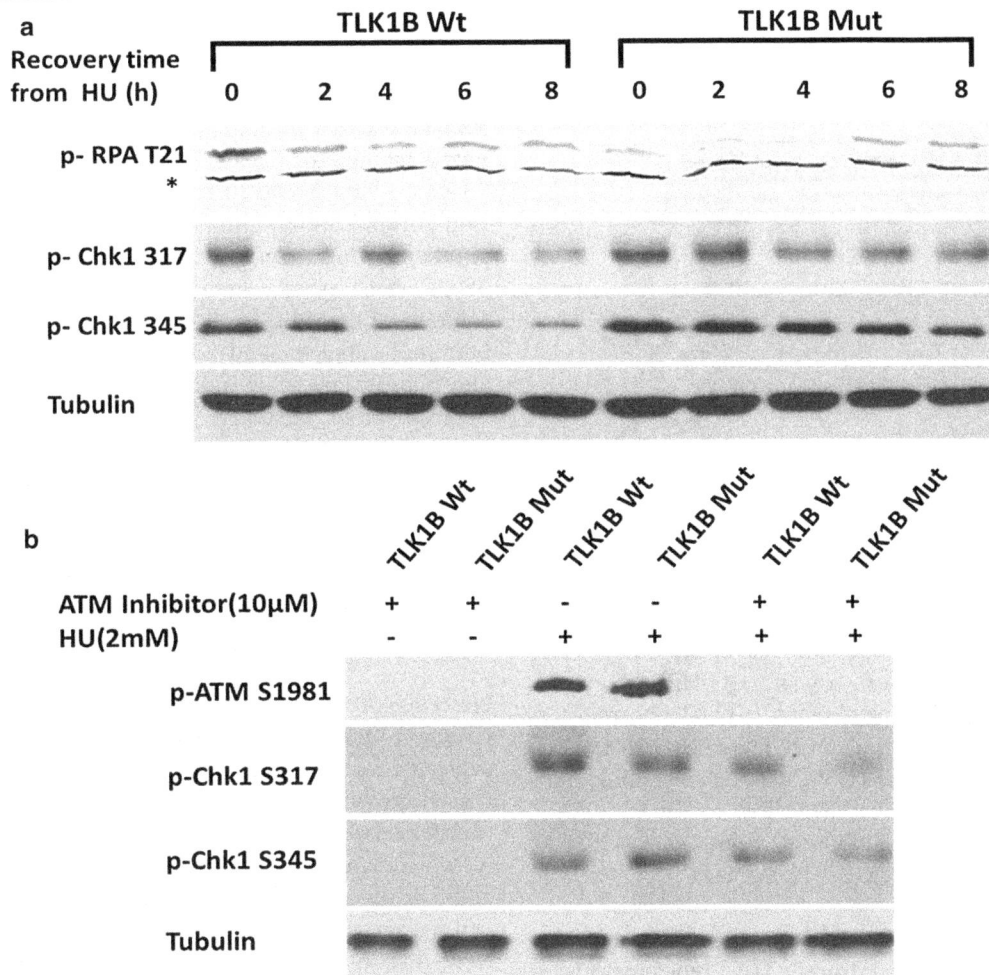

**Fig. 9** Chk1 activation in Wt and Mut TLK1B expressing cells after treatment and recovery from HU **a** Mut TLK1B expressing cells show reduced phosphorylation of RPA at T21. During recovery from HU, phosphorylation of Chk1 at S317 and S345 persists for 8 h in cells expressing Mut TLK1B, in contrast to Wt expressing cells. **b** Treatment of cells with KU-55933 leads to a reduction of Chk1 phosphorylation in the Mut TLK1B expressing cells

## Conclusions

Our collective work has demonstrated that TLK1B acts as a chaperone for Rad9. In the absence of TLK1 kinase activity, i.e., in presence of DNA damage or when a kinase-dead protein is used, we showed that TLK1B promoted the association of Rad9 with a DSB and presumably also SSB. Once the damage is repaired, the kinase activity recovers and then TLK1B phosphorylates Rad9 at S328, promoting its dissociation from 9-1-1 and export to the cytoplasm, thereby mediating the deactivation of the DDR. In conclusion transient inhibition of the kinase after DNA damage is crucial in retaining 9-1-1 at damage sites until repair is complete, and mutations in the TLK1 gene which can activate the kinase may potentially cause accumulation of DSBs.

## Methods

Hydroxyurea (Catalog No. H8627), doxorubicin hydrochloride (Catalog No. D1515), thioridazine hydrochloride (Catalog No. T9025), KU-55933 (Catalog No SML1109) and G418 disulfate salt (Catalog No. A1720) were purchased from Sigma. Dulbecco's modified Eagle's medium (DMEM) was obtained from Life Technologies (Catalog No. 12100-046), fetal bovine serum (FBS) was obtained from Atlanta Biologicals (Catalog No. 900108). Enhanced chemiluminescence solution was obtained from Thermo-Scientific (Catalog number No. 32106).

Antibodies used in this study were: rabbit α-Actin (Ab1801, Abcam), rabbit α-Rad9 phospho-328 (AP3225a, Abgent), rabbit α-Chk1 phospho-317 (AP3070a, Abgent), rabbit α-Chk1 phospho-345 (sc-17922, Santa Cruz Biotechnology), goat α-Rad9 (sc-10465, Santa Cruz

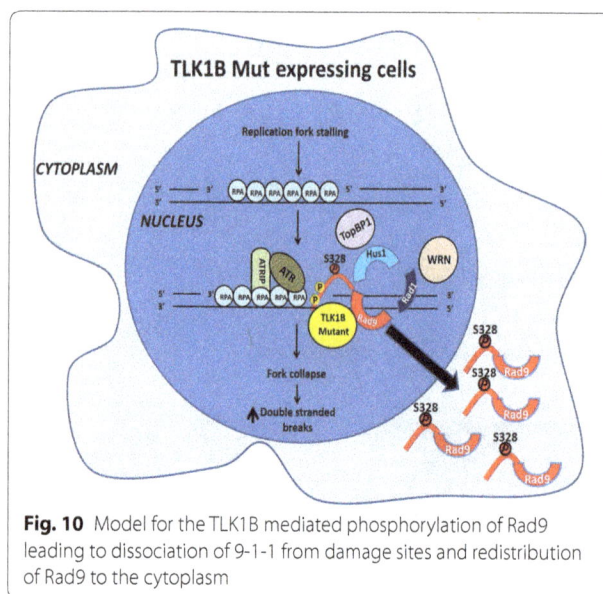

**Fig. 10** Model for the TLK1B mediated phosphorylation of Rad9 leading to dissociation of 9-1-1 from damage sites and redistribution of Rad9 to the cytoplasm

Biotechnology), mouse α-Rad9 (sc-8324, Santa Cruz Biotechnology), donkey α-goat IgG-HRP (sc-2020, Santa Cruz Biotechnology), rabbit α-TLK1 phospho-695 (4121S, Cell Signaling), α-rabbit IgG-HRP (7074S, Cell Signaling), α-mouse IgG-HRP (7076, Cell Signaling), rabbit α-TLK1 (GTX102891, GeneTex), mouse α- H2A.X phospho-139 (05-636, Millipore), mouse α-Flag (F1804, Sigma) and Rabbit α-Flag (F7425, Sigma).

### Cell culture

HEK293 cells, obtained from ATCC repository, were maintained in DMEM supplemented with 10 % FBS and 1 % penicillin–streptomycin at 37 °C in a humidified incubator with 5 % $CO_2$. For ATM inhibitor experiment, cells were pretreated with 10 μM KU-55933 for 2 h. 2 mM HU was then added into the media for 16 h. For inhibition of TLK cells were pretreated with 10 μM Thioridazine hydrochloride (THD) for 2 h before addition of 2 mM HU.

### Cell cycle analysis

HEK293 cells were seeded in T-25 flasks at a density of $2 \times 10^5$ cells/flask. Cells were treated with 2 mM HU for 16 h. They were briefly washed with 1× PBS and fresh media was added to the cells. Cells were allowed to recover for indicated time points. Cells were washed with 1× PBS and trypsinized. Cell suspensions were centrifuged at 1000 rpm for 5 min, and pellets were fixed with ethanol and stained with 50 μg/ml propidium iodide (Sigma, Catalog No. P4170). Percentages of cells within each of the cell cycle compartments (G0/G1, S, or G2/M) were determined using a FACS Calibur flow cytometer (Becton–Dickinson).

### Chromatin-bound fractionation

To isolate chromatin, ~$10^7$ cells cells were resuspended in buffer A (10 mM HEPES, [pH 7.9], 10 mM KCl, 1.5 mM $MgCl_2$, 0.34 M sucrose, 10 % glycerol and 1 mM DTT) supplemented with halt EDTA free protease and phosphatase inhibitors (Life Technologies, Catalog No. 78441). Triton X-100 (0.1 %) was added, and the cells were incubated for 5 min on ice. Cytosolic proteins were separated from nuclei by centrifugation (4 min, 1300×$g$). Nuclei were washed once in solution A, and then lysed in solution B (3 mM EDTA, 0.2 mM EGTA and 1 mM dithiothreitol) supplemented with halt EDTA free protease and phosphatase inhibitors for 30 min. Insoluble chromatin was then separated from soluble nuclear proteins by centrifugation (4 min, 1700×$g$), washed once in solution B, and collected by centrifugation (1 min, 10,000×$g$). The final chromatin pellet was resuspended in SDS sample buffer. Samples were sonicated for 15 s. Aliquots of each fraction were separated on sodium dodecyl sulfate–polyacrylamide (SDS-PAGE) gels and blotted onto polyvinylidene difluoride (PVDF) membranes.

### Western blot analysis

Cells were lysed in 1X SDS sample buffer. Lysates were sonicated for 15 s and heated at 100 °C for 5 min. Proteins were separated on 6–12 % SDS-PAGE gels and transferred to PVDF membranes (Millipore). Membranes were incubated with PBS containing 0.05 % Tween 20 and 5 % non-fat dry milk to block non-specific binding and were incubated with primary antibodies; membranes were then incubated with appropriate secondary antibodies conjugated to horseradish peroxidase. Immunoreactive bands were visualized using chemiluminescence reagent.

### Construction of mammalian expression vectors and generation of stable cell lines

We had cDNA of the human TLK1B cloned into the BK-shuttle vector [9]. To sub-clone TLK1B into pIRES2 vector the TLK1B ORF was amplified by PCR with the primers: 5′-GTACCGGAATTCAAAATTATTCAGACTGATCTC-3′; 5′-TAATTAGGATCCTGGAGGAAAGTCAGTAAGTAATTA-3′, containing an EcoRI and a BamHI tail respectively. The TLK1B PCR product was sub-cloned in the plasmid pIRES2-EGFP, which was cut with the same enzymes. TLK1B S457A mutant was generated from pIRES2-EGFP using the QuikChange Site-Directed Mutagenesis Kit (Stratagene, Catalog No. 200518) with the following primer: 5′-GAGAAGATCAAAT**G**CTTCAGGAAACCTACAC-3′.

The PCR product was transformed in bacteria, and the presence of the nucleotide substitution was confirmed by DNA sequencing. To generate stable cell lines, HEK293 cells were transfected using Lipofectamine 3000 (Life Technologies, L3000001) as per the manufacturer's protocol, and stably transfected cells were selected with G418 (500 µg/ml). The pIRES2-EGFP permits the translation of both the TLK1B gene cloned into the multiple cloning site and EGFP from the single bicistronic mRNA. After 30 days of selection in G418, cells expressing high levels of GFP were sorted by flow-cytometry.

We had cDNA of the human Flag-tagged Rad9 cloned into an episomal pREP10 vector. Stable expression of this vector requires expression of full length EBNA-1. To generate stable cell lines we transfected HEK293 c-18 cells (ATCC CRL-10852) which stably express full length EBNA-1. Under G418 selection HEK293 c-18 cells stably express EBNA-1 which is required to maintain pREP10 vector episomally. Maintenance of pREP10 vector requires hygromycin selection.

Rad9 S328D and T355D mutant was generated using the QuikChange Site-Directed Mutagenesis Kit (Stratagene, Catalog No. 200518) with the following primers: 5'-CTGCCCTCCATTTCCCTT**GAC**CCTGGCCCCCAG-3' (Rad9 S328D)5'- CAGTACAGTGCCTGGG**GAT**CCCCCACCCAAGAAGTTC-3' (Rad9 T355D).

The PCR product was transformed in bacteria, and the presence of the nucleotide substitutions were confirmed by DNA sequencing. To generate stable cell lines, HEK293 c-18 cells were transfected using Lipofectamine 3000 as per the manufacturer's protocol, and cells were selected under G418 (250 µg/ml) and hygromycin (50 µg/ml) challenge for 30 days.

## Immunofluorescence

Cells were grown on culture slides. To remove soluble proteins and fix chromatin-bound proteins, cells were pre-extracted in buffer1 (1 % Triton X-100, 10 mM HEPES pH 7.4, 10 mM NaCl and 3 mM MgCl$_2$) supplemented with halt EDTA free protease and phosphatase inhibitors for 5 min at 4 °C. Cells were then fixed in 4 % paraformaldehyde for 10 min at 4 °C and then treated in buffer 2 (0.5 % Triton X-100, 20 mM HEPES pH 7.4, 50 mM NaCl, 3 mM MgCl$_2$ and 300 mM sucrose) supplemented with halt EDTA free protease and phosphatase inhibitors for 5 min at 4 °C. Cells were then blocked for 1 h in SuperBlock solution.

To look at Rad9 cellular localization cells were fixed with methanol and then rehydrated with PBS and blocked for 1 h in SuperBlock solution. Staining with primary antibody was performed overnight at 4 °C in blocking solution, whereas species specific fluorescein/Texas red conjugated secondary antibody (Vector Labs) was applied for 1 h at RT, followed by counterstaining

with DAPI. All the primary antibodies were used at a 1:250 dilution, whereas the secondary antibodies were employed at a 1:500 dilution. Fluorescence images were captured using a Zeiss Axioskop 2 microscope.

## Immunoprecipitation

For co-immunoprecipitation (CoIP) experiment, cells were lysed in the lysis buffer (1 % Triton X-100, 0.5 % Na-Doxycholate, 150 mM NaCl, 2.5 mM MgCl2, 1 mM EGTA, 1 mM EDTA, 20 mM Tris/HCl pH 8.0) supplemented with halt EDTA free protease and phosphatase inhibitors. Cell lysates were sonicated for 20 cycles using Diagenode Bioruptor and then incubated with 250 U of benzonase. In total 1.5 mg of cell lysate was precleared with protein A/G Sepharose beads and then incubated overnight at 4 °C with rabbit polyclonal anti-Rad9 (4 µg; Santa Cruz Biotech) or anti-Rabbit IgG, and tumbled with 70 µl of protein A/G Sepharose beads for 4 h at 4 °C. After extensive washing in CoIP buffer, proteins were eluted by boiling treatment in 2X electrophoresis sample buffer prior to Western blotting analysis.

## Comet assay

Neutral comet assay was performed using comet assay kit from TREVIGEN (Catalog# 4250-050-K) following manufacturer's instructions. DNA was stained with SYBR Gold and fluorescence images were captured using a Zeiss Axioskop 2 microscope. Tail moments were quantified for each cell using ImageJ OpenComet plugin. 50 comet images were measured for each treatment.

## Statistics

Statistical analysis of data was done using one way analysis of variance (ANOVA) with Sigma Stat statistical software. A $p$ value of 0.05 or less was considered significant.

## Additional files

**Additional file 1: Figure S1.** Immunofluorescence was performed as described in methods using the α-Rad9 Ab to detect the cellular distribution of Rad9 in the overexpressed Wt and mut TLK1B expressing cells.

**Additional file 2: Figure S2.** A) Distribution of p-Rad9 (T355) in the cytoplasmic fraction in Wt and Mut TLK1B expressing cells. Note that p-Rad9 (T355) shows an increased cytoplasmic accumulation in Mut TLK1B expressing cells after treatment with HU. It should be noted that Immobilon Western HRP Substrate (Millipore catalog# WBKLS0500) was used which provides high sensitivity. B) Chromatin fractionation assay was performed to examine the cellular distribution of the overexpressed flag tagged Rad9 (T355D) mutant after treatment with 2 mM HU for 16 h and allowing them to recover at indicated timepoints. α-Flag Ab detects specifically the cellular distribution of overexpressed flag tagged Rad9 (T355D) mutant.

## Authors' contributions

SA and ADB designed the experiments. SA performed the experiments. SA and ADB analyzed the data and wrote the manuscript. Both authors read and approved the final manuscript.

## Acknowledgements

This work was supported by a Grant from the Feist Weiller Cancer Center from LSUHSC, Shreveport. We want to thank Dr. Scott Davey for the kind gift of p-Rad9(T355) antibody. We wish to thank Dr. Abhijit Rath for technical help.

## Competing interests

The authors declare that they have no conflict of interests.

## References

1. Roe J, Rivin C, Sessions R, Feldmann K, Zambryski P. The Tousled gene in *A. thaliana* encodes a protein kinase homolog that is required for leaf and flower development. Cell. 1993;75(5):939–50.
2. Lario LD, Ramirez-Parra E, Gutierrez C, Spampinato CP, Casati P. Anti-silencing function1 proteins are involved in ultraviolet-induced DNA damage repair and are cell cycle regulated by E2F transcription factors in Arabidopsis. Plant Physiol. 2013;162(2):1164–77.
3. Sen S, De Benedetti A. TLK1B promotes repair of UV-damaged DNA through chromatin remodeling by Asf1. BMC Mol Biol. 2006;7:37.
4. Wang Y, Liu J, Xia R, Wang J, Shen J, Cao R, Hong X, Zhu JK, Gong Z. The protein kinase TOUSLED is required for maintenance of transcriptional gene silencing in Arabidopsis. EMBO Rep. 2007;8(1):77–83.
5. Han Z, Riefler GM, Saam JR, Mango SE, Schumacher JM. The C. elegans Tousled-like kinase contributes to chromosome segregation as a substrate and regulator of the Aurora B kinase. Curr Biol. 2005;15(10):894–904.
6. Sunavala-Dossabhoy G, Li Y, Williams B, De Benedetti A. A dominant negative mutant of TLK1 causes chromosome missegregation and aneuploidy in normal breast epithelial cells. BMC Cell Biol. 2003;4:16.
7. Li Z, Gourguechon S, Wang CC. Tousled-like kinase in a microbial eukaryote regulates spindle assembly and S-phase progression by interacting with Aurora kinase and chromatin assembly factors. J Cell Sci. 2007;120(21):3883–94.
8. Sillje H, Nigg E. Identification of human Asf1 chromatin assembly factors as substrates of Tousled-like kinases. Curr Biol. 2001;11(13):1068–73.
9. Li Y, DeFatta R, Anthony C, Sunavala G, De Benedetti A. A translationally regulated Tousled kinase phosphorylates histone H3 and confers radioresistance when overexpressed. Oncogene. 2001;20(6):726–38.
10. Sunavala-Dossabhoy G, De Benedetti A. Tousled homolog, TLK1, binds and phosphorylates Rad9; tlk1 acts as a molecular chaperone in DNA repair. DNA Repair. 2009;8:87–102.
11. Carrera P, Moshkin Y, Gronke S, Sillje H, Nigg E, Jackle H, Karch F. Tousled-like kinase functions with the chromatin assembly pathway regulating nuclear divisions. Genes Dev. 2003;17(20):2578–90.
12. Han Z, Saam J, Adams H, Mango S, Schumacher J. The *C. elegans* Tousled-like kinase (TLK-1) has an essential role in transcription. Curr Biol. 2003;13:1921–9.
13. Sunavala-Dossabhoy G, Balakrishnan S, Sen S, Nuthalapaty S, De Benedetti A. The radioresistance kinase TLK1B protects the cells by promoting repair of double strand breaks. BMC Mol Biol. 2005;6:19.
14. Canfield C, Rains J, De Benedetti A. TLK1B promotes repair of DSBs via its interaction with Rad9 and Asf1. BMC Mol Biol. 2009;10:110.
15. Broustas CG, Lieberman HB. Contributions of Rad9 to tumorigenesis. J Cell Biochem. 2012;113(3):742–51.
16. Weiss RS, Enoch T, Leder P. Inactivation of mouse Hus1 results in genomic instability and impaired responses to genotoxic stress. Genes Dev. 2000;14(15):1886–98.
17. Hopkins KM, Auerbach W, Wang XY, Hande MP, Hang H, Wolgemuth DJ, Joyner AL, Lieberman HB. Deletion of mouse rad9 causes abnormal cellular responses to DNA damage, genomic instability, and embryonic lethality. Mol Cell Biol. 2004;24(16):7235–48.
18. Bao S, Lu T, Wang X, Zheng H, Wang LE, Wei Q, Hittelman WN, Li L. Disruption of the Rad9/Rad1/Hus1 (9-1-1) complex leads to checkpoint signaling and replication defects. Oncogene. 2004;23(33):5586–93.
19. Ellison V, Stillman B. Biochemical characterization of DNA damage checkpoint complexes: clamp loader and clamp complexes with specificity for 5' recessed DNA. PLoS Biol. 2003;1(2):E33.
20. Majka J, Binz SK, Wold MS, Burgers PM. Replication protein A directs loading of the DNA damage checkpoint clamp to 5'-DNA junctions. J Biol Chem. 2006;281(38):27855–61.
21. Parrilla-Castellar ER, Arlander SJ, Karnitz L. Dial 9-1-1 for DNA damage: the Rad9-Hus1-Rad1 (9-1-1) clamp complex. DNA Repair (Amst). 2004;3(8–9):1009–14.
22. Sancar A, Lindsey-Boltz LA, Unsal-Kacmaz K, Linn S. Molecular mechanisms of mammalian DNA repair and the DNA damage checkpoints. Annu Rev Biochem. 2004;73:39–85.
23. Pichierri P, Nicolai S, Cignolo L, Bignami M, Franchitto A. The RAD9-RAD1-HUS1 (9.1.1) complex interacts with WRN and is crucial to regulate its response to replication fork stalling. Oncogene. 2012;31(23):2809–23.
24. Oshima J. The Werner syndrome protein: an update. BioEssays. 2000;22(10):894–901.
25. Salk D. In vitro studies of Werner syndrome cells: aberrant growth and chromosome behavior. Basic Life Sci. 1985;35:419–26.
26. Fukuchi K, Martin GM, Monnat RJ Jr. Mutator phenotype of Werner syndrome is characterized by extensive deletions. Proc Natl Acad Sci U S A. 1989;86(15):5893–7.
27. Gebhart E, Bauer R, Raub U, Schinzel M, Ruprecht KW, Jonas JB. Spontaneous and induced chromosomal instability in Werner syndrome. Hum Genet. 1988;80(2):135–9.
28. Machwe A, Lozada E, Wold MS, Li GM, Orren DK. Molecular cooperation between the Werner syndrome protein and replication protein A in relation to replication fork blockage. J Biol Chem. 2011;286(5):3497–508.
29. Groth A, Lukas J, Nigg EA, Sillje HH, Wernstedt C, Bartek J, Hansen K. Human Tousled like kinases are targeted by an ATM- and Chk1-dependent DNA damage checkpoint. EMBO J. 2003;22(7):1676–87.
30. Kelly R, Davey SK. Tousled-like kinase-dependent phosphorylation of Rad9 plays a role in cell cycle progression and G2/M checkpoint exit. PLoS One. 2013;8(12):e85859.
31. Ronald S, Awate S, Rath A, Carroll J, Galiano F, Dwyer D, Kleiner-Hancock H, Mathis JM, Vigod S, De Benedetti A. Phenothiazine inhibitors of TLKs affect double-strand break repair and DNA damage response recovery and potentiate tumor killing with radiomimetic therapy. Gene Cancer. 2013;4(1–2):39–53.
32. Byrnes K, De Benedetti A, Holm N, Luke J, Nunez J, Chu Q, Meschonat C, Abreo F, Johnson L, Li B. Correlation of TLK1B in elevation and recurrence in doxorubicin-treated breast cancer patients with high eIF4E overexpression. J Am Coll Surg. 2007;204(5):925–33.
33. Medhurst AL, Warmerdam DO, Akerman I, Verwayen EH, Kanaar R, Smits VA, Lakin ND. ATR and Rad17 collaborate in modulating Rad9 localisation at sites of DNA damage. J Cell Sci. 2008;121(Pt 23):3933–40.
34. Polo SE, Jackson SP. Dynamics of DNA damage response proteins at DNA breaks: a focus on protein modifications. Genes Dev. 2011;25(5):409–33.
35. Zou L, Cortez D, Elledge SJ. Regulation of ATR substrate selection by Rad17-dependent loading of Rad9 complexes onto chromatin. Genes Dev. 2002;16(2):198–208.
36. Roos-Mattjus P, Hopkins KM, Oestreich AJ, Vroman BT, Johnson KL, Naylor S, Lieberman HB, Karnitz LM. Phosphorylation of human Rad9 is required for genotoxin-activated checkpoint signaling. J Biol Chem. 2003;278(27):24428–37.
37. Wu X, Shell SM, Zou Y. Interaction and colocalization of Rad9/Rad1/Hus1 checkpoint complex with replication protein A in human cells. Oncogene. 2005;24(29):4728–35.
38. Pichierri P, Ammazzalorso F, Bignami M, Franchitto A. The Werner syndrome protein: linking the replication checkpoint response to genome stability. Aging. 2011;3(3):311–8.
39. Savic V, Yin B, Maas NL, Bredemeyer AL, Carpenter AC, Helmink BA, Yang-Iott KS, Sleckman BP, Bassing CH. Formation of dynamic gamma-H2AX domains along broken DNA strands is distinctly regulated by ATM

and MDC1 and dependent upon H2AX densities in chromatin. Mol Cell. 2009;34(3):298–310.

40. Bakkenist CJ, Kastan MB. DNA damage activates ATM through inter-molecular autophosphorylation and dimer dissociation. Nature. 2003;421(6922):499–506.

41. Kumar R, Horikoshi N, Singh M, Gupta A, Misra HS, Albuquerque K, Hunt CR, Pandita TK. Chromatin modifications and the DNA damage response to ionizing radiation. Front Oncology. 2012;2:214.

42. Deem AK, Li X, Tyler JK. Epigenetic regulation of genomic integrity. Chromosoma. 2012;121(2):131–51.

43. Olson E, Nievera CJ, Klimovich V, Fanning E, Wu X. RPA2 is a direct down-stream target for ATR to regulate the S-phase checkpoint. J Biol Chem. 2006;281(51):39517–33.

44. Sakasai R, Shinohe K, Ichijima Y, Okita N, Shibata A, Asahina K, Teraoka H. Differential involvement of phosphatidylinositol 3-kinase-related protein kinases in hyperphosphorylation of replication protein A2 in response to replication-mediated DNA double-strand breaks. Genes Cells. 2006;11(3):237–46.

45. Vassin VM, Anantha RW, Sokolova E, Kanner S, Borowiec JA. Human RPA phosphorylation by ATR stimulates DNA synthesis and prevents ssDNA accumulation during DNA-replication stress. J Cell Sci. 2009;122(Pt 22):4070–80.

46. Kastan MB, Lim DS. The many substrates and functions of ATM. Nat Rev Mol Cell Biol. 2000;1(3):179–86.

47. Abraham RT. Cell cycle checkpoint signaling through the ATM and ATR kinases. Genes Dev. 2001;15(17):2177–96.

48. Shiloh Y. ATM and related protein kinases: safeguarding genome integrity. Nat Rev Cancer. 2003;3(3):155–68.

49. Bartek J, Lukas J. Chk1 and Chk2 kinases in checkpoint control and cancer. Cancer Cell. 2003;3(5):421–9.

50. Gatei M, Sloper K, Sorensen C, Syljuasen R, Falck J, Hobson K, Savage K, Lukas J, Zhou BB, Bartek J, et al. Ataxia-telangiectasia-mutated (ATM) and NBS1-dependent phosphorylation of Chk1 on Ser-317 in response to ionizing radiation. J Biol Chem. 2003;278(17):14806–11.

51. Sorensen CS, Syljuasen RG, Falck J, Schroeder T, Ronnstrand L, Khanna KK, Zhou BB, Bartek J, Lukas J. Chk1 regulates the S phase checkpoint by coupling the physiological turnover and ionizing radiation-induced accelerated proteolysis of Cdc25A. Cancer Cell. 2003;3(3):247–58.

52. Liu S, Opiyo SO, Manthey K, Glanzer JG, Ashley AK, Amerin C, Troksa K, Shrivastav M, Nickoloff JA, Oakley GG. Distinct roles for DNA-PK, ATM and ATR in RPA phosphorylation and checkpoint activation in response to replication stress. Nucleic Acids Res. 2012;40(21):10780–94.

53. Lin YF, Shih HY, Shang Z, Matsunaga S, Chen BP. DNA-PKcs is required to maintain stability of Chk1 and Claspin for optimal replication stress response. Nucleic Acids Res. 2014;42(7):4463–73.

54. Vidal-Eychenie S, Decaillet C, Basbous J, Constantinou A. DNA structure-specific priming of ATR activation by DNA-PKcs. J Cell Biol. 2013;202(3):421–9.

55. Wolfort R, De Benedetti A, Nuthalapaty S, Yu H, Chu Q, Li B. Up-regulation of TLK1B by eIF4E overexpression predicts cancer recurrence in irradiated patients with breast cancer. Surgery. 2006;140:161–9.

56. Sunavala-Dossabhoy G, Fowler M, De Benedetti A. Translation of the radioresistance kinase TLK1B is induced by gamma-irradiation through activation of mTOR and phosphorylation of 4E-BP1. BMC Mol Biol. 2004;5:1.

57. Lieberman HB, Bernstock JD, Broustas CG, Hopkins KM, Leloup C, Zhu A. The role of RAD9 in tumorigenesis. J Mol Cell Biol. 2011;3(1):39–43.

58. Zhu A, Zhang C, Lieberman H. Rad9 has a functional role in human prostate carcinogenesis. Cancer Res. 2008;68(5):1267–74.

59. Pandita RK, Sharma GG, Laszlo A, Hopkins KM, Davey S, Chakhparonian M, Gupta A, Wellinger RJ, Zhang J, Powell SN, et al. Mammalian Rad9 plays a role in telomere stability, S- and G2-phase-specific cell survival, and homologous recombinational repair. Mol Cell Biol. 2006;26(5):1850–64.

60. Pandita TK. Role of mammalian Rad9 in genomic stability and ionizing radiation response. Cell Cycle. 2006;5(12):1289–91.

61. St Onge RP, Besley BD, Pelley JL, Davey S. A role for the phosphorylation of hRad9 in checkpoint signaling. J Biol Chem. 2003;278(29):26620–8.

62. Ammazzalorso F, Pirzio LM, Bignami M, Franchitto A, Pichierri P. ATR and ATM differently regulate WRN to prevent DSBs at stalled replication forks and promote replication fork recovery. EMBO J. 2010;29(18):3156–69.

# 12

# Identification of G-quadruplex structures that possess transcriptional regulating functions in the *Dele* and *Cdc6* CpG islands

Daniyah H. Bay[1,2†], Annika Busch[1,3†], Fred Lisdat[3], Keisuke Iida[4], Kazunori Ikebukuro[5], Kazuo Nagasawa[5], Isao Karube[1] and Wataru Yoshida[1*]

## Abstract

**Background:** G-quadruplex is a DNA secondary structure that has been shown to play an important role in biological systems. In a previous study, we identified 1998 G-quadruplex-forming sequences using a mouse CpG islands DNA microarray with a fluorescent-labeled G-quadruplex ligand. Among these putative G-quadruplex-forming sequences, G-quadruplex formation was verified for 10 randomly selected sequences by CD spectroscopy and DMS footprinting analysis. In this study, the biological function of the 10 G-quadruplex-forming sequences in the transcriptional regulation has been analyzed using a reporter assay.

**Results:** When G-quadruplex-forming sequences from the *Dele* and *Cdc6* genes have been cloned in reporter vectors carrying a minimal promoter and the luciferase gene, luciferase expression is activated. This has also been detected in experiments applying a promoterless reporter vector. Mutational analysis reveals that guanine bases, which form the G-tetrads, are important in the activation. In addition, the activation has been found to decrease by the telomestatin derivative L1H1-7OTD which can bind to the G-quadruplex DNA. When *Dele* and *Cdc6* CpG islands, containing the G-quadruplex-forming sequence, have been cloned in the promoterless reporter vector, the luciferase expression is activated. Mutational analysis reveals that the expression level is decreased by mutation on *Dele* G-quadruplex; however, increased by mutation on *Cdc6* G-quadruplex.

**Conclusion:** *Dele* and *Cdc6* G-quadruplex formation is significant in the transcriptional regulation. *Dele* and *Cdc6* G-quadruplex DNA alone possess enhancer and promotor function. When studied in more complex CpG islands *Dele* G-quadruplex also demonstrates promotor activity, whereas *Cdc6* G-quadruplex may possess a dual function of transcriptional regulation.

**Keywords:** *Cdc6*, *Dele*, G-quadruplex, Transcriptional regulation, 7OTD

## Background

Although genomic DNA usually adopts the canonical double helix structure [1–3], formation of non-canonical DNA structures, such as G-quadruplex (G4) DNA and i-motif, have also been found in human genomic DNA [2, 4, 5]. G4 is a DNA secondary structure that consists of two or more planar guanine tetrads and is stabilized by Hoogsteen hydrogen bonds along with a monovalent cation [3, 6–10]. DNA strand orientation and length of the loop sequences connecting the guanine runs provide a structural diversity to G4 forms, such as parallel, antiparallel, or a mixed structure [2, 9–12]. The characteristic topologies and existence of G4 in significant locations, such as in the telomeres and several promoters, suggest potential functions of G4 structures in gene regulation [13].

G4-forming sequences in promoter regions have been found to play a functional role in the suppression of proto-oncogenes [2, 14], such as *c-MYC* [15–18], *BCL-2*

---

*Correspondence: yoshidawtr@stf.teu.ac.jp
†Daniyah H. Bay and Annika Busch contributed equally to this work
[1] School of Bioscience and Biotechnology, Tokyo University of Technology, 1404-1 Katakuramachi, Hachioji, Tokyo 192-0982, Japan
Full list of author information is available at the end of the article

[19], *VEGF* [20], and *RET* [21]. Several studies on the G4-forming sequence within the nuclease hypersensitivity element III$_1$ (NHEIII$_1$) of the *c-MYC* promoter have reported that mutating the G4-forming sequence destabilizes the G4 structure, resulting in a three-fold increase in basal transcriptional activity of the *c-MYC* promoter [15, 16]. Binding of the *c-MYC* G4 to the cationic porphyrin TMPyP4 resulted in repression of the promoter activity [17]. Nucleolin has been identified as the *c-MYC* G4 binding protein that represses *c-MYC* expression [18]. The *BCL-2* gene contains a GC-rich region upstream of the P1 promoter that has been shown to be critically involved in the regulation of the *BCL-2* gene expression. It has been demonstrated that three individual G4 structures can be formed in this GC-rich region [19, 22]. As for *VEGF*, the stabilization of G4 DNA by quindoline derivatives represses gene transcription and consequently causes angiogenesis inhibition [20]. In addition, G4 ligands, such as TMPyP4 and telomestatin, have been shown to stabilize the G4 structure of the *RET* proto-oncogene promoter and lead to the repression of gene expression [21].

On the other hand, there are few examples of the transcriptional activation by G4 formation. A recent study of p32 G4 structure that is located in the P1 promoter of the *Bcl-2* gene has demonstrated that reduced transcription activity in the mutated vectors compared to the native vectors [23]. This result implicated a transcriptional activation by the G4 structure. Furthermore, in the insulin-linked polymorphic region (ILPR), G4 DNA is formed by a two-repeats of consensus sequence [24, 25], and the G-quartet formation was observed to activate transcription, where single/double mutation in the sequence has reduced promoter activity [26–28]. In addition, the G4-forming *c-myb* GGA repeat region has shown to play a contrast role of both a transcriptional repressor and an activator, where one or two deletions of $(GGA)_4$ motifs have increased *c-myb* promoter activity, while the deletion of all three regions has eliminated the promoter activity [29]. The involvement of G4 structures in regulating the transcription process suggests that they may hold the key to new therapeutic approaches in numerous areas of human disease, including cancer [30].

A genome-wide in silico analysis has demonstrated that sequences with the potential to form G4 motifs are enriched near transcription start sites, in the telomeres, ribosomal DNA, immunoglobulin heavy-chain switch regions, and CpG islands (CGIs) [31–33], suggesting widespread regulatory influence of the G4 motifs [34]. In a previous study, we identified 1998 G4-forming sequences in mouse CGIs using a mouse CGI microarray with a fluorescent-labeled G4 ligand L1Cy5-7OTD [35]. Among the identified G4-forming sequences, CD

spectroscopy and DMS footprinting analysis were performed on 10 randomly selected G4-forming sequences and the G4 structure formation was confirmed. In this study, our aim was to analyze the biological function of the 10 G4 DNAs the in transcriptional regulation.

## Results

### *Dele* and *Cdc6* G4-forming sequences activate reporter gene expression

The 10 G4 DNA sequences from *Jard2*, *Foxa2*, *Med4*, *Chd4*, *Ntpcr*, *Bmi1*, *Wt1*, *Sp130*, *Cdc6*, and *Dele* genes (Table 1; Additional file 1) have been cloned into the luciferase reporter vector with a minimal promoter containing a TATA-box promoter element. The sequences were 42–50 nucleotides long. The successful plasmid construction has been verified by gel shift experiments and sequencing of the reporter vector.

The *Dele* G4-forming sequence was found on two divergently overlapping genes, *Dele* and *1700086O06Rik*, located on the mouse genome; therefore, *Dele* G4 has been cloned into the luciferase vector in the forward and reverse direction (*Dele*-F G4 and *Dele*-R G4). The reporter vectors have been transfected into NIH3T3 cells and expressed luciferase activity is measured after 48 h of cultivation. In order to reduce experimental variability the measured enzyme activity has been normalized with respect to the pGL4.74 vector which is coding for Renilla luciferase. The relative luciferase activity with and without G-quadruplex forming sequences is evaluated for a minimum of three transfections. The results of this set of

**Table 1 DNA sequences used in this study**

| Name | Sequences (5′–3′) |
| --- | --- |
| *Jard2* | GTGAGGCTAGG**GGG**T**GG**TGG**TGGTGGG**GGTGAGGAAGGGAAAGAT |
| *Dele* | ATAGCGCCAGT**GGGTGGG**CTTAGATCTGGGAA**GGG**C**GGG**ACAGAG |
| *Foxa2* | GTCCAGGAAGGCTAGA**GGTGGGGGGG**C**GGG**TACCGGTGAAGGGAG |
| *Chd4* | TAAAGAGGA**GGG**T**GG**CGGTAGTGGA**GGGGGGGG**GTTGGAGTTGGTT |
| *Ntpcr* | CTTGTGTGTC**GGG**AA**GGGGGGGGGGGG**GAGCGTTGGAAACGCATGC |
| *Med4* | ACTT**GGG**TA**GG**C**GGG**CTT**GG**GAGGCTCCGTTGGACGTGGGGTCTA |
| *Bmi1* | CACTCTTTTT**GGGG**TT**GGG**ACTGA**GG**T**GG**C**GG**TCACGCGAGGATC |
| *Wt1* | AGTA**GGG**AGCTTT**GG**AAT**GAGGG**ATTAACACTTT**GGGGG**ACTTAGTC |
| *Sp130* | AG**GGG**TA**GG**TT**GGG**T**GG**TAAGAGGTGGTAAGCGGAGCGGCTGCTG |
| *Cdc6* | T**GGGG**AGGCT**GGG**T**GG**AGGACAAAGTAGAAATAAAAATACGGAAGTAGAT |

Guanine runs, which form the G-quadruplex structures, are shown in bold and mutation sites are underlined experiments are compiled in Fig. 1 and show rather high expression values for *Cdc6*, *Dele*-F and *Dele*-R.

Mutant vectors have also been constructed and investigated in order to verify that really the G4-forming sequence affects the transcriptional activity. Mutant-type DNAs have been designed by replacing one of the guanine triplets to thymines (Additional file 1). These guanines were reported to be strongly involved in G4 formation [35], which can be verified by CD measurements in this study (see below).

A comparative analysis of expression levels with the G4-forming sequences and the mutant sequences demonstrates a noticeable differences for *Foxa2*, *Med4*, *Chd4*, *Ntpcr*, *Cdc6*, *Dele*-F and, *Dele*-R, but no significant difference in *Jard2*, *Bmi1*, *Wt1*, and *Sp130* (Fig. 1; Additional file 2). These results indicate that these DNA sequences possess an enhancer activity on the luciferase reporter vector, which is connected to the possibility to form specific G4 secondary structures. We focused on *Cdc6* and *Dele* G4 DNAs because the G4 DNAs showed a remarkable high activation in wild-type G4 DNAs. Moreover in regards to the statistical analysis, the t-test between wild-type and mutant-type demonstrated that *Cdc6* and *Dele* showed more significant difference ($P < 0.0001$) compared to *Foxa2*, *Med2* ($P < 0.01$) and *Chd4*, *Ntpcr* ($P < 0.001$); therefore, we performed detail analysis of the *Cdc6* and *Dele* G4 DNAs.

It has been reported that enhancers frequently do not only interact with promoters but also promotor–promotor interaction is feasible, indicating that promoter can also work as enhancer for other gene promoters [36]. To investigate whether *Dele* and *Cdc6* G4 DNAs also possess

a promoter activity, they have been cloned into a promoterless vector and then a reporter assay is performed. It is found that *Dele*-F, *Dele*-R and *Cdc6* G4 DNAs activate luciferase expression in this system. Furthermore, the activation of protein expression is clearly decreased by thymine mutations in the G4 region (Fig. 2; Additional file 3). These results indicate that *Dele* and *Cdc6* G4 DNAs have a role in transcriptional activation, both as a promoter and enhancer. In order to perform this function, the formation of secondary structures seems to be essential.

In order to evaluate the function of *Dele* and *Cdc6* G4 DNAs in CpG islands (CGIs) that may contain more regulatory elements, the 696-bp DNA fragment containing 475-bp *Cdc6* CGI and 555-bp DNA fragment containing 477-bp *Dele* CGI have been cloned into the promoterless vector to analyze the transcriptional activities. The results demonstrate that the luciferase activities have increased more than one order of magnitude by the cloning of *Cdc6*, *Dele*-F and *Dele*-R CGI DNAs (Fig. 3; Additional file 4). The luciferase activities are higher than that of G4 alone cloning vectors, indicating that the CGIs contain cis-regulatory elements outside of the G4 forming regions.

The transcriptional activity of the mutant-type of *Cdc6* CGI DNA has been found higher than that of the wild-type. This result suggests that *Cdc6* G4 may possess a dual function of transcriptional regulation, such as *c-myc* G4 [16], swinging in both ways as an activator and as a suppressor under the effect of transcriptional factors.

The transcriptional activity of the mutant-type of *Dele*-F CGI DNA has decreased compared to the wild-type.

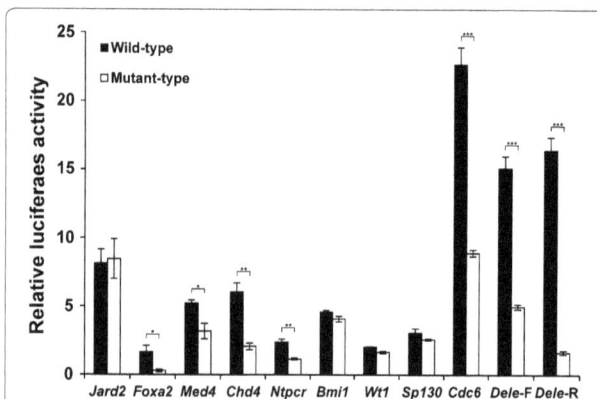

**Fig. 1** Luciferase reporter assay for evaluation of the enhancer activities of G4 DNA sequences. The G4-forming sequences were cloned into the pGL4.23 vector containing minimal promoter. *Black bars* represent the wild-types and *white bars* represent the mutant-types. Luciferase activities relative to the pGL4.23 vector are shown (mean ± SD, n = 3). Wild- and mutant-types samples t-test differences: *P < 0.01, **P < 0.001, ***P < 0.0001

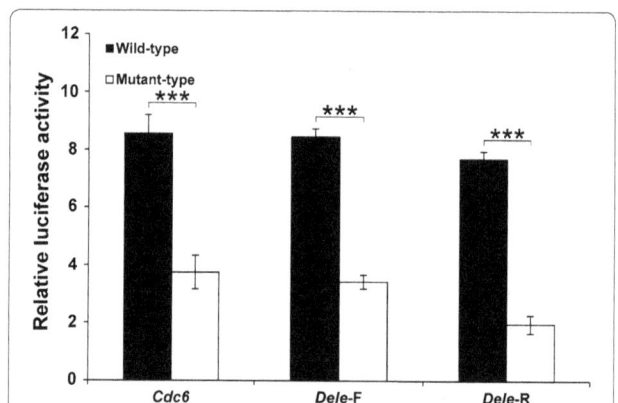

**Fig. 2** Reporter assay for evaluation of the promoter activities of *Cdc6*, *Dele*-F, and *Dele*-R G4 DNAs. The G4-forming sequences were cloned into the pGL4.10 vector not containing any promoter. *Black bars* represent the wild-types and *white bars* represent the mutant-types. Luciferase activities relative to the pGL4.10 vector are shown (mean ± SD, n = 3). Wild- and mutant-types samples t-test differences: ***P < 0.0001

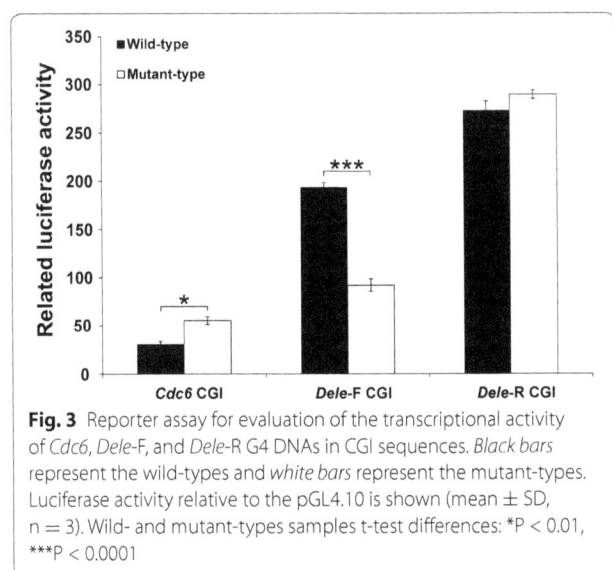

**Fig. 3** Reporter assay for evaluation of the transcriptional activity of Cdc6, Dele-F, and Dele-R G4 DNAs in CGI sequences. *Black bars* represent the wild-types and *white bars* represent the mutant-types. Luciferase activity relative to the pGL4.10 is shown (mean ± SD, n = 3). Wild- and mutant-types samples t-test differences: *P < 0.01, ***P < 0.0001

In contrast to *Dele*-F, the transcriptional activity of the *Dele*-R CGI mutant-type remains similar to that of the wild-type. Two divergently overlapping genes, *Dele* and *1700086O06Rik*, are located on the *Dele* CGI, suggesting that the *Dele* CGI may contain at least two regulatory sequences. Our results indicate that the *Dele* CGI contains a promoter for *1700086O06Rik*, an insulator, and a promoter for *Dele* (that also possesses enhancer activity). The insulator sequence may have a blocking activity that prevents promoter-enhancer interaction [37]; therefore, the enhancer activity of the *Dele* G4 for *1700086O06Rik* promoter has not been detected.

### G4 ligand suppresses the transcriptional activation of *Dele* and *Cdc6* G4 DNAs

Ligands which can bind to G-quadruplex structures can have different effect on the secondary DNA structure. Distortion or stabilization is often found depending on the binding mode of the ligand. Here the effect of a telomestatin derivative L1H1-7OTD has been investigated with respect to the effect of the G4 ligand on the transcriptional activity of the G4 DNA. This ligand can bind to the top G-tetrad structure through π-stacking and electrostatic interaction [38, 39].

In a first set of experiments, reporter assays with *Dele*-F, *Dele*-R and *Cdc6* G4 DNA reporter vectors carrying a minimal promoter have been performed in the presence of L1H1-7OTD. The vectors have been transfected into NIH3T3 cells and then the medium has been changed to the medium containing the G4 ligand 1 day after the transfection. After the 1 day cultivation, the luciferase activity has been measured. In this assay, results with L1H1-7OTD-treated and -untreated cells are normalized

with L1H1-7OTD-treated and -untreated controls, respectively. As a result, there has been no significant difference of luciferase activity in the presence and the absence of the ligand (Additional file 5). One reason for this could be that the G4 ligand can not easily enter the cells. Therefore in a second set of experiments, the vectors have been mixed with the G4 ligand first and then transfected into NIH3T3 cells in the presence of the G4 ligand in the medium. The results show an inhibition of the luciferase expression with the wild-types of the G4 ligand-treated *Dele*-F and *Cdc6* G4 DNAs, whereas no significant changes have been detected with all the mutant vectors (Fig. 4; Additional file 6). Although CD measurements show a stabilizing effect of the L1H1-7OTD binding to the G quadruplex (see below), this has no beneficial effect on the activation of protein expression. In this moment one can only hypothesize that the ligand binding to the top of the G4 structure modifies the interaction of the DNA structure with respective transcription factors by influencing the dynamics, the charge and the steric conditions.

### CD spectroscopy analyses of *Dele* and *Cdc6* G4 structures

CD spectra analysis has been performed to confirm the secondary structure formation of the wild- and mutant-type of *Dele* and *Cdc6* G4 DNA (Fig. 5; Additional file 7). The CD spectra of the wild-types show a positive cotton effect at around 264 nm for *Dele* G4 DNA, and at around 262 nm for *Cdc6* G4 DNA, with a negative effect at around 242 nm for *Dele* G4 DNA, and at around 240 nm for *Cdc6* G4 DNA. Thus, the G quadruplex formation can be confirmed under these experimental conditions.

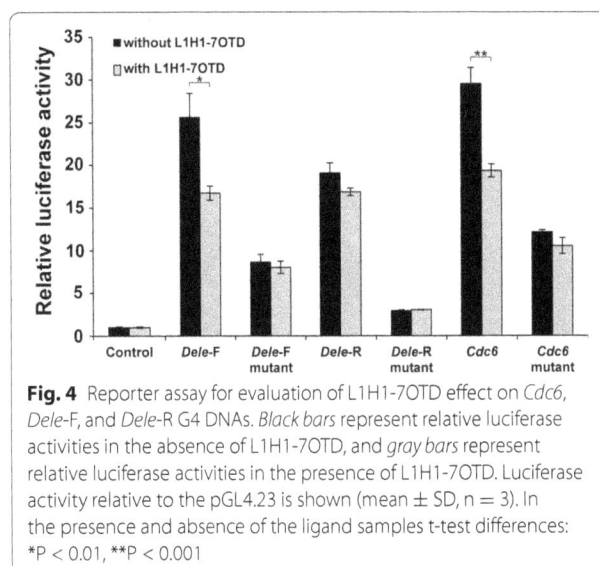

**Fig. 4** Reporter assay for evaluation of L1H1-7OTD effect on *Cdc6*, *Dele*-F, and *Dele*-R G4 DNAs. *Black bars* represent relative luciferase activities in the absence of L1H1-7OTD, and *gray bars* represent relative luciferase activities in the presence of L1H1-7OTD. Luciferase activity relative to the pGL4.23 is shown (mean ± SD, n = 3). In the presence and absence of the ligand samples t-test differences: *P < 0.01, **P < 0.001

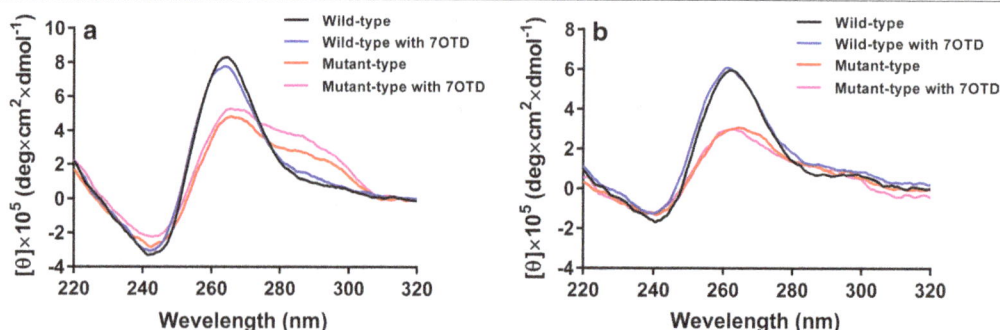

**Fig. 5** CD spectra of the *Dele* G4 (**a**) and *Cdc6* G4 DNAs (**b**). The wild-type (*black*), mutant-type (*red*), wild-type with L1H1-7OTD (*blue*), and mutant-type with L1H1-7OTD (*pink*) were analyzed at 25 °C in TK buffer (50 mM Tris–HCl, 100 mM KCl, pH 7.5)

In the presence of the G4 ligand (L1H1-7OTD), these spectra do not change much, indicating that the formation of parallel-type G4 structures in the wild-types of *Dele* and *Cdc6* G4 DNAs is not disturbed by the presence of the ligand (Additional files 8, 9) [40]. CD melting analysis reveals that the $T_m$ values are 70 °C for *Dele* G4 DNA and 77 °C for *Cdc6* G4 DNA in the absence of the G4 ligand, while the $T_m$ values increase to 72 °C for *Dele* G4 DNA, and to 81 °C for *Cdc6* G4 DNA by addition of the G4 ligand. These are clear arguments that the ligand binding to the G4 structure of the *Dele* and *Cdc6* increases the stability of the G4s at least under the chosen experimental conditions (Fig. 6; Additional file 10). The findings are in agreement with an NMR study on a structurally similar macrocyclic compound L2H2-6M(2) OTD, which has been found to bind to the top G-tetrad structure through π-stacking and electrostatic interaction. Here the folding topology did not change upon ligand binding [41].

CD spectra of the mutant-type sequences (mutating one G-run to T-run) that may form transient secondary structures, show only low molar ellipticity at around 260 nm and a shoulder between 280 and 300 nm, which are not related to the characteristic G4 structures (Fig. 5; Additional files 8, 9). Additionally in the presence of the G4 ligand, the CD spectra of the mutant-type sequences are not affected. These results strongly support the G4 formation of the two wild-type sequences of *Dele* and *Cdc6* and also verify the disability of the selected mutants to form such defined secondary structures. In consequence, the conclusions, which have been taken from the comparative analysis of mutated and wild type sequences, seem to be valid.

## Discussion

Recently, the death ligand signal enhancer (DELE) has been identified as a binding protein for the death-associated protein 3 (DAP 3), which is induced by various stimuli to regulate cell apoptosis. Stable expression of DELE induces apoptosis, whereas the knockdown of DELE has rescued HeLa cells from apoptosis induction [42]. It has also been reported that the cell-division-cycle 6 (CDC6) protein is essential for DNA replication, and the down-regulation of the *Cdc6* gene causes inhibition in cell growth accompanied by an increase in cell apoptosis [43]. Abnormal apoptosis is related to many diseases involving

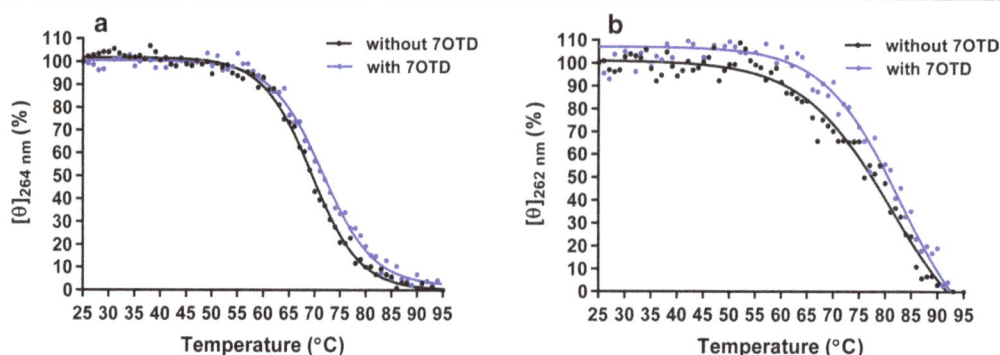

**Fig. 6** Circular dichroism melting curves of wild-types of *Dele* G4 (**a**) and *Cdc6* G4 DNAs (**b**) at 264 and 262 nm, respectively, with (*blue*) and without the ligand (*black*) in TK buffer (50 mM Tris–HCl, 100 mM KCl, pH 7.5)

atrophy [44], such as Parkinson's disease [45], with excessive cell death leading to tissue and organ damage. In contrast, some oncogenic mutations disrupt apoptosis, leading to tumor progression [46] or metastasis. Therefore, further analysis of *Dele* and *Cdc6* G4 DNAs may contribute to elucidate the apoptosis mechanism.

## Conclusions

The reporter assay for G4-forming DNA sequences has demonstrated that *Dele* and *Cdc6* G4 DNA may have the role of promoter and enhancer i.e., activating the transcription. This can be concluded from comparative experiments with mutant DNA structures. CD spectra verify the G4 formation of the wild type sequences and the disability of the studied mutants to form this secondary structure. The activation of transcription is inhibited by the telomestatin derivative L1H1-7OTD. CD spectra analysis demonstrates that binding of L1H1-7OTD stabilizes the G4 structures, but will also influence interaction with transcription factors. While for *Dele-F* G4 DNA transcriptional activation has also been verified in experiments with this sequence in CpG islands (CGIs), the *Cdc6* G4 DNA results here in suppression of protein expression. This indicates that *Cdc6* G4 may perform a dual functional role in the transcriptional regulation. In summary the results obtained suggest that *Dele* and *Cdc6* G4 structures are formed under physiological conditions in the cell and play a role in regulating transcription. Consequently, the study contributes to the elucidation of mechanisms of *Dele* and *Cdc6* gene regulation.

## Methods

### Plasmid construction

The G4 DNAs and mutant-type DNAs (Additional file 1) were cloned in the *Sfi*I site of the pGL4.23 [luc2/minP] or pGL4.10 [luc2] (Promega Corporation, Madison, WI, USA) and then transformed into *E. coli* DH5α (TOYOBO, Osaka, Japan). The plasmids were purified by PureYield Plasmid Miniprep System (Promega Corporation, Madison, WI, USA). *E. coli* HST04 *dam−/dcm−* (Takara, Tokyo, Japan) was transformed by the plasmids to prepare DNA methylation-free plasmids. All plasmids were sequenced using a 3730xl DNA analyzer (Thermo Fisher Scientific, Waltham, MA, USA). The *Dele* and *Cdc6* CGIs were amplified from C57BL/6 mouse genomic DNA by PCR. The PCR primers are shown in Additional file 11. The PCR products were purified by Wizard SV Gel and PCR Clean-Up System (Promega Corporation, Madison, WI, USA) and then digested by *Sfi*I (NEB, Ipswich, MA, USA). The products were cloned in the *Sfi*I site of pGL4.10 [luc2] and the plasmids were prepared as described above. To construct the mutant-types vectors, the site-directed mutagenesis was performed using KOD

-plus- mutagenesis kit (TOYOBO, Osaka, Japan) according to manufacturer's protocol (Additional file 11).

### Cell culture

NIH3T3 cells (RCB1862, RIKEN BRC) were cultured in dulbecco modified eagle medium (DMEM) medium (Sigma-Aldrich, St. Louis, Missouri, USA) containing 10% fetal bovine serum (Sigma-Aldrich, St. Louis, Missouri, USA), 1× penicillin–streptomycin-L-glutamine solution (Wako, Tokyo, Japan) at 37 °C in 5% $CO_2$.

### Reporter assay

The NIH3T3 cells were transfected with 100 ng of the firefly luciferase reporter vector and 100 ng of the Renilla luciferase control vector by Lipofectamine 3000 (Thermo Fisher Scientific, Waltham, MA, USA) according to the manufacturer's protocol. After 48 h, cells were harvested, and luciferase activities were measured using the Dual-Luciferase Reporter Assay System (Promega Corporation, Madison, WI, USA) and SPARK 10 M microplate reader (TECAN, Männedorf, Switzerland). The ratio of firefly luciferase activity to Renilla luciferase activity was utilized to calculate the firefly luciferase expression level in each cell and then these results were normalized with that of the pGL4.10 or pGL4.23 luciferase reporter vector. All reporter assays were performed in triplicate.

In the experiment to analyze the G4 ligand effect, two methods were used. In the first method, the 100 ng of the firefly luciferase reporter vector, and 100 ng of the Renilla control vector were transfected into NIH3T3 cells by Lipofectamine 3000. After 24 h, the medium was changed to one with 10 µM G4 ligand L1H1-7OTD. After 24 h, the luciferase activity was measured as described above. In the second method, the culture medium was changed to one with 10 µM G4 ligand L1H1-7OTD prior to transfection. The NIH3T3 cells were transfected with 100 ng of the firefly luciferase reporter vector, 100 ng of the Renilla control vector, and 10 µM of the G4 ligand L1H1-7OTD by Lipofectamine 3000, after which the luciferase activity was measured as described above. The luciferase expression level in the presence or absence of L1H1-7OTD was normalized with that of the pGL4.23 luciferase reporter vector in the presence or absence of L1H1-7OTD, respectively. All reporter assays were performed in triplicate.

### CD spectra measurement

Wild-types and mutant-types of the *Dele* and *Cdc6* G4 oligonucleotides were purchased from Macrogen, South Korea (Additional file 7). The oligonucleotides were dissolved as to create stock solutions (100 µM) in distilled water. Prior to use, all oligonucleotides were diluted to 15 µM in TK buffer (50 mM Tris–HCl, 100 mM KCl, pH 7.5). The oligonucleotides were denatured at 95 °C for 3 min and then allowed to cool to room temperature for 30 min. After heat-treatment, the oligonucleotides were diluted to 10 μM in the presence or absence of 10 μM L1H1-7OTD, and then incubated for 10 min before measurement. The CD spectra were measured with a J-1500 CD Spectrometer (JASCO, Tokyo, Japan) at 220 to 320 nm using a 1 mm path-length cuvette at every 5 °C from 20 to 95 °C. The baseline of each spectrum was corrected for signal contributions by the buffer with and without the G4 ligand. In the CD melting analysis, molar ellipticities were measured concomitantly at 1 °C intervals for wild-types of *Dele* and *Cdc6* G4 DNAs, at 264 nm and 262 nm, respectively. To determine $T_m$ values, the molar ellipticity at 25 °C was set as 100%, and the molar ellipticity at 95 °C was set as 0%. Using Ghraph-Pad Prism7 software, curve fitting was performed to yield normalized molar ellipticities. $T_m$ values were recorded as temperatures equivalent to 50% of the normalized molar ellipticity.

## Additional files

**Additional file 1.** G4 DNA sequences used in the reporter assay.

**Additional file 2.** Luciferase reporter assay results of G4 DNA sequences cloned in pGL4.23 vector.

**Additional file 3.** Luciferase reporter assay results of G4 DNA sequences cloned in pGL4.10 vector.

**Additional file 4.** Luciferase reporter assay results of CGI DNA sequences cloned in pGL4.10 vector.

**Additional file 5.** Luciferase reporter assay results with or without L1H1-7OTD for G4 DNA sequences cloned in pGL4.23 vector.

**Additional file 6.** Luciferase reporter assay results with or without L1H1-7OTD for G4 DNA sequences cloned in pGL4.23 vector.

**Additional file 7.** G4 DNA sequences used in the CD Spectroscopy.

**Additional file 8.** CD spectra results of *Dele* G4 DNA wild or mutant-type sequences with or without L1H1-7OTD.

**Additional file 9.** CD spectra results of *Cdc6* G4 DNA wild-type sequence with or without L1H1-7OTD.

**Additional file 10.** Detailed molar ellipticity of wild-types of *Dele* and *Cdc6* G4 DNAs in the absence and the presence of L1H1-7OTD.

**Additional file 11.** Primer sequences used for CGI vectors construction and mutagenesis.

## Abbreviations
7OTD: 7 oxazole telomestatin derivative; CD: circular dichroism; Cdc6: cell-division-cycle 6; CpG: cytosine-phosphate-guanine; Dele: death ligand signal enhancer; G4: G-quadruplex.

## Authors' contributions
WY conceived and designed the experiments. DHB and AB performed the experiments. DHB, AB, FL, Kli, Klk, KN, IK, and WY analyzed the data. DHB, FL and WY wrote the manuscript. All authors read and approved the final manuscript.

## Author details
[1] School of Bioscience and Biotechnology, Tokyo University of Technology, 1404-1 Katakuramachi, Hachioji, Tokyo 192-0982, Japan. [2] Biology Department, Umm Al-Qura University, Makkah, Kingdom of Saudi Arabia. [3] Biosystems Technology, Institute of Applied Life Sciences, Technical University of Applied Sciences Wildau, Wildau, Germany. [4] Graduate School of Science and Engineering, Saitama University, c/o Saitama Cancer Center, Saitama, Japan. [5] Department of Biotechnology and Life Science, Tokyo University of Agriculture and Technology, Tokyo, Japan.

## Acknowledgements
We thank Dr. Tomohiko Yamazaki (National Institute for Materials Science, Japan) for the kind gift of NIH3T3 cells.

## Competing interests
The authors declare that they have no competing interests.

## Funding
This study was supported by Tokyo University of Technology. The funding body had no role in the design of the study, collection, analysis, interpretation of data and in writing the manuscript.

## References
1. Watson JD, Crick FH. Molecular structure of nucleic acids; a structure for deoxyribose nucleic acid. Nature. 1953;171:737–8.
2. Qin Y, Hurley LH. Structures, folding patterns, and functions of intra-molecular DNA G-quadruplexes found in eukaryotic promoter regions. Biochimie. 2008;90:1149–71.
3. Hardin CC, Watson T, Corregan M, Bailey C. Cation-dependent transition between the quadruplex and Watson-Crick hairpin forms of d(CGCG3GCG). Biochemistry. 1992;31:833–41.
4. Phan AT, Mergny JL. Human telomeric DNA: G-quadruplex, i-motif and Watson-Crick double helix. Nucleic Acids Res. 2002;30:4618–25.
5. Simone R, Fratta P, Neidle S, Parkinson GN, Isaacs AM. G-quadruplexes: emerging roles in neurodegenerative diseases and the non-coding transcriptome. FEBS Lett. 2015;589:1653–68.
6. Gellert M, Lipsett MN, Davies DR. Helix formation by guanylic acid. Proc Natl Acad Sci USA. 1962;48:2013–8.
7. Williamson JR, Raghuraman MK, Cech TR. Monovalent cation-induced structure of telomeric DNA: the G-quartet model. Cell. 1989;59:871–80.
8. Parkinson GN, Lee MP, Neidle S. Crystal structure of parallel quadruplexes from human telomeric DNA. Nature. 2002;417:876–80.
9. Duquette ML, Handa P, Vincent JA, Taylor AF, Maizels N. Intracellular transcription of G-rich DNAs induces formation of G-loops, novel structures containing G4 DNA. Genes Dev. 2004;18:1618–29.
10. Schlachter C, Lisdat F, Frohme M, Erdmann VA, Konthur Z, Lehrach H, et al. Pushing the detection limits: the evanescent field in surface plasmon resonance and analyte-induced folding observation of long human telomeric repeats. Biosens Bioelectron. 2012;31:571–4.
11. Maizels N, Gray LT. The G4 genome. PLoS Genet. 2013;9:e1003468.
12. Agrawal P, Hatzakis E, Guo K, Carver M, Yang D. Solution structure of the major G-quadruplex formed in the human VEGF promoter in K+: insights into loop interactions of the parallel G-quadruplexes. Nucleic Acids Res. 2013;41:10584–92.
13. Bochman ML, Paeschke K, Zakian VA. DNA secondary structures: stability and function of G-quadruplex structures. Nat Rev Genet. 2012;13:770–80.

14. Brooks TA, Kendrick S, Hurley LH. Making sense of G-quadruplex and i-motif functions in oncogene promoters. FEBS J. 2010;277:3459–69.

15. Siddiqui-Jain A, Grand CL, Bearss DJ, Hurley LH. Direct evidence for a G-quadruplex in a promoter region and its targeting with a small molecule to repress c-MYC transcription. Proc Natl Acad Sci USA. 2002;99:11593–8.

16. Borgognone M, Armas P, Calcaterra NB. Cellular nucleic-acid-binding protein, a transcriptional enhancer of c-Myc, promotes the formation of parallel G-quadruplexes. Biochem J. 2010;428:491–8.

17. Seenisamy J, Rezler EM, Powell TJ, Tye D, Gokhale V, Joshi CS, et al. The dynamic character of the G-quadruplex element in the c-MYC promoter and modification by TMPyP4. J Am Chem Soc. 2004;126:8702–9.

18. Gonzalez V, Guo K, Hurley L, Sun D. Identification and characterization of nucleolin as a c-myc G-quadruplex-binding protein. J Biol Chem. 2009;284:23622–35.

19. Wang XD, Ou TM, Lu YJ, Li Z, Xu Z, Xi C, et al. Turning off transcription of the *bcl-2* gene by stabilizing the *bcl-2* promoter quadruplex with quindoline derivatives. J Med Chem Soc. 2010;53:4390–8.

20. Wu Y, Zan LP, Wang XD, Lu YJ, Ou TM, Lin J, et al. Stabilization of VEGF G-quadruplex and inhibition of angiogenesis by quindoline derivatives. Biochim Biophys Acta. 2014;1840:2970–7.

21. Guo K, Pourpak A, Beetz-Rogers K, Gokhale V, Sun D, Hurley LH. Formation of pseudosymmetrical G-quadruplex and i-motif structures in the proximal promoter region of the RET oncogene. J Am Chem Soc. 2007;129:10220–8.

22. Dexheimer TS, Sun D, Hurley LH. Deconvoluting the structural and drug-recognition complexity of the G-quadruplex-forming region upstream of the bcl-2 P1 promoter. J Am Chem Soc. 2006;128:5404–15.

23. Sun H, Xiang J, Shi Y, Yang Q, Guan A, Li Q, et al. A newly identified G-quadruplex as a potential target regulating Bcl-2 expression. Biochim Biophys Acta. 2014;1840:3052–7.

24. Dhakal S, Yu Z, Konik R, Cui Y, Koirala D, Mao H. G-quadruplex and i-motif are mutually exclusive in ILPR double-stranded DNA. Biophys J. 2012;102:2575–84.

25. Connor AC, Frederick KA, Morgan EJ, McGown LB. Insulin capture by an insulin-linked polymorphic region G-quadruplex DNA oligonucleotide. J Am Chem Soc. 2006;128:4986–91.

26. Verma A, Yadav VK, Basundra R, Kumar A, Chowdhury S. Evidence of genome-wide G4 DNA-mediated gene expression in human cancer cells. Nucleic Acids Res. 2009;37:4194–204.

27. Catasti P, Chen X, Moyzis RK, Bradbury EM, Gupta G. Structure-function correlations of the insulin-linked polymorphic region. J Mol Biol. 1996;264:534–45.

28. Timmer CM, Michmerhuizen NL, Witte AB, Van Winkle M, Zhou D, Sinniah K. An isothermal titration and differential scanning calorimetry study of the G-quadruplex DNA-insulin interaction. J Phys Chem B. 2014;118:1784–90.

29. Palumbo SL, Memmott RM, Uribe DJ, Krotova-Khan Y, Hurley LH, Ebbinghaus SW. A novel G-quadruplex-forming GGA repeat region in the c-myb promoter is a critical regulator of promoter activity. Nucleic Acids Res. 2008;36:1755–69.

30. Balasubramanian S, Neidle S. G-quadruplex nucleic acids as therapeutic targets. Curr Opin Chem Biol. 2009;13:345–53.

31. Huppert JL, Balasubramanian S. G-quadruplexes in promoters throughout the human genome. Nucleic Acids Res. 2007;35:406–13.

32. Zhao Y, Du Z, Li N. Extensive selection for the enrichment of G4 DNA motifs in transcriptional regulatory regions of warm blooded animals. FEBS Lett. 2007;581:1951–6.

33. Eddy J, Maizels N. Conserved elements with potential to form polymorphic G-quadruplex structures in the first intron of human genes. Nucleic Acids Res. 2008;36:1321–33.

34. Verma A, Halder K, Halder R, Yadav V, Rawal P, Thakur RK, et al. Genome-wide computational and expression analyses reveal G-quadruplex DNA motifs as conserved cis-regulatory elements in human and related species. J Med Chem. 2008;51:5641–9.

35. Iida K, Nakamura T, Yoshida W, Tera M, Nakabayashi K, Hata K, et al. Fluorescent-ligand-mediated screening of G-quadruplex structures using a DNA microarray. Angew Chem Int Ed Engl. 2013;52:12052–5.

36. Li G, Ruan X, Auerbach RK, Sandhu KS, Zheng M, Wang P, et al. Extensive promoter-centered chromatin interactions provide a topological basis for transcription regulation. Cell. 2012;148:84–98.

37. Yoshida W, Tomokawa J, Inaki M, Kimura H, Onodera M, Hata K, Nakabayashi K. An insulator element located at the cyclin B1 interacting protein 1 gene locus is highly conserved among mammalian species. PLoS ONE. 2015;10:e0131204.

38. Iida K, Nagasawa K. Macrocyclic polyoxazoles as G-quadruplex ligands. Chem Rec. 2013;13:539–48.

39. Tera M, Iida K, Ishizuka H, Takagi M, Suganuma M, Doi T, et al. Synthesis of a potent G-quadruplex-binding macrocyclic heptaoxazole. ChemBioChem. 2009;10:431–5.

40. Karsisiotis AI, Hessari NM, Novellino E, Spada GP, Randazzo A, Webba da Silva M. Topological characterization of nucleic acid G-quadruplexes by UV absorption and circular dichroism. Angew Chem Int Ed Engl. 2011;50:10645–8.

41. Chung WJ, Heddi B, Tera M, Iida K, Nagasawa K, Phan AT. Solution structure of an intramolecular (3 + 1) human telomeric G-quadruplex bound to a telomestatin derivative. J Am Chem Soc. 2013;135:13495–501.

42. Harada T, Iwai A, Miyazaki T. Identification of DELE, a novel DAP3-binding protein which is crucial for death receptor-mediated apoptosis induction. Apoptosis. 2010;15:1247–55.

43. Feng CJ, Lu XW, Luo DY, Li HJ, Guo JB. Knockdown of Cdc6 inhibits proliferation of tongue squamous cell carcinoma Tca8113 cells. Technol Cancer Res Treat. 2013;12:173–81.

44. Zhang G, Liu K, Ling X, Wang Z, Zou P, Wang X, et al. DBP-induced endoplasmic reticulum stress in male germ cells causes autophagy, which has a cytoprotective role against apoptosis in vitro and in vivo. Toxicol Lett. 2016;245:86–98.

45. Rasheed MZ, Tabassum H, Parvez S. Mitochondrial permeability transition pore: a promising target for the treatment of Parkinson's disease. Protoplasma. 2017;254:33–42.

46. Zakraoui O, Marcinkiewicz C, Aloui Z, Othman H. Lebein, a snake venom disintegrin, suppresses human colon cancer cells proliferation and tumor-induced angiogenesis through cell cycle arrest, apoptosis induction and inhibition of VEGF expression. Mol Carcinog. 2017;56:18–35.

# 13

# RNA polymerase II depletion promotes transcription of alternative mRNA species

Lijian Yu[1†], Mayuri Rege[2†], Craig L. Peterson[2*] and Michael R. Volkert[1*]

## Abstract

**Background:** Cells respond to numerous internal and external stresses, such as heat, cold, oxidative stress, DNA damage, and osmotic pressure changes. In most cases, the primary response to stress is transcriptional induction of genes that assist the cells in tolerating the stress and facilitate the repair of the cellular damage. However, when the transcription machinery itself is stressed, responding by such standard mechanisms may not be possible.

**Results:** In this study, we demonstrate that depletion or inactivation of RNA polymerase II (RNAPII) changes the preferred polyadenylation site usage for several transcripts, and leads to increased transcription of a specific subset of genes. Surprisingly, depletion of RNA polymerase I (RNAPI) also promotes altered polyadenylation site usage, while depletion of RNA polymerase III (RNAPIII) does not appear to have an impact.

**Conclusions:** Our results demonstrate that stressing the transcription machinery by depleting either RNAPI or RNAPII leads to a novel transcriptional response that results in induction of specific mRNAs and altered polyadenylation of many of the induced transcripts.

**Keywords:** Transcription, RNA polymerase depletion, Transcriptional stress, Altered polyadenylation preference, mRNA induction

## Background

UV damage causes lesions in the genome that interfere with cellular DNA metabolism and lead to cell death. UV damage also induces a DNA repair response that triggers a cascade of events, including transcriptional activation of a large number of DNA damage response genes that facilitate DNA repair, and trigger cell cycle arrest until DNA repair is complete [1, 2]. In contrast, the presence of UV lesions in DNA also causes RNA polymerase arrest, blocking elongation and thus, antagonizes genome-wide transcription [3–5]. The arrest of RNAP II initiates a DNA repair response known as transcription coupled repair (TCR). RNA polymerase and associated proteins serve as the DNA damage recognition factor for TCR, recruiting nucleotide excision repair proteins

to preferentially repair DNA damage located in the template strand of actively transcribing DNA [4]. The arrest of multiple RNA polymerase complexes at DNA damage sites throughout the genome also reduces the number of functionally active and free, RNA polymerase molecules available.

The *RPB2* gene locus has served as an ideal system to investigate effects of UV damage in the cell [3, 6–8]. Given the thorough investigation of transcription of this gene after UV treatment, our recent work on recovery after UV damage at *RPB2* revealed an unexpected transcriptional response. Prior to UV treatment, the *RPB2* mRNA terminated at a polyA site that is 58 nt downstream from the stop codon. Shortly after UV treatment, a distal polyA site 345 nt downstream of the stop codon is preferentially used, resulting in a longer transcript. Moreover, the abundance of this *RPB2* long form increased markedly over the first 60 min after UV. This increase was not due to stabilization of the longer *RPB2* transcript, as the half-lives of both the short and long form of *RPB2* transcripts were comparable. Thus, UV treatment induced transcription of the *RPB2* gene and the mRNA

*Correspondence: craig.peterson@umassmed.edu;
michael.volkert@umassmed.edu
[†]Lijian Yu and Mayuri Rege contributed equally to this work
[1] Microbiological and Physiological Systems, University of Massachusetts Medical School, 55 Lake Avenue North, Worcester, MA 01655, USA
[2] Program in Molecular Medicine, University of Massachusetts Medical School, 373 Plantation Street, Worcester, MA 01605, USA

that was produced preferentially utilized a distal polyA site [9]. Given that UV treatment generally inhibits transcription genome-wide until the damage is repaired, this serendipitous observation was quite remarkable.

In this study we extend this work to test whether production of the *RPB2* long form is a general hallmark of transcriptional stress. We also examine if other genes show a similar response, and if depletion or inactivation of other RNA polymerases serves as an inducer of the response. We find that inactivation of RNAPII or nuclear depletion of either RNAPI or RNAPII triggers transcriptional changes similar to the changes seen after UV treatment. Thus it appears that treatments that reduce the level of free or active transcription complexes cause a type of transcriptional stress that triggers induction of specific genes and modulation of polyadenylation (polyA) site usage.

## Results

### Depletion of RNA polymerase II induces the long form of *RPB2* mRNA

UV damage has both positive and negative effects on transcription. It triggers a UV induced DNA-damage response that stimulates transcription of genes required for DNA repair and cellular recovery while the presence of lesions in the template DNA stalls transcription throughout the genome [3–5]. Our previous study demonstrated several additional changes in transcription after UV treatment. Specifically, the polyA site preference at the *RPB2* gene shifts from production of a 4010 nt mRNA to a longer 4297 nt mRNA, and transcription of the long form is induced dramatically [9]. One possibility is that this transcriptional change is due to a direct response to UV damage. Alternatively, induction of the long form of *RPB2* may be due to the general inhibition of transcription that results from UV treatment [3–5]. As an initial test of the latter possibility, the *rpb1-1* temperature sensitive allele of the gene encoding the largest subunit of RNAPII (Rpb1/Rpo21) was used to rapidly inactivate RNAPII at 37 °C [10], and the abundance of *RPB2* mRNA species was analyzed by Northern analysis. When transcription is inhibited by incubating the *rpb1-1* mutant at 37 °C, changes in *RPB2* expression are observed that are similar to those seen after UV treatment. The expression of the 4297 nt long *RPB2* mRNA increases after Pol II inactivation, while levels of the shorter 4010 nt *RPB2* mRNA decrease (Fig. 1a). As would be expected from a genome-wide inactivation of the transcriptional machinery, levels of other control mRNAs, such as *SUB1* and *YRA1* decline. This indicates that the inactivation of RNAPII results in the expected inhibition of general transcription, but with the striking exception that transcription of the long *RPB2* mRNA was

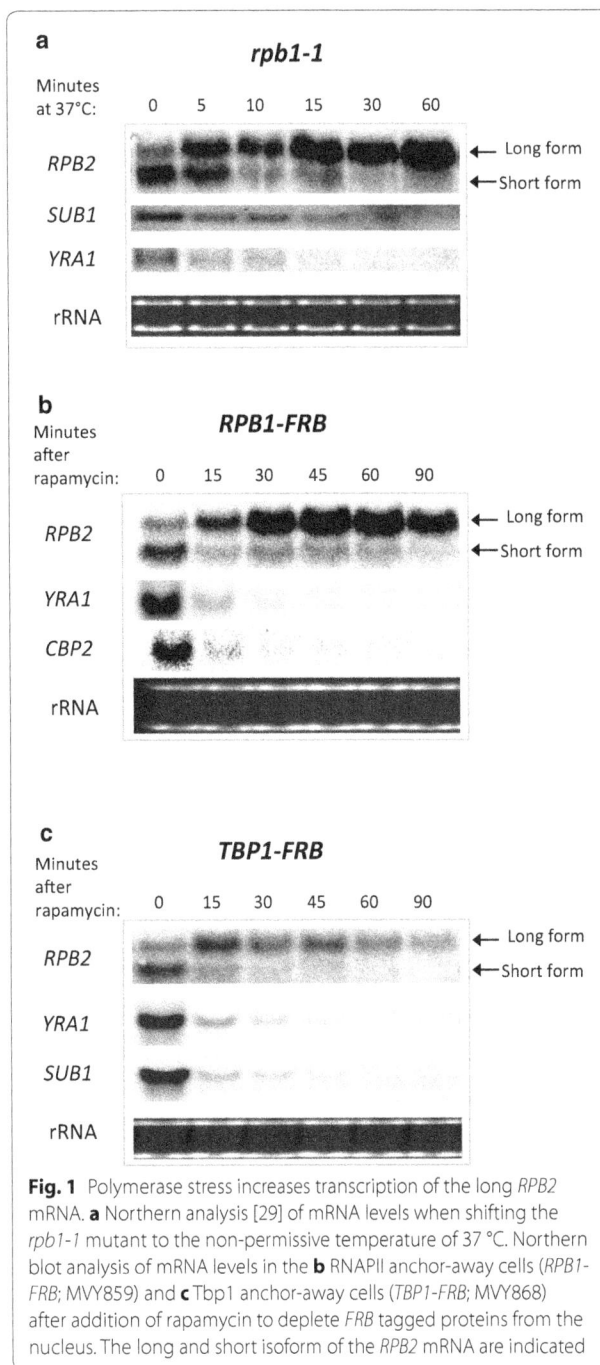

Fig. 1 Polymerase stress increases transcription of the long *RPB2* mRNA. **a** Northern analysis [29] of mRNA levels when shifting the *rpb1-1* mutant to the non-permissive temperature of 37 °C. Northern blot analysis of mRNA levels in the **b** RNAPII anchor-away cells (*RPB1-FRB*; MVY859) and **c** Tbp1 anchor-away cells (*TBP1-FRB*; MVY868) after addition of rapamycin to deplete *FRB* tagged proteins from the nucleus. The long and short isoform of the *RPB2* mRNA are indicated

not repressed at the restrictive temperature but rather appears to be induced.

To test if simply depleting RNAPII from the nucleus is sufficient to induce transcription of the long form of *RPB2*, we employed the anchor-away technique, which is a rapamycin-inducible system that rapidly depletes nuclear proteins to the cytoplasm [11]. A previous study used the anchor away method to demonstrate that

RNAPII can be depleted from genes, and RNAPII transcription is reduced within 15 min of rapamycin addition. Likewise, we find that nuclear depletion of the largest subunit of RNAPII, Rpb1, leads to a rapid reduction in *CBP2* and *YRA1* mRNAs (*RPB1-FRB* strain, Fig. 1b). A similar reduction in transcription is also seen for the short *RPB2* mRNA (Fig. 1b). In contrast, the long *RPB2* mRNA is not inhibited, but instead its synthesis increases over time after depleting RNAPII. Thus, it appears that depletion of actively transcribing RNAP II complexes from the nucleus triggers the observed changes in polyA site preference and induction of *RPB2* transcription.

As an independent means to deplete RNAPII from target genes, the anchor away strategy was used to deplete the general RNAPII transcription factor, TATA binding protein 1 (*Tbp1*). Tbp1 depletion leads to the expected decrease in expression of several genes, including the short form of *RPB2* (*TBP1-FRB* strain, Fig. 1c). In contrast, Tbp1 depletion led to an increase in expression of the long *RPB2* mRNA, although expression of the long form of *RPB2* slowly declined with time. Thus, like the case for depletion or inactivation of RNAPII, depletion of Tbp1 results in an induction of the long form of *RPB2*.

### Induction of the long form of *RPB2* is not due to increased RNAPII occupancy

The induction of the long form of *RPB2* mRNA following inactivation or depletion of RNAPII raises the possibility that the residual RNAPII is preferentially retained, or accumulates, at the *RPB2* locus. To test this possibility, chromatin immunoprecipitation (ChIP) was used to measure RNAPII levels at the *RPB2* locus in *RPB1-FRB* cells treated with either DMSO or rapamycin for one hour. Contrary to our expectation, rapamycin addition led to a rapid decrease in RNAPII occupancy throughout the *RPB2* coding region, similar to other genes (Fig. 2b; Additional file 1: Figure S1). Thus, either the residual RNAPII remaining at the *RPB2* locus is actively engaged in transcription, or other RNA polymerases substitute for RNAPII when it is depleted or inactivated. To determine if other polymerases induce this response and to test whether other RNAPs are required for the increased expression of the large *RPB2* transcript, we examined the effects of depleting or inactivating RNAPI or RNAPIII.

### Depletion of RNAPI promotes expression of the long form of *RBP2*

The anchor away strategy was used to test if depletion of either RNAPI or RNAPIII might also trigger the change in *RPB2* transcription. Surprisingly, depletion of the Rpa190 subunit of RNAPI led to an increase in the level of the long *RPB2* mRNA, while the short form of *RPB2*

was decreased only slightly (Fig. 3a). In contrast, depletion of the Rpc160 subunit of RNAPIII had no impact on either the short or long forms of *RPB2* (Fig. 3b). Thus, depletion of either RNAPI or RNAPII causes increased expression of the long form of *RPB2*, whereas depletion of RNAPIII has no effect.

Since RNAPI and RNAPIII share multiple subunits with RNAPII, we tested the formal possibility that RNAPI or RNAPIII might substitute for RNAPII at the *RPB2* gene after RNAPII depletion. However, ChIP analysis of the Rpa190 subunit of RNAPI and the Rpc34 subunit of RNAPIII did not detect an increase in their levels at the *RPB2* locus following depletion of RNAPII (Fig. 2c, d). Several functional studies were also used to probe whether RNAPI or RNAPIII might play a more direct role in *RPB2* transcription under conditions where RNAPII is depleted or inactivated. To test if RNAPI plays a role in gene expression after transcriptional stress, we used a strain that carries a deletion of the *RPA135* gene (*rpa135Δ*) that encodes an essential subunit of RNAPI. This strain is viable as it also harbors a plasmid-borne copy of the 35S rRNA gene expressed from a RNAPII-dependent promoter. The wild type *RPB1* gene was then replaced with the *rpb1-1* allele, producing a strain that lacks RNAPI and expresses the temperature-sensitive allele of RNAPII. After shifting the *rpb1-1 rpa135Δ* double mutant to the non-permissive temperature, the long form of *RPB2* was induced and the short form was eliminated. These results eliminate the possibility that RNAPI is required for transcription of the long form of *RPB2* mRNA when RNAPII is inactivated (Fig. 3c).

Similarly, we tested whether RNAPIII is required for eliciting the *RPB2* transcriptional response. As an initial test, we depleted RNAPIII by anchoring away the Rpc160 subunit and then induced transcriptional stress by UV treatment. As observed previously for a wild type strain, the long form of *RPB2* is preferentially produced after UV damage even when RNAPIII is depleted (Fig. 3d). As a second strategy, RNAPIII and RNAPII were simultaneously depleted by the anchor away strategy. In this case, the results are similar to those seen when RNAPII alone is depleted (compare Fig. 3e with Fig. 1b), demonstrating that RNAPIII depletion does not alter the response to transcriptional stress. These data indicate that neither RNAPI nor RNAPIII play a direct role in the transcriptional response at *RPB2*.

### Analysis of genome-wide changes in transcription after RNAPII depletion

To examine genome-wide changes in transcriptional profiles following depletion of RNAPII, we re-analyzed the datasets of Geisberg et al. [12] who used RNAPII depletion to measure the half-lives of mRNA isoforms

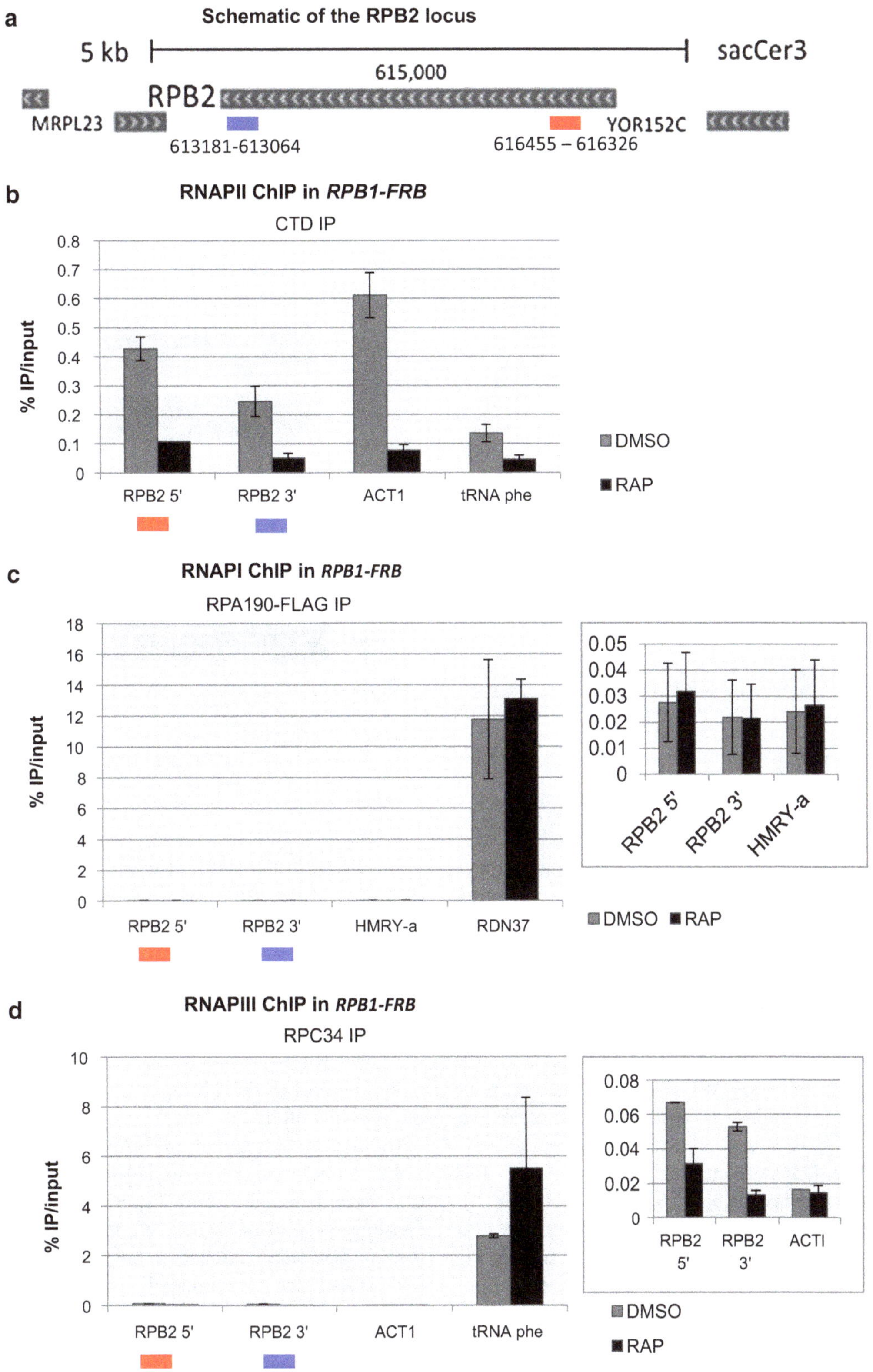

**Fig. 2** Chromatin Immunoprecipitation of RNA polymerase subunits at the *RPB2* locus after RNAPII depletion. **a** The schematic indicates locations of the primers used for chromatin immunoprecipitation at the 5′ and 3′ end of the RPB2 locus. IP was performed using the **b** anti-CTD antibody for RNAPII. **c** anti-FLAG antibody for the Rpa190-FLAG subunit of RNAPI and **d** anti-Rpc34. for RNAPIII. *RDN37* and tRNA phe are known targets of RNAPI and RNAPIII respectively and thus were used as positive controls

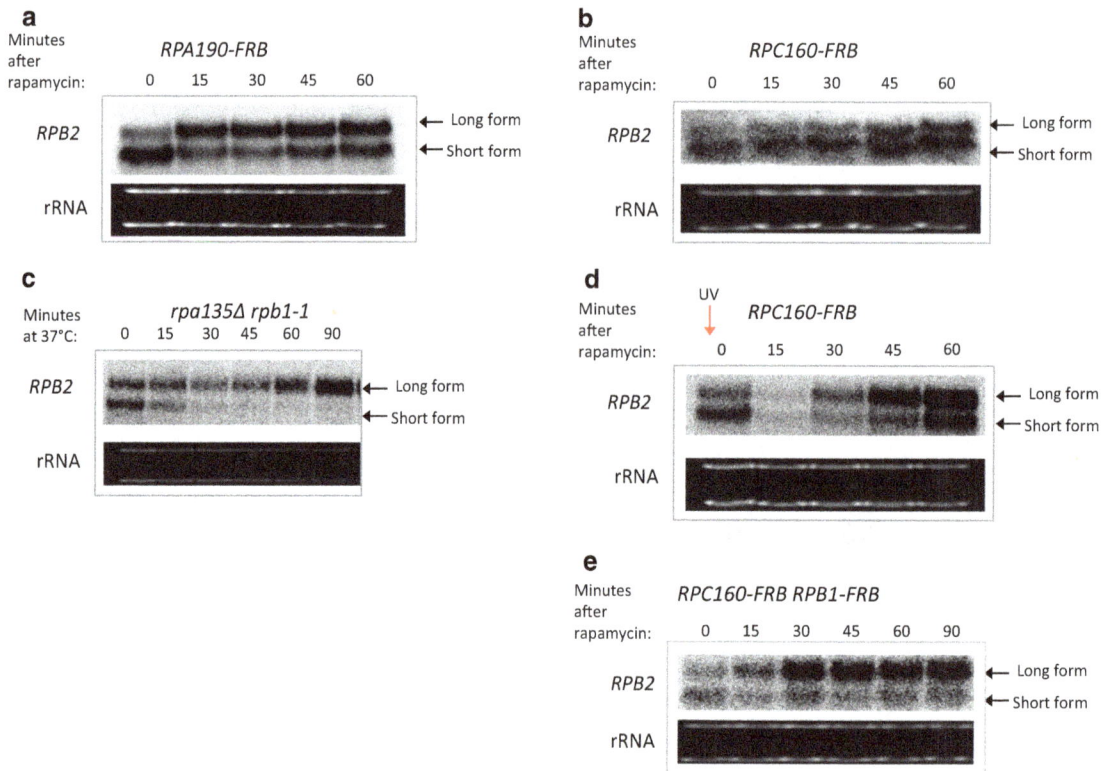

**Fig. 3** Depletion of RNAPI but not RNAPIII induces the long *RPB2* mRNA. **a** Nothern analysis [29] of the *RPB2* mRNA isoforms after **a** RNAPI anchor-away (*RPA190*-FRB; MVY860), **b** RNAPIII anchor-away (*RPC160-FRB*; MVY862). **c** Northern analysis of the *RPB2* mRNA isoforms after heat inactivation of RNAPII in the *rpa135Δ rpb1-1* (MVY851). **d** Northern analysis of the *RPB2* mRNA isoforms after RNAPIII anchor-away (*RPC160-FRB*; MVY862), except that cells are UV irradiated (70 J/m$^2$) before adding rapamycin. **e** Northern analysis of the *RPB2* mRNA isoforms after depletion of RNAPII and RNAPIII simultaneously (*RPA190-FRB RPB1-FRB*; MVY872)

with different polyA tails. We obtained their dataset and analyzed which mRNA species increase after depleting RNAPII. After clustering and annotating synonymous reads in each dataset, we obtained the absolute reads for all mRNA species at each time point in the two experiments ("clustered.absolute.reads.A.txt" and "clustered. absolute.reads.B.txt" in Additional file 4). Geisberg et al. [12] found 7 long-lived mRNAs with half lives >100 min (Additional file 2: Table S1). We used the median reads of these 7 mRNAs to establish the baseline of fluctuations in RNA sequencing reads and adjusted the absolute reads to this baseline ("clustered.normalized.reads.A.txt" and "clustered.normalized.reads.B.txt" in Additional file 4). Figure 4a shows the normalized sequencing reads of 100 random transcripts after RNAPII depletion in Experiment A. Most mRNAs decline after RNAPII depletion. However, unlike other transcripts, the *DDR2* mRNA with a 275 bp 3′UTR shows a continuous increase in abundance following RNAPII depletion, indicative of continuous RNA synthesis despite transcriptional inhibition. The increase in *DDR2* expression was confirmed by RT-PCR analysis of RNA isolated from cells depleted for

either RNAPI or RNAPII (Fig. 5). We then looked at the dynamics of the *RPB2* mRNA isoforms in the two independent experiments (Fig. 4b, c). A long *RPB2* transcript with a 344 bp 3′UTR increases in abundance at 40 min after RNAPII depletion and then declines in both experiments. The 345th nucleotide after the *RPB2* stop codon is an adenine which is likely to be a part of the polyA tail of the *RPB2* mRNA, thus the 344 bp 3′UTR mRNA is essentially identical to the 345 bp 3′UTR mRNA identified in our previous study.

To identify additional mRNAs that are induced by RNAPII depletion, we searched for induced transcripts that were detected in both experiments A and B in the Geisburg et al. dataset [12]. We found 66 genes (70 isoforms) that were induced following RNAPII depletion (Additional file 4 "induced.normalized.common.txt"). The gene selection criteria were stringent in order to identify only those genes that are clearly induced. *RPB2* is not represented in this list due to the transient nature of its induction and the apparently high level of expression measured at time 0 in experiment B (Fig. 4b, c). Most genes have all of their mRNA isoforms repressed

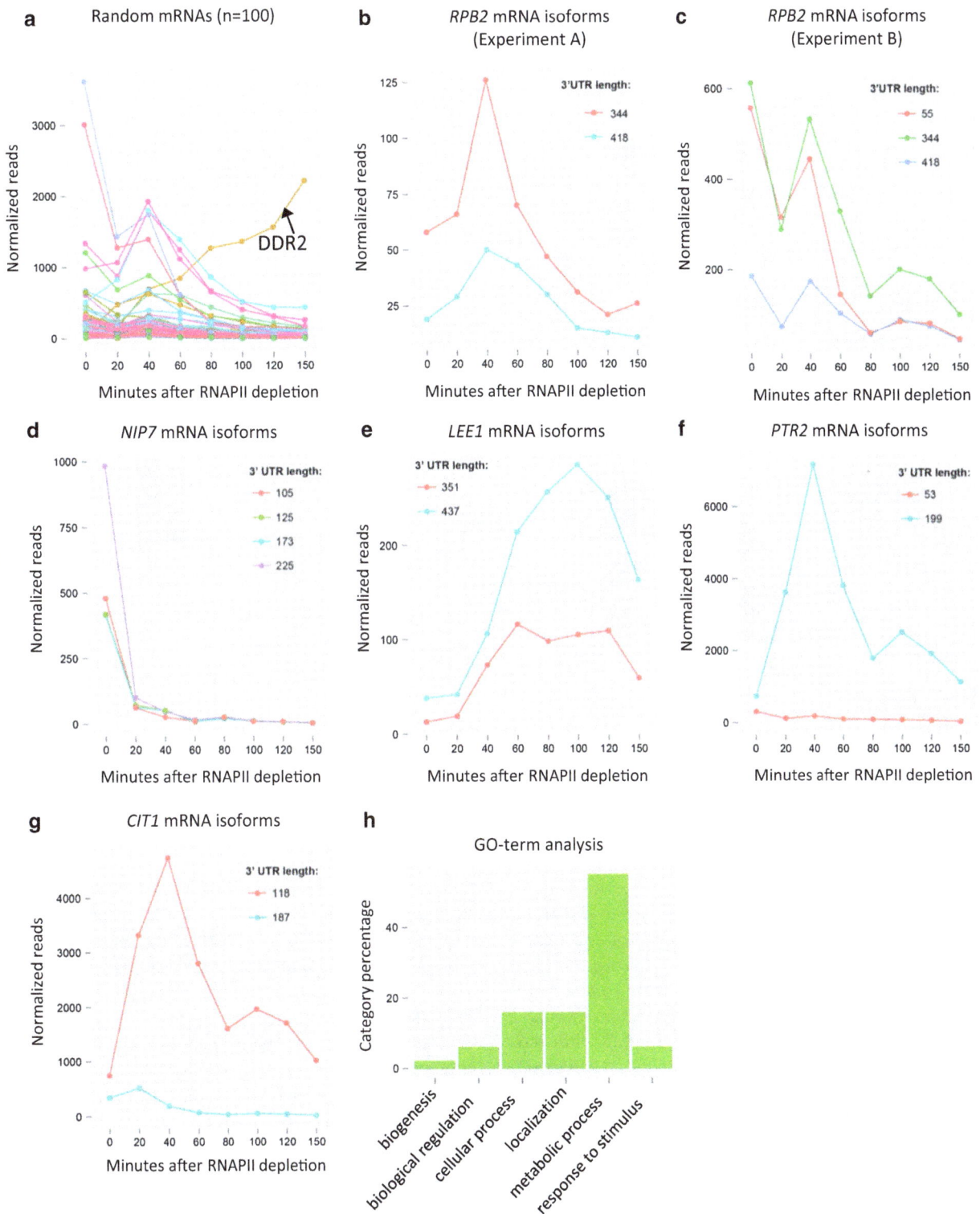

**Fig. 4** Analysis of Direct RNA Sequencing (DRS) data for mRNA isoforms after RNAPII depletion. **a** Normalized sequencing reads of 100 random transcripts in Experiment A. **b** Normalized sequencing reads of the *RPB2* mRNAs in Experiment A. **c** Normalized sequencing reads of the *RPB2* mRNAs in Experiment B. Normalized sequencing reads for the **d** *NIP7*, **e** *LEE1*, **f** *PTR2* and **g** *CIT1* genes, which represent the distinct patterns of mRNA isoform changes observed after RNAPII. **h** Gene ontology enrichment analysis of the 66 genes that were induced in both Experiment A and Experiment B after RNAPII depletion. Dataset for Experiment A and Experiment B were downloaded from NCBI (GSE52286)

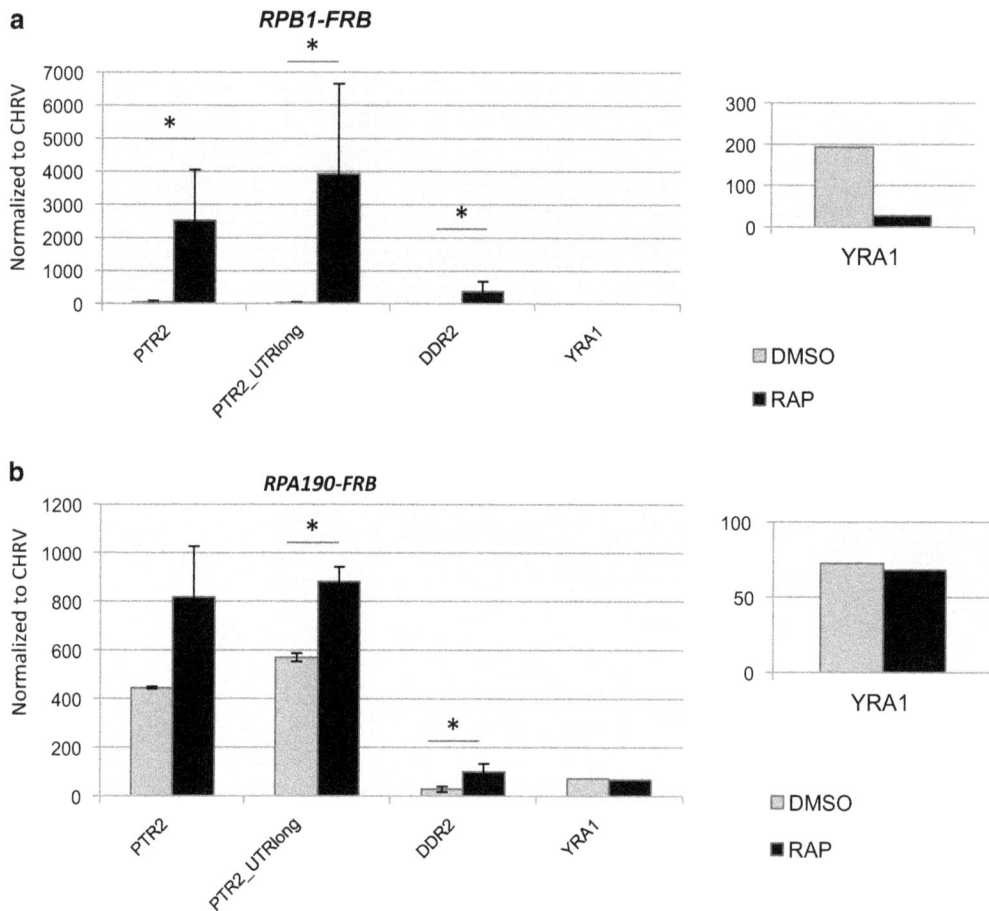

**Fig. 5** Depletion of RNAPI or RNAPII induces expression of a subset of mRNAs. RT-PCR analysis of RNA isolated from either the *RPB1-FRB* strain following nuclear depletion (+RAP) **a** or the *RPA190-FRB* strain following depletion (+RAP) **b** RNA levels were normalized to RNA from an intergenic region on chromosome V

after RNAPII depletion (~80 % in both experiments), as represented by the *NIP7* gene (Fig. 4d). However, among the 66 genes induced by transcriptional stress, a subset of genes (10–20 % depending on the experiment) have all isoforms induced and behave similarly to the *LEE1* gene (Fig. 4e). Most other genes (80–90 %), exemplified by *PTR2*, have only one isoform preferentially induced, and show results similar to those seen for *RPB2* (Fig. 4f). Similar to the case of *DDR2*, an increase in *PTR2* mRNA levels is observed after either RNAPI or RNAPII depletion (Fig. 5). Additionally, in the case where two or more isoforms are alternatively expressed, the induced isoform is not always the long form. For example, the *CIT1* gene normally transcribes the long form and the shorter form is induced after RNAPII depletion (Fig. 4g). To characterize whether the genes induced by RNAPII depletion had any obvious functional relationship, we used PANTHER (pantherdb.org) to classify their molecular functions. As shown in Fig. 4h, most genes encode products involved

in metabolic functions (Additional file 4 "induced genes.annotations.xlsx"). In addition, we also performed a search for common sequence motifs within the 5′ or 3′ UTRs, but no unique motifs were identified.

## Discussion

Recent studies indicate that transcription of protein coding genes produces a heterogeneous set of mRNAs that show alternative 3′ end formation and polyadenylation [9, 12–17]. Changes in 3′ end formation can have dramatic impacts on mRNA stability, transport, and resulting protein expression [14, 18]. A number of regulatory mechanisms have been identified that control alternative 3′ end formation, including alterations in the levels of transcription initiation or elongation [19–22]. For instance, studies in both Drosophila and budding yeast have found that slowing the elongation rate of RNAPII favors the selection of proximal termination sites, whereas enhanced elongation rates favor 3′ end formation at more distal positions [19, 21]. Such studies have led to a model whereby there is a kinetic competition between RNAPII elongation and 3′ end selection. Recently, we reported that transcription of the yeast *RPB2* gene produces mRNAs with alternative sites of polyadenylation and 3′ end formation. Furthermore, we found that a promoter-distal polyA site is preferentially used following recovery from UV damage. Here, we find that depletion of the RNAPII machinery is sufficient to induce the preferential use of the distal polyA site at the *RPB2* locus, and furthermore, a genome-wide analysis of transcript levels identifies a larger set of transcripts that are induced following RNAPII depletion. Surprisingly, we also find that depletion of RNAPI also leads to changes in 3′ end formation at *RPB2*, suggesting a more general response to transcriptional stress.

In mammalian cells, the efficiency of polyA site usage and transcription termination correlates with transcriptional activity [23]. For instance, highly expressed genes tend to utilize more proximal polyA, whereas more lowly expressed genes choose alternative, distal polyA sites. The choice of mRNA cleavage and polyA addition sites appears to be due to an increased propensity of RNAPII to pause at a proximal polyA site under conditions of high transcriptional activity [23]. Consistent with these correlative studies, we found that the depletion of RNAPII alters 3′ end usage at a subset of yeast genes. Using the anchor away system, RNAPII was rapidly removed from the nucleus, leading to efficient depletion of RNAPII from coding regions. As expected, depletion of RNAPII leads to greatly decreased densities of RNAPII at gene coding regions as well as the shutdown of a large percentage of the yeast transcriptome. However, we identified ~70 genes whose expression continues under conditions of RNAPII depletion, and many of these mRNAs also show a switch in 3′ end formation. In several cases, more distal polyA sites are used, though examples of a switch to a more proximal site was also identified.

The *RPB2* gene is an example of a gene that continues to be expressed following RNAPII depletion, and a long form of *RPB2* is induced under these conditions. This long form of *RPB2* is also induced during recovery from UV damage. Importantly, our previous study demonstrated that the long form of *RPB2* does not have an altered decay rate, indicating that continued expression is due to transcription [9]. Remarkably, RNAPII density is dramatically decreased at *RPB2* during the anchor away protocol, eliminating a simple model in which genes such as *RPB2* compete more effectively for a limiting pool of RNAPII. These lower levels of RNAPII are consistent with a model in which RNAPII density impacts alternative polyA site usage.

How can *RPB2* mRNA levels be maintained at normal or even higher levels if RNAPII levels are decreased? We entertained the possibility that other RNA polymerases might substitute for RNAPII during the depletion conditions, but co-depletion of RNAPI or RNAPIII with RNAPII did not eliminate the induction of the long form of *RPB2*. These results indicate that the residual levels of RNAPII are responsible for the continued expression of the long form of *RPB2*. This raises the possibility that these bound RNAPII molecules are engaged in a more productive mode of transcription. Interestingly, many genes that encode subunits of RNAPII are increased in expression after inactivation of the RNA exosome [24]. Furthermore, we previously reported that yeast genes with a lower density of RNAPII are not as likely to be targeted by the RNA exosome, consistent with the idea that the transcription of some genes during RNAPII depletion may be more productive [24]. Interestingly, genotoxic stress has been shown to inhibit the nonsense mediate decay pathway [25]. One possibility is that polymerase depletion may also lead to a similar type of stress that down regulates decay of a subset of transcripts. Since the induced, long *RPB2* isoform does not appear to have an altered half-life [9], such a decay mechanism may operate co-transcriptionally.

Laferte and colleagues [26] reported that increased levels of RNAPI activity not only induce expression of the 35S rRNA but also the RNAPII-dependent expression of ribosomal protein genes. They suggested the possibility of communication between RNAPI and RNAPII that would ensure coordinated expression levels of ribosomal components. Remarkably, we find that depletion of RNAPI leads to a RNAPII-dependent induction of the *RPB2* gene and a change in the selection of the 3′ end. This response occurs rapidly (<60′), so it likely occurs prior to any decreases in translational capacity. We also find that the *PTR2* and *DDR2* transcripts are also increased following either RNAPI or RNAPII depletion, though the magnitude of the transcriptional induction is much larger when RNAPII is depleted. We favor a model proposed by Lafrete [26] and colleagues in which transcription of 35S rRNA or the structure of the nucleolus is monitored in some way by the RNAPII machinery leading to an altered transcriptional activity.

## Conclusions

We demonstrate that depletion of RNAPII and RNAPI, but not RNAPIII results in a transcriptional stress response that triggers the induction of specific mRNAs and that many of the mRNAs induced also exhibit a change in the polyadenylation site that is preferentially used. Thus, transcriptional stress response must result in modification of the transcriptional machinery to cause these changes.

## Methods

### Yeast strains and plasmids

Yeast strains and plasmids used in this study are listed in Table 1. Strain construction details are described below. Primer sequences are listed in Additional file 3: Table S2.

To construct strain MVY851 (*rpa135Δ rpb1-1*), we mated MVY845 and MVY451, sporulated the diploid cells, and dissected the tetrads onto YEP-galactose. Spore colonies that grew on Leucine-dropout Galactose plates (*rpa135::LEU2*) and could not grow at 37 °C (*rpb1-1*), were further tested as follows. To test *rpa135::LEU2*, we PCR amplified the genomic DNA with primer RPA135-f (TCAAACTTACTTCAGCTGTCTCG) and primer RPA135-r (GGCTTCCAAACCCTTTAGGT), digested the PCR fragment with restriction enzyme XbaI and separated the digested DNA on agarose gels. The *RPA135* wild type strain had 3 bands of 1957 bp, 1101 bp, and 886 bp, the *rpa135::LEU2* mutant clone had a 3.7 kb band. To confirm the *rpb1-1* allele, we PCR amplified the genomic DNA with RPB1f (GTTGATCATCCGTTGTCGTG) and RPB1-4955r (GGTGACGTTGGGCTGTAACT), and sequenced with primer RPB1r (CATTTGAACCCGAATCAACC). A G≥A point mutation at bp 4310 of the *RPB1* gene was confirmed in the temperature sensitive clone.

To construct strain MVY860 (*RPA190-FRB::KANMX6*), we PCR amplified the *KanMX6* gene from pFA6a-FRB-KanMX6 using primers RPA190aaf (GAACAATGTT GGTACGGGTTCATTTGATGTGTTAGCAAAGGTT CCAAATGCGGCTcggatccccgggttaattaa) and RPA190aar (CTCCTTCAAATAAACTAATATTAAATCGTAATAATT ATGGGACCTTTTGCCTGCTTgaattcgagctcgtttaaac)

to produce DNA fragment R1, PCR amplified a 403 bp fragment from the *RPA190* gene using primers RPA190-5586f (TGTGGGATCAAGAGGCATTT), RPA190-5589r (CGCATTTGGAACCTTTGCTA) to produce DNA fragment A1, PCR amplified a 394 bp fragment from the *RPA190* gene using primers RPA190-5999f (CAG-GCAAAAGGTCCCATAAT) and RPA190-6393r (ATTG-GCCGTTCCTTCAAATA) to produce DNA fragment A2, then assembled purified DNA fragments A1, A2 and R1 using overlap extension PCR (one cycle of at 95 °C for 2 min, then 13 cycles at 95 °C for 20 s, 55 °C 20 s, and 72 °C 30 s, followed by one cycle at 72 °C for 4 min) to produce the DNA molecule F1. F1 was then used as a template in which we PCR amplified an 2870 bp DNA using primers RPA190-5600-SalI (GCGC-GTCGAC-GCATTTA TCGACGTTGATGGT) and RPA190-6309r-SacII (GGCA-CCGCGG- CCAGCACGTAATCGAGAAAA). The resultant 2870 bp PCR product was then digested the with SalI and SacII and inserted between the SalI and SacII sites of the pBluescript SK vector plasmid to produce pMV1374. pMV1374 was then digested with SacII, SalI, and DrdI to release the 2768 bp fragment containing sequences homolgous to sequences flanking *RPA190* but containing the *FRB-KANMX6* cassette. The gel purified 2768 bp fragment was then transformed into strain MVY858 and selected for clones that are resistant to G418 and sensitive to Rapamycin. The resulting clones were confirmed by PCR amplifying the genomic DNA using primers FRB-r (CCTTCATGCCACATCTCATG) and RPA190-5528f (CCAGAACCTGAAAACGGAAA) and looking for amplification of the signature 564 bp PCR product.

### Table 1 Yeast strains used in this study

| Strain | Original name, genotype (annotation) | Notes | References |
|---|---|---|---|
| MVY150 | W303-1B, MATα *ade2-1 trp1-1 can1-100 leu2-3, 112 his3-11,15 ura3-1* | | [7] |
| MVY451 | SYY9, MATa *ade2-1 his3-11, 15 leu2-3, 112 trp1-1 ura3-1 can1-100 rpb1-1* (derived from W303) | | [27] |
| MVY845 | NOY408-1a, MATa *rpa135::LEU2 ade2-1 ura3-1 trp1-1 leu2-3,112 his3-11 can1-100* carrying pNOY102 (*GAL7-35S rDNA URA3*) | | [28] |
| MVY851 | MATα rpa135::LEU2 *ade2-1 ura3-1 trp1-1 leu2-3,112 his3-11 can1-100* carrying pNOY102 (*GAL7-35S rDNA URA3*) *rpb1-1* | | This study |
| MVY858 | HHYy168, MATa *tor1-1 fpr1::NAT RPL13A-2xFKBP12::TRP1* | | [11] |
| MVY859 | HHY170, MATα *tor1-1 fpr1::NAT RPL13A-2xFKBP12::TRP1 RPO21-FRB::KANMX6* | | [11] |
| MVY860 | MVY858 with *RPA190-FRB::KANMX6* | | This study |
| MVY862 | MVY858 with *RPC160-FRB::KANMX6* | | This study |
| MVY868 | MVY858 with *TBP1-FRB::KANMX6* | | This study |
| MVY872 | MVY859 with *RPA190-FRB-HIS3MX* | | This study |
| MVY874 | MVY859 with *RPC160-FRB-HIS3MX* | | This study |
| MVY879 | MVY872 with *RPC160-FRB-LEU2* | | This study |
| CY1940 | MVY858 with *RPA190* FLAG | Pol I AA | This study |
| CY2032 | MVY859 with *RPA190* FLAG | Pol I AA Pol II AA | This study |

To construct strain MVY862 (*RPC160-FRB::KANMX6*), we PCR amplified the *KanMX6* gene from pFA6a-FRB-KanMX6 using primers RPC160aaf (TCCTAAGCGATG TCTATTTGAAAGTCTCTCAAATGAGGCAGCTTTA AAAGCGAACcggatccccgggttaattaa) and RPC160aar (TGACTGTGGTAGAAAAATAATACAAATGCTATAA AAAAGTTTAAAAACGACTACTgaattcgagctcgtttaaac), transformed the strain MVY858 with the purified PCR fragment, and selected the transformants on YPD plates containing 200ug/ml G418, and confirmed the RPC160-FRB::KANMX6 construct by PCR amplifying the characteristic 806 bp fragment using primers KanMX-905 (TTTGATGACGAGCGTAATGG) and RPC160 + 233r (TCAGCTTGTGAGTGCATACCA).

To construct strain MVY872 (*RPB1-FRB::KANMX6 RPA190-FRB-HIS3MX)*, we first replaced the KanMX6 marker in pMV1374 with the *HIS3MX6* marker with the following steps: digestion of pMV1374 with BamHI and EcoRI and purification of the 3.3 kb vector, digestion of the pFA6a-*FRB-HIS3MX6* with BamHI and EcoRI to release the 1.9 kb insert containing the HIS3MX6 gene. The vector and insert were then ligated to obtain pMV1377 (pBluescriptSK with *RPA190-FRB-HIS3MX6*). This plasmid was then sequenced using T7/T3 primers to confirm its structure. Then we digested pMV1377 with SacII and SalI and DrdI to release the 2655 bp *RPA190-FRB-HIS3MX6* fragment and used the fragment to transform MVY859 and selecting on the Histidine-dropout minimum plate. Clones were confirmed by PCR amplification with primers FRBr (CCTTCATGCCACATCTCATG) and RPA190-5528f (CCAGAACCTGAAAACGGAAA) with expected amplicon of size 600 bp.

To construct strain MVY874 (*RPB1-FRB::KANMX6 RPC160-FRB-HIS3MX*), we first cloned the *RPC160-FRB::KANMX6* construct from MVY862 into pBluescriptSK to make plasmid pMV1379 with the following steps, PCR amplification was performed using the genomic DNA of strain MVY862 as template and primers RPC160-4859-SalIf (ACGC-GTCGAC-GATGACGGCA AGAGGGAAT) and RPC160-5866-SacIIr (GGCA-CCGCG G-TTGCCTCTCAATGCCCATA). The resultant amplicon was then ligated into the pBluescriptSK vector with SalI and SacII to construct plasmid pMV1379. It structure was confirmed by DNA seuquencing using T7/T3 primers. Plasmids pMV1379 and pFA6a-FRB-HIS3MX6 were then digested with BamHI and PmeI and ligated to swap the HIS3MX6 and KanMX6 markers to make pMV1382 (RPC160-FRB-HIS3MX6) sequencing using T7/T3 primers confirmed its structure. pMV1382 was then digested with SacII, SalI, and DrdI to release the 2953 bp fragment, containing RPC160 homologous sequences flanking the FRB-HIS3MX6 cassette, which was used to transform strain MVY859 and selecting on

Histidine dropout plate for His + clones. The presence of the *RPC160-FRB-HIS3MX* construct was confirmed by amplifying the signature 700 bp fragment using primers FRBr (CCTTCATGCCACATCTCATG) and RPC160-4796f (GCGCTTCCTGATGTTGTTGT).

To construct strain MVY879 (MVY872 with *RPC160-FRB-LEU2*), we first replaced the KanMX6 marker in pMV1379 with the *LEU2* gene from pRS315 using the following step: PCR amplification of the *LEU2* gene from plasmid pRS315 using primers LEU2-BglIIf (GCGC-AGATCT-ACCCTCGAGGAGAACTTCTAGTA TATC) and LEU2-PmeIr (GTAC-GTTTAAAC-TCGACT ACGTCGTAAGGCC). The amplicon was digested with BglII and PmeI and inserted into the vector pMV1379 between the cloning sites BglII and PmeI to make the plasmid pMV1381. Its structure was confirmed by sequencing using primers T7/T3. Plasmid pMV1381 was then digested with SalI and SacII to release the *RPC160-FRB-LEU2* fragment (3872 bp), which was used to transform the yeast strain MVY872. Transformants were selected on Leucine-dropout plates and positive clones are confirmed by PCR amplifying the signature 4 kbp fragment using primers RPC160+233r (TCAGCTTGTGAGTGCATACCA) and RPC160-4796f (GCGCTTCCTGATGTTGTTGT).

### Anchor-away and UV irradiation

Yeast cells were cultured in YPD with 200 µg/ml of G418 overnight, centrifuged and diluted in fresh YPD to OD = 0.1 and cultured for another 6 h. If cells were to be UV irradiated, cells were collected and resuspended in PBS and irradiated with UV at $1.71 \text{ J/m}^2$/s for 42 s (70 J/ $m^2$), then resuspended in YPD and cultured at 30 °C in the dark. Rapamycin was added to the culture to a final concentration of 8 µg/ml to induce anchor-away.

### Heat shock at 37 °C

Heat shock induction was performed by culturing cells at 23 °C, then mixing the cells with an equal volume of YPD at 51 °C to immediately raise the incubation temperature to 37 °C. The cultures were then maintained at 37 °C.

### Northern blot analysis

Procedures of yeast RNA extraction (hot phenol method) and Northern analysis are as described previously [29]. Northern blot images were acquired on a BAS-2500 Image Scanner (Fujifilm) and processed using Multi Gauge 3.0 (Fujifilm)

### RT-PCR analyses of RNA

Yeast were grown in yeast extract peptone (YEP) media with 2 % glucose at 30 °C at an OD600 of 0.8, ensuring equal cell number and cell pellets were snap-frozen in liquid nitrogen until RNA extraction. All samples in one biological replicate were processed together for RNA extraction to minimize variations across mutants and a minimum of two biological replicates were performed for each sample. The cell pellets were thawed on ice and 200 μL of lysis buffer (50 mM Tris pH7-7.4, 130 mM NaCl, 5 mM EDTA, 5 % SDS), 200 μL of Phenol (pH 4.0): Chloroform: Isoamylalcohol (PCI, 25:24:1) and 200 μl of glass beads (by volume) were added before vortexing in the cold room for 20 min at maximum speed. Samples were spun for 15 min at 4 °C at >13,000 rpm to separate the organic and aqueous phases. The upper aqueous phase was transferred to a fresh precooled 1.7 ml tube; an equal amount of PCI was added before shaking the tubes vigorously, followed by centrifugation at maximum speed. This step was repeated once more and then the aqueous layer was mixed with an equal amount of Chloroform: Isoamyl alcohol (CI, 24:1) to remove any residual phenol. The tubes were shaken and spun as above. Phase Lock Heavy tubes (5 Prime) were prepared with a short pre-spin (30 s, 3000 rpm) and the upper aqueous layer from the RNA samples added along with an equal quantity CI. The phases were mixed by inverting the tubes vigorously, taking care to avoid any vortexing, and spun at >13,000 rpm for 2 min. The upper layer was transferred to fresh precooled 1.7 ml tubes containing 1/20th the volume of 3 M sodium acetate (pH 4.2), 2 volumes of 100 % ethanol and inverted. RNA was precipitated at −20 °C for 30 min, spun at 15 min at >13000 rpm and supernatant discarded. The pellet was rinsed with 80 % ethanol, spun for 2 min and the supernatant discarded. The pellets were allowed to air dry by inverting for ~30 min, resuspended in 50–100 μl of DEPC water and quantified using a Nanodrop. Ideally, A260/280 for RNA ~2 and A260/230 ~2, and it is important to note that lower ratios might indicate organic contamination. Integrity of the RNA was confirmed on a 2 % Agarose gel that showed a light smear with 2 bands for the high molecular weight ribosomal subunits and 3 bands for the low molecular weight RNAs. Total RNA was treated for at least 1 h with TURBO DNase (TURBO DNA free kit, Ambion #1907) by incubation for 30 min in the 37 °C water bath in 50 μL reactions with ≤25 μg of RNA in each tube. Care was taken to avoid sample agitation as DNase is extremely heat labile. 10X Inactivation reagent was added after the incubation and frequently tapped to ensure uniform mixing. Samples were centrifuged for 2 min at >13,000 rpm and the supernatant was used for the RT-PCR. Total RNA of 60 ng per well

was determined to amplify signal in the linear range for most sample/primer sets. Primers were designed to be specific using Primer3 qRT-PCR settings and melt-curve analysis was performed in each run to confirm unique product amplification. The One-step RT-PCR reaction mix contained 2X Sybr Green PCR mix from Invitrogen (SYBR Green, ROX, dNTPs), Superscript RT III, primers). In addition to using equal RNA amounts, signal from all primer sets were normalized to a non-transcribed region on CHRV.

### Analysis of the DRS dataset

Direct RNA sequencing (DRS) dataset for Experiment A and Experiment B were downloaded from NCBI (GSE52286). Reference yeast genome sequence file (sacCer3, 2008 June assembly) was downloaded from yeastgenome.org. Yeast gene annotations (sgdGene2008.txt and sgdToName2008.txt) were downloaded from UCSC Table browser. The sequence analysis was done using custom-written R programs, which are available upon request. For transcripts that were undetectable at only some of the 8 time points, we set the read counts as 1 to avoid missing values. Nearby sequencing reads (distance less than 15 base pairs) were clustered to the most abundant isoforms. We set the minimum total reads of the clustered absolute transcripts in all 8 experiments to 300 reads and set the maximum 3′UTR length as 1000 bp. Induced transcripts were defined as transcripts whose normalized reads following Pol II depletion were at least twofold higher than at time 0 for at least 4 time points in the 7 time series in both Experiment A and Experiment B. To search for sequence motifs, Multiple Em for Motif Elicitation (MEME; http://meme-suite.org/tools/meme) was used to probe an 800 bp 5′UTR from the 70 induced gene isoforms as well as 70 random genes. MEME was also used to probe sequences from the protein stop codon to the polyA site.

### ChIP experiments

Yeast strains were grown in rich media with 2 % glucose at 30 °C and either DMSO or Rapamycin (8 μg/ml final concentration) was added for 60 min before fixation with 1.2 % formaldehyde. Cells were quenched with 2.5 M glycine, centrifuged, rinsed with cold water and stored at −80 °C until chromatin preparation. Chromatin preparation, immunoprecipitation and DNA extraction were performed as described in [30]. The anti-CTD antibody (1 μL for 100μL chromatin) was used to immunoprecipitate Pol II. The anti-Rpc34 antibody (1 μL for 100μL chromatin) was used to immunoprecipitate Pol

III. The Pol I subunit *RPA190* was C-terminally tagged with a FLAG tag and a anti-FLAG antibody used for immunoprecipitaion.

## Additional files

**Additional file 1:Figure S1.** Confirmation of RPB2 Northern blot data by qRT-PCR. The schematic in panel A indicates locations of the primers that amplify the long RPB2 mRNA (purple; amplifies only the 4297 nt long form) and the short RPB2 mRNA (green; amplifies both 4010 nt and 4297 nt forms). Panel B presents the RT- PCR analysis of RNA isolated from the RPB1-FRB strain following nuclear depletion (+RAP). RNA levels were normalized to RNA from an intergenic region on chromosome V as done in Figure 5.

**Additional file 2:Table S1.** mRNAs with long half-lives from Geisberg et al. [12]. The seven long lived mRNA species identified by Geisberg et al. that were used in this study to establish the baseline and the relevant features of each mRNA are listed in Supplementary Table.

**Additional file 3:Table S2.** Primer sequences. Sequences of all primers used in this study and the experiment in which they were used are listed in Supplementary Table S2.

**Additional file 4:** Gene expression data files. Text files of all absolute and normalized reads, as described in the text, are contained within the gene expression data files.

## Abbreviations
RNAPI: RNA polymerase I; RNAPII: RNA polymerase II; RNAPIII: RNA polymerase III; TCR: transcription coupled repair; polyA: polyadenylation; *Tbp1*: TATA binding protein 1; ChIP: chromatin immunoprecipitation; YEP: yeast extract peptone; DRS: direct RNA sequencing.

## Authors' contributions
LY and MRV initiated the study. All authors participated in experimental strategy and design. LY and MR conducted the experiments. All authors participated in data analysis, interpretation of the results, and writing of the manuscript. All authors read and approved the final manuscript.

## Acknowledgements
We thank Shuyun Dong, the Laemmli laboratory, and the Nomura laboratory for yeast strains. We also acknowledge the kind gifts from several colleagues: anti-CTD antibody from David Bentley (Univ. of Colorado, Denver), anti-FLAG antibody from Anthony Imbalzano (Univ. of Massachusetts, Worcester) and anti-Rpc34 antibody from Steve Hahn (FHCC, Seattle).

## Competing interests
The authors declare that they have no competing interests.

## Funding
This work was funded by NIH Grants GM49650 to CLP and CA100122 to MRV. The funding agencies played no role in the design, execution, interpretation of the experiments, or writing of the manuscript.

## References
1. Elledge SJ. Cell cycle checkpoints: preventing an identity crisis. Science. 1996;274(5293):1664–72.
2. Gasch AP, Huang M, Metzner S, Botstein D, Elledge SJ, Brown PO. Genomic expression responses to DNA-damaging agents and the regulatory role of the yeast ATR homolog Mec1p. Mol Biol Cell. 2001;12(10):2987–3003.
3. Mao P, Meas R, Dorgan KM, Smerdon MJ. UV damage-induced RNA polymerase II stalling stimulates H2B deubiquitylation. Proc Natl Acad Sci USA. 2014;111(35):12811–6.
4. Hanawalt PC, Spivak G. Transcription-coupled DNA repair: two decades of progress and surprises. Nat Rev Mol Cell Biol. 2008;9(12):958–70.
5. Somesh BP, Reid J, Liu WF, Sogaard TM, Erdjument-Bromage H, Tempst P, Svejstrup JQ. Multiple mechanisms confining RNA polymerase II ubiquitylation to polymerases undergoing transcriptional arrest. Cell. 2005;121(6):913–23.
6. Sweder KS, Hanawalt PC. Preferential repair of cyclobutane pyrimidine dimers in the transcribed strand of a gene in yeast chromosomes and plasmids is dependent on transcription. Proc Natl Acad Sci USA. 1992;89:10696–700.
7. van Gool AJ, Verhage R, Swagemakers SM, van de Putte P, Brouwer J, Troelstra C, Bootsma D, Hoeijmakers JH. RAD26, the functional S. cerevisiae homolog of the Cockayne syndrome B gene ERCC6. EMBO J. 1994;13(22):5361–9.
8. Li S, Ding B, LeJeune D, Ruggiero C, Chen X, Smerdon MJ. The roles of Rad16 and Rad26 in repairing repressed and actively transcribed genes in yeast. DNA Repair. 2007;6(11):1596–606.
9. Yu L, Volkert MR. UV damage regulates alternative polyadenylation of the RPB2 gene in yeast. Nucleic Acids Res. 2013;41(5):3104–14.
10. Kim TS, Liu CL, Yassour M, Holik J, Friedman N, Buratowski S, Rando OJ. RNA polymerase mapping during stress responses reveals widespread nonproductive transcription in yeast. Genome Biol. 2010;11(7):R75.
11. Haruki H, Nishikawa J, Laemmli UK. The anchor-away technique: rapid, conditional establishment of yeast mutant phenotypes. Mol Cell. 2008;31(6):925–32.
12. Geisberg JV, Moqtaderi Z, Fan X, Ozsolak F, Struhl K. Global analysis of mRNA isoform half-lives reveals stabilizing and destabilizing elements in yeast. Cell. 2014;156(4):812–24.
13. Velten L, Anders S, Pekowska A, Jarvelin AI, Huber W, Pelechano V, Steinmetz LM. Single-cell polyadenylation site mapping reveals 3′ isoform choice variability. Mol Syst Biol. 2015;11(6):812.
14. Gupta I, Clauder-Munster S, Klaus B, Jarvelin AI, Aiyar RS, Benes V, Wilkening S, Huber W, Pelechano V, Steinmetz LM. Alternative polyadenylation diversifies post-transcriptional regulation by selective RNA-protein interactions. Mol Syst Biol. 2014;10:719.
15. Tian B, Hu J, Zhang H, Lutz CS. A large-scale analysis of mRNA polyadenylation of human and mouse genes. Nucleic Acids Res. 2005;33(1):201–12.
16. Ozsolak F, Kapranov P, Foissac S, Kim SW, Fishilevich E, Monaghan AP, John B, Milos PM. Comprehensive polyadenylation site maps in yeast and human reveal pervasive alternative polyadenylation. Cell. 2010;143(6):1018–29.
17. Ulitsky I, Shkumatava A, Jan CH, Subtelny AO, Koppstein D, Bell GW, Sive H, Bartel DP. Extensive alternative polyadenylation during zebrafish development. Genome Res. 2012;22(10):2054–66.
18. Moore MJ. From birth to death: the complex lives of eukaryotic mRNAs. Science. 2005;309(5740):1514–8.
19. Hazelbaker DZ, Marquardt S, Wlotzka W, Buratowski S. Kinetic competition between RNA Polymerase II and Sen1-dependent transcription termination. Mol Cell. 2013;49(1):55–66.
20. Ji Z, Luo W, Li W, Hoque M, Pan Z, Zhao Y, Tian B. Transcriptional activity regulates alternative cleavage and polyadenylation. Mol Syst Biol. 2011;7:534.

21. Moreira A. Integrating transcription kinetics with alternative polyadenylation and cell cycle control. Nucleus. 2011;2(6):556–61.

22. Pinto PA, Henriques T, Freitas MO, Martins T, Domingues RG, Wyrzykowska PS, Coelho PA, Carmo AM, Sunkel CE, Proudfoot NJ, et al. RNA polymerase II kinetics in polo polyadenylation signal selection. EMBO J. 2011;30(12):2431–44.

23. Pendlebury A, Frayling IM, Santibanez Koref MF, Margison GP, Rafferty JA. Evidence for the simultaneous expression of alternatively spliced alkylpurine N-glycosylase transcripts in human tissues and cells. Carcinogenesis. 1994;15(12):2957–60.

24. Rege M, Subramanian V, Zhu C, Hsieh TH, Weiner A, Friedman N, Clauder-Munster S, Steinmetz LM, Rando OJ, Boyer LA, et al. Chromatin dynamics and the RNA exosome function in concert to regulate transcriptional homeostasis. Cell Rep. 2015;13(8):1610–22.

25. Popp MW, Maquat LE. Attenuation of nonsense-mediated mRNA decay facilitates the response to chemotherapeutics. Nat Commun. 2015;6:6632.

26. Laferte A, Favry E, Sentenac A, Riva M, Carles C, Chedin S. The transcriptional activity of RNA polymerase I is a key determinant for the level of all ribosome components. Genes Dev. 2006;20(15):2030–40.

27. He F, Li X, Spatrick P, Casillo R, Dong S, Jacobson A. Genome-wide analysis of mRNAs regulated by the nonsense-mediated and 5′ to 3′ mRNA decay pathways in yeast. Mol Cell. 2003;12(6):1439–52.

28. Nogi Y, Yano R, Nomura M. Synthesis of large rRNAs by RNA polymerase II in mutants of Saccharomyces cerevisiae defective in RNA polymerase I. Proc Natl Acad Sci USA. 1991;88(9):3962–6.

29. He F, Jacobson A. Identification of a novel component of the nonsense-mediated mRNA decay pathway by use of an interacting protein screen. Genes Dev. 1995;9(4):437–54.

30. Bennett G, Papamichos-Chronakis M, Peterson CL. DNA repair choice defines a common pathway for recruitment of chromatin regulators. Nat Commun. 2013;4:2084.

# 14

# Signal transducer and activator of transcription STAT5 is recruited to *c-Myc* super-enhancer

Sophia Pinz[†], Samy Unser[†] and Anne Rascle[*]

## Abstract

**Background:** *c-Myc* has been proposed as a putative target gene of signal transducer and activator of transcription 5 (STAT5). No functional STAT5 binding site has been identified so far within the *c-Myc* gene locus, therefore a direct transcriptional regulation by STAT5 remains uncertain. *c-Myc* super-enhancer, located 1.7 Mb downstream of the *c-Myc* gene locus, was recently reported as essential for the regulation of *c-Myc* gene expression by hematopoietic transcription factors and bromodomain and extra-terminal (BET) proteins and for leukemia maintenance. *c-Myc* super-enhancer is composed of five regulatory regions (E1–E5) which recruit transcription and chromatin-associated factors, mediating chromatin looping and interaction with the *c-Myc* promoter.

**Results:** We now show that STAT5 strongly binds to *c-Myc* super-enhancer regions E3 and E4, both in normal and transformed Ba/F3 cells. We also found that the BET protein bromodomain-containing protein 2 (BRD2), a co-factor of STAT5, co-localizes with STAT5 at E3/E4 in Ba/F3 cells transformed by the constitutively active STAT5-1*6 mutant, but not in non-transformed Ba/F3 cells. BRD2 binding at E3/E4 coincides with *c-Myc* transcriptional activation and is lost upon treatment with deacetylase and BET inhibitors, both of which inhibit STAT5 transcriptional activity and *c-Myc* gene expression.

**Conclusions:** Our data suggest that constitutive STAT5 binding to *c-Myc* super-enhancer might contribute to BRD2 maintenance and thus allow sustained expression of *c-Myc* in Ba/F3 cells transformed by STAT5-1*6.

**Keywords:** STAT5, *c-Myc*, BET, BRD2, Super-enhancer, Chromatin

## Background

*c-Myc* is a master regulator of essential biological processes such as cell proliferation, survival, differentiation, metabolism, angiogenesis and pluripotency establishment and maintenance [1, 2]. *c-Myc* is found overexpressed in most human cancers and is a hallmark of tumor initiation and maintenance [3, 4]. Characterizing the regulatory mechanisms of *c-Myc* gene expression is therefore fundamental for a better understanding of its deregulation in cancer and to possibly identify novel therapies aiming at controlling its expression. *c-Myc* gene transcription is regulated by multiple transcription factors via responsive elements located within both its promoter and remote enhancer regions [1, 5–9]. A number of reports, including those from our lab, provided evidence that *c-Myc* expression is regulated by signal transducer and activator of transcription 5 (STAT5) [10–13]. STAT5 is an essential regulator of cell differentiation, proliferation and survival [11, 14, 15] and is frequently constitutively activated in cancer. STAT5 constitutive activation results in *c-Myc* overexpression, increased cell proliferation and reduced cell apoptosis, and is as such an important player in cancer initiation and progression [11, 16–20]. Among molecules that suppress STAT5 activity, inhibitors of tyrosine kinases, of deacetylases and of bromodomain and extra-terminal (BET) proteins represent promising therapeutic agents, either alone or in combination [21–28]. We and others showed that expression

---
*Correspondence: anne.rascle@klinik.uni-regensburg.de
[†]Sophia Pinz and Samy Unser contributed equally to this work
Stat5 Signaling Research Group, Institute of Immunology, University of Regensburg, 93053 Regensburg, Germany

of STAT5 target genes, including *c-Myc*, is reduced upon treatment with deacetylase inhibitors (trichostatin A, suberoylanilide hydroxamic acid, apicidin, valproic acid, sodium butyrate) or with the BET inhibitor (+)-JQ1 [12, 29–31]. Further elucidating the process of STAT5-induced expression of *c-Myc* and of its inhibition by clinically relevant therapeutic agents is critical for improving cancer therapy.

Among the evidence of a direct regulation of *c-Myc* by STAT5, an elegant study by Lord et al. demonstrated that *c-Myc* expression in response to IL-2 and IL-3 is dependent on the transactivation domain of STAT5 [10]. Accordingly, *c-Myc* expression is upregulated in cells expressing constitutively active STAT5, including in the BCR-ABL-transformed human leukemic cell line K562 [11–13, 32]. Constitutive activation of STAT5 by the oncogenic tyrosine kinase BCR-ABL contributes to K562 cell transformation [33, 34]. Overexpression of *c-Myc* in K562 cells is inhibited by the BCR-ABL tyrosine kinase inhibitor imatinib but also by the deacetylase inhibitor trichostatin A (TSA), which we showed to inhibit STAT5-mediated transcription [12, 31, 32]. In the same line, overexpression of *c-Myc* in Ba/F3 cells expressing the constitutively active STAT5 mutant 1*6 [35] is repressed by TSA [12, 31]. In apparent contradiction with a direct regulation a *c-Myc* by STAT5, we found that STAT5 knock-down in Ba/F3 cells did not affect IL-3-induced *c-Myc* gene expression [13]. Two other acknowledged direct target genes of STAT5, *Bcl-x* and *Id-1* [10, 13, 36–38], remained equally unaffected upon STAT5 knock-down in Ba/F3 cells [13]. Although these observations might be interpreted as an indication that these genes are not regulated by STAT5, they might also reflect an unconventional mechanism of regulation by STAT5, possibly not as sensitive to the partial (60 %) knock-down generated upon siRNA transfection [13]. Interestingly, in contrast to most classical STAT5 target genes (e.g., *Cis*, *Osm*, *Spi2.1*, ...) which display STAT5 responsive elements within their promoter region [13], functional STAT5 binding sites have been identified outside the promoter regions of *Bcl-x* and *Id-1*, notably within *Bcl-x* first intron [36, 38] and within *Id-1* enhancer located several kb downstream of the *Id-1* gene [37]. These observations raise the possibility that regulation of *c-Myc* expression by STAT5 might be likewise unconventional, possibly involving distal elements. In support of this proposition, we previously attempted and failed to detect STAT5 binding along the *c-Myc* gene [13], including at GAS elements present in its promoter and known to mediate transcriptional response to other STAT family members [39–41].

Recently, several research groups described the role of a 3′ super-enhancer in the regulation of *c-Myc* gene expression in hematopoietic cells and its importance in

*c-Myc* overexpression in leukemic cells [5–8, 42]. *c-Myc* super-enhancer is located 1.7 Mb downstream of the *c-Myc* coding region. It consists of five enhancer regions (E1–E5) with multiple binding sites for transcription factors. These transcription factors recruit transcriptional co-factors, in particular the BET bromodomain protein BRD4 and the SwItch/Sucrose non-fermentable (SWI/SNF) protein BRG1. These chromatin-associated factors mediate *c-Myc* transcription via long-range chromatin looping and selective interaction with the *c-Myc* promoter [5–7]. On the other hand, we recently demonstrated that STAT5 transcriptional activity is regulated by BET proteins, including BRD2 [31]. We found that BRD2 is recruited along STAT5 at the proximal promoter and transcription start site of the conventional STAT5 target gene *Cis*, both upon IL-3-induced STAT5 activation and in cells expressing and transformed by constitutively active STAT5. BRD2 association with the *Cis* gene is lost upon treatment with the BET inhibitor (+)-JQ1 (thereafter referred to as JQ1 [43]), resulting in transcriptional inhibition [31]. Interestingly, deacetylase inhibitors also repress STAT5-mediated transcription by displacing BRD2 and preventing recruitment of the transcriptional machinery [12, 31]. Our data therefore suggest that STAT5 recruits BRD2 which, in turn, supports transcriptional activation by assisting the assembly of the pre-initiation complex [12, 31]. Given that STAT5-induced expression of *c-Myc* is also inhibited by JQ1 [30, 31, 44], and given the absence of STAT5 binding activity within *c-Myc* promoter and gene locus, we tested the hypothesis that STAT5 binds to the 3′ super-enhancer and recruits BRD2 proteins to regulate *c-Myc* gene expression.

## Results

### *c-Myc* expression is induced by STAT5 and repressed by inhibitors of deacetylases and of BET proteins in Ba/F3-derived cells

This study was conducted in the mouse IL-3-dependent pro-B cell line Ba/F3 [45] and in its transformed IL-3-independent counterpart Ba/F3-1*6, which expresses the constitutively active mutant STAT5A-1*6 [35]. Whereas STAT5 phosphorylation, binding to DNA and transcriptional activity are induced by IL-3 in Ba/F3 and Ba/F3-WT cells, they are constitutive in Ba/F3-1*6 cells in the absence of IL-3 (Fig. 1a, b and [11, 12, 29, 31, 35]), hence mimicking the situation found in cancer cells. As such, Ba/F3 (or Ba/F3-WT) and Ba/F3-1*6 cells represent ideal experimental models to study and compare STAT5-mediated transcription in normal and cancer cells respectively. Similarly to that of the STAT5 target genes *Cis* and *Osm* [12, 31], expression of *c-Myc* was induced by IL-3 in cells expressing wild-type STAT5 while it was upregulated in cells expressing the STAT5A-1*6 mutant

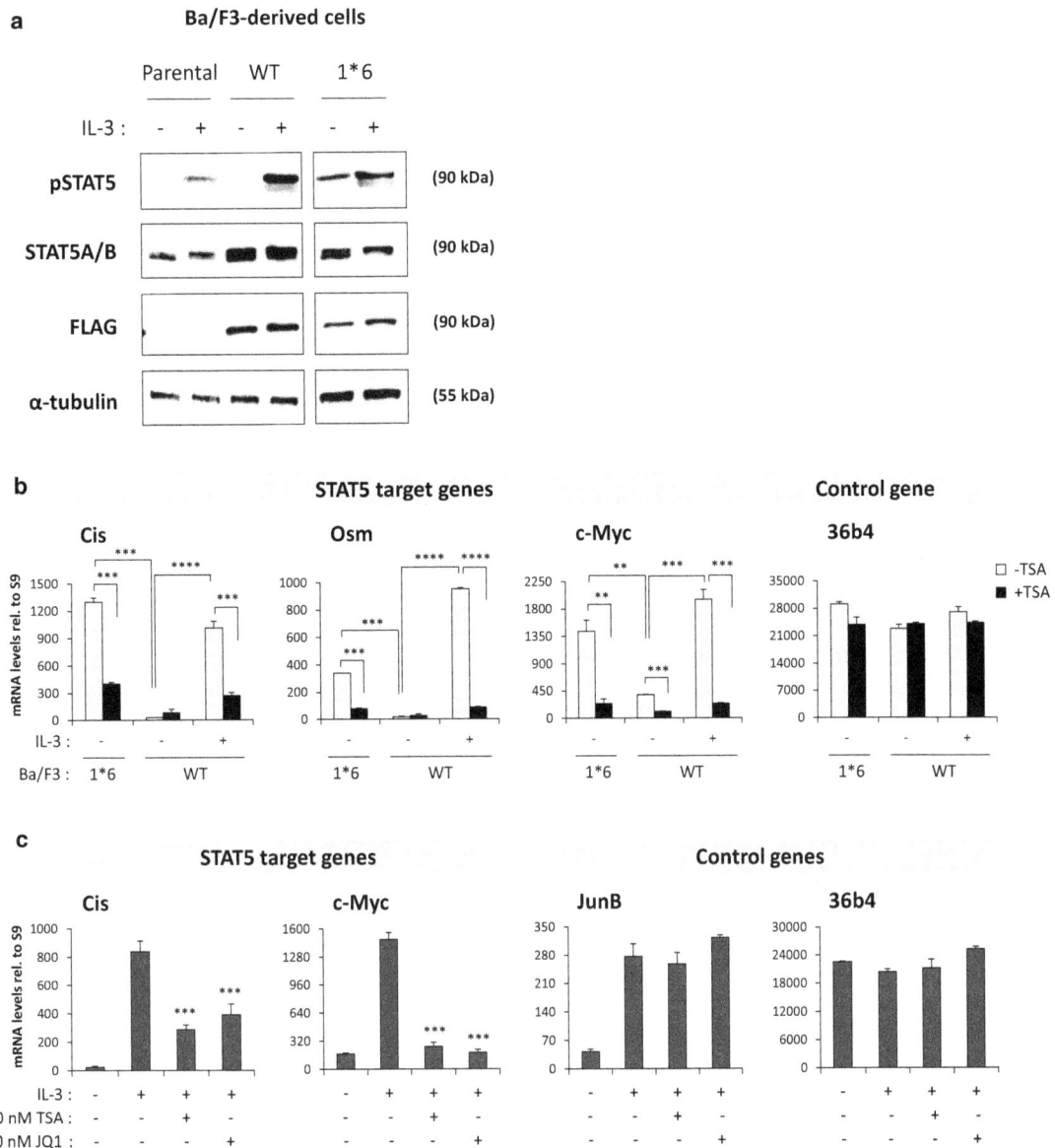

**Fig. 1** *c-Myc* gene expression in Ba/F3-derived cell lines. **a** STAT5 protein and phosphorylation levels in the parental Ba/F3 cells and in Ba/F3 cells stably expressing STAT5A-WT (Ba/F3-WT) and STAT5A-1*6 (Ba/F3-1*6). Ba/F3 and Ba/F3-WT cells, which grow in IL-3-containing medium, were withdrawn from IL-3 for 11 h and stimulated with IL-3 for 15 min. IL-3-independent Ba/F3-1*6 cells were stimulated with IL-3 in parallel. Brij whole-cell protein lysates (20 μg) were analysed by Western blot using antibodies specific for phosphorylated STAT5 (pSTAT5), total STAT5A and STAT5B (STAT5A/B), transgenic STAT5A-WT and STAT5A-1*6 (FLAG), and α-tubulin as a loading control. STAT5A/B signal in parental Ba/F3 cells corresponds to endogenous STAT5 protein levels, while the signals detected in the stable cell lines Ba/F3-WT and Ba/F3-1*6 represent both endogenous (STAT5A and STAT5B) and transgenic (STAT5A-WT or -1*6) proteins. **b** STAT5-mediated expression of *c-Myc* in Ba/F3-WT and -1*6 cells. Rested Ba/F3-WT cells stably expressing wild-type STAT5A were pre-treated 30 min with 200 nM TSA and further stimulated with IL-3 for 30 min. Ba/F3-1*6 cells expressing constitutively active STAT5A-1*6 were treated in parallel with 200 nM TSA for 60 min. Expression of STAT5 target genes (*Cis, Osm, c-Myc*) and of the housekeeping gene *36b4* was analysed by RT-qPCR. Expression of *c-Myc*, like that of *Cis* and *Osm*, was induced by STAT5A-WT and by constitutively active STAT5A-1*6, in an IL-3-dependent and -independent manner respectively. STAT5A-WT- and STAT5A-1*6-mediated expression of *c-Myc*, *Cis* and *Osm* was inhibited by the deacetylase inhibitor trichostatin A (TSA). Student's t tests were employed to compare on the one hand IL-3-induced (WT) or 1*6-induced gene expression to the unstimulated WT control, and on the other hand TSA-treated to the vehicle control in each condition; **$P < 0.01$, ***$P < 0.001$, ****$P < 0.0001$; a $P$ value $<0.05$ was considered statistically significant. **c** Expression of STAT5 target genes is inhibited by both deacetylase (TSA) and BET (JQ1) inhibitors. Ba/F3 cells were pre-treated with 20 nM TSA or 500 nM JQ1 for 30 min and stimulated with IL-3 for 60 min. Expression of STAT5 target genes (*Cis, c-Myc*) and of control genes (IL-3-dependent MAPK target gene *JunB* and housekeeping gene *36b4*) was analysed by RT-qPCR, as above. IL-3-induced expression of *Cis* and *c-Myc*, but not that of *JunB*, was inhibited by TSA and JQ1. One-way ANOVA with Dunnett's multiple comparison test was used to assess differences between TSA- or JQ1-treated conditions and the vehicle-treated IL-3-stimulated control; ***$P < 0.001$; a $P$ value $<0.05$ was considered statistically significant

in the absence of IL-3 (Fig. 1b). In both cell lines, expression of *Cis*, *Osm* and *c-Myc* was inhibited by the deacetylase inhibitor trichostatin A (TSA) (Fig. 1b), in agreement with our previous reports that deacetylase inhibitors (notably TSA, valproic acid, apicidin, sodium butyrate and clinically relevant SAHA) prevent STAT5-mediated transcription [12, 31]. Likewise, STAT5-mediated expression of *Cis* and *c-Myc* in Ba/F3 cells was inhibited by the BET inhibitor JQ1 (Fig. 1c), as already reported [6, 30, 31, 44]. Expression of the IL-3-inducible but STAT5-independent gene *JunB* was not inhibited by TSA or JQ1 (Fig. 1c) while expression of the housekeeping gene *36b4* remained unaffected in all conditions (Fig. 1b, c).

### STAT5 binds to *c-Myc* super-enhancer

Ba/F3 and Ba/F3-1*6 cells were used to investigate transcription factor recruitment along the *c-Myc* gene locus in normal and STAT5-transformed cells respectively. Chromatin immunoprecipitation (ChIP) assays were conducted in Ba/F3 and Ba/F3-1*6 cells using STAT5-specific antibodies. Immunoprecipitated genomic DNA was analysed by quantitative PCR using primers specific for the *c-Myc* gene locus—including known STAT binding sites—and for its 3′ super-enhancer (E1 to E5). Primers specific for regions encompassing the STAT5 binding sites of the *Cis* and *Osm* genes were investigated as controls (Fig. 2a). Upon IL-3 stimulation of Ba/F3 cells, STAT5 was specifically detected at the STAT5 binding sites present within the proximal promoter of the *Cis* and *Osm* genes (Fig. 2b). With the exception of a very weak STAT5 binding at a previously described STAT1 binding site within *c-Myc* promoter [40], no STAT5 binding was detected within the *c-Myc* locus. By contrast, a strong signal was detected specifically at the E3 and E4 regions of *c-Myc* super-enhancer (Fig. 2b). Interestingly, a similar binding pattern was found in Ba/F3-1*6 cells (Fig. 2c), demonstrating that both IL-3-induced wild-type STAT5 and constitutively activated STAT5-1*6 bind to *c-Myc* super-enhancer elements E3 and E4. In support of STAT5 binding to E3 and E4, sequence analysis of the *c-Myc* super-enhancer revealed the presence of clusters of putative STAT5 binding sites within E3 and E4, both consensus (TTCNNNGAA) and non-consensus [46], conserved between mouse and human genomes at the same position (Fig. 3). No conserved putative STAT5 binding sites were identified within E1, E2 and E5 (not shown).

### BRD2 co-localizes with constitutively active STAT5 at *c-Myc* super-enhancer

Given that BRD4 is recruited at *c-Myc* super-enhancer in hematopoietic and leukemic cells [5, 6], and since we recently showed the implication of BRD2 in STAT5-mediated transcription in Ba/F3 and Ba/F3-1*6 cells [31],

we investigated the recruitment of BRD2 along the *c-Myc* gene and super-enhancer by chromatin immunoprecipitation (Fig. 4). In cells expressing STAT5A-1*6, BRD2 bound at E3 and E4, but not at E1, E2 and E5 of *c-Myc* super-enhancer (Fig. 4a). Unexpectedly, BRD2 was also detected at the reported STAT1 binding site within the *c-Myc* promoter. As previously described [31], BRD2 was found proximate to the transcription start site of the *Cis* gene (Fig. 4a). BRD2 binding at *Cis* and *c-Myc* was lost upon treatment with the BET inhibitor JQ1 (Fig. 4a), demonstrating BRD2 binding specificity. Interestingly, BRD2 binding was reduced at *Cis* transcription start site and abolished at *c-Myc* E3/E4 upon TSA treatment, but not at the *c-Myc* promoter (STAT1) location. We previously showed that TSA-induced inhibition of STAT5 transcriptional activity correlates with BRD2 loss at the STAT5 target gene *Cis* [31]. We therefore show here that, similarly, BRD2 binding at E3/E4—but not at STAT1 element within its promoter—along with STAT5, strictly correlates with *c-Myc* transcriptional activation. Our data thus strongly suggest that BRD2 association with *c-Myc* super-enhancer is involved in STAT5-mediated transcription of *c-Myc*. The presence of BRD2 around the STAT1 binding site and its insensitivity to TSA, suggest that BRD2 is recruited in a STAT5-independent manner to the *c-Myc* promoter, possibly by other transcription factors [5–7], and that BRD2 association at this site is not implicated in STAT5-mediated transcription of the *c-Myc* gene.

To better characterize the role of STAT5 in BRD2 recruitment to STAT5 target genes, we next addressed whether STAT5A-1*6 and BRD2 physically interact. Co-immunoprecipitation experiments were carried out on nuclear lysates from formaldehyde-crosslinked Ba/F3-1*6 cells using STAT5A- and BRD2-specific antibodies and following the ChIP protocol (Additional file 1: Fig. S1A–C). Upon immunoprecipitation (IP), samples were analysed by Western blot (Additional file 1: Fig. S1B) and quantitative PCR as a control (Additional file 1: Fig. S1C). Recruitment efficiency and specificity of STAT5A-1*6 and BRD2 to the *Cis* gene was comparable to that observed before (Additional file 1: Fig. S1C). No co-immunoprecipitation of STAT5A-1*6 and BRD2 was detected in Western blot, following neither STAT5A nor BRD2 IP (Additional file 1: Fig. S1B). It should be noted that, while STAT5A was strongly immunoprecipitated under the ChIP experimental conditions, BRD2 was poorly pulled-down in the same conditions (Additional file 1: Fig. S1B), probably explaining the weak signals usually detected by qPCR following BRD2 ChIP (Additional file 1: Fig. S1C; Fig. 4a), but also making it difficult to evidence a co-immunoprecipitation with STAT5A. The weak immunoprecipitation efficiency observed in BRD2

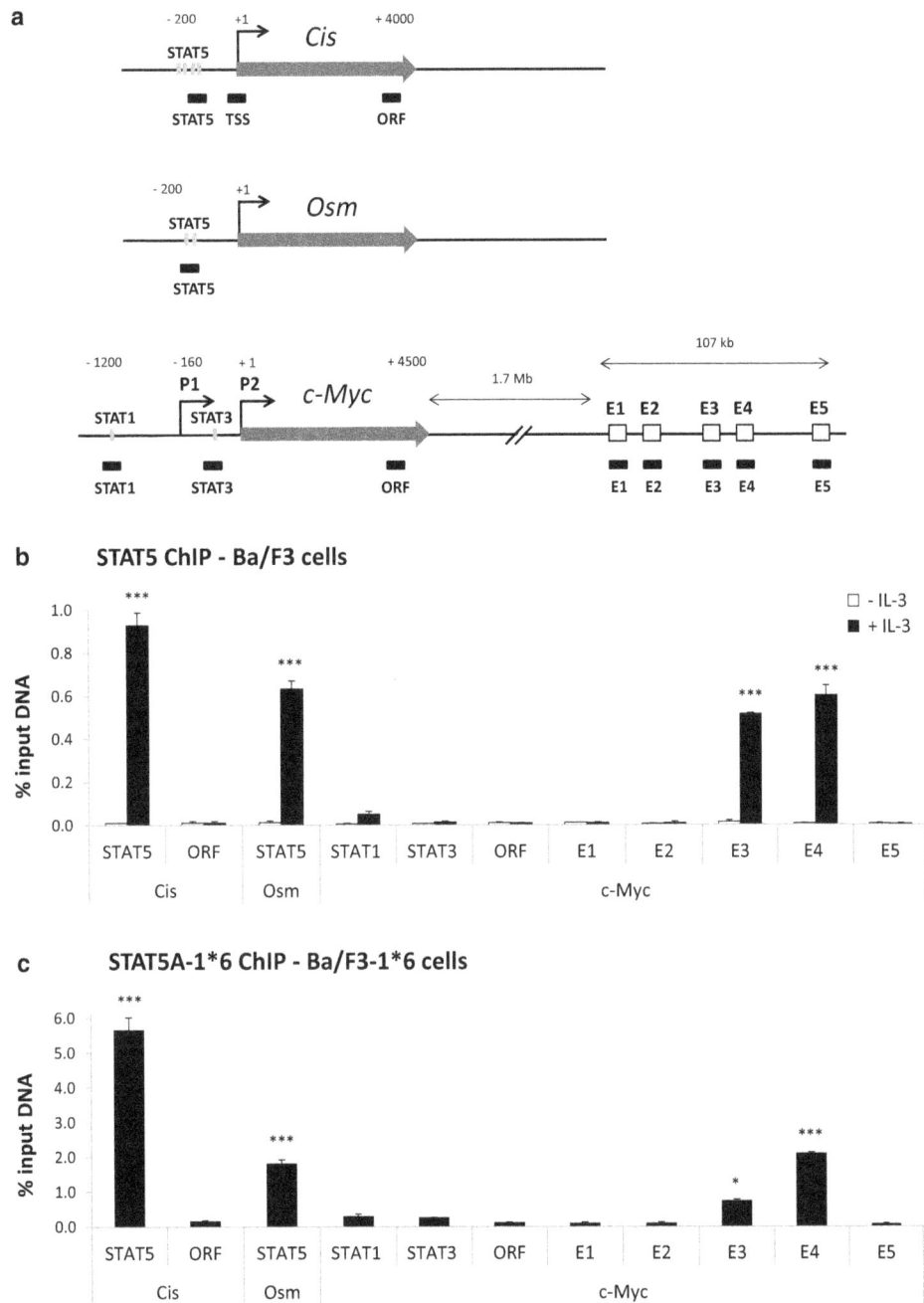

**Fig. 2** STAT5 specifically binds to the E3 and E4 regions of *c-Myc* super-enhancer. **a** Schematic representation of the mouse STAT5 target genes *Cis*, *Osm* and *c-Myc* and of the qPCR amplicons analysed following chromatin immunoprecipitation. Nucleotide positions are relative to the respective transcription start sites (TSS). Functional STAT binding sites within the individual promoter regions are indicated as *grey bars*. P1 and P2 designate *c-Myc* dual promoters, P2 being predominant in normal cells. E1 to E5 symbolise the five domains of *c-Myc* 3′ super-enhancer. *Black boxes* underneath the respective genes represent the qPCR amplicons. Primers amplifying regions within *Cis* and *c-Myc* open reading frame (ORF) were used as controls. **b** Chromatin immunoprecipitation (ChIP) was performed with whole-cell lysates from Ba/F3 cells (conventional ChIP protocol), either unstimulated or stimulated 30 min with IL-3, using antibodies specific for STAT5A + STAT5B (STAT5 ChIP). Immunoprecipitated genomic DNA (gDNA) was analysed by qPCR using primers shown in **a**. One-way ANOVA with Dunnett's post test was used to evaluate IL-3-induced STAT5 enrichment at the various loci compared to the signal detected at the *c-Myc* "ORF" region, used as a reference and background control; ***$P < 0.001$; a $P < 0.05$ was considered statistically significant. **c** ChIP was conducted with nuclear lysates (alternative ChIP protocol) from Ba/F3-1*6 cells grown in the absence of IL-3, using antibodies specific for STAT5A, which recognize STAT5A-1*6 mutant. Immunoprecipitated gDNA was analysed by qPCR, as in **b**. One-way ANOVA with Dunnett's multiple comparison test was used to evaluate STAT5 enrichment at the various loci vs. *c-Myc* "ORF" region, used as a reference and background control; *$P < 0.05$, ***$P < 0.001$

## *c-Myc* - E3

```
h TACATCCTTACAACAGAACCCTCTCT-GGAAATGAGACAATTATTTCTTTCTAGTGGGGTTGCAGAGAGTAGAGGAAGAGAGTTGGAAAGGAGAAGGAGAGCTAGT
m CATCTTCTTGCAACAGTACCCAGACTTGGAAACAAGACAATTATTT-TTTCTAA-GGGGCTTGCAGAGAGTAGAGAAAGAAAGCTGGAAAGGAGAAGGGGAGTCAGG

h GGATGGGAAACCAACAAAATGTGGTTCATTTTCAGGAAGCTGCGAGGCTGGATCTTTCAGAACAGCTCATATTTTTTCGAGATGCATTTTTTTTGAGGTCCCCCACT
m GGACAGGAGAACCACACAGTGTGGTTCCTTTTCAGGAAGCAGCCGGGCTGGATCTCCAAGGGCAACTCCTGTTTTCTCAAGATGCATTTTCTTTGAGGTCCCCCACT

h CCGGATACCCCAGTGCCCTCCTTTCAGCCTTGGCCCTTGCATCACCACCAGCCAGGGGAAGGAGCCGTGGCATGGCAGTGGTCACAGTTCTCTTGGATAAGCTTTA C
m CCGGATACCCCAGTGCCCTCCTTTCAGCCCCAGCCTTTGCATCACCACCAGCCAGGGGAAGGAGCCGTGGCATAGCGGCGGTCACAGTTCACTTGGATAAACTCTA C

h CGTAATCTTCATAAACCTCAATGCTCAAGGTGCAGAGGGAGATTGGAGGAAACTTGTTTTTCCGTTGGTTTCTCTCTGACTGCTGCGGCTAATCATTCCTAGAACAG
m CATAATCTTCAGAAACCTCAATGCACAAGGTGCAGAGGAAGATTGGAGGAAACTTGTTTTTCCGTTGGTTTCTCTGTGACTGCTGTGGCCAGTCATTCCTAGAACAG

h GAAGCTGGAGAAATGTAGGTAGGAGACACTCTTCTGACAAGTCAGGTTGCCTCACCCTTCCAGGGCACAACTGATTAGGAAAAGCTTCAGGGTTACATAACCT
m GAAGCTGGGGAAATTTAGGTAGGAGACATTATTCTGACCAGTCAGGTTTCTTCACCAGTCTAGGGTACAACTGATTAGGAAAAGCCTCC---TTGCATAACCT
```

## *c-Myc* - E4

```
h AGGGGTGGGCTGTGGAAAGACACAAGGGT-GGATAAGTCTGAATGGACCCC-CCCTCCAATGGTCACACCCC------------CCCCAGCCCCAATTCAGTTAAAG
m AGAAGTGGTTTGTGGAAAGACAAGAGGGTTGGACAAGTGTGAATGGAGGCCGCCCTCCAATGGTCACACCCGAGCCCACCCCCACCCCAGCCTCAATTCAGTTAAAG

h ATTGCAGGAAGTGGTTTGTCCGTTTAGAAGTCTTCACAGCTCTGAGGCCCTTCTGAGAAATTAACTTGCAGTTCTCTAAAGATAGATTGCGCAAAATGGGAGGGGGC
m ATTGCAGGAAGTGGTTTGTCCGTTTAGAAGTCATCGCAGCCCTGAGGCCCTTCTGAGAAATGAACTTGCAGTTCTCTAAAGATAGATTGTGCAAAACAGAAGGGGGT

h A-GCCCCCAAGGACCCTCCAACAGCAATTCTCCTGAGGACTTCCAGAGACCTCTGCCAGTGCCCCCACCCCCACCCCGCCCCCTT-GAGCAGCTGCCTGTCCTTGGA
m AAGCCCCCAAGGACCCTCCAACAGCAGCTCTC-TGGAGACTGT-AGAAATCGCCTTGGGGGCCTCCAGGCCCACCCCACTCCTTTAGAGCAGCTGTCA-TCCCCGA A

h TGGACCAAATCAACACCCGACTCTTGTGCAAGGCTCGGGTTTCATG-TCCCCAGTGAACAAATTCTGTTCTGAAGAGGCAGTGGGGTGGGGAGGCTGAGGCGGGGAG
m TGGTCCAAATCAACACCTCACTCTTGTGCAAAGCTCAGGTTTCTGAATATCTAGCTAGCCAATTTTGTTCTGAGGAGGC----GGCAGCGGCTTTTAAAGGGGGGCA

h GGGACATCAAAAACAGGGGGAGGGGAGGGGAGGATGGTGGGGGTGGGGTTTTCAGCTGCTCATGTCATTTATTTTCCACGGAATCTTAACTGTAATTTCAAACTGCA
m GGG-------------GTGGGGAGTAGTTGAAGACTGTGGGGGGAGGGTTCTCAGTTGTTCCTGTCACTCATTTTCCATGCATTCC-AACTGTAATTTCAAACGACC

h TCTGGGTTTTCTGAAACTATGTGATTTGCCATGAGGTTCACAGAGAGTTTA
m TCTGGGTTGTCTGAAACCATGTGATTTGACACGAGGTTAACCGAGAGTTTA
```

**Fig. 3** Sequence alignment of the mouse and human enhancer regions E3 and E4 located 3′ of the *c-Myc* gene. Consensus STAT5 binding sites (TTCNNNGAA) that are conserved between mouse and human sequences are designated as *white boxes*. Non-consensus STAT5 binding sites (usually TTCNNNNAA or TTNNNNGAA) are *underlined* [46]. E3 and E4 contain each one consensus and a cluster of 8–9 non-consensus putative STAT5 binding motifs that are well-conserved in both mouse and human genomes at the same position. Such pattern of conserved motifs was absent from E1, E2 and E5 elements (not shown)

ChIP is likely the consequence of the experimental conditions used (formaldehyde-mediated crosslinking and/or IP buffer composition). Indeed, conventional immunoprecipitation from nuclear lysates of non-crosslinked cells (Additional file 1: Fig. S1D) and using mild buffer conditions resulted in efficient BRD2 immunoprecipitation (Additional file 1: Fig. S1E). However, no co-immunoprecipitation of STAT5A was observed in these conditions. Altogether, these experiments suggest that STAT5A-1*6 and BRD2 do not directly interact but are rather co-recruited at the chromatin level.

Finally, we investigated BRD2 association with the *c-Myc* gene and its super-enhancer in IL-3-stimulated Ba/F3 cells. While BRD2 was recruited upon IL-3 stimulation at the transcription start site of the *Cis* gene, as previously described [31], no BRD2 enrichment above the IgG background was detected along the *c-Myc* gene and downstream enhancer in the same conditions (Fig. 4b). This observation suggests that, by contrast to STAT5A-1*6, IL-3-induced wild-type STAT5 might not be efficient in recruiting and/or stabilizing BRD2 at *c-Myc* enhancer.

## Discussion

This study identified functional STAT5 binding sites possibly regulating *c-Myc* transcription. We showed that STAT5 specifically binds to enhancer regions E3 and E4 of *c-Myc* 3′ super-enhancer. Putative STAT5 binding sites within E3/E4 are organised in clusters of consensus and non-consensus binding motifs. Similar functional STAT5 binding sites have been described at other STAT5 target genes [13]. Binding of wild-type STAT5 to E3/E4 was induced by IL-3 in normal Ba/F3 cells while STAT5A-1*6 binding was constitutive and independent of IL-3 in transformed Ba/F3-1*6 cells. Interestingly, ChIP-Seq assays performed in human HEL leukemia cells revealed that $JAK2^{V617F}$-induced constitutive active STAT5 [47, 48] is bound at *c-Myc* enhancer region E3 (ArrayExpress

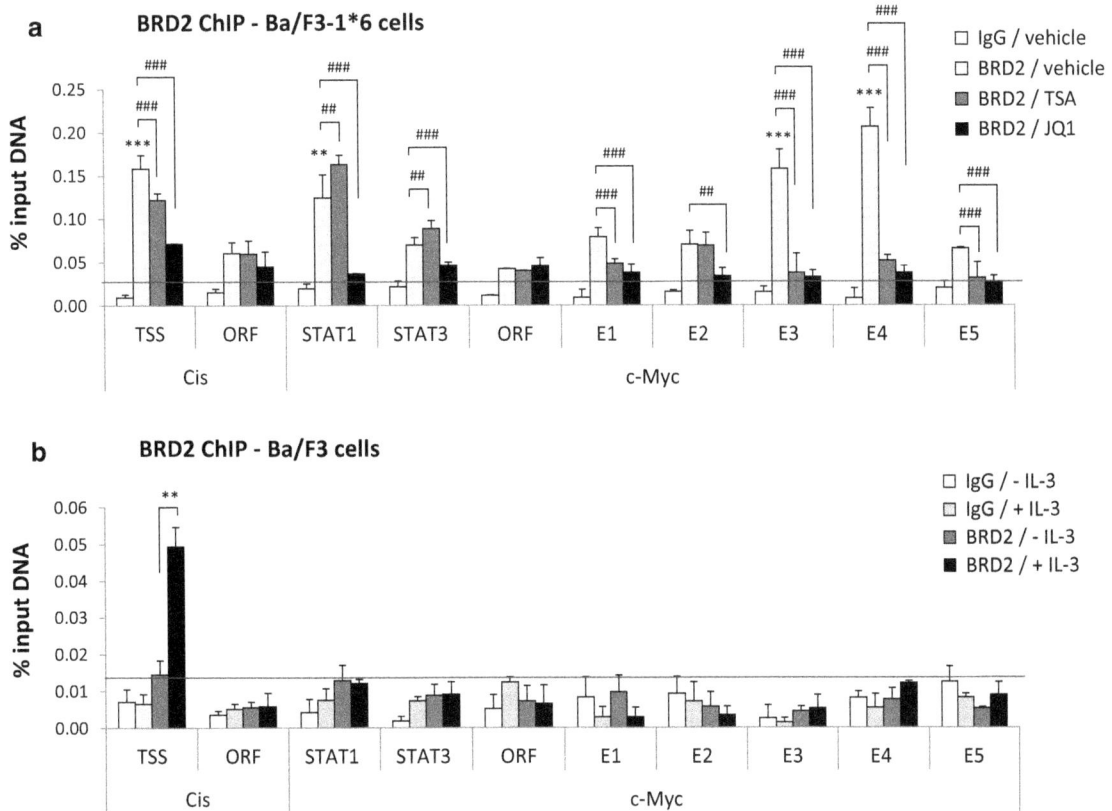

**Fig. 4** BRD2 co-localizes with STAT5 at E3/E4 in Ba/F3 cells transformed by constitutive active STAT5A-1*6. **a** Ba/F3-1*6 cells were treated for 60 min with 200 nM TSA, 1 μM JQ1 or 0.02 % DMSO (vehicle). Nuclei isolation and ChIP was performed following the alternative protocol, and using BRD2-specific antibodies or the same amount of rabbit IgG as a control. qPCR primers are depicted in Fig. 2a. Only the vehicle-treated IgG control is shown. TSA- and JQ1-treated IgG controls exhibited similar background levels (data not shown), as previously reported [31]. BRD2 binds strongly to the *Cis* transcription start site (TSS), the *c-Myc* promoter region (STAT1) and enhancer regions E3 and E4. BRD2 association is lost upon JQ1 treatment and is reduced at *Cis* TSS and *c-Myc* E3/E4 upon treatment with the deacetylase inhibitor TSA. Background cut-off (*horizontal line*) was defined as the mean of the signal generated by the IgG negative controls plus 2x the standard deviation (SD) of the IgG control (mean IgG + 2x SD). BRD2 enrichment at the various loci in the vehicle-treated condition was evaluated using One-way ANOVA with Dunnett's multiple comparison vs. *c-Myc* "ORF" region, used as a reference; of note, BRD2 signal intensity at the *c-Myc* "ORF" region is not reduced upon JQ1 treatment, suggesting that it corresponds to background BRD2 signal; **$P < 0.01$, ***$P < 0.001$; a $P < 0.05$ was considered statistically significant. One-way ANOVA with Dunnett's post test was used to assess differences between TSA- or JQ1-treated conditions and the vehicle-treated control; ##$P < 0.01$, ###$P < 0.001$; a $P < 0.05$ was considered statistically significant. **b** Ba/F3 cells were stimulated with IL-3 for 30 min and BRD2 ChIP was conducted with whole-cell lysates using BRD2-specific antibodies or rabbit IgG, following the conventional ChIP protocol [12, 31]. Genomic DNA was analysed as in **a**. Only background signals were detected along the *c-Myc* gene locus and downstream super-enhancer in Ba/F3 cells, while BRD2 was recruited in response to IL-3 at the transcription start site (TSS) of the *Cis* gene. Background cut-off (*horizontal line*) was defined as described in A (mean IgG + 2x SD). Student's t test was used to monitor IL-3-induced BRD2 recruitment at *Cis* TSS; **$P < 0.01$; a $P < 0.05$ was considered statistically significant

accession number E-MTAB-1096; [49]), providing another evidence of STAT5 constitutive binding at *c-Myc* super-enhancer in leukemia cells.

Constitutive binding of STAT5 at E3/E4 in Ba/F3-1*6 cells coincided with that of BRD2, suggesting that STAT5 might play a role in BRD2 maintenance at *c-Myc* super-enhancer in transformed cells. By contrast, BRD2 was not detected by ChIP at the *c-Myc* locus in IL-3-stimulated Ba/F3 cells. We cannot exclude at this point that this absence of specific signal is due to a detection problem. Alternatively, it might be the consequence of transient STAT5 binding in Ba/F3 cells, as opposed to constitutive

STAT5 binding in Ba/F3-1*6 cells which might be necessary for BRD2 maintenance at *c-Myc* super-enhancer. We could not evidence an interaction between STAT5A-1*6 and BRD2 in co-immunoprecipitation assays. Therefore, our data suggest that constitutive binding of STAT5A-1*6 at E3/E4 might—directly or indirectly—assist and/or stabilize BRD2 association with *c-Myc* super-enhancer. Furthermore, it remains possible that other BET protein(s) are involved in the regulation of *c-Myc* by wild-type STAT5 in Ba/F3 cells. This proposition is supported by our previous observation that BRD2 was detected at the *Cis* but not at the *Osm* promoter in IL-3-stimulated Ba/

F3 cells [31]. This is also in agreement with the implication of BRD4 in long-range regulation of *c-Myc* transcription via its 3′ super-enhancer [5, 6]. Interestingly, it was shown that multiple transcription factors, the BET protein BRD4 and the SWI/SNF component BRG1 are recruited to *c-Myc* 3′ enhancer and contribute to *c-Myc* gene transcription [5, 6]. Notably, BRG1 was proposed to maintain transcription factor occupancy at the enhancer region and to facilitate interactions with the *c-Myc* promoter [5]. Whether such a mechanism of stabilisation by BRG1 also takes place in STAT5-mediated transcription remains to be shown. Maintenance of STAT5 occupancy at *c-Myc* super-enhancer would be an attractive explanation for the absence of effect of STAT5 (partial) knockdown on *c-Myc* gene expression [13].

## Conclusions

We showed that constitutive binding of STAT5 and maintenance of BRD2 at E3/E4 in transformed Ba/F3-1*6 cells correlated with transcriptional activation of *c-Myc*. Furthermore, BRD2 binding at E3/E4 was lost upon TSA- and JQ1-mediated inhibition of *c-Myc* expression, both of which inhibit STAT5-induced transcription [31]. Our data therefore suggest that constitutive binding of STAT5A-1*6 contributes to BRD2 maintenance at *c-Myc* super-enhancer in transformed Ba/F3-1*6 cells, which in turn might be implicated in *c-Myc* overexpression. In support of a role of STAT5 in *c-Myc* overexpression in leukemia via *c-Myc* super-enhancer, E3 was recently shown to display enhancer activity in K562 leukemia cells, using a luciferase reporter assay [5]. This finding nicely fits with our previous finding that *c-Myc* expression in K562 cells is dependent on BCR-ABL-induced constitutive active STAT5 [32]. Our model is hence in line with the recently suggested leukemia maintenance function attributed to *c-Myc* 3′ super-enhancer, via the recruitment of BET proteins by hematopoietic transcription factors [5, 6, 8]. We propose that the transcription factor STAT5 might play a similar role in *c-Myc* overexpression, in leukemia exhibiting constitutive STAT5 activation. Further functional assays will be necessary to verify this proposition.

## Methods
### Chemicals
Dimethyl sulfoxide (DMSO) and trichostatin A (TSA) were purchased from SIGMA (D-2650 and T-8552 respectively). (+)-JQ1 (BPS Bioscience #27401)—hereafter abbreviated to JQ1—was purchased from BIOMOL GmbH. TSA and JQ1 were dissolved in DMSO at a final concentration of 1 mM (TSA) and 5 mM (JQ1). DMSO was used as vehicle control. Its final concentration was adjusted to 0.02 % in all conditions.

### Cells
All cell lines were cultivated at 37 °C under 5 % $CO_2$ in a humidified incubator. The interleukin-3 (IL-3)-dependent mouse pro-B cell line Ba/F3 (a kind gift from Jacqueline Marvel, IFR 128 BioSciences Gerland-Lyon Sud, France [45]) was grown in RPMI 1640 (PAN-Biotech P04-16500) supplemented with 10 % heat-inactivated fetal calf serum (FCS; PAN-Biotech), penicillin/streptomycin (100 U/mL penicillin, 100 μg/mL streptomycin; PAN-Biotech) and 2 ng/ml rmIL-3 (ImmunoTools). No ethical approval was required to obtain and use the Ba/F3 parental cell line. Ba/F3-derived cell lines were generated according to German GenTSV (genetic engineering safety regulations; authorization AZ.55.1-8791.7.52). The IL-3-independent Ba/F3-1*6 cell line (clone F7) stably expressing the FLAG-tagged constitutively active mouse STAT5A-1*6 mutant [35] has been described [32], and was grown in RPMI 1640 supplemented with 10 % heat-inactivated FCS, penicillin/streptomycin and 600 μg/ml G418 (SIGMA A-1720). The IL-3-dependent Ba/F3-WT cell line (clone A7) expressing FLAG-tagged wild-type (WT) mouse STAT5A was generated by electroporating Ba/F3 cells with a pcDNA3-based expression vector allowing expression of a mSTAT5A-WT-FLAG fusion protein. Stably transfected cells were selected in IL-3-containing medium in the presence of 800 μg/mL G418 (PAA). Individual clones were isolated and characterized to verify mSTAT5A-WT transgene expression and proper IL-3-dependent activation of STAT5A-WT. Ba/F3-WT clone A7 was used for this study and served as a control for potential adverse effects of FLAG-tagged protein overexpression. Ba/F3-WT cells were maintained in RPMI 1640 supplemented with 10 % heat-inactivated FCS, penicillin/streptomycin, 600 μg/ml G418 (SIGMA A-1720) and 2 ng/ml rmIL-3. For cytokine stimulation of Ba/F3 and Ba/F3-WT cells, cells were washed twice in RPMI 1640 and rested in RPMI 1640, 10 % FCS, penicillin/streptomycin for 6–12 h before addition of 5 ng/ml IL-3 for 15–120 min, as indicated. For inhibitor treatment of Ba/F3 and Ba/F3-WT cells, rested cells were pre-treated 30 min with the respective compound or with DMSO (vehicle) prior to IL-3 stimulation. Ba/F3-1*6 cells were treated with inhibitors or vehicle for 60 min.

For co-immunoprecipitation assays from non-crosslinked cells, previously described Ba/F3-tet-on-1*6 cells conditionally expressing STAT5A-1*6 in the presence of doxycycline were used [31]. Briefly, Ba/F3-tet-on-1*6 cells were grown for 9 h in the presence of 1 μg/ml doxycycline in IL-3-free medium. Cells were harvested for nuclear fractionation and western blot analysis.

### Gene expression analysis by RT-qPCR
Following inhibitor and cytokine treatments, cells were harvested, cDNA synthesized and quantitative PCR

performed as previously described [31]. Nucleotide sequence of the qPCR primers used in this study have been published [12, 13, 31]. Data were normalized to mouse S9 ribosomal mRNA and expressed as relative mRNA levels, like previously reported [12, 31]. Data are mean ± SD of the quantitative PCR performed in either duplicate or triplicate, and are representative of at least two independent experiments.

## Chromatin immunoprecipation (ChIP)

Chromatin immunoprecipitation (ChIP) was carried out from either whole-cell (conventional protocol) or nuclear (alternative protocol) lysates, following reported procedures [12, 31, 50]. Chromatin Immunoprecipitation from nuclear lysates yields stronger signals than from whole cell lysates [50]. Antibodies used were as follows: STAT5A (Santa Cruz Biotechnology sc-1081), STAT5A + B (Santa Cruz Biotechnology sc-835), BRD2 (Bethyl A302-583A) and IgG from rabbit serum (SIGMA I-5006). Antibody concentrations used were as reported [12, 31]. Co-precipitated genomic DNA was measured by quantitative PCR. Mouse Cis- and Osm-specific primers have been described [31]. Mouse c-Myc-specific forward and reverse primers were the following, respectively: STAT1, TTTATTCTAGGGTCTCTGCAGGC and GAAAACCCGGACTTCCCAG; STAT3, CCCTCCTGC CTCCTGAAGG and CAGGATCCCTCCCCTCCC; ORF, AACAACCGCAAGTGCTCCAG and GTCGTTT TCCTCCGTGTCTGAG; E1, ACGCTCAGAGTGCTT TCCAT and GGTGGTGTGGGGTGACTAATAT; E2, GTGGGAGGGACTGAAATGGAG and TGGGCAAAG CTAGAGGCAGAT; E3, GAACAGGAAGCTGGGGAA AT and TGCAAGGAGGCTTTTCCTAA; E4, CACCCC AGCCTCAATTCAGT and GCTGCGATGACTT CTAAACGG; E5, GCAACAGCAAGAACCAGTGA and TGCTTCTCCTGAACCACCTT. Results are expressed as percentage (%) of input DNA. Data are mean ± SD of the quantitative PCR performed in either duplicate or triplicate, and are representative of at least two independent experiments.

## Protein analysis by immunoprecipitation and Western blot

Western blots were performed as described [32], using the following antibodies and respective dilutions: pSTAT5 (#9351, Cell Signaling Technology; 1:1000), STAT5A (L-20, sc-1081, Santa-Cruz Biotechnology; 1:1000), STAT5A/B (C-17, sc-835, Santa-Cruz Biotechnology; 1:1000), FLAG (M2, SIGMA F-1804; 1:500), Brd2 (Bethyl A302–583A; 1:2000), α-tubulin (DM1A, sc-32293, Santa-Cruz Biotechnology; 1:200), HDAC1 (Millipore 05-100; 1:1000), Anti-Rabbit IgG-Peroxidase (SIGMA A0545; 1:10,000), Anti-Mouse IgG-Peroxidase (SIGMA A8924; 1:10,000). HDAC1 and α-tubulin were

used as nuclear and cytosolic protein markers respectively to control the quality of the nuclear fractionation, like previously reported [31].

Brij whole-cell protein lysis and nuclear fractionation from non-crosslinked cells were conducted as previously described [31]. Immunoprecipitations from non-crosslinked Ba/F3-tet-on-1*6 cells was performed using 250 μg nuclear protein lysate diluted 1:10 in Brij buffer (10 mM Tris–HCl pH 7.5, 150 mM NaCl, 2 mM EDTA pH 8.0, 0.875 % Brij 97, 0.125 % NP40, 10 mM NaF, 1 mM $Na_3VO_4$, 10 μg/ml leupeptin, 10 μg/ml aprotinin, 0.5 mM phenylmethylsulfonyl fluoride) and 1.2 μg BRD2 antibody (Bethyl A302–583A) or 1.2 μg rabbit IgG (SIGMA I-5006) as a negative control. Immunoprecipitations were conducted for 5 h and immunocomplexes were collected using protein A-Sepharose beads. Consecutive to washing in Brij buffer, beads were boiled in 60 μl Laemmli buffer. Half of the eluted bead fraction was analyzed by Western blot in parallel to input lysate (3 % or 8 μg nuclear proteins) and immunoprecipitation supernatants (same volume as input lysate).

Nuclear-enriched protein lysate preparation and immunoprecipitations (IP) from formaldehyde-crosslinked Ba/F3-1*6 cells were carried out following the alternative ChIP protocol described above [31, 50], using $7 \times 10^6$ cells for the nuclei preparation and 2.4 μg antibody (rabbit IgG, STAT5A or BRD2) per IP. Subsequent to the last washing step, ~20 % beads were further processed for genomic DNA isolation and qPCR analysis, following the ChIP protocol. The remaining beads were boiled 10 min in 50 μl Laemmli buffer and ~40 % of the eluted bead fraction (20 μl) was analysed by Western blot in parallel to input lysate (3 %) and immunoprecipitation supernatants (same volume as input lysate).

## Additional file

**Additional file 1: Fig. S1.** Protein interaction between STAT5A-1*6 and BRD2 cannot be evidenced in co-immunoprecipitation assays. Nuclear lysates from formaldehyde-crosslinked (A-C) or non-crosslinked (D, E) STAT5A-1*6-expressing cells were prepared as described in the Methods section. Nuclear protein enrichment was verified by Western blot using antibodies specific for the nuclear and cytosolic proteins HDAC1 and α-tubulin respectively, and STAT5A-1*6 expression was monitored using the FLAG antibody (A, D). Immunoprecipitations (IP) were performed as described in the "Methods" section using the indicated antibodies. Input (In), immunoprecipitation supernatants (SN) and eluted bead fractions (B) were analysed by immunoblot (IB) using the indicated antibodies (B, E). In panel B, arrow points to BRD2 and (*) indicates a non-specific signal associated with the bead fractions. Bead samples from the IP experiment shown in panel B (crosslinked cells) were further processed for ChIP analysis by qPCR, using the Cis-specific primers depicted in Fig. 2a (C). In panel C, background cut-off (dotted line) was defined as in legend to Fig. 4 (mean IgG background + 2x SD). One-way ANOVA with Dunnett's multiple comparison test was used to evaluate BRD2 and STAT5 enrichment at the STAT5 binding site (STAT5) and transcription start site (TSS) of the Cis gene, in comparison to the "ORF" region, used as a reference and background control; ***$P < 0.001$; a $P$ value $< 0.05$ was considered statistically significant.

## Abbreviations

Ba/F3: mouse pro-B IL-3-dependent cell line; Ba/F3-1*6: IL-3-independent BaF3 cell line stably expressing constitutive active STAT5A-1*6; BCR-ABL: tyrosine kinase oncogenic fusion; BET: bromodomain and extra-terminal protein; BRD2 and BRD4: bromodomain-containing protein 2 and 4 respectively, members of the BET protein family; BRG1: Brahma-related gene-1 encoding an ATPase and component of the SWI/SNF complex; DMSO: dimethyl sulfoxide; GAS element: Gamma interferon activation site (STAT-responsive) element; IL-2 and IL-3: interleukin-2 and -3; (+)-JQ1: inhibitor of BET protein function; K562: human leukemic cell line; SAHA: suberoylanilide hydroxamic acid; STAT5: signal transducer and activator of transcription 5; STAT5A-1*6: constitutively active STAT5A mutant; SWI/SNF: SWItch/Sucrose Non-Fermentable, chromatin remodelling complex; TSA: trichostatin A.

## Authors' contributions

AR designed and performed experiments, and drafted the manuscript. SP and SU performed experiments and revised the manuscript. All authors read and approved the final manuscript.

## Acknowledgements

We thank Jacqueline Marvel for providing the Ba/F3 cell line and Krystina Beer for her technical contribution while completing an internship in our group. We are grateful to Joachim Griesenbeck for critically reading the manuscript.

## Competing interests

The authors declare that they have no competing interests.

## Funding

This work was supported by the Deutsche Forschungsgemeinschaft [RA 2010/2-1 to A.R.], the Deutsche Krebshilfe [109750 to A.R.], and institutional research funds [Foerderlinie C to A.R.; Frauenfoerderung, Bayerisches Programm zur Realisierung der Chancengleichheit fuer Frauen in Forschung und Lehre, to S.P.]. The funding bodies had no role in the design of the study, in collection, analysis, and interpretation of data and in writing the manuscript.

## References

1. Hoffman B, Amanullah A, Shafarenko M, Liebermann DA. The proto-oncogene c-myc in hematopoietic development and leukemogenesis. Oncogene. 2002;21:3414–21.
2. Chappell J, Dalton S. Roles for MYC in the establishment and maintenance of pluripotency. Cold Spring Harb Perspect Med. 2013;3:a014381.
3. McKeown MR, Bradner JE. Therapeutic strategies to inhibit MYC. Cold Spring Harb Perspect Med. 2014;4:a014266.
4. Gabay M, Li Y, Felsher DW. MYC activation is a hallmark of cancer initiation and maintenance. Cold Spring Harb Perspect Med. 2014;4:a014241.
5. Shi J, Whyte WA, Zepeda-Mendoza CJ, Milazzo JP, Shen C, Roe J-S, et al. Role of SWI/SNF in acute leukemia maintenance and enhancer-mediated Myc regulation. Genes Dev. 2013;27:2648–62.
6. Roe J-S, Mercan F, Rivera K, Pappin DJ, Vakoc CR. BET bromodomain inhibition suppresses the function of hematopoietic transcription factors in acute myeloid leukemia. Mol Cell. 2015;58:1028–39.
7. Sotelo J, Esposito D, Duhagon MA, Banfield K, Mehalko J, Liao H, et al. Long-range enhancers on 8q24 regulate c-Myc. Proc Natl Acad Sci USA. 2010;107:3001–5.
8. Yashiro-Ohtani Y, Wang H, Zang C, Arnett KL, Ballis W, I lo Y, et al. Long-range enhancer activity determines Myc sensitivity to Notch inhibitors in T cell leukemia. Proc Natl Acad Sci USA. 2014;111:E4946–53.
9. Sengupta D, Kannan A, Kern M, Moreno MA, Vural E, Stack B, et al. Disruption of BRD4 at H3K27Ac-enriched enhancer region correlates with decreased c-Myc expression in Merkel cell carcinoma. Epigenetics. 2015;10:460–6.
10. Lord JD, McIntosh BC, Greenberg PD, Nelson BH. The IL-2 receptor promotes lymphocyte proliferation and induction of the c-myc, bcl-2, and bcl-x genes through the trans-activation domain of Stat5. J Immunol Baltim Md. 1950;2000(164):2533–41.
11. Nosaka T, Kawashima T, Misawa K, Ikuta K, Mui AL, Kitamura T. STAT5 as a molecular regulator of proliferation, differentiation and apoptosis in hematopoietic cells. EMBO J. 1999;18:4754–65.
12. Rascle A, Johnston JA, Amati B. Deacetylase activity is required for recruitment of the basal transcription machinery and transactivation by STAT5. Mol Cell Biol. 2003;23:4162–73.
13. Basham B, Sathe M, Grein J, McClanahan T, D'Andrea A, Lees E, et al. In vivo identification of novel STAT5 target genes. Nucleic Acids Res. 2008;36:3802–18.
14. Wakao H, Gouilleux F, Groner B. Mammary gland factor (MGF) is a novel member of the cytokine regulated transcription factor gene family and confers the prolactin response. EMBO J. 1994;13:2182–91.
15. Grimley PM, Dong F, Rui H. Stat5a and Stat5b: fraternal twins of signal transduction and transcriptional activation. Cytokine Growth Factor Rev. 1999;10:131–57.
16. Nosaka T, Kitamura T. Pim-1 expression is sufficient to induce cytokine independence in murine hematopoietic cells, but is dispensable for BCR-ABL-mediated transformation. Exp Hematol. 2002;30:697–702.
17. Liu CB, Itoh T, Arai K, Watanabe S. Constitutive activation of JAK2 confers murine interleukin-3-independent survival and proliferation of BA/F3 cells. J Biol Chem. 1999;274:6342–9.
18. Gesbert F, Griffin JD. Bcr/Abl activates transcription of the Bcl-X gene through STAT5. Blood. 2000;96:2269–76.
19. Valentino L, Pierre J. JAK/STAT signal transduction: regulators and implication in hematological malignancies. Biochem Pharmacol. 2006;71:713–21.
20. Ren S, Cai HR, Li M, Furth PA. Loss of Stat5a delays mammary cancer progression in a mouse model. Oncogene. 2002;21:4335–9.
21. Romanski A, Schwarz K, Keller M, Wietbrauk S, Vogel A, Roos J, et al. Deacetylase inhibitors modulate proliferation and self-renewal properties of leukemic stem and progenitor cells. Cell Cycle Georget Tex. 2012;11:3219–26.
22. Kosan C, Ginter T, Heinzel T, Krämer OH. STAT5 acetylation: mechanisms and consequences for immunological control and leukemogenesis. JAK-STAT. 2013;2:e26102.
23. Pietschmann K, Bolck HA, Buchwald M, Spielberg S, Polzer H, Spiekermann K, et al. Breakdown of the FLT3-ITD/STAT5 axis and synergistic apoptosis induction by the histone deacetylase inhibitor panobinostat and FLT3-specific inhibitors. Mol Cancer Ther. 2012;11:2373–83.
24. Evrot E, Ebel N, Romanet V, Roelli C, Andraos R, Qian Z, et al. JAK1/2 and Pan-deacetylase inhibitor combination therapy yields improved efficacy in preclinical mouse models of JAK2V617F-driven disease. Clin Cancer Res. 2013;19:6230–41.
25. Wang Y, Fiskus W, Chong DG, Buckley KM, Natarajan K, Rao R, et al. Cotreatment with panobinostat and JAK2 inhibitor TG101209 attenuates JAK2V617F levels and signaling and exerts synergistic cytotoxic effects against human myeloproliferative neoplastic cells. Blood. 2009;114:5024–33.
26. Nguyen T, Dai Y, Attkisson E, Kramer L, Jordan N, Nguyen N, et al. HDAC inhibitors potentiate the activity of the BCR/ABL kinase inhibitor KW-2449 in imatinib-sensitive or -resistant BCR/ABL+ leukemia cells in vitro and in vivo. Clin Cancer Res. 2011;17:3219–32.
27. Fiskus W, Sharma S, Qi J, Shah B, Devaraj SGT, Leveque C, et al. BET protein antagonist JQ1 is synergistically lethal with FLT3 tyrosine kinase inhibitor (TKI) and overcomes resistance to FLT3-TKI in AML cells expressing FLT-ITD. Mol Cancer Ther. 2014;13:2315–27.
28. Fiskus W, Sharma S, Qi J, Valenta JA, Schaub LJ, Shah B, et al. Highly active combination of BRD4 antagonist and histone deacetylase inhibitor against human acute myelogenous leukemia cells. Mol Cancer Ther. 2014;13:1142–54.

29. Rascle A, Lees E. Chromatin acetylation and remodeling at the Cis promoter during STAT5-induced transcription. Nucleic Acids Res. 2003;31:6882–90.

30. Liu S, Walker SR, Nelson EA, Cerulli R, Xiang M, Toniolo PA, et al. Targeting STAT5 in hematological malignancies through inhibition of the bromodomain and extra-terminal (BET) bromodomain protein BRD2. Mol Cancer Ther. 2014;13:1194–205.

31. Pinz S, Unser S, Buob D, Fischer P, Jobst B, Rascle A. Deacetylase inhibitors repress STAT5-mediated transcription by interfering with bromodomain and extra-terminal (BET) protein function. Nucleic Acids Res. 2015;43:3524–45.

32. Pinz S, Unser S, Brueggemann S, Besl E, Al-Rifai N, Petkes H, et al. The synthetic α-bromo-2′,3,4,4′-tetramethoxychalcone (α-Br-TMC) inhibits the JAK/STAT signaling pathway. PLoS One. 2014;9:e90275.

33. de Groot RP, Raaijmakers JA, Lammers JW, Jove R, Koenderman L. STAT5 activation by BCR-Abl contributes to transformation of K562 leukemia cells. Blood. 1999;94:1108–12.

34. Nieborowska-Skorska M, Wasik MA, Slupianek A, Salomoni P, Kitamura T, Calabretta B, et al. Signal transducer and activator of transcription (STAT)5 activation by BCR/ABL is dependent on intact Src homology (SH)3 and SH2 domains of BCR/ABL and is required for leukemogenesis. J Exp Med. 1999;189:1229–42.

35. Onishi M, Nosaka T, Misawa K, Mui AL, Gorman D, McMahon M, et al. Identification and characterization of a constitutively active STAT5 mutant that promotes cell proliferation. Mol Cell Biol. 1998;18:3871–9.

36. Silva M, Benito A, Sanz C, Prosper F, Ekhterae D, Nuñez G, et al. Erythropoietin can induce the expression of bcl-x(L) through Stat5 in erythropoietin-dependent progenitor cell lines. J Biol Chem. 1999;274:22165–9.

37. Xu M, Nie L, Kim S-H, Sun X-H. STAT5-induced Id-1 transcription involves recruitment of HDAC1 and deacetylation of C/EBPbeta. EMBO J. 2003;22:893–904.

38. Nelson EA, Walker SR, Alvarez JV, Frank DA. Isolation of unique STAT5 targets by chromatin immunoprecipitation-based gene identification. J Biol Chem. 2004;279:54724–30.

39. Kiuchi N, Nakajima K, Ichiba M, Fukada T, Narimatsu M, Mizuno K, et al. STAT3 is required for the gp130-mediated full activation of the c-myc gene. J Exp Med. 1999;189:63–73.

40. Ramana CV, Grammatikakis N, Chernov M, Nguyen H, Goh KC, Williams BR, et al. Regulation of c-myc expression by IFN-gamma through Stat1-dependent and -independent pathways. EMBO J. 2000;19:263–72.

41. Grigorieva I, Grigoriev VG, Rowney MK, Hoover RG. Regulation of c-myc transcription by interleukin-2 (IL-2). Identification of a novel IL-2 response element interacting with STAT-4. J Biol Chem. 2000;275:7343–50.

42. Uslu VV, Petretich M, Ruf S, Langenfeld K, Fonseca NA, Marioni JC, et al. Long-range enhancers regulating Myc expression are required for normal facial morphogenesis. Nat Genet. 2014;46:753–8.

43. Filippakopoulos P, Qi J, Picaud S, Shen Y, Smith WB, Fedorov O, et al. Selective inhibition of BET bromodomains. Nature. 2010;468:1067–73.

44. Mertz JA, Conery AR, Bryant BM, Sandy P, Balasubramanian S, Mele DA, et al. Targeting MYC dependence in cancer by inhibiting BET bromodomains. Proc Natl Acad Sci USA. 2011;108:16669–74.

45. Palacios R, Steinmetz M. Il-3-dependent mouse clones that express B-220 surface antigen, contain Ig genes in germ-line configuration, and generate B lymphocytes in vivo. Cell. 1985;41:727–34.

46. Soldaini E, John S, Moro S, Bollenbacher J, Schindler U, Leonard WJ. DNA binding site selection of dimeric and tetrameric Stat5 proteins reveals a large repertoire of divergent tetrameric Stat5a binding sites. Mol Cell Biol. 2000;20:389–401.

47. Quentmeier H, MacLeod RA, Zaborski M, Drexler HG. JAK2 V617F tyrosine kinase mutation in cell lines derived from myeloproliferative disorders. Leukemia. 2006;20:471–6.

48. Bar-Natan M, Nelson EA, Walker SR, Kuang Y, Distel RJ, Frank DA. Dual inhibition of Jak2 and STAT5 enhances killing of myeloproliferative neoplasia cells. Leukemia. 2012;26:1407–10.

49. Dawson MA, Foster SD, Bannister AJ, Robson SC, Hannah R, Wang X, et al. Three distinct patterns of histone H3Y41 phosphorylation mark active genes. Cell Rep. 2012;2:470–7.

50. Pinz S, Rascle A. Assessing HDAC function in the regulation of signal transducer and activator of transcription 5 (STAT5) activity using chromatin immunoprecipitation (ChIP). Methods Mol Biol. 2016 **(in press)**.

# 15

# Positive selection and functional divergence of farnesyl pyrophosphate synthase genes in plants

Jieying Qian[1†], Yong Liu[2†], Naixia Chao[1], Chengtong Ma[1], Qicong Chen[1], Jian Sun[1] and Yaosheng Wu[1*]

## Abstract

**Background:** Farnesyl pyrophosphate synthase (FPS) belongs to the short-chain prenyltransferase family, and it performs a conserved and essential role in the terpenoid biosynthesis pathway. However, its classification, evolutionary history, and the forces driving the evolution of FPS genes in plants remain poorly understood.

**Results:** Phylogeny and positive selection analysis was used to identify the evolutionary forces that led to the functional divergence of FPS in plants, and recombinant detection was undertaken using the Genetic Algorithm for Recombination Detection (GARD) method. The dataset included 68 FPS variation pattern sequences (2 gymnosperms, 10 monocotyledons, 54 dicotyledons, and 2 outgroups). This study revealed that the FPS gene was under positive selection in plants. No recombinant within the FPS gene was found. Therefore, it was inferred that the positive selection of FPS had not been influenced by a recombinant episode. The positively selected sites were mainly located in the catalytic center and functional areas, which indicated that the 98S and 234D were important positively selected sites for plant FPS in the terpenoid biosynthesis pathway. They were located in the FPS conserved domain of the catalytic site. We inferred that the diversification of FPS genes was associated with functional divergence and could be driven by positive selection.

**Conclusions:** It was clear that protein sequence evolution via positive selection was able to drive adaptive diversification in plant FPS proteins. This study provides information on the classification and positive selection of plant FPS genes, and the results could be useful for further research on the regulation of triterpenoid biosynthesis.

**Keywords:** Biological evolution, Farnesyl pyrophosphate synthase, Positive selection, Terpenoid biosynthesis

## Background

Triterpenoids are a large class of plant secondary metabolites. They enable plants to withstand pathogens and pests [1, 2]. Many different plant species synthesize triterpenoid saponins during normal growth and development [3]. In clinical medicine, it has been shown that triterpene saponins have anti-tumor, anti-inflammatory, and anti-viral activities. They also help lower cholesterol and elevate immunity [4–11]. Generally, the biosynthetic

pathway for terpenoids can be divided into four or five stages. These are the formation of IPP (isopentenyl diphosphate, C5 unit), GPP (geranyl diphosphate, C10 unit), FPP (farnesyl diphosphate, C15 unit), squalene (C30 unit), 2, 3-oxidosqualene, and triterpenoid [3, 12, 13]. Farnesyl pyrophosphate synthase (FPS) catalyzes FPP formation. FPS has been widely found in lower green algae up to higher eudicot plants and has been cloned from various plants [14–22]. However, its origin, evolution, and structural and functional divergence remain poorly understood.

Farnesyl pyrophosphate synthase belongs to the short-chain prenyltransferase family [23] and it accelerates the head-to-tail condensation reaction of dimethylallyl pyrophosphate (DMAPP) with two molecules of

*Correspondence: wuyaosheng03@sina.com
†Jieying Qian and Yong Liu are co-first authors
[1] Key Laboratory of Biological Molecular Medicine Research of Guangxi Higher Education, Department of Biochemistry and Molecular Biology, Guangxi Medical University, Nanning, Guangxi, People's Republic of China
Full list of author information is available at the end of the article

isopentenyl pyrophosphate (IPP) to form FPP [24], which is the precursor of all sesquiterpenes and triterpenoids [25]. FPS provides substrate FPP to squalene synthase and sesquiterpene synthase [15]. Squalene synthase plays a role in steroid and triterpenoid synthesis, which are involved in cell membrane system building. Sesquiterpene synthase plays a role in the synthesis of cyclic sesquiterpene compounds [26]. FPS mainly affects sesquiterpene compounds [22] and then squalene synthase (SS) primarily controls downstream triterpenoid synthesis [27–29]. The large FPS functional diversity suggests that it may be subject to positive Darwinian selection. The conserved domains 90–104 (LVLDDIMDSSH-TRRG) and 225–237 (MGTYFQVQDDYLD) of *Panax notoginseng* FPS (*PnFPS*) have important effects on the catalytic activity of isopentenyl pyrophosphate synthase (Trans-IPPS) in downstream products [30]. However, it is not known how the FPS genes evolved and functionally diverged, or whether positive selection is associated with the two important functional domains. Furthermore, it remains unclear what the evolutionary relationships are between some essential catalytic sites. In this study, we analyzed nucleotide and amino acid residue divergence in the FPS genes from 68 species of land plants. Likelihood methods that utilized the site-model, branch-model, and branch-site model were used to investigate potential positive selection patterns for plant FPS.

## Results
### Origins of the FPS genes during plant evolution
A rooted maximum-likelihood (ML) phylogenetic tree based on codon alignment was produced by the Bayesian method in order to explore the origin and evolutionary history of FPS genes among plants. The FPS cDNA sequences from 68 species were used to reconstruct a phylogenetic tree. In addition, we used the Bayesian posterior probability (PP) to evaluate all clade supports. The analysis revealed that the FPS genes mainly fell into one of three general groups: gymnosperms (A), monocotyledons (B), and dicotyledons (C) (Fig. 1). The monocotyledons FPS isoforms are a highly supported monophyletic group and are thus separated from the dicot isoforms. The dicotyledons group contains representatives from all of the available dicots, including verified FPS sequences from *Panax notoginseng*, *Panax ginseng*, *Gynostemma pentaphyllum*, etc. The gymnosperm FPS also formed a

separate cluster that was closest to the monocots. The phylogeny showed that FPS genes consist of several distinct branch clusters, indicating that the formation of the paralogous lineages occurred before divergence of the individual species [31], and that *Chlamydomonas reinhardtii (CrFPS)* and *Huperzia serrate (HsFPS)* were outgroups of the assigned lineages. In plants, gene evolution leading to functional divergence plays a crucial role in the diversification of biochemical metabolites [32]. These findings were consistent with previous studies on the phylogenetic classification of terrestrial plants. Thus, the terrestrial plant phylogenetic tree for FPS genes may reflect the genetic relationships among different species. Based on the lineages of the tree, we inferred that the metabolites produced by different species varied as the accompanying metabolic pathway diverged. Plant FPS is located at a branch point of the terpenoid synthesis pathway and is responsible for directing carbon flow away from the central portion of the isoprenoid pathway [30]. Two types of terpenoids occurred. These were tetracyclic and pentacyclic triterpenoids. For example, ginsenoside, the main component of ginseng, is a dammarane tetracyclic triterpenoid. The oleanane-type pentacyclic triterpenoids are the most widespread, and hitherto most extensively studied compounds in the family Araliaceae, family Cucurbitaceae, and family Leguminosae.

### Detection of recombinant episodes
We were able to detect positive selection pressures using the evolutionary phylogenetic tree. However, recombination can have a profound impact on the evolutionary process [33] and can adversely affect the power and accuracy of phylogenetic reconstruction, molecular clock inference, and the detection of positively-selected sites [34–36]. Therefore, the recombination factor must be considered before performing positive selection analysis. In our study, Mafft software was used to align the 68 FPS sequences and convert the format to fasta. The aligned sequences file was used by the Genetic Algorithm for Recombination Detection (GARD) and Recombination Detection Program (RDP) methods to detect the recombinant events. The GARD and RDP analysis found no recombinant within the FPS genes. Therefore, it was inferred that the positive selection of FPS has not been influenced by a recombinant episode.

(See figure on next page.)
**Fig. 1** Phylogenetic tree of terrestrial plant FPS. The phylogenetic tree of plant FPSs was constructed through the Bayesian analyses. Posterior probabilities are labeled above branches. *Chlamydomonas reinhardtii (CrFPS)* and *Huperzia serrate (HsFPS)* were used as outgroups. The clades of gymnosperms, monocotyledons and dicotyledons were labeled as *A*, *B* and *C*, respectively. The *numbers* indicate the Bayesian probabilities for each phylogenetic clade. Posterior probability values were to only show the pp values smaller than 1.0 with the tree

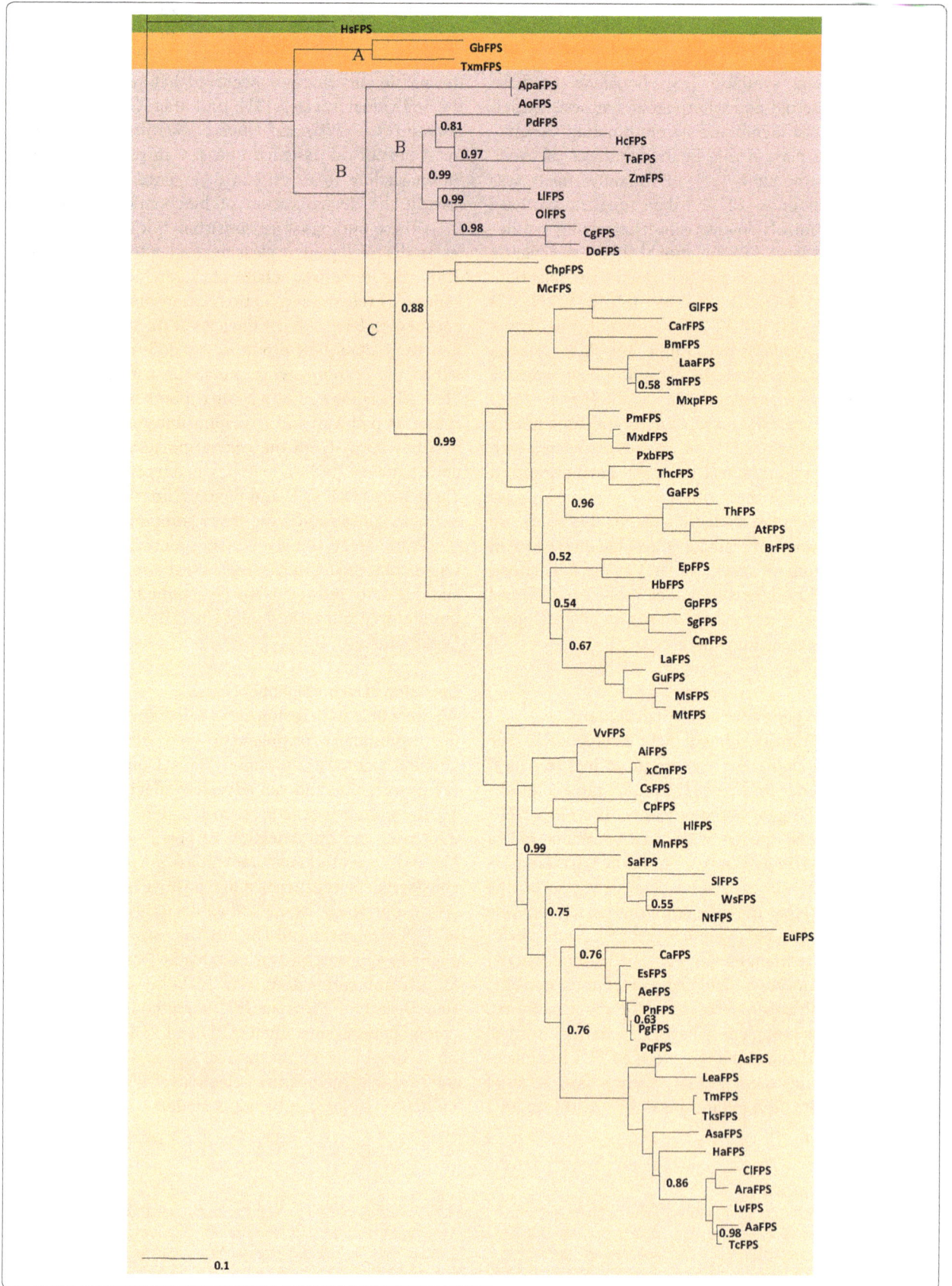

## Positively selected sites in the FPS family and their putative biological significance

The site-specific model, the branch model, the branch-site model, and PAML package version 4.4 were used to detect the selective pressure on the FPS family in plants. After removing the gaps, all the amino acids sites were analyzed using the CodeML program. In the site model, none of the positive selection sites was detected by the M0 vs. M3 or M2a vs. M1a model. However, the alternative models, M3 and M8, may fit the data significantly better than the null models, M0 and M7 (for M3 vs. M0, $2\Delta L = 2715.02$, $p < 0.001$; for M8 vs. M7, $2\Delta L = 9346.66$, $p < 0.001$), but only M8 identified several sites with an $\omega$ value significantly greater than 1. Therefore, at the PP > 95% level, 39 amino acid sites were identified as being under positive selection by M8 (Table 1), including 28 positive selection sites with a PP > 99% (Table 1) and 11 sites as potential targets of positive selection with a PP > 0.95 (1M, 2S, 6T, 10E, 29D, 111L, 125L, 176S, 195S, 310K, and 326A). Positive selection may only happen during specific stages of evolution or in specific branches, which means that positive selection may only affect some branches. Therefore, we used a branch-specific model to detect positive selection. The branch model suggested that the free ratio model was significantly higher than the one ratio model ($2\Delta lnL = 256.64$, $p = 0.00$), which indicated that there was heterogeneous selection among branches. The selective pressure on the different branches and sites was investigated by using the branch-site model to directly search for the positively-selected amino acid sites. Branch-site model was used to search for amino acid sites that underwent positive selection in branches a, b, and c, and then fixed the three branches as foreground branches in the branch-site model. According to the likelihood ratio test (LRT) for the branch-site (Table 1), comparisons of BSa1 vs. BSa0-fix ($2\Delta lnL = 10.56$, $p = 0.0012$), BSb1 vs. BSb0-fix ($2\Delta lnL = 10.12$, $p = 0.01$), and BSc1 vs. BSc0-fix ($2\Delta lnL = 9.98$, $p = 0.01$), were significantly different. Naive Empirical Bayes (NEB) analysis and Bayes Empirical Bayes (BEB) analysis were undertaken, but the BEB analysis showed the posteriori probability of the positive selection sites better than the NEB analysis. The positive pressure computation showed that there were three amino acid sites (98S, 148D, 234D) in the branch with a $p < 0.01$ for BSa1 vs. BSa0-fix, which were considered to have undergone positive selection. The analysis showed that (1) FPS genes suffered from positive selection during the plant evolutionary process; and (2) some representative positively-selected sites were located in the catalytic region. These features suggested that positive selection sites located in the functional domain of FPS are important components of the FPS functional structure.

## Protein structural characteristics of FPS in plants

In addition to the above-mentioned phylogenetic and the positive selection FPS analysis, we also conducted detailed structural studies based on the two-dimensional model containing the protein sequence alignment of the FPS in several important medicinal herbs, such as *Panax ginseng* (*PgFPS*), *Panax quinquefolium* (*PqFPS*), *Gynostemma pentaphyllum* (*GpFPS*), *Panax notoginseng* (*PnFPS*), and *Eleutherococcus senticosus* (*EsFPS*). *PnFPS* was used as the reference sequence. These FPSs shared a high level of sequence similarity in the coding region. The structure of the FPS members is highly conserved. The conserved sites (shaded) and the functional areas are shown in Fig. 2. The observations suggested that these areas may undergo positive Darwinian selection or an increase in the fixation of neutral mutations due to the relaxation of functional constraints. We mapped these sites onto the model as well as their sequence alignments. The results showed that the distribution of these sites was largely disordered, but a few sites were concentrated in some special FPS spatial locations.

## Distributions of possible positive selection sites on FPS three dimensional structures

We predicted the positive selection sites using the BEB method. Thirty-nine sites were identified as positively selected at a BEB posterior probability threshold of 95% in the site-model. In order to draw positive selected sites onto a plant FPS three-dimensional model, we first built an energy-minimized model using a homology modeling approach [37]. We took the protein structure of *Panax notoginseng* as an example and analyzed the relationship between positive selection sites and functional sites. The PDB data was produced in Swiss model (http://swiss-model.expasy.org/), where the highest sequence similarity identified in the PSI-BLAST analysis corresponded to the FPS. We mapped three positively selected sites (98S, 148D, and 234D) and tested them in the branch-site model. Other important positively selected sites tested in the site model were mapped onto the surface of the three-dimensional structure by Pymol (http://PyMOLwiki.org). As shown in Fig. 3, positively-selected 59K and 60L were relatively adjacent to the acylated 46G site in the spatial structure (Fig. 3a: involved in N-myristoylation site 46G), and 302D was near to the protein kinase C phosphorylation site in the spatial structure (Fig. 3b: Involved in the protein kinase C phosphorylation site). In Fig. 3c, positively selected site 98S was close to the chemical binding site 97D. Furthermore, in the 111L and 250T active sites, positively-selected site 176S was significantly related to the active sites (Fig. 3d: involved in active site lid residues 111L and 250T). In the highly conserved domain, positive selection sites 98S, and 234D were located in the

**Table 1  Positive selection sites of FPS tested through the site model, branch model and branch-site model**

| Model | Estimate of parameters | lnL | LRT pairs | df | 2ΔlnL | P value | Positive selection sites |
|---|---|---|---|---|---|---|---|
| *Site model* | | | | | | | |
| M0:one ratio | $\omega = 0.11676$ | −30867.39 | | | | | |
| M3:discrete | $p_0 = 0.49134$, $p_1 = 0.296497$, $p_2 = 0.21217$, $\omega_0 = 0.01740$, $\omega_1 = 0.15430$, $\omega_2 = 0.50703$ | −29509.88 | M0/M3 | 3 | 2715.02 | 0 | none |
| M1a:neutral | $p_0 = 0.75883$, $p_1 = 0.24117$, $\omega_0 = 0.07661$, $\omega_1 = 1.00000$ | −30122.51 | M1a/M2a | 2 | 0 | 1 | none |
| M2a:selection | $p_0 = 0.75883$, $p_1 = 0.15391$, $p_2 = 0.08726$, $\omega_0 = 0.07661$, $\omega_1 = 1.0000$, $\omega_2 = 1.00000$ | −30122.51 | | | | | |
| M7:beta | $p = 0.37262$, $q = 1.74173$ | −29440.75 | | | | | |
| M8:beta&ω | $p_0 = 1.0000$, $p = 0.28555$, $q = 1.02636$, ($p_1 = 0.00000$), $\omega_2 = 2.36785$ | −34114.08 | M7/M8 | 2 | 9346.66 | 0 | 3D, 7R, 14V, 21N, 25F, 27F, 34W, 47K, 59K, 60L, 65K, 98S, 99S, 181P, 207S, 213K, 233D, 293E, 275F, 286A, 251D, 252I, 270E, 302D, 305A, 309S, 336G, 342Q (all were[a]) |
| *Branch model* | | | | | | | |
| Model 0:(one-ratio) | $\omega = 0.11676$ | −30867.39 | M0/Free model | 135 | 258.64 | 0 | none |
| Free model | $\omega_a = 1.0131$, $\omega_b = 1.3249$, $\omega_c = 540.6926$ | −30738.07 | | | | | |
| *Branch-site model* | | | | | | | |
| BSa1 | $p_0 = 0.00006$, $p_1 = 0.00001$, $p_{2a} = 0.85238$, $p_{2b} = 0.14755$, $b\omega_0 = 0.07506$, $\omega_1 = 1.00000$, $\omega_{2a} = 0.07506$, $\omega_{2b} = 1.00000$, $f\omega_0 = 0.07506$, $\omega_1 = 1.00000$, $\omega_{2a} = 1.00000$, $\omega_{2b} = 1.00000$ | −25664.61 | BSa1/BSa0-fix | 1 | 10.56 | 0 | 98S, 148D, 234D |
| BSa0_fix | $p_0 = 0.81966$, $p_1 = 0.14204$, $p_{2a} = 0.03264$, $p_{2b} = 0.00566$, $b\omega_0 = 0.07521$, $\omega_1 = 1.00000$, $\omega_{2a} = 0.07521$, $\omega_{2b} = 1.00000$, $f\omega_0 = 0.07521$, $\omega_1 = 1.00000$, $\omega_{2a} = 1.00000$, $\omega_{2b} = 1.00000$ | −25669.89 | | | | | none |
| BSb1 | $p_0 = 0.02224$, $p_1 = 0.00386$, $p_{2a} = 0.82997$, $p_{2b} = 0.14393$, $b\omega_0 = 0.07528$, $\omega_1 = 1.00000$, $\omega_{2a} = 0.07528$, $\omega_{2b} = 1.00000$, $f\omega_0 = 0.07528$, $\omega_1 = 1.00000$, $\omega_{2a} = 1.00000$, $\omega_{2b} = 1.00000$ | −25663.83 | BSb1/BSb0-fix | 1 | 10.12 | 0 | none |
| BSb0_fix | $p_0 = 0.10932$, $p_1 = 0.01896$, $p_{2a} = 0.74286$, $p_{2b} = 0.12886$, $b\omega_0 = 0.07528$, $\omega_1 = 1.00000$, $\omega_{2a} = 0.07528$, $\omega_{2b} = 1.00000$, $f\omega_0 = 0.07528$, $\omega_1 = 1.00000$, $\omega_{2a} = 8.15518$, $\omega_{2b} = 8.15518$, $f\omega_0 = 0.07528$, $\omega_1 = 1.00000$, $\omega_{2a} = 8.15518$, $\omega_{2b} = 8.15518$ | −25668.89 | | | | | |
| BSc1 | $p_0 = 0.03778$, $p_1 = 0.00655$, $p_{2a} = 0.81443$, $p_{2b} = 0.14124$, $b\omega_0 = 0.07526$, $\omega_1 = 1.00000$, $\omega_{2a} = 0.07526$, $\omega_{2b} = 1.00000$, $f\omega_0 = 0.07526$, $\omega_1 = 1.00000$, $\omega_{2a} = 1.00000$, $\omega_{2b} = 8.15518$ | −25664.75 | BSc1/BSc0-fix | 1 | 9.98 | 0 | none |

**Table 1 Continued**

| Model | Estimate of parameters | lnL | LRT pairs | df | 2ΔlnL | P value | Positive selection sites |
|-------|------------------------|-----|-----------|----|-------|---------|--------------------------|
| ScO_fix | p0 = 0.01751, p1 = 0.00303, p2a = 0.83486, p2b = 0.14461, bω0 = 0.07526, ω1 = 1.00000, ω2a = 0.07526, ω2b = 1.00000, fω0 = 0.07526, ω1 = 1.00000, ω2a = 1.00000, ω2b = 1.00000 | −25669.74 | | | | | |

Selection analysis by site model was performed using CodeML implemented in PAML. Significant tests at 1% cut off

*lnL* log-likelihood values, *LRT* likelihood ratio test, *ω2* average dN/dS ratio for sites subject to positive selection, *p and q* shape parameters for the beta distribution of *ω, p0, p1, and p2* proportions of codons subject to purifying selection, neutral evolution, and positive selection, respectively, *df* degrees of freedom, *2ΔlnL* twice the log-likelihood difference of the model compared

[a] Posterior probability >99%

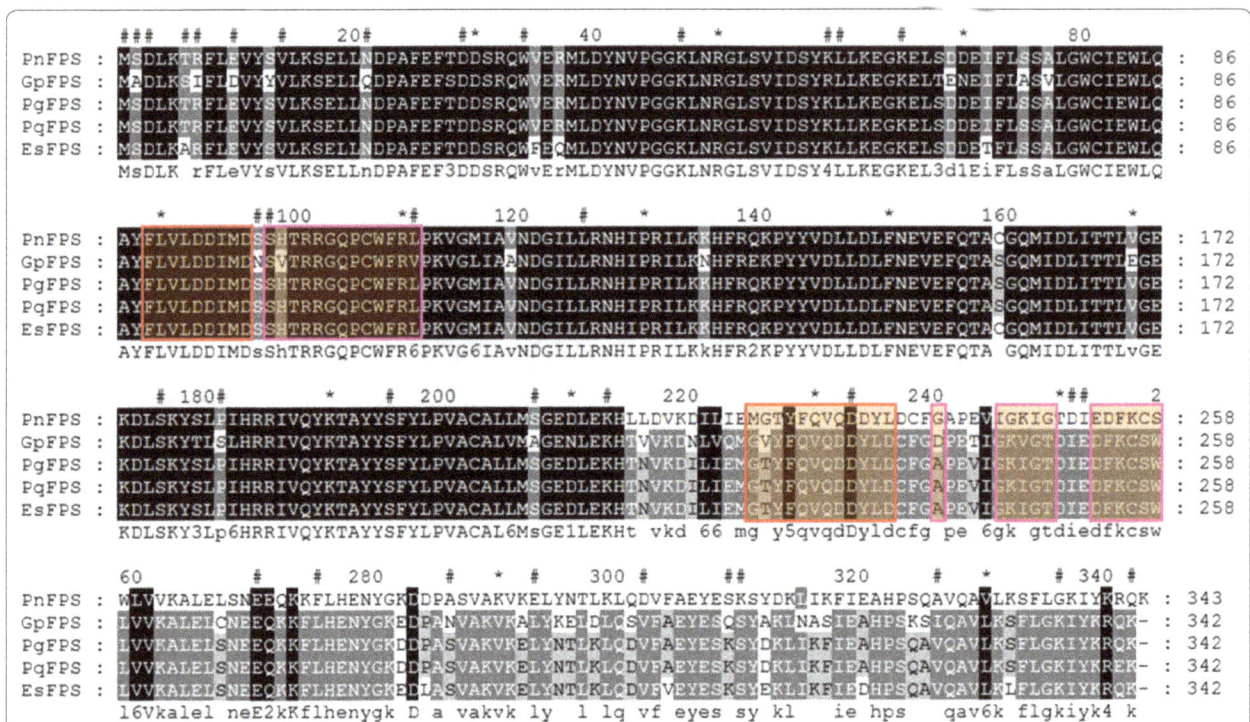

**Fig. 2** Multi-alignment of the amino acid sequences of partial terrestrial plant FPS. *PnFPS*, *PgFPS*, *EsFPS*, *PqFPS*, and *GpFPS* represent farnesyl pyrophosphate synthase cloned from *Panax notoginseng, Panax ginseng, Eleutherococcus senticosus, Panax quinquefolium,* and *Gynostemma pentaphyllum,* respectively. The positive selection sites for *FPS* in the above five common medicinal plants were marked and displayed through GeneDoc (http://www.nrbsc.org/gfx/genedoc). *PnFPS* was used as the reference sequence. The conserved sites were shaded. *Hash symbol* positive selection site; *red box* conserved sites of trans-isoprenyl diphosphate synthases (Trans IPPS); *carmine box* active site lid residues

important DDXX (XX)D aspartate-rich domains (Fig. 3e: positive selection sites tests in the branch-site model). Positive selection sites 207S and 213K were close to the substrate-$Mg^{2+}$ binding sites 247K and 251D (Fig. 3f: involved in substrate-$Mg^{2+}$ binding site 247K and 251D). All of these positively-selected sites may be key amino acids for this important functional region.

## Discussion

FPS plays a vital role in the isoprenoid biosynthesis pathway. The reaction catalyzed by FPS is considered the rate-limiting step and determines the flow fate of farnesyl diphosphate [15, 22, 30]. In this study, we reported the molecular evolution of positive selection sites in plant FPS genes for the first time. The gene expression analysis showed that FPS genes could increase terpenoid accumulation in plants [15, 38, 39]. In our study, we combined molecular phylogenetic analysis, putative biological significance, and protein structure analysis to clarify the evolutionary mechanisms. However, how FPS improves the triterpenoid content in the biosynthesis pathway is still not clear, and their biological roles in many species are also poorly understood.

As the number of FPS gene sequences cloned in our laboratory and collected from the database increased, it became more feasible to explore the evolutionary relationships and the functional diversity of the FPS family. In this study, 68 sequences were used for phylogenetic reconstruction by Bayesian methods. The phylogenetic analysis showed that FPS gene formation occurred before the divergence of individual species. The phylogenetic tree allowed us to investigate FPS evolution and to further understand the relationship between FPS structure and function in plants. These results are consistent with the phylogenetic classification of terrestrial plants and similar to the functional divergence analysis. The phylogenetic analysis clearly showed how FPS was classified, which may affect its functional divergence.

Positive selection is the retention and spread of advantageous mutations throughout a population and has long been considered synonymous with protein functional shifts [40]. Previous research found that positively-selected genes are more likely to interact with each other than genes not under positive selection [41].

In the evolutionary history of many microorganisms, positive selection and homologous recombination are

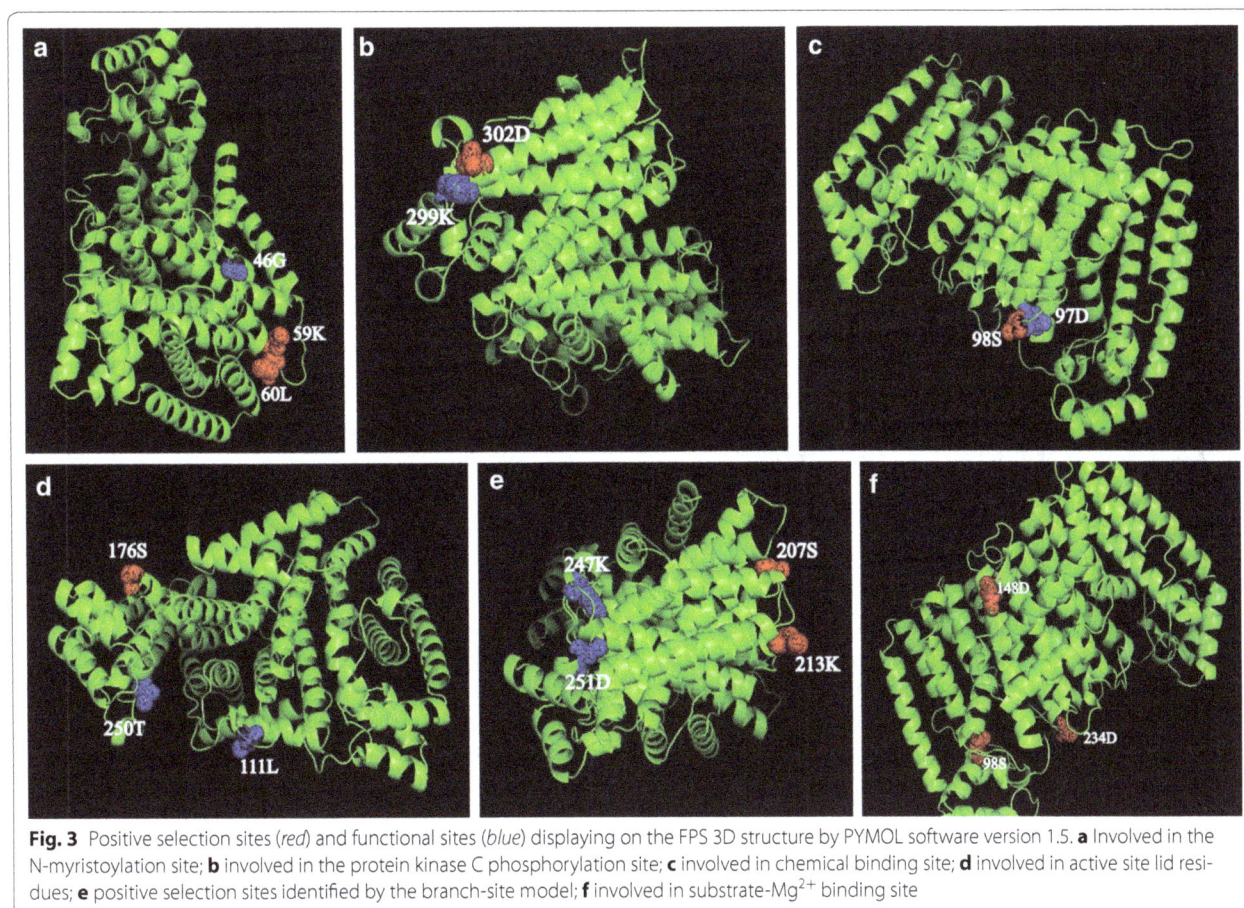

**Fig. 3** Positive selection sites (*red*) and functional sites (*blue*) displaying on the FPS 3D structure by PYMOL software version 1.5. **a** Involved in the N-myristoylation site; **b** involved in the protein kinase C phosphorylation site; **c** involved in chemical binding site; **d** involved in active site lid residues; **e** positive selection sites identified by the branch-site model; **f** involved in substrate-Mg$^{2+}$ binding site

two indispensable forces that drive adaptation to new niches. Therefore, before undertaking the positive selection analysis, we detected potential recombination events in order to assure the accuracy of any positive selections found. GARD found no evidence of recombination, which meant that the positive selections detected were statistically reliable. The selection events on coding sequences could affect gene expression regulation. Therefore, it is vital to detect positively-selected sites on the FPS ORF in order to get a further insight into the relationship between its structure and function. Site model, branch model, and branch-site model were used to detect positive selection among pre-specified groups. The ω values from the site model analysis did not fit the data well enough to describe the variability under selection pressure across amino acid sites. However, the branch model results showed that the ω ratios varied among clades, which meant that this model could be used to evaluate some sites in specific clades of the FPS phylogenic tree. Using molecular adaptive evolution and the positive selection principle to search corresponding functional sites can provide valuable reference information for FPSs that influence the regulation of synthetic triterpenoids.

About 20 years ago, several structural FPS genes from *Homo sapiens, Rattus rattus, Callus gallus, Saccharomyces cerevisiae, Escherichia coli,* and *Bacillus stearothermophilus* were identified and characterized, and five regions with highly conserved residues and sequence comparisons revealed two conserved DDXX(XX)D aspartate-rich domains [42], which were considered to be binding sites for the diphosphate moieties in IPP and allylic substrates. Now, many plant FPS genes have been cloned and identified too [14, 18, 19, 43, 44]. As shown in the space structure of *PnFPS* in Fig. 2, the positively-selected 59K site is overlapped in protein kinase C phosphorylation sites and 207S coincides with casein kinase II phosphorylation sites. Positions 90–104 (LVLDDIMDSSHTRRG) and 225–237 (MGTYFQVQDDYLD) in *PnFPS* contain the isopentenyl pyrophosphate synthase (Trans-IPPS) conserved domain of the catalytic site, and positive selection sites 98S and 99S are in the conserved domains. The first aspartate-rich region is an FPS chain length determination (CLD) region for the consecutive condensations of isopentenyl diphosphate with allylic diphosphates. A conversion analysis of archaeal geranylgeranyl pyrophosphate synthase (GGPS) to FPS inferred that the archaeal

GGPSs had evolved into type I and type II FPSs in eukaryotes and prokaryotes, respectively, and that the conserved CLD region made significant differences to some important FPS functions [45]. It was predicted that the region around the first aspartate-rich motif was essential for the product specificity of all FPP synthases and that the aromatic amino acid on the fifth amino acid before the first aspartate-rich motif (DDXX (XX)D, FARM) had been replaced. In this study, the positive selection sites 98S and 99S in plant FPS were found to be close to the first conserved motif (DDIMD). Therefore, 98S and 99S might be important sites that affect the biochemical function of plant FPS. Moreover, the site 59K coincided with protein kinase C phosphorylation, which indicated that 59K might undergo positive selection, so we inferred that this site could be related to protein tyrosine phosphorylation. A mutation in this site might change the downstream reactions during secondary metabolite biosynthesis. 207S also underwent a positive pressure that corresponded to the casein kinase II phosphorylation sites. These sites may be associated with protein kinase phosphorylation and acylation, and site-directed mutagenesis experiments would confirm this. Positive selection site 234D was located in the functional domains of the 225–237 amino acids (MGTYFQVQDDYLD). This was tested in the branch-site model, which showed better than any other model that they had important and potential positive selection functions during evolution. Furthermore, positive-selection site 98S contained casein kinase II phosphorylation and chemical binding sites, such as $Mg^{2+}$ binding site, which are relatively close in the space structure. It could be deduced that the 98S located in the highly conserved aspartate-rich region is the more important functional site. To further characterize the relationship between functional divergence and the site-specific evolution of amino acids, some potential amino acid sites associated with positive selection were chosen and mapped to the sequence alignment and the 3D structural model. The results showed that the functional divergence of the 98S site occurred during the site-specific evolution of amino acids, which suggested that 98S site-specific evolution was closely related to functional divergence in the FPS family.

## Conclusions

This study is the first large-scale evolutionary analysis of FPS in land plants. It explores the relationship between the molecular evolution of positive selection sites and their roles in plant FPS. Our results indicate: (1) FPS genes in plants appeared very early, and could be traced back to the bryophyte divergence to pteridophyte, which then evolved into gymnospermae, monocotyledonae, and dicotyledoneae; and (2) a number of signals for positive selection exist in plant FPSs. Thirty-nine positively selected sites in the site model and three positively selected sites in the branch-site model were detected, respectively. Furthermore, 98S was detected by both models and was located in the catalytic center. Therefore, 98S was considered the most significant site for plant FPS during the terpenoid synthesis process. 234D, which was detected in the branch-site model and was located in the functional domains, may provide an important reference for exploring further functional sites for FPS in the triterpenoid biosynthesis pathway. (3) The diversification of FPS genes among terrestrial plants could be attributed to functional divergence, which probably improves the activity of the enzymes in the triterpenoid biosynthesis pathway when plants adapt to terrestrial environments. This study provides useful information for further research on the regulation of triterpenoid biosynthesis.

## Methods
### Sequence data
In our study, plant FPS gene sequences contain two parts. FPS sequences in *Panax notoginseng* (GenBank accession AAY53905) and *Gynostemma pentaphyllum* (GenBank accession KJ917160) were cloned by our laboratory using rapid-amplification of cDNA ends (RACE) technology, and other cDNA sequences for FPS genes were collected from existing databases. The amino acid sequences were downloaded from GenBank at the National Center for Biotechnology Information (NCBI) (http://www.ncbi.nlm.nih.gov/) and the UniProt databases (http://www.uniprot.org/) (information about the total FPS sequences is shown in Additional file 1, downloaded before 2015-06). Then, BLAST and PSI-BLAST searches against the non-redundant database of FPS genomes at UniProt and NCBI were conducted. Only the full-length coding sequences were utilized in the final analysis. All partial, putative, redundant, and incomplete CDs were eliminated from our original sequences. In addition, each corresponding protein was matched to CDs. The final data included 68 sequences from terrestrial plants. These consisted of 2 gymnospermae, 10 monocotyledons, 54 dicotyledons, and *Chlamydomonas reinhardtii (CrFPS) and Huperzia serrate (HsFPS)* as outgroups.

### Sequence alignment
Multiple sequence alignments were performed using MUSCLE software [46] with the default parameters (http://www.ebi.ac.uk/Tools/msa/muscle/) to align the sequences of the proteins after the exclusion of poorly aligned positions, gap positions, and highly divergent regions. Then the CDs sequences were rearranged according to their amino acid alignment. The aligned amino acids and rearranged CDs were entered into

EMBL web tool PAL2NAL [47] (http://www.bork.embl.de/pal2nal/), which can form multiple codon alignments from matching amino acid sequences. The nucleotide sequences after PAL2NAL alignment were then converted to the nexus format using MEGA4.0 software [48] for phylogenetic analysis.

## Phylogenetic analysis

Phylogenetic trees were generated using MrBayes version 3.1.2 software [49, 50]. Before the MrBayes tree could be constructed, we had to modify the parameters in the nexus file using PAUP* version 4.0 [51] and Modeltest version 3.7 [52] to produce test outfiles that could be used to obtain a list of the best settings for these parameter types. The Akaike Information Criterion (AIC) [53] in PAUP* version 4.0 was used to evaluate the estimate of the most appropriate model for amino acid substitution during the tree-building analysis. ML [54] optimizations and distance methods were evaluated by the PhyML program in PAUP* version 4.0. Then the likelihood settings were obtained from the best-fit model (GTR + I + G) selected by AIC [55] in Modeltest 3.7. It comprises three important commands that can be used to specify the evolutionary model (lset), the prior knowledge (prset), the generation time, and the sampling frequency (mcmc). The parameters added and modified in the nexus file for tree reconstruction were as follows: Statefreqpr = dirichlet (0.2722, 0.2343, 0.2413, and 0.2522), revmatpr = dirichlet (1.4781, 3.1597, 1.1667, 1.1255, 4.6277, and 1.0000), shapepr = fixed (2.2202), pinvarpr = fixed (0.0054), unlinkshape = (3), mcmcp ngen = 10,000,000, and samplefreq = 10,000; mcmc. There were 10 million generations with sampling every 10 thousand generations [56, 57]. After completing the MrBayes analysis, the first 250,000 generations were discarded from every run. The remaining data were used to compute the phylogenetic trees and to determine the posterior probabilities at the different nodes. When all the parameters had been completely modified, we used MrBayes to construct the phylogenetic tree [50].

## Detection of recombination events

According to previous research, LRT can lead to the false detection of positive selection in the presence of a recombination event [58]. Although recombination between species may occur in animals and plants, the sequence divergence is generally too low for phylogeny-based likelihood methods to be useful [59]. Recombination events may affect the detection of the positively-selective evidence. Therefore, we first tested for recombination signals between sequences involved in the alignment of FPS genes. The GARD approach [33] was applied to screen multiple sequence alignments for evidence of phylogenetic incongruence, and to identify the number and location of breakpoints and sequences involved in putative recombination events [34]. RDP software was also used to detect recombination events in FPS.

## Positively-selected sites and putative biological significance

To explore the selection pressure, we performed a strict statistical analysis using the CodeML program in the PAML version 4 software [60] using branch model, site model, and branch-site model [61] in a run based on the non-synonymous (dN) and synonymous (dS) nucleotide substitution rate ratio (dN/dS) or ω. Four files needed to be entered into CodeML: the nuc file, the treeview file, the corresponding ctl file, and the CodeML application program. The nuc file was produced from a DAMBE format conversion using PAML. If ω > 1, then there was a positive selection on some branches or sites, but the positive selection sites may occur in very short episodes or on only a few sites during the evolution of duplicated genes; ω < 1 suggests a purifying selection (selective constraints); and ω = 1 indicates neutral evolution. The parameter estimates (ω) and likelihood scores [62] were calculated for three pairs of models. These were M0 (one-ratio) vs. M3 (discrete), M1a (nearly-neutral) vs. M2a (positive-selection), and M7 (beta) vs. M8 (beta&ω) [50]. In these models, M0 assumed a constant ω ratio for all FPS coding sites; M3 allowed for three discrete classes of ω within the gene that was contrasted with LRT against the M0 model where the ω ratio was averaged over all gene sites; and M1a allowed for two classes of ω sites: negative sites with $\omega_0 < 1$ estimated from our data; and neutral sites with $\omega_1 = 1$, whereas M2a added a third class with $\omega_2$ possibly >1 estimated from our data. M7 was a null model in which ω was assumed to be beta-distributed among sites and M8 was an alternative selection model that allowed an extra category of positively selected sites [63]. The LRT [64] was used to compare the fit to the data of two nested models, which measured the statistical significance of each pair of nested models. The twice the log likelihood difference between each pair models (2ΔL) follows a Chi square distribution with the number of degrees of freedom equal to the difference in the number of free parameters. Therefore, we can get a p value for this LRT [65]. A significantly higher likelihood of the alterative model compared to the null model suggests positive selection. Generally, all positive selection sites were calculated by the M8 model, which provided some useful information for the branch-specific and branch-sites analysis. These site models might not detect positive selection affecting only a few sites along a few lineages after a duplication event, so we also implemented the branch model to select the statistically significant "foreground branch" under positive selection. This was achieved by comparing the fit to the data of the "one-ratio" model (M0) with the "free ratios" model (FR), where the rate parameters were estimated independently in each lineage. All other branches in the tree were "background" branches. The background branches share the same distribution of ω values among sites, whereas different values can apply to the foreground branch. Then, the branch-site model were applied, which further estimated the different dN/dS values among the significant branches detected by the branch model and among sites [66]. Finally, a Bayes empirical Bayes (BEB) [64] approach was then used to calculate the posterior probabilities that a site comes from the site class with $\omega > 1$, which, when implemented in PAML4, were used to identify sites under positive selection or purifying selection in the foreground group with significant LRTs [64]. Each branch group was labeled as a foreground group as well.

## Positive selections in the protein sequences and structure analysis

All the protein sequences for FPS were aligned with Clustal W and displayed through GeneDoc (http://www.nrbsc.org/gfx/genedoc), which enabled us to examine the possible mechanisms driving the structural evolution of the FPS family in the triterpenoid biosynthesis pathway. First, the functional areas in the model are composed of positively-selected sites and post-translational modification sites, such as the highly conserved aspartate rich region located between positions 100 and 104 (DDSKD), protein kinase C phosphorylation sites, casein kinase II phosphorylation sites, N-myristoylation sites, amidation sites, and the conservative catalytic sites for isopentenyl pyrophosphate synthase (Trans-IPPS). Secondly, the positive selection sites related to the above functional sites were marked according to the experimental results. Thirdly, the transmembrane predictions and the post-translational modification sites above were determined with predict protein [67, 68] and TMHMM2.0 [69]. An estimate for the prediction accuracy was based on the confidence score for the modeling. Finally, PYMOL software version 1.5 [70] (http://www.pymol.org/) was used to predict the potential impact that those positive selected sites may have on the overall structure and function of the protein. Furthermore, the functional areas and relevant positive-selection sites identified in the evolutionary analysis were built into three dimensional graphic models and are shown by the highlighted parts.

## Abbreviations
FPS: farnesyl pyrophosphate synthase; SS: squalene synthase; ML: maximum-likelihood; GARD: Genetic Algorithm for Recombination Detection; RDP: Recombination Detection Program; PP: posterior probability; RACE: rapid amplification of cDNA ends; BEB: Bayes empirical Bayes; LRT: likelihood ratio test; dN: non-synonymous; dS: synonymous; AIC: Akaike Information Criterion.

## Authors' contributions
JYQ carried out the molecular genetic studies, drafted the manuscript, and performed the statistical analysis; YSW conceived, designed the study, and revised the manuscript. YL was involved in the experimental analysis of data and results; NXC and CTM collected and sorted the sequence data; QCC participated in the sequence alignment; JS participated in data analysis, coordination and helped to draft the manuscript. All authors read and approved the final manuscript.

## Author details
[1] Key Laboratory of Biological Molecular Medicine Research of Guangxi Higher Education, Department of Biochemistry and Molecular Biology, Guangxi Medical University, Nanning, Guangxi, People's Republic of China. [2] Schools of Pharmacy, Guangdong Medical University, Dongguan, Guangdong, People's Republic of China.

## Acknowledgements
We are grateful for the grants from the National Natural Science Foundation of China. We appreciate the significant critical opinion and the revision suggestions of the manuscript from two peer reviewers and the editor of BMC Molecular Biology, which greatly improved the manuscript.

## Competing interests
The authors declare that they have no competing interests.

## Funding
This study was supported by grants from the National Natural Science Foundation of China (31260069). The funding body had no role in design of the study, manuscript preparation, or the decision to submit the manuscript for publication.

## References
1. Sparg SG, Light ME, van Staden J. Biological activities and distribution of plant saponins. J Ethnopharmacol. 2004;94:219–43.
2. Suzuki H, Achnine L, Xu R, Matsuda SP, Dixon RA. A genomics approach to the early stages of triterpene saponin biosynthesis in Medicago truncatula. Plant J. 2002;32:1033–48.
3. Haralampidis K, Trojanowska M, Osbourn AE. Biosynthesis of triterpenoid saponins in plants. Adv Biochem Eng Biotechnol. 2002;75:31–49.
4. Zong J, Wang R, Bao G, Ling T, Zhang L, Zhang X, Hou R. Novel triterpenoid saponins from residual seed cake of Camellia oleifera Abel. show anti-proliferative activity against tumor cells. Fitoterapia. 2015;104:7–13.
5. Yang P, Li X, Liu YL, Xu QM, Li YQ, Yang SL. Two triterpenoid glycosides from the roots of Camellia oleifera and their cytotoxic activity. J Asian Nat Prod Res. 2015;17:800–7.
6. Wu C, Zhang RL, Li HY, Hu C, Liu BL, Li YL, Zhou GX. Triterpenoid saponins from the root bark of Schima superba and their cytotoxic activity on B16 melanoma cell line. Carbohydr Res. 2015;413:107–14.

7. Liu C, Sun H, Wang W T, Zhang J B, Cheng A W, Guo X, Sun J Y. A new triterpenoid saponin from *Gleditsia sinensis* and its antiproliferative activity. Nat Prod Res. 2015: 1–6.
8. Zhao M, Ma N, Qiu F, Hai WL, Tang HF, Zhang Y, Wen AD. Triterpenoid saponins from the roots of *Clematis argentilucida* and their cytotoxic activity. Planta Med. 2014;80:942–8.
9. Mu LH, Bai L, Dong XZ, Yan FQ, Guo DH, Zheng XL, Liu P. Antitumor activity of triterpenoid saponin-rich *Adisia gigantifolia* extract on human breast adenocarcinoma cells in vitro and in vivo. Biol Pharm Bull. 2014;37:1035–41.
10. Koneri RB, Samaddar S, Ramaiah CT. Antidiabetic activity of a triterpenoid saponin isolated from *Momordica cymbalaria* Fenzl. Indian J Exp Biol. 2014;52:46–52.
11. Zhang W, Yao MN, Tang HF, Tian XR, Wang MC, Ji LJ, Xi MM. Triterpenoid saponins with anti-myocardial ischemia activity from the whole plants of *Clematis tangutica*. Planta Med. 2013;79:673–9.
12. Zheng X, Xu H, Ma X, Zhan R, Chen W. Triterpenoid saponin biosynthetic pathway profiling and candidate gene mining of the *Ilex asprella* root using RNA-Seq. Int J Mol Sci. 2014;15:5970–87.
13. Huang L, Li J, Ye H, Li C, Wang H, Liu B, Zhang Y. Molecular characterization of the pentacyclic triterpenoid biosynthetic pathway in *Catharanthus roseus*. Planta. 2012;236:1571–81.
14. Kim OTAJ, Hwang SJ, Hwang B. Cloning and expression of a farnesyl diphosphate synthase in *Centella asiatica* (L.) Urban. Mol Cells. 2005;19:294–9.
15. Zhao YJ, Chen X, Zhang M, Su P, Liu YJ, Tong YR, Wang XJ, Huang LQ, Gao W. Molecular cloning and characterisation of farnesyl pyrophosphate synthase from *Tripterygium wilfordii*. PLoS ONE. 2015;10:e0125415.
16. Mekkriengkrai DST, Hirooka K, Sakdapipanich J, Tanaka Y, Fukusaki E, Kobayashi A. Cloning and characterization of farnesyl diphosphate synthase from the rubber-producing mushroom *Lactarius chrysorrheus*. Biosci Biotechnol Biochem. 2004;68:2360–8.
17. Kojima NSW, Viroonchatapan E, Suh DY, Iwanami N, Hayashi T, Sankaw U. Geranylgeranyl diphosphate synthases from *Scoparia dulcis* and *Croton sublyratus*. cDNA cloning, functional expression, and conversion to a farnesyl diphosphate synthase. Chem Pharm Bull (Tokyo). 2000;48:1101–3.
18. Sanmiya KIT, Matsuoka M, Miyao M, Yamamoto N. Cloning of a cDNA that encodes farnesyl diphosphate synthase and the blue-light-induced expression of the corresponding gene in the leaves of rice plants. Biochim Biophys Acta. 1997;1350:240–6.
19. Pan ZHL, Backhaus RA. Cloning, characterization, and heterologous expression of cDNAs for farnesyl diphosphate synthase from the guayule rubber plant reveals that this prenyltransferase occurs in rubber particles. Arch Biochem Biophys. 1996;332:196–204.
20. Matsushita YKW, Charlwood BV. Cloning and analysis of a cDNA encoding farnesyl diphosphate synthase from *Artemisia annua*. Gene. 1996;172:207–9.
21. Delourme DLF, Karst F. Cloning of an *Arabidopsis thaliana* cDNA coding for farnesyl diphosphate synthase by functional complementation in yeast. Plant Mol Biol. 1994;26:1867–73.
22. Lan JB, Yu RC, Yu YY, Fan YP. Molecular cloning and expression of *Hedychium coronarium* farnesyl pyrophosphate synthase gene and its possible involvement in the biosynthesis of floral and wounding/herbivory induced leaf volatile sesquiterpenoids. Gene. 2013;518:360–7.
23. Dhar MK, Koul A, Kaul S. Farnesyl pyrophosphate synthase: a key enzyme in isoprenoid biosynthetic pathway and potential molecular target for drug development. New Biotechnol. 2013;30:114–23.
24. Lange BM, Rujan T, Martin W, Croteau R. Isoprenoid biosynthesis: the evolution of two ancient and distinct pathways across genomes. Proc Natl Acad Sci USA. 2000;97:13172–7.
25. Szkopinska A, Plochocka D. Farnesyl diphosphate synthase; regulation of product specificity. Acta Biochim Pol. 2005;52:45–55.
26. McGarvey DJ, Croteau R. Terpenoid metabolism. Plant Cell. 1995;7:1015–26.
27. Wang JR, Lin JF, Guo LQ, You LF, Zeng XL, Wen JM. Cloning and characterization of squalene synthase gene from *Poria cocos* and its up-regulation by methyl jasmonate. World J Microbiol Biotechnol. 2014;30:613–20.
28. Kim TD, Han JY, Huh GH, Choi YE. Expression and functional characterization of three squalene synthase genes associated with saponin biosynthesis in *Panax* ginseng. Plant Cell Physiol. 2011;52:125–37.
29. Lee MH, Jeong JH, Seo JW, Shin CG, Kim YS, In JG, Yang DC, Yi JS, Choi YE. Enhanced triterpene and phytosterol biosynthesis in *Panax* ginseng overexpressing squalene synthase gene. Plant Cell Physiol. 2004;45:976–84.
30. Dhar MK, Koul A, Kaul S. Farnesyl pyrophosphate synthase: a key enzyme in isoprenoid biosynthetic pathway and potential molecular target for drug development. N Biotechnol. 2013;30:114–23.
31. Danielson PB, Alrubaian J, Muller M, Redding JM, Dores RM. Duplication of the POMC gene in the paddlefish (*Polyodon spathula*): analysis of gamma-MSH, ACTH, and beta-endorphin regions of ray-finned fish POMC. Gen Comp Endocrinol. 1999;116:164–77.
32. Huang Y, Wang X, Ge S, Rao GY. Divergence and adaptive evolution of the gibberellin oxidase genes in plants. BMC Evol Biol. 2015;15:207.
33. Kosakovsky Pond SL, Posada D, Gravenor MB, Woelk CH, Frost SD. GARD: a genetic algorithm for recombination detection. Bioinformatics. 2006;22:3096–8.
34. Posada D, Crandall KA. The effect of recombination on the accuracy of phylogeny estimation. J Mol Evol. 2002;54:396–402.
35. Posada D. Unveiling the molecular clock in the presence of recombination. Mol Biol Evol. 2001;18:1976–8.
36. Shriner D, Nickle DC, Jensen MA, Mullins JI. Potential impact of recombination on sitewise approaches for detecting positive natural selection. Genet Res. 2003;81:115–21.
37. Landgraf R, Xenarios I, Eisenberg D. Three-dimensional cluster analysis identifies interfaces and functional residue clusters in proteins. J Mol Biol. 2001;307:1487–502.
38. Schmidt A, Gershenzon J. Cloning and characterization of isoprenyl diphosphate synthases with farnesyl diphosphate and geranylgeranyl diphosphate synthase activity from Norway spruce (*Picea abies*) and their relation to induced oleoresin formation. Phytochemistry. 2007;68:2649–59.
39. Yokoyama T, Ostermann A, Mizuguchi M, Niimura N, Schrader TE, Tanaka I. Crystallization and preliminary neutron diffraction experiment of human farnesyl pyrophosphate synthase complexed with risedronate. Acta Crystallogr F Struct Biol Commun. 2014;70:470–2.
40. Morgan CC, Shakya K, Webb A, Walsh TA, Lynch M, Loscher CE, Ruskin HJ, O'Connell MJ. Colon cancer associated genes exhibit signatures of positive selection at functionally significant positions. BMC Evol Biol. 2012;12:114.
41. Vamathevan JJ, Hasan S, Emes RD, Amrine-Madsen H, Rajagopalan D, Topp SD, Kumar V, Word M, Simmons MD, Foord SM, Sanseau P, Yang Z, Holbrook JD. The role of positive selection in determining the molecular cause of species differences in disease. BMC Evol Biol. 2008;8:273.
42. Chen A, Kroon PA, Poulter CD. Isoprenyl diphosphate synthases: protein sequence comparisons, a phylogenetic tree, and predictions of secondary structure. Protein Sci. 1994;3:600–7.
43. Hefner JKR, Croteau R. Cloning and functional expression of a cDNA encoding geranylgeranyl diphosphate synthase from *Taxus canadensis* and assessment of the role of this prenyltransferase in cells induced for taxol production. Arch Biochem Biophys. 1998;360:62–74.
44. Xiang L, Zhao K, Chen L. Molecular cloning and expression of *Chimonanthus praecox* farnesyl pyrophosphate synthase gene and its possible involvement in the biosynthesis of floral volatile sesquiterpenoids. Plant Physiol Biochem. 2010;48:845–50.
45. Ohnuma S, Hirooka K, Ohto C, Nishino T. Conversion from archaeal geranylgeranyl diphosphate synthase to farnesyl diphosphate synthase. Two amino acids before the first aspartate-rich motif solely determine eukaryotic farnesyl diphosphate synthase activity. J Biol Chem. 1997;272:5192–8.
46. Edgar RC. MUSCLE: multiple sequence alignment with high accuracy and high throughput. Nucleic Acids Res. 2004;32:1792–7.
47. Suyama M, Torrents D, Bork P. PAL2NAL: robust conversion of protein sequence alignments into the corresponding codon alignments. Nucleic Acids Res. 2006;34:W609–12.
48. Kumar S, Tamura K, Nei M. MEGA3: integrated software for molecular evolutionary genetics analysis and sequence alignment. Brief Bioinform. 2004;5:150–63.
49. Ronquist F, Huelsenbeck JP. MrBayes 3: Bayesian phylogenetic inference under mixed models. Bioinformatics. 2003;19:1572–4.
50. Yang Z, Wong WS, Nielsen R. Bayes empirical bayes inference of amino acid sites under positive selection. Mol Biol Evol. 2005;22:1107–18.
51. Posada D. Using MODELTEST and PAUP* to select a model of nucleotide substitution. Curr Protoc Bioinformatics. 2003; Chapter 6: Unit 6 5.

52. Posada DCK. MODELTEST: testing the model of DNA substitution. Bioinformatics. 1998;14:817–8.

53. Galtier N. Maximum-likelihood phylogenetic analysis under a covarion-like model. Mol Biol Evol. 2001;18:866–73.

54. Bielawski JP, Yang Z. Maximum likelihood methods for detecting adaptive evolution after gene duplication. J Struct Funct Genomics. 2003;3:201–12.

55. Rokas A, Nylander JA, Ronquist F, Stone GN. A maximum-likelihood analysis of eight phylogenetic markers in gallwasps (Hymenoptera: Cynipidae): implications for insect phylogenetic studies. Mol Phylogenet Evol. 2002;22:206–19.

56. Huelsenbeck JP, Ronquist F. MRBAYES: Bayesian inference of phylogenetic trees. Bioinformatics. 2001;17:754–5.

57. Wang P, Liao Z, Guo L, Li W, Chen M, Pi Y, Gong Y, Sun X, Tang K. Cloning and functional analysis of a cDNA encoding *Ginkgo biloba* farnesyl diphosphate synthase. Mol Cells. 2004;18:150–6.

58. Anisimova M, Bielawski JP, Yang Z. Accuracy and power of bayes prediction of amino acid sites under positive selection. Mol Biol Evol. 2002;19:950–8.

59. Anisimova M, Bielawski JP, Yang Z. Accuracy and power of the likelihood ratio test in detecting adaptive molecular evolution. Mol Biol Evol. 2001;18:1585–92.

60. Yang Z. PAML: a program package for phylogenetic analysis by maximum likelihood. Comput Appl Biosci. 1997;13:555–6.

61. Guindon S, Rodrigo AG, Dyer KA, Huelsenbeck JP. Modeling the site-specific variation of selection patterns along lineages. Proc Natl Acad Sci USA. 2004;101:12957–62.

62. Wong WS, Yang Z, Goldman N, Nielsen R. Accuracy and power of statistical methods for detecting adaptive evolution in protein coding sequences and for identifying positively selected sites. Genetics. 2004;168:1041–51.

63. Tang Y, Wei Y, He W, Wang Y, Zhong J, Qin C. GATA transcription factors in vertebrates: evolutionary, structural and functional interplay. Mol Genet Genomics. 2014;289:203–14.

64. Nielsen R, Yang Z. Likelihood models for detecting positively selected amino acid sites and applications to the HIV-1 envelope gene. Genetics. 1998;148:929–36.

65. Whelan SGN. Distributions of statistics used for the comparison of models of sequence evolution in phylogenetics. Mol Biol Evol. 1999;19:1292.

66. Zhang J, Nielsen R, Yang Z. Evaluation of an improved branch-site likelihood method for detecting positive selection at the molecular level. Mol Biol Evol. 2005;22:2472–9.

67. Bairoch A, Bucher P, Hofmann K. The PROSITE database, its status in 1997. Nucleic Acids Res. 1997;25:217–21.

68. Rost B, Yachdav G, Liu J. The PredictProtein server. Nucleic Acids Res. 2004;32:W321–6.

69. Moller S, Croning MD, Apweiler R. Evaluation of methods for the prediction of membrane spanning regions. Bioinformatics. 2001;17:646–53.

70. Simmons AD, Nguyen TK, Follis JL, Ribes-Zamora A. Using a PyMOL activity to reinforce the connection between genotype and phenotype in an undergraduate genetics laboratory. PLoS ONE. 2014;9:e114257.

# 16

# Confirmation of translatability and functionality certifies the dual endothelin 1/VEGFsp receptor (DEspR) protein

Victoria L. M. Herrera[1,2], Martin Steffen[3], Ann Marie Moran[1,2], Glaiza A. Tan[1,2], Khristine A. Pasion[1,2], Keith Rivera[4], Darryl J. Pappin[4] and Nelson Ruiz-Opazo[1,2]*

## Abstract

**Background:** In contrast to rat and mouse databases, the NCBI gene database lists the human dual-endothelin1/VEGFsp receptor (DEspR, formerly *Dear*) as a unitary transcribed pseudogene due to a stop [TGA]-codon at codon#14 in automated DNA and RNA sequences. However, re-analysis is needed given prior single gene studies detected a tryptophan [TGG]-codon#14 by manual Sanger sequencing, demonstrated DEspR translatability and functionality, and since the demonstration of actual non-translatability through expression studies, the standard-of-excellence for pseudogene designation, has not been performed. Re-analysis must meet UNIPROT criteria for demonstration of a protein's existence at the highest (protein) level, which a priori, would override DNA- or RNA-based deductions.

**Methods:** To dissect the nucleotide sequence discrepancy, we performed Maxam–Gilbert sequencing and reviewed 727 RNA-seq entries. To comply with the highest level multiple UNIPROT criteria for determining DEspR's existence, we performed various experiments using multiple anti-DEspR monoclonal antibodies (mAbs) targeting distinct DEspR epitopes with one spanning the contested tryptophan [TGG]-codon#14, assessing: (a) DEspR protein expression, (b) predicted full-length protein size, (c) sequence-predicted protein-specific properties beyond codon#14: receptor glycosylation and internalization, (d) protein-partner interactions, and (e) DEspR functionality via DEspR-inhibition effects.

**Results:** Maxam–Gilbert sequencing and some RNA-seq entries demonstrate two guanines, hence a tryptophan [TGG]-codon#14 within a compression site spanning an error-prone compression sequence motif. Western blot analysis using anti-DEspR mAbs targeting distinct DEspR epitopes detect the identical glycosylated 17.5 kDa pull-down protein. Decrease in DEspR-protein size after PNGase-F digest demonstrates post-translational glycosylation, concordant with the consensus-glycosylation site beyond codon#14. Like other small single-transmembrane proteins, mass spectrometry analysis of anti-DEspR mAb pull-down proteins do not detect DEspR, but detect DEspR-protein interactions with proteins implicated in intracellular trafficking and cancer. FACS analyses also detect DEspR-protein in different human cancer stem-like cells (CSCs). DEspR-inhibition studies identify DEspR-roles in CSC survival and growth. Live cell imaging detects fluorescently-labeled anti-DEspR mAb targeted-receptor internalization, concordant with the single internalization-recognition sequence also located beyond codon#14.

**Conclusions:** Data confirm translatability of DEspR, the full-length DEspR protein beyond codon#14, and elucidate DEspR-specific functionality. Along with detection of the tryptophan [TGG]-codon#14 within an error-prone compression site, cumulative data demonstrating DEspR protein existence fulfill multiple UNIPROT criteria, thus refuting its pseudogene designation.

**Keywords:** DEspR, Pseudogene, DEspR protein–protein interactions

---

*Correspondence: nruizo@bu.edu
[1] Whitaker Cardiovascular Institute, Boston University School of Medicine, 700 Albany Street, Boston, MA 02118, USA
Full list of author information is available at the end of the article

## Background

In contrast to rat and mouse databases listing Dear as a gene, DNA and RNA sequence databases list the human Dear gene or the dual-endothelin1/VEGFsp receptor (DEspR) as a pseudogene [1] (Additional file 1: Figure S1). Automated DNA sequence databases report a stop codon [TGA] instead of tryptophan [TGG] at codon#14 reported in single gene study [2]. The current NCBI pseudogene annotation updated in May 2016 and referenced in other sites is discrepant with the single research group single-gene studies of human DEspR showing expression in human kidney via immunohistochemistry using a polyclonal anti-DEspR antibody, and functional studies of human DEspR expressed in permanent Cos1 cell transfectants detecting the predicted protein size by Western blot analysis as well as binding to DEspR-ligands (endothelin-1 and VEGFsp) [2]. The NCBI pseudogene annotation is also discrepant with the single gene study demonstrating DEspR-specific functional roles in cancer and putative regulation at the splicing level with detection of both unspliced and spliced DEspR RNA in human tumor cells by allele-specific amplification-refractory mutation system (ARMS) methodology [3].

Experimental clarification is warranted since the basis for the NCBI pseudogene annotation, automated DNA/RNA-sequencing, is known to have reproducible systematic sequencing errors, regardless of technology [4]. Occurrences of, hence risks for, systematic errors eliminate the a priori assumption that multiple occurrences negate errors. More specifically, systematic errors in high throughput DNA sequencing has been observed to occur "even in overlapping paired reads from high-coverage data, approximately one in 1000 bp, and are highly replicable across experiments" [4]. Moreover, given discrepancies among methodologies, the Sanger sequencing is the final determinant of sequence discrepancy since "any difference from the Sanger sequence is defined as a sequencing error" [5].

Further support for the need for scientific clarification is found in the GENCODE Pseudogene Resource which states that "the definition of a pseudogene is based on the presence of specific characteristics such as premature stop codon, coding sequence frame shift, truncation, or disabling insertion/deletion—*unless evidence (transcriptional, functional, publication) shows that the locus represents a protein-coding gene*" [6]. Concordant with Pei et al. [6], Kageyama et al. [7] explicitly states that "before a particular transcript can be determined to be a long noncoding RNA (or 'transcribed pseudogene'), there must be somewhat convincing evidence for its lack of translatability." Hence, deductions from automated sequence databases need to be evaluated in the context of experimental evidence for translatability and functionality.

To address the need for scientific clarification of DEspR (Dear) as a gene or pseudogene, we therefore upheld established standards that (1) "*the translatability of the candidate can be validated with specific antibodies against amino acid sequences predicted from the ORF,*" [7] and that (2) "*an assessment of ... a protein's molecular activity by biochemical methods should be the final certification of an active gene product*" [7]. These perspectives are codified in the Central Protein Resource UNIPROT criteria for translatability: "evidence for existence of a protein at the protein level, such as via antibody detection, is the highest level of evidence" [8].

Here we confirm the existence of the DEspR gene product at the nucleotide and protein level. We show that both Sanger and Maxam–Gilbert sequencing detect two G's for a tryptophan [TGG]-codon#14. We also determine that some RNA-seq entries also contain two G's but with an extra A or TA, as the region in question spans a canonical Yamakawa compression motif [9]. Monoclonal anti-DEspR antibodies to different DEspR epitopes detect the identical glycosylated 17.5 kDa pull-down protein from membrane-bound proteins from human tumor cells and the Cos1-DEspR + permanent transfectants cells. More importantly, DEspR molecular activity, protein–protein interactions, protein-specific properties and functionality are shown, and found to play key roles in cancer stem cell anoikis resistance and growth.

## Results

### Confirmation of two G's and compression site

To determine whether the two G's are detected forming tryptophan [TGG]-codon#14, as observed in Sanger sequencing (Additional file 2: Figure S2), we performed manual Maxam–Gilbert sequencing using 8 % denaturing polyacrylamide sequencing gels on $^{32}$P end-labeled PCR-amplified cDNA electrophoresed at three different fixed wattages (Fig. 1a). Clearly, the two G's are noted with the corresponding C bands that accompany most G nucleotides. Interestingly, the compression in the gel run is consistently observed in Sanger sequencing and in Maxam–Gilbert sequencing, and contains the compression motif found in 68 % of sequencing errors with compression [9] (Fig. 1a; Additional file 2: Figure S2). This compression results in the "slippage' of the two G's with a T nucleotide, thus leaving a space in the gel read. Additionally, the compression region is within a 3-nt stem-loop structure that could render this region prone to sequencing errors which are not present in rat and mouse DEspR gene sequences, both of which also contain the identical tryptophan [TGG]-codon#14 (Additional file 2: Figure S2).

Importantly, queries against subsets of the NCBI sequence read archive (SRA) database reveals DEspR-specific exon sequences which would not be expected for the

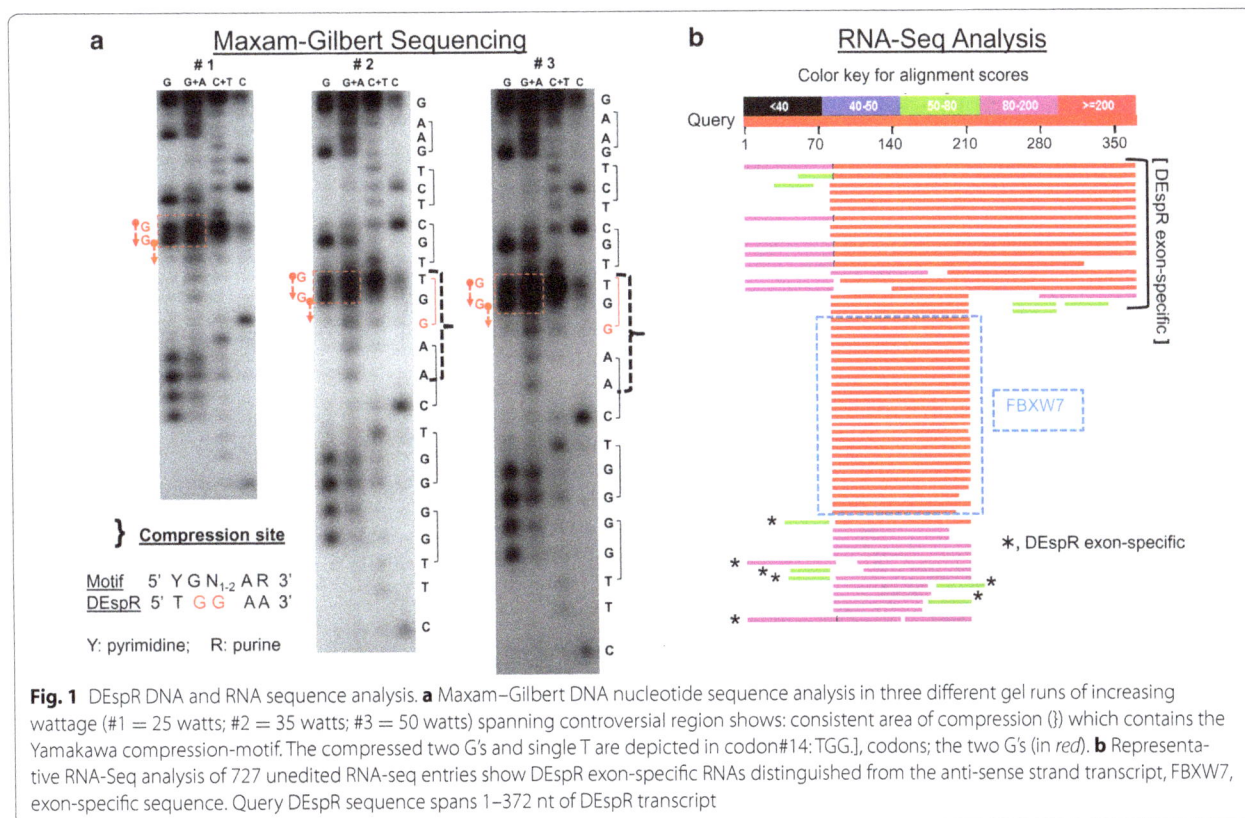

**Fig. 1** DEspR DNA and RNA sequence analysis. **a** Maxam–Gilbert DNA nucleotide sequence analysis in three different gel runs of increasing wattage (#1 = 25 watts; #2 = 35 watts; #3 = 50 watts) spanning controversial region shows: consistent area of compression (}) which contains the Yamakawa compression-motif. The compressed two G's and single T are depicted in codon#14: TGG.], codons; the two G's (in *red*). **b** Representative RNA-Seq analysis of 727 unedited RNA-seq entries show DEspR exon-specific RNAs distinguished from the anti-sense strand transcript, FBXW7, exon-specific sequence. Query DEspR sequence spans 1–372 nt of DEspR transcript

FBXW7 transcript in the opposite strand (Fig. 1b). Notably, five entries show the two G's out of 124 data entries in said file (Additional file 3: Figure S3), and that these two G's are associated with an insertion of a "T–A' or an "A" (Additional file 3: Figure S3). These observations show that the two G's are indeed detected and are located within a problematic sequencing region, concordant with observations in manual sequencing gels of nucleotide compression spanning a known sequence motif for compression [9]. These data show the questioned two G's in support of tryptophan [TGG] codon in 5/124 sequences similar to manual sequencing runs. Given that Sanger-sequencing is the accepted standard final determinant of nucleotide discrepancies [5], these experimental data support the need for clarification and demonstration of DEspR protein expression.

### Analysis of translatability and protein-specific properties

In order to clarify the existence of DEspR protein complying with established UNIPROT criteria for detection of protein by antibody made from deduced amino acid sequences, we performed anti-DEspR mAb pull-down and subsequent Western blot and mass spectrometry analyses of pull-down products from cell membranes. Membranes were isolated from human glioblastoma

tumor cells and U87 cancer stem-like cells (CSC) which we isolated and characterized for CSC tumor initiating properties [3]. We performed pull-down experiments using an anti-human DEspR specific antibody, 5g12e8 mAb (Fig. 2a), which detected DEspR on prior Western blot analysis of CSC membrane proteins [3].

Analysis of pull-down products by mass spectrometry (MS) (Table 1; Additional file 4: Table S1) revealed that the DEspR-protein interacts with several proteins involved in intracellular trafficking, angiogenesis, and/or cancer: vimentin, Gal-3, Gal-1 and TMED10. Although MS analyses did not detect DEspR (Table 1), Western blot analyses of pull-down products detected DEspR protein bands at ~17.5 and ~12.5 kDa, larger than the deduced amino acid sequence predicting ~10 kDa, and distinct from the other proteins in the pull-down-complex: Rab-1b (22 kDa), Galectin-1 (14 kDa) and TMED10 (25 kDa) (Fig. 2b). Data indicate that DEspR is likely glycosylated given its consensus glycosylation site [N-F-S-G] (Fig. 2a), and clearly distinct from the other proteins in the pull-down complex as evidenced by size, relative abundance, and the lack of antibody cross-reactivity (Fig. 2b). Glycosylation of DEspR is confirmed after peptide-N-glycosidase F (PNGase F) treatment of pull-down proteins showing decrease in size of DEspR upon Western blot

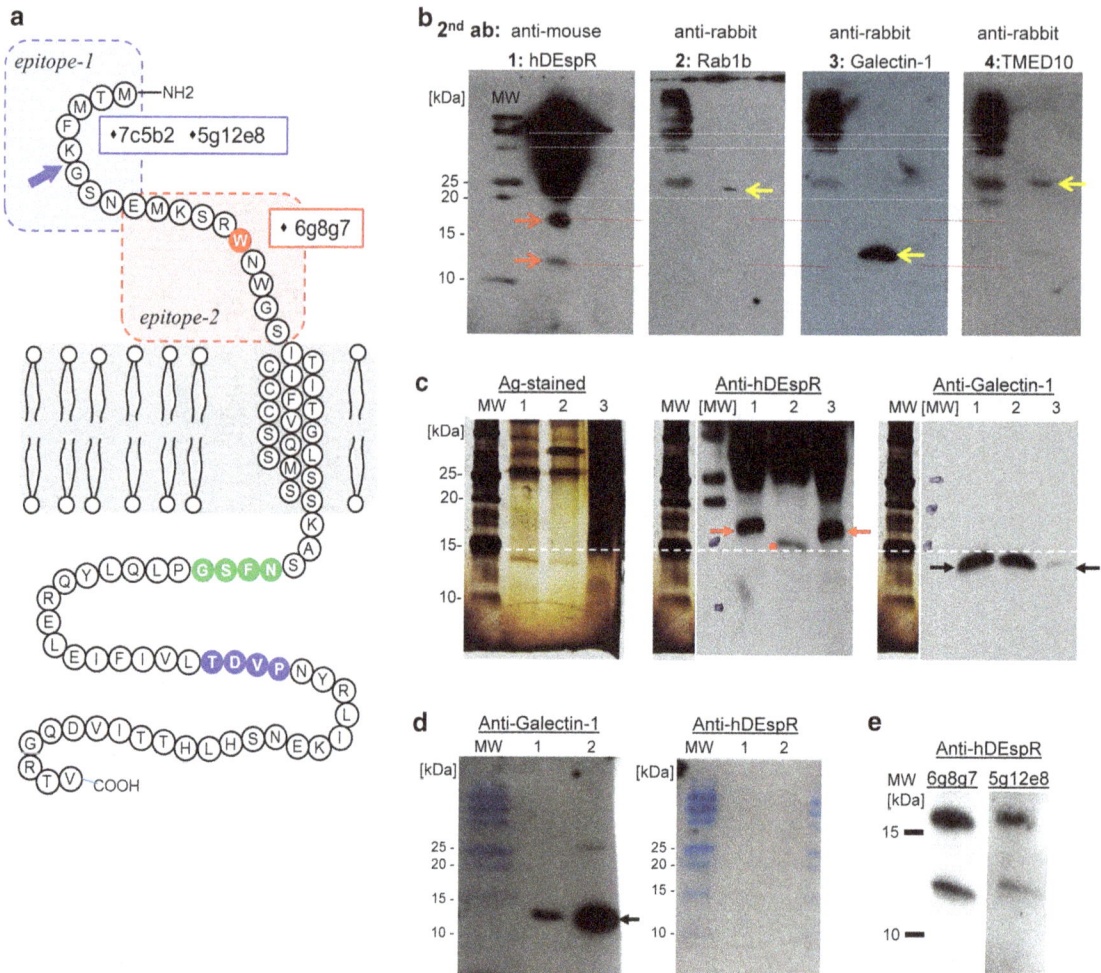

**Fig. 2** Analysis of DEspR translatability. **a** Schematic diagram of DEspR protein and mAb-epitopes. Two distinct peptides (*epitope-1, epitope-2*) in the extracellular domain were used to develop murine monoclonal antibodies (mAbs). Two high-affinity mAbs target human-specific epitope-1: 7c5b2, 5g12e8; and one high-affinity mAb targets the pan-species reactive epitope-2, identical in human, monkey, and rat. Epitope-2 spans the putative ligand binding domain [24]. The 5g12e8 mAb was used in pull-down experiments; 5g12e8 and 6g8g7 were used in Western blot analyses, 7c5b2 mAb was used in FACs analysis, immunostaining, and internalization assays, and all three were used in functional inhibition assays. The contested tryptophan (W)-aa#14 (*red*); consensus glycosylation site sequence: (*green*, N-F-S-G), known internalization recognition sequence: (*blue*: T-D-V-P). A *blue arrow* marks the splice junction between exon1 and exon2, i.e. between amino-acids G and K (aa#5-#6). **b** Sequential Western blot analyses of pull-down proteins from glioblastoma U87 membrane proteins using different antibodies specific for proteins identified by mass spectrometry analysis of pull-down protein-products. The identical blot was sequentially probed, stripped of antibody, confirmed as stripped, then re-probed in the following order: #1: anti-hDEspR-5g12e8 mouse mAb, #2: anti-Rab1b rabbit polyclonal Ab (pAb); #3: anti-Galectin-1 rabbit pAb; #4: anti-TMED10 rabbit pAb. Molecular weight markers are noted. DEspR bands are ~17.5 and 12.5 kDa. Expected sizes are detected for Rab1b: 22 kDa, Galectin-1: 14 kDa, and TMED10: 25 kDa. **c** *Panel*-1 shows silver-stained gel of pull-down protein products using 5g12e8 mAb from membrane proteins isolated from: (1) glioblastoma U87 CSCs, (2) PNGase-digested sample of pull-down proteins from U87 CSCs, (3) permanent transfectants DEspR-positive Cos1-cells. *Panel*-2 shows Western blot analysis using anti-DEspR 5g12e8 mAb showing DEspR band (*lane 1*), smaller DEspR + band after PNGase digest-samples (*lane 2*), and identically-sized DEspR band in DEspR-positive Cos1-cell permanent transfectants showing appropriate splicing and translatability of DEspR-minigene transfected into Cos-1 cells (*lane 3*). *Panel*-3 Western blot analysis of different wells in the same gel run probed with anti-Galectin-1 pAb showing distinct sized protein bands, thus confirming DEspR-specific bands are not Galectin-1 protein bands, and that Galectin-1 is not glycosylated as reported. **d** Western blot analysis of Galectin-1 recombinant protein. *Panel 1* overlay of gel-image and western blot image showing detection of Galectin-1 recombinant protein at expected size 14 kDa. *Panel 2* overlay of gel-image and western blot image probed with anti-DEspR 5g12e8 mAb showing non-cross reactivity of anti-hDEspR mAb with Galectin-1. **e** Sequential western blot analysis of 5g12e8-pull-down proteins from U87 CSCs probed first with 6g8g7 (*left panel*), and subsequently with 5g12e8 after 'stripping' (*right panel*), detects identical protein bands. This confirms that 6g8g7 epitope is on the same protein as 5g12e8 epitope, thus corroborating DEspR protein existence

**Table 1 Proteins pulled-down with anti-DEspR mAb 5g12e8 from U87 CSC membranes**

| Accession | % | #SM | #P | #AAs | Mass | Score | Description |
|---|---|---|---|---|---|---|---|
| P08670 | 55.8 | 36 | 31 | 466 | 53,619 | 177.59 | *Vimentin (VIM)* |
| P60709 | 34.4 | 11 | 9 | 375 | 41,709 | 66.79 | *Actin, cytoplasmic 1 (ACTB)* |
| P68371 | 41.6 | 13 | 12 | 445 | 49,799 | 56.53 | Tubulin beta-4B chain (TUBB4B) |
| P07437 | 35.1 | 11 | 10 | 444 | 49,638 | 47.84 | *Tubulin beta chain (TUBB)* |
| Q13885 | 29.0 | 10 | 9 | 445 | 49,874 | 42.11 | Tubulin beta-2A chain (TUBB2A) |
| P17931 | 25.6 | 7 | 5 | 250 | 26,136 | 39.26 | Galectin-3 (LGALS3) |
| P09382 | 49.6 | 7 | 6 | 135 | 14,706 | 38.90 | *Galectin-1 (LGALS1)[a]* |
| Q13509 | 25.6 | 8 | 8 | 450 | 50,400 | 32.86 | Tubulin beta-3 chain (TUBB3) |
| Q9BUF5 | 15.7 | 6 | 5 | 446 | 49,825 | 21.46 | Tubulin beta-6 chain (TUBB6) |
| P61006 | 13.0 | 2 | 2 | 207 | 23,653 | 12.01 | Ras-related protein Rab-8A (RAB8A) |
| C9J8S3 | 15.0 | 2 | 2 | 160 | 18,015 | 11.01 | *Ras-related protein Rab-7a (RAB7A)* |
| P61026 | 6.0 | 1 | 1 | 200 | 22,526 | 9.59 | *Ras-related protein Rab-10 (RAB10)* |
| P51153 | 5.9 | 1 | 1 | 203 | 22,759 | 7.60 | Ras-related protein Rab-13 (RAB13) |
| H0YNE9 | 6.4 | 1 | 1 | 188 | 21,854 | 7.51 | Ras-related protein Rab-8B (Fragment) (RAB8B) |
| G3V2K7 | 7.2 | 1 | 1 | 153 | 16,893 | 5.96 | *Transmembrane emp24 domain-containing protein 10 (TMED10)[a]* |
| Q9Y3B3-2 | 5.3 | 1 | 1 | 188 | 21,219 | 4.85 | Transmembrane emp24 domain-containing protein 7 (TMED7) |
| Q9BVK6 | 3.8 | 1 | 1 | 235 | 27,260 | 4.36 | Transmembrane emp24 domain-containing protein 9 (TMED9) |
| F8WBC0 | 34.4 | 1 | 1 | 32 | 3501 | 2.90 | *Ras-related protein Rap-1b (Fragment) (RAP1B)* |

Proteins in italic face letters are proteins detected in three independent pull-down experiments

% the percentage of the protein sequence covered by identified peptides; #SM the number of peptide spectrum matches; #P the total number of distinct peptide sequences identified in the protein group; #AAs number of amino acids; Mass, mass in Daltons; Score, the sum of the ion scores of all peptides that were identified

[a] Proteins verified by Western blot analysis of pull-down proteins; Rab-1b was detected in two independent experiments and verified by Western blot analysis (Fig. 2b)

analysis (Fig. 2c, middle panel, lane 2). These data validate the canonical glycosylation site beyond the purported pseudogene stop codon#14, and provides a mechanism for DEspR-Galectin-1 and DEspR-Galectin-3 binding as observed in the pull-down products.

However, because of the abundance of Galectin-1 in the pull-down products, we performed more Western blot analyses to eliminate the possibility that anti-DEspR antibody cross reacts with Galectin-1 and vice versa. As shown in Fig. 2c, the anti-human DEspR 5g12e8 mAb detects the 17.5 kDa glycosylated DEspR in the pull-down products from U87-CSC cell membrane proteins and from the membrane proteins isolated from DEspR-positive Cos1-permanent transfectant cells. These DEspR + Cos1-permanent transfectants cells were previously shown to express human DEspR by immunostaining, and to competitively bind ligands and anti-DEspR (7c5b2) mAb [2]. After PNGase F digestion, 5g12e8 also detected the smaller, hence deglycosylated DEspR (Fig. 2c, middle panel). In contrast, anti-Galectin-1 mAb detected a 14 kDa protein band with markedly different size and expression levels to DEspR in all three lanes (Fig. 2c, right panel).

To further confirm this distinction, we performed double Western blot analyses of Galectin-1 recombinant protein. As shown in Fig. 2d, anti-Galectin-1 mAb detected

human Galectin-1, in contrast to the anti-human DEspR 5g12e8 mAb which did not cross react to both 100 and 500 ng of purified recombinant human Galectin-1 on Western blot analysis, thus confirming that DEspR in the pull-down product is distinct from other pull-down products (Fig. 2b) especially Galectin-1 (Fig. 2c). Additionally, since Galectin-1 is known not to be glycosylated [10] as confirmed in Fig. 2c, the PNGase F-treated sample showing a decrease in size cannot, therefore, be Galectin-1.

To further corroborate the detection of human DEspR protein in human tumor cells, we performed sequential probing of the identical Western blot with two different anti-human DEspR mAbs targeting different epitopes. As shown in Fig. 2e, the 6g8g7 mAb targeting the epitope spanning the disputed tryptophan [TGG]-codon#14 (Fig. 2a) detects both the glycosylated 17.5 kDa and the less glycosylated 12.5 kDa DEspR bands. After removal of the 6g8g7 antibody and demonstration of no residual signals, the Western blot was re-probed with the 5g12e8 mAb used in the pull-down experiment and which targets a human DEspR-specific epitope spanning a peptide present only if correctly spliced (Fig. 2a). This 5g12e8 probed western blot detected the identical protein bands (Fig. 2e), thus showing that two different anti-hDEspR mAbs 6g8g7 and 5g12e8, which target different epitopes, detect the same protein bands representing spliced and glycosylated DEspR.

## Protein–protein interactions: analysis of co-localization of DEspR and Galectin-1

Since both DEspR and Galectin-1 are "pulled-down" consistently (n = 6), we next determined whether they are co-localized in tumors. We performed double immunostaining of xenograft tumors derived from U87-CSCs [3] using human specific anti-human DEspR mAb 7c5b2 and human-specific anti-Galectin-1 antibody. As shown in Fig. 3a, DEspR and Galectin-1 are indeed co-localized in tumors, and located more in the expanding tumor zone. Additionally, as shown in Fig. 3b, DEspR and Galectin-1 are co-expressed in tumor cells that have invaded through the xenograft tumor cap and adhered onto a microvessel in the surrounding subcutaneous tissue of the host nude rat. The detection of co-localization is concordant with the pull-down of both DEspR and Galectin-1 in multiple independent experiments (Table 1).

## Analysis of functionality at multiple levels certifies DEspR protein

While we showed DEspR functionality in human tumor cells by showing that DEspR inhibition via the 7c5b2 mAb inhibited angiogenesis, tumor growth and invasiveness [3], the NCBI pseudogene designation persists (updated in May 2016). In addition to manual nucleotide DNA sequencing (Fig. 1a; Additional file 2: Figure S2), review of RNA-Seq sequence entries (Fig. 1b; Additional file 3: Figure S3), and detection of DEspR translatability or expression (Figs. 2, 3), to provide further evidence against the pseudogene designation, we further studied DEspR functionality since demonstration of functionality of a protein certifies that protein's existence [7]. To ascertain functionality, we used different anti-human DEspR mAbs whose epitopes are depicted in Fig. 2a: 7c5b2, 5g12e8 and 6g8g7 (Fig. 4), and tested for DEspR in different cancer tissue types.

As shown in Fig. 4, we studied DEspR-inhibition effects on anoikis resistant cancer stem-like cells (CSCs) isolated from human U87 glioblastoma tumor cell line [3], human MDA-MB-231 triple negative breast cancer cell line, human H460 non-small cell lung cancer cell line and human Panc1 pancreatic ductal adenocarcinoma cell line [3]. To first demonstrate that DEspR protein is expressed on the different CSCs, we performed FACS analysis using AF568-fluorescently labeled 7c5b2 anti-DEspR mAb (Fig. 2a). Figure 4a shows that, compared to isotype control, fluorescently labeled 7c5b2-mAb specifically detected DEspR on the cell membrane in ~50 % of U87-CSCs, ~60 % of MB-231-CSCs, ~71 % of H460-CSCs and ≥55 % of Panc1-CSCs in the suspension culture conditions used.

To demonstrate the functionality of DEspR proteins present on the cell membrane, in vitro inhibition studies were performed using the 7c5b2, 5g12e8 and 6g8g7 anti-human DEspR mAbs which target different DEspR epitopes (Fig. 2a). As shown in Fig. 4b, all three mAbs significantly inhibited CSC growth in the four cancer tissue type-CSCs tested (P < 0.0001, One Way ANOVA with Tukey's multiple comparisons test), in contrast to non-treated contemporaneous controls respectively. We note that CSC growth is representative of survival in suspension culture, proliferation and ability to form stem-cell like spheroids in suspension culture conditions. We also note some variations in the inhibition of CSC growth in vitro by the different mAbs, with 7c5b2 an 6g8g7 performing better than 5g12e8 (Fig. 4), but that all mAbs significantly inhibit survival, proliferation, and spheroid formation (CSC-growth) of U87-CSCs, MDA-MB-231-CSCs, H460-CSCs and Panc1-CSCs.

To further demonstrate DEspR functionality, we performed live-cell imaging of Panc1 cells using AF568-fluorescently labeled 7c5b2 mAb (Fig. 5) given the single consensus internalization-recognition sequence [T-D-V-P] in the deduced DEspR amino acid protein located beyond the contested stop codon vs amino acid #14-W (Fig. 2a). Demonstration of internalization would indicate translation and expression of the DEspR protein beyond the purported [TGA] stop codon#14, thus refuting the presence of the stop codon. As shown in Fig. 5a, time series of live-cell imaging shows increasing amounts of intracellular internalization beginning around 15–30 min and incrementally increasing up to 75 min after addition of fluorescent 7c5b2 anti-DEspR mAb. Higher magnification confirms intracellular fluorescence accumulation with fluorescently labeled 7c5b2 but not with isotype control IgG2b-AF568 (Fig. 5b), thus indicating AF568-labeled 7c5b2-DEspR-mediated internalization rather than non-specific endocytosis. Demonstration of internalization affirms DEspR protein functionality, and further refutes the presence of the stop codon.

## Discussion

### Stop [TGA] codon vs tryptophan [TGG] codon within a sequence-compression region

The detection of the two G's to make tryptophan [TGG]-codon#14 in manual sequencing Sanger-dideoxy sequencing and Maxam and Gilbert sequencing, along with confirmation of sequence compression at the site within a Yamakawa compression-motif whether sequencing the sense strand (Maxam–Gilbert) or the antisense strand (Sanger dideoxy sequence), strongly suggests that the NCBI designation of human Dear/DEspR as a pseudogene based on stop codon#14 is erroneous. Following the accepted standard that manual Sanger sequencing defines

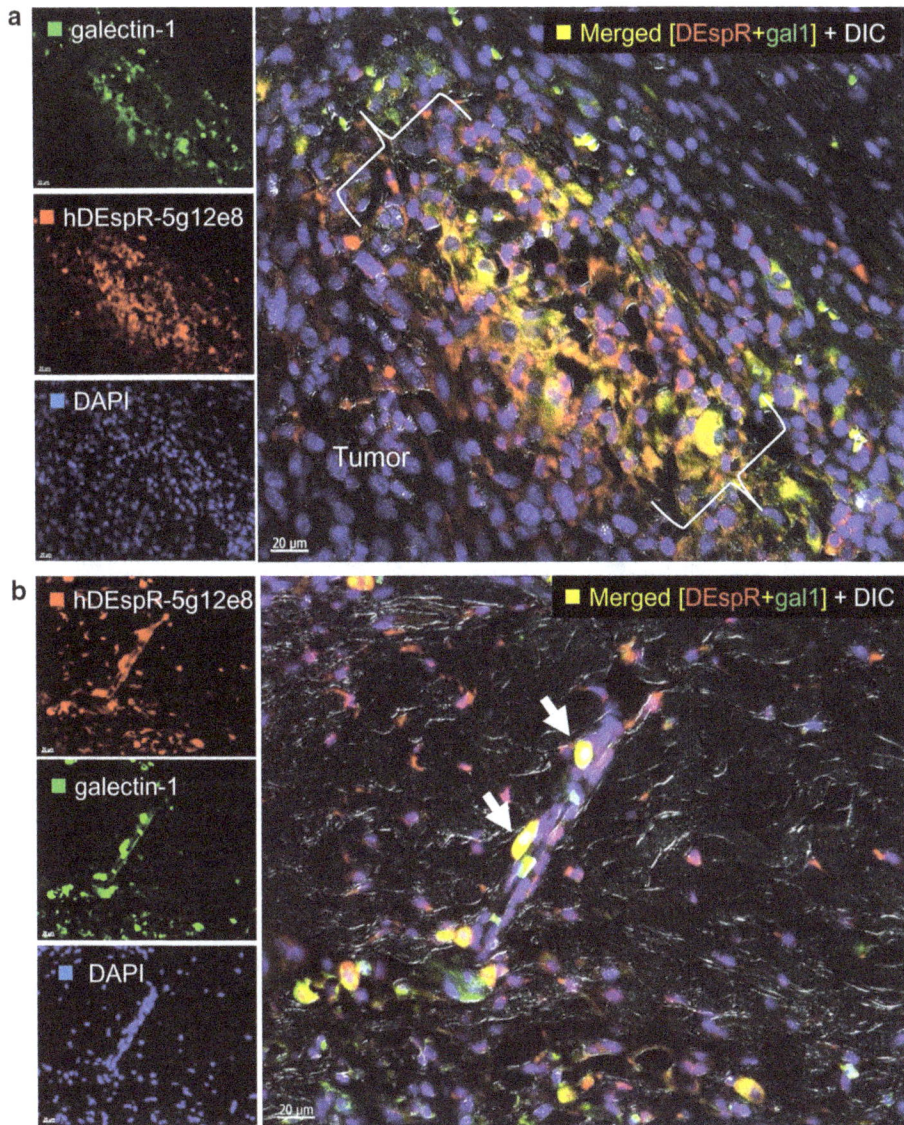

**Fig. 3** Dual-fluorescence co-immunostaining analysis of DEspR and Galectin-1 expression. **a** Analysis of the expanding tumor zone of a U87-CSC xenograft tumor invading through the tumor fibrous cap. Co-localization (*yellow, yellow dotted circle*) of increased human-DEspR expression (*red dotted circle*) in invasive U87 tumor cells and Galectin-1 (*green dotted circle*) is observed in the invasive front. {}, invasive tumor front. Host subcutaneous tissue is to the *upper right* corner. **b** Co-localization of hDEspR and Galectin-1 is detected in tumor cells adhering to the outer wall of a microvessel in the subcutaneous tissue demonstrating invasive nature of U87 in xenograft tumors and homing to microvessels

the correct sequence rather than the automated sequence, the single gene analysis data indicates that the TGG codon is present rather than the TGA stop codon reported in automated DNA sequencing databases. This interpretation is supported by the fact that systematic sequencing errors are documented to occur with both position- and sequence-specificity even in "next generation sequencing technologies at rates greater than prior technologies" [4]. Existence of the tryptophan[TGG]-codon#14 is independently supported by detection of the DEspR protein by

mAbs 5g12e8 and 7c5b2 whose peptide epitopes exist only if the DEspR RNA is spliced, and in the case of 6g8g7, only if the contested tryptophan[TGG]-codon#14 is present (Fig. 2a). Notably, 5g12e8 and 6g8g7 detect DEspR-protein as a glycosylated 12.5–17.5 kDa protein band, much greater than the 1.5 kDa predicted from a 13 amino acid peptide should there be a stop codon at codon#14. Furthermore, functional validation of the glycosylation and internalization recognition sequences beyond codon#14 (Figs. 2, 5) experimentally refute a stop codon at codon#14.

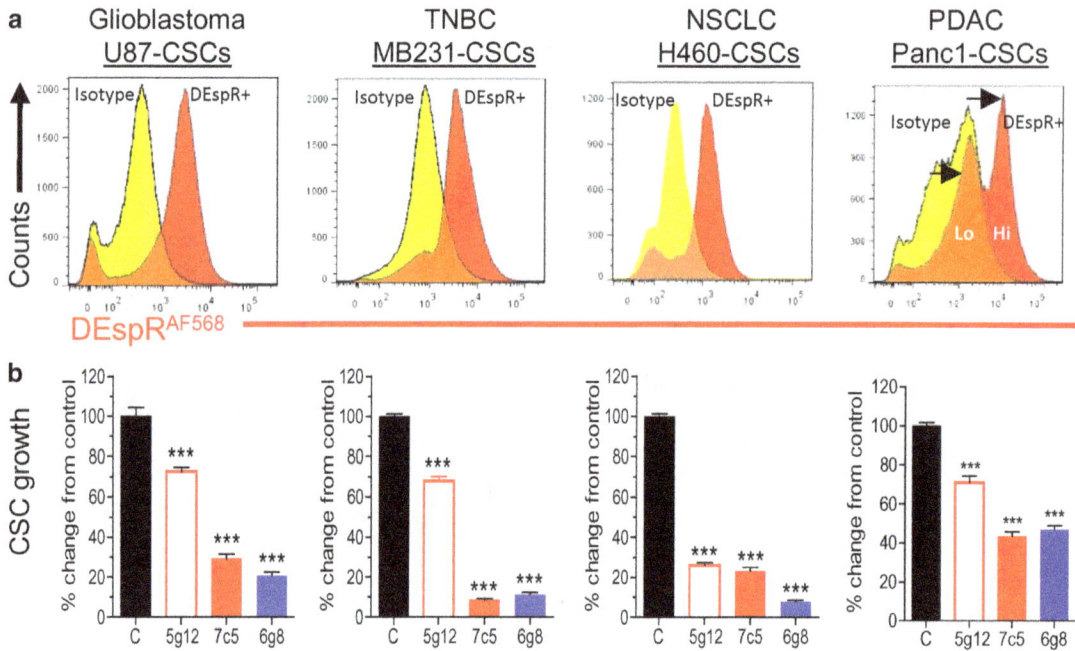

**Fig. 4** Demonstration of DEspR protein and functionality in different human tumor cell lines by anti-DEspR mAbs. **a** FACs analysis of different cancer tissue type CSCs: glioblastoma, triple negative breast cancer (TNBC), non-small cell lung cancer (NSCLC), and pancreatic ductal adenocarcinoma (PDAC) using AF-568 labeled murine anti-DEspR mAbs compared with AF568-IgG2b murine isotype control. In Panc1 CSCs, low and high DEspR+ CSCs are detected. **b** Multiple murine anti-DEspR mAbs (5g12e8, 7c5b2, 6g8g7) inhibit CSC growth in suspension culture determined by the number of live CSCs after 5 days of incubation with murine anti-DEspR mAbs, compared with corresponding control non-treated CSCs. Comparative analysis is presented using % change from respective controls; ***, One Way ANOVA with Tukey's multiple comparisons test $P < 0.0001$. Epitope 1 murine mAbs: 5g12, 5g12e8, 7c5, 7c5b2; Epitope 2 murine mAbs: 6g8, 6g8e8

**Insights regarding other high throughput database reports**

The non-detection of DEspR sequences in normal human tissue RNA-Seq databases is unsurprising since DEspR is not expressed or minimally expressed in normal tissues. Moreover, detection of unspliced DEspR-specific RNA as the dominant species [3] does not indicate non-translation, since regulation at the splicing level has recently been described for granulocyte differentiation whereby the dominant RNA species in the cell is the unspliced form [11, 12], similar to observations for detection of both spliced and unspliced DEspR-transcripts by ARMS-PCR [3].

Moreover, since DEspR overlaps with a larger transcript, FBXW7, on the opposite strand [3], RNA-seq databases that specifically exclude double stranded DNA [13] will also exclude RNA–RNA sequences such as DEspR-FBXW7 hybrids. The process-exclusion of RNA–RNA hybrids in RNA-Seq entries in current methodologies is due to the standard of excellence requiring elimination of all double stranded DNA in RNA-Seq libraries [13], which inadvertently will also exclude RNA–RNA hybrids. More specifically, since RNA–RNA hybrids can form in standard RNA isolation methods with phenol extraction [14], and can form during RNA-seq library

preparation using primer annealing reactions at temperatures lower than the melting temperatures (Tm) of RNA–RNA hybrids, and given process-exclusion of any dsDNA or RNA–RNA or dsDNA-RNA hybrids, there is an inadvertent bias introduced against the detection of DEspR-specific transcripts due to RNA–RNA hybrids formed between DEspR transcript and exon#5 (or #6 depending on dataset numbering) of FBXW7 transcript on the opposite strand.

In parallel, the non-detection in proteomic databases is concordant with difficulties in detecting membrane proteins with single transmembrane domains that are less than 150 kDa [15], especially if glycosylated, since PNGase F treatments do not necessarily ensure detection by MS [16]. The current non-detection of DEspR transcripts in transcriptomics could also be due to the exclusion of RNA–RNA hybrids, and in primer pair-specific amplification due to the absence of DEspR-specific primers given the non-recognition of the DEspR gene by NCBI.

**Confirmation of translatability**

Demonstration of translatability through different experimental designs and using anti-human-DEspR mAbs

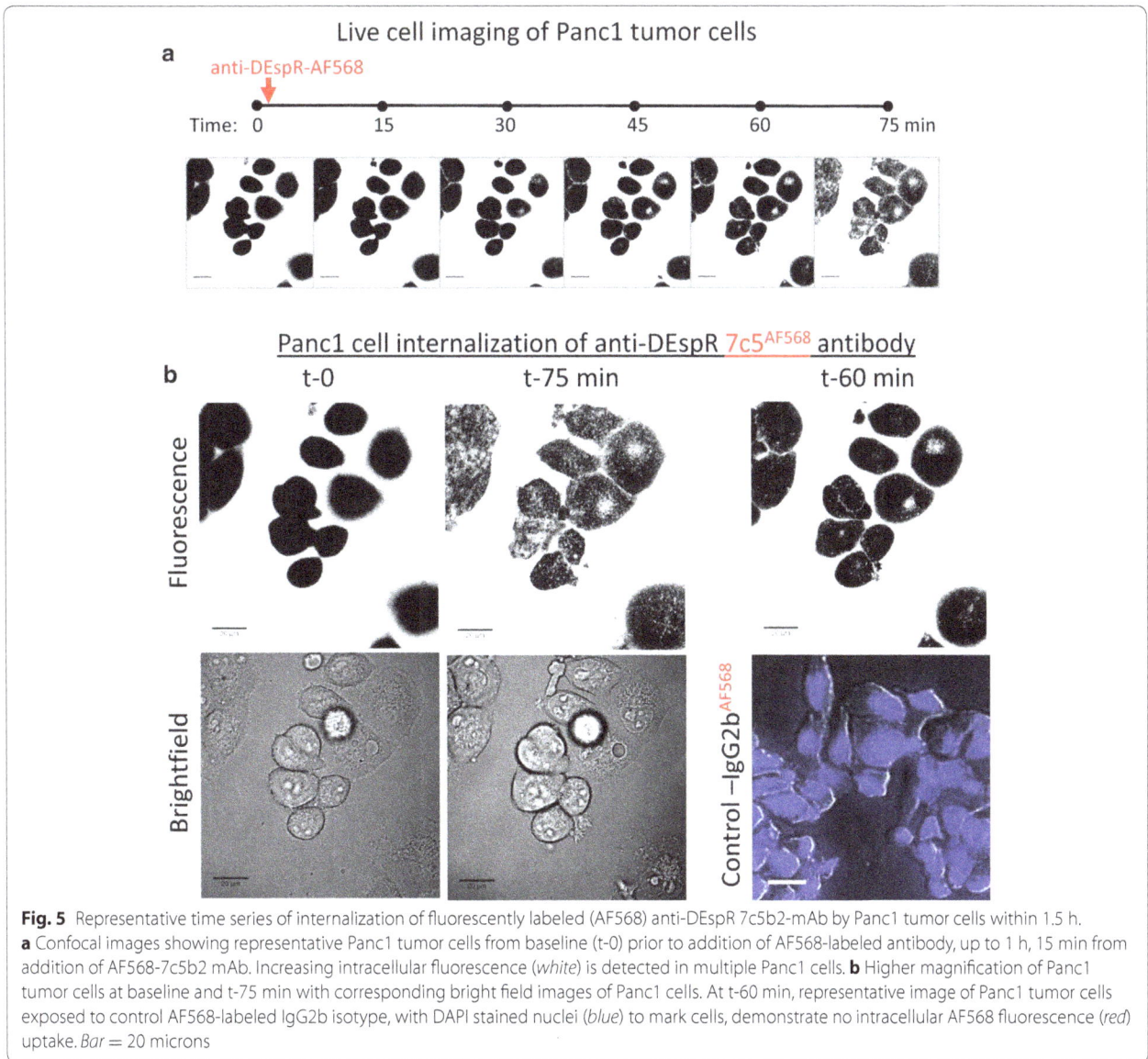

**Fig. 5** Representative time series of internalization of fluorescently labeled (AF568) anti-DEspR 7c5b2-mAb by Panc1 tumor cells within 1.5 h. **a** Confocal images showing representative Panc1 tumor cells from baseline (t-0) prior to addition of AF568-labeled antibody, up to 1 h, 15 min from addition of AF568-7c5b2 mAb. Increasing intracellular fluorescence (*white*) is detected in multiple Panc1 cells. **b** Higher magnification of Panc1 tumor cells at baseline and t-75 min with corresponding bright field images of Panc1 cells. At t-60 min, representative image of Panc1 tumor cells exposed to control AF568-labeled IgG2b isotype, with DAPI stained nuclei (*blue*) to mark cells, demonstrate no intracellular AF568 fluorescence (*red*) uptake. *Bar* = 20 microns

specific for different epitopes substantially fulfill UNI-PROT criteria for determining a protein's existence at the highest level such as detection of protein by antibody, demonstration of protein–protein interactions, demonstration of post-translational modifications, and demonstration of protein functionality [8].

Pull-down protein analysis experiments confirm the existence of DEspR via direct detection of the protein in human tumor cells by Western blot analysis of DEspR using multi-epitope anti-DEspR mAbs. The detection of the identical protein band by Western blot 'walking' with distinct anti-DEspR mAbs targeting two different epitopes, 5g12e8 mAb binding to DEspR epitope-1 that spans a splice junction, and 6g8g7 binding to DEspR

epitope 2 that spans tryptophan[TGG]-codon#14, demonstrate the existence of the DEspR protein. If the stop codon were indeed present at codon#14 instead of tryptophan, there would simply be a 13 amino acid peptide with 1.5 kDa MW rather than the glycosylated 17.5 and 12.5 kDa protein bands, or the non-glycosylated DEspR-sequence predicted ~10 kDa protein.

Moreover, if there were indeed a stop codon at codon#14, DEspR would not be glycosylated nor internalized as the consensus sequence motifs for these sites are beyond the reported stop codon#14. The detection of the identical-sized protein on Western blot analysis of DEspR-positive Cos1 cell permanent transfectants and U87 tumor cells corroborates the protein product of the transfected plasmid DEspR mini-gene construct previously reported, and corroborates splicing of the unspliced DEspR cDNA mini-gene construct [2].

## Detection of DEspR by Western blotting but not by mass spectrometry analysis

We note that although DEspR was detected consistently in denatured conditions via Western blot analysis of membrane protein pull-down products using 5g12e8 and 6g8g7 anti-DEspR mAbs, MS did not detect DEspR-specific peptides. However, non-detection by MS does not negate DEspR protein existence for several reasons. First, based on deduced amino acid sequence predicting a single transmembrane integral membrane ~10 kDa protein that is glycosylated to 17.5/12.5 kDa, non-detection by peptide mass fingerprinting methods using techniques such as MALDI-TOF MS analysis is not surprising given that only 204 integral membrane proteins were detected on mass spectrometry in a study of rat endothelial cells at the National Center for Proteomics Research, and that no VEGF receptor-2 (151 kDa) was detected in this study [17]. Using 6718 as total number of membrane proteins [18], Mirza et al. [17] detected 3 % integral membrane proteins on MS, while Peng et al. [19] detected 301 integral membrane proteins via SDS-PAGE shotgun proteomics or 4.5 % of 6718 total membrane proteins, and highest count using SCX-RPLC-MS/MS (MudPIT) strategy detected 876 integral membrane proteins or 13 % of 6718 integral membrane proteins in murine NK cells [20]. Notably, of the ones detected by MS using an improved method that detects more membrane proteins using a centrifugal proteomic reactor, *all the proteins detected were >150 kDa and ≥2 transmembrane domains* [15].

Other factors could also account for non-detection by MS with current methodologies. As stated by Bensalem et al. [21], peptide mass fingerprinting of membrane proteins, using techniques such as MALDI-TOF MS, remains a significant challenge for at least three reasons: (1) membrane proteins are naturally present at low levels, (2) many detergents strongly inhibit proteases and have deleterious effects on MALDI spectra, and (3) despite the presence of detergent, membrane proteins are unstable and often aggregate [21]. Additionally, glycosylation of DEspR also impedes current proteomic detection as reported by Cao et al. [16] stating that "deglycosylation of plasma membrane proteins by treatment with PNGase-F did not yield detection of additional hydrophobic proteins". Therefore, peptide mass fingerprinting MS analysis of PNGase-treated pull-down proteins which did not identify DEspR does not negate the existence of DEspR protein. These observations by others account for the non-detection by MS of the single transmembrane

receptor <150 kDa DEspR protein, thus excluding MS non-detection as counter evidence for the existence of DEspR protein, especially given multiple other experimental evidence that fulfill UNIPROT criteria for determining a protein [8].

## Demonstration of protein interactions, protein-specific properties and functionality

Aside from ligand-specific activated DEspR signaling [3], detection of DEspR-protein interactions in the DEspR-Galectin-1 complex, along with other pull-down protein partners, complies with a UNIPROT criterion for ascertaining existence of a protein as a gene product through the demonstration of protein-interactions [8]. The distinct functionality of DEspR vs Galectin-1 in terms of anoikis resistance indicates DEspR protein-specific functions, rather than the potential counter-argument that DEspR functionality is merely due to the known Galectin-1 functions.

More specifically, DEspR-specific inhibition by blocking antibodies induces anoikis in four cancer tissue type CSCs, corroborating earlier observations for Panc1 and U87 CSCs [3]. In contrast, Galectin-1 is pro-anoikis [22]. Given these two observations, we hypothesize that glycosylated DEspR plays a role in anoikis resistance by binding to Galectin-1 and inhibiting its pro-anoikis function, much like glycosylation of p16-INK4a de-induces Galectin-1 pro-anoikis functions through upregulation of Galectin-3 [22]. Confirmation of the DEspR-Galectin-1 complex is demonstrated by co-localization on double-immunostaining experiments, thus further strengthening evidence for translatability and functionality by the detection of protein–protein interactions following UNIPROT criteria [8]. Distinguishing the DEspR protein band by size and quantity from its pull-down partners identified from MS by their corresponding antibodies in serial sequential probing of the identical western blot, corroborates their detection by MS, as well as eliminates the potential of erroneous cross-reactivity pull-down results.

More importantly, given that demonstration of protein functionality certifies a protein's existence [7, 8], the demonstration of DEspR expression in >50 % of CSCs in four different cancer tissue types by FACS analysis and DEspR functionality via blocking antibody inhibition of CSC growth in all four different cancer tissue type CSCs—glioblastoma, triple negative breast cancer, non-small cell lung cancer and pancreatic ductal adenocarcinoma, provide compelling evidence of the existence of DEspR protein based on UNI-PROT criteria and standards in the field [7, 8].

Furthermore, the demonstration of internalization of DEspR protein via live cell imaging is consistent with a consensus internalization recognition sequence downstream to amino acid#14. This not only confirms translatability of the DEspR protein beyond 13 amino acids, thus refuting the stop codon presence, but also demonstrates a protein-specific property, i.e. receptor internalization and expands DEspR's multi-functionality. Concordantly, these new data on functionality are supported by previous findings of DEspR roles in tumor angiogenesis, invasiveness, growth, as well as CSC survival in adverse conditions [3].

## Conclusions

Altogether, in the context of an error prone site at codon#14, multiple protein assays independently and collectively demonstrate the existence of the DEspR protein, and analyses that show DEspR protein properties (protein–protein interactions, glycosylation, internalization) and functional roles in multiple cancer tissue type CSCs, collectively certify the DEspR protein based on multiple UNIPROT criteria and demonstrate that the human DEspR gene is not a pseudogene. Given that fulfillment of a single UNIPROT criterion for a protein's existence is sufficient, fulfillment of multiple criteria present compelling experimental evidence, at the highest protein level, that functional human DEspR protein exists.

## Methods

### Maxam–Gilbert sequencing

We performed Maxam–Gilbert sequencing essentially as described [23]. The hDEspR cDNA was 5′-end labelled with polynucleotide kinase using gamma-$^{32}$P ATP. Chemical treatments for Guanine (G), Guanine + Adenine (G + A), Cytosine + Thymine (C + T) and Cytosine (C) cleavages were done as described [23]. Cleavage products from the four reactions were size separated by 8 % denaturing polyacrylamide gel electrophoresis and fragments visualized by autoradiography. Electrophoresis was performed at three different fixed wattages: 25 watts (#1), 35 watts (#2) and 50 watts (#3).

### Cell lines and antibody development and characterization by ELISA

Verified glioblastoma U87 MG (cat# ATCC HTB-14), triple negative breast cancer MDA-MB-231 (cat# ATCC HTB-26), non-small cell lung cancer NCI-H460 (cat# ATCC HTB-177) and pancreatic cancer Panc-1 (cat# ATCC CRL-1469) cell lines were obtained from ATCC. Isolation and propagation of U87, MB-231, H460 and Panc1 cancer stem cell-like cells (CSCs) was done as described [3]. CSCs were maintained and expanded through passage-5 in complete MammoCult® medium (Stem Cell Technologies, BC, Canada) containing 0.5 % Methylcellulose (Stem Cell Technologies, BC, Canada) in

100 mm ultra-low attachment plates in 5 % CO2 humidified incubator at 37 °C. Testing for increased tumorigenicity was performed in vivo at passage-5. Monoclonal antibody development was custom performed by Pro-Mab Biotechnologies (Richmond, CA). For 7c5b2 [3] and 5g12e8 mAbs we used a nine amino-acid peptide, $M_1$TMFKGSN $E_9$ at the amino-terminal end of hDEspR [2] and a ten amino-acid peptide, $E_9$MKSRWNWGS$_{18}$, for 6g8g7 mAb as antigens respectively. Screening of hybridoma supernatants and characterization of monoclonal antibodies were performed by ELISA using corresponding antigenic peptides. Serial dilutions of primary antibodies were incubated at 37 °C for 1 h. The wells were then incubated with horse radish peroxidase (HRP)-labeled anti-IgG (Sigma) at 37 °C for 1 h. Reactions were analyzed at 450 nm after addition of 3,3′5,5′-tetramethylbenzidine substrate at 37 °C for 10 min.

### Magnetic bead immunoprecipitation (pull-down) of DEspR protein-complex

U87 CSC membranes, Cos1-hDEspR permanent cell transfectant membranes and U87 xenograft tumor (in nude rats) membranes were isolated by differential centrifugation as described [2]. Antibody coupling to magnetic Dynabeads M-450 Epoxy (Invitrogen) was performed as per manufacturer's instructions using 200 μg of anti-hDEspR 5g12e8 mAb per 1 mL of beads (4 × 10$^8$ beads). For target binding 100 μL of 5g12e8-coupled beads (4 × 10$^7$ beads) and 1.3 mg membrane protein were incubated for 2 h at 4 °C in 1 ml ice-cold PBS containing 2 mM EDTA. To capture mAb-bound protein, the tube was placed in a magnet for 2 min, the supernatant discarded and the beads washed five times with 1 mL PBS buffer containing 2 mM EDTA and 0.2 % Tween-20 at room temperature. This washed magnetic bead-'pull-down'-protein complex was then analyzed using different methods. (a) For SDS-PAGE (18 % polyacrylamide) the beads were resuspended in 20 μL of 1X Laemmle sample buffer and proteins denatured at 65 °C × 30 min in order to size-separate and assess size(s) of protein(s) 'pulled-down'. (b) For PNGase F (SIGMA) treatment of pull-down proteins to assess glycosylation states of proteins pulled-down, the beads were resuspended in 20 μL of a buffer containing 20 mM ammonium bicarbonate pH 8.0, 1 % SDS, 100 mM beta-mercaptoethanol and incubated for 30 min at 65 °C. The tube was then placed in a magnet for 2 min. The supernatant was collected to which 20 μL of 0.25 M KCl was added to eliminate SDS by precipitation. To lower KCl concentration the supernatant (40 μL) was collected, diluted 10X with 20 mM ammonium bicarbonate pH 8.0 and concentrated by using a Microcon YM-3 centrifugal filter (Millipore). The concentrated sample was treated with PNGase F (2.5 units) in a final volume of 20 μL for 3 h at 37 °C. Gels were stained either with QC Colloidal Coomassie Stain (Bio-Rad) or Silver (BioRad Silver Stain Plus Kit) following manufacturer's instructions.

## Western blot analysis

Western blot analysis was done essentially as described [2] using pull-down proteins extracted from $2 \times 10^7$ 5g12e8-coupled Dynabeads. Serial Western blot analyses of proteins pulled down—hDEspR, hGalectin-1, hRab-1b and hTMED10—were done sequentially by stripping and re-probing the same blot in the following order: 1st, human-specific DEspR mAb (5g12e8 40 μg/ml, secondary anti-mouse IgG at 1:20,000), 2nd, Rab-1b polyclonal antibody (pAb) (Santa Cruz Biotechnology, cat # sc-599, at 1 μg/ml, secondary anti-rabbit IgG at 1:20,000), 3rd, human-specific Galectin-1 (Abcam, cat # ab25138 at 0.4 μg/ml, secondary anti-rabbit IgG 1:20,000) and 4th, pAb TMED10 (Abcam, cat # ab72666, at 10 μg/ml, secondary anti-rabbit IgG 1:20,000).

Serial Western blot analyses of U87 tumor cell membrane and Cos1-hDEspR cell membrane pull-downs were done sequentially by stripping and re-probing the same blot in order: 1st, hDEspR mAb (5g12e8 40 μg/ml, secondary anti-mouse IgG at 1:20,000) and then 2nd, hGalectin-1 (Abcam, cat#ab25138, at 0.4 μg/ml, secondary anti-rabbit IgG 1:20,000).

Serial western blot of U87 tumor membrane pull-down was also reacted sequentially by stripping and re-probing the same blot using first the following in order: 1st, hDEspR mAb (6g8g7 40 μg/ml, secondary anti-mouse IgG at 1:20,000), and 2nd, hDEspR mAb (5g12e8 40 ug/ml, secondary anti-mouse IgG at 1:20,000).

Immunoreactive proteins were detected by chemiluminescence using the ECL Western Detection kit (GE Healthcare) as per manufacturer's specifications.

## Mass spectrometry analysis

Mass spectrometry of pull-down proteins was custom-performed by Creative Proteomics (Shirley, NY). Briefly, proteins were reduced for 40 min with 5 mM dithiothreitol in 25 mM NH4HCO3 at room temperature and alkylated for 40 min with 15 mM iodoacetamide in 25 mM NH4HCO3 in the dark. After washed and dehydrated, the alkylated samples were digested overnight at 37 °C with trypsin in a 1:50 enzyme-to-substrate ratio (Promega, V5113). Following digestion, the peptide mixtures were acidified with trifluoroacetic (TFA) to 1 %, and desalted by home-made C18 tips. Finally, the desalted peptide samples were dried and dissolved in 10 μL of 0.1 % formic acid in water and subjected to nanoLC-MS/MS analysis in a Q Exactive mass spectrometer. The raw MS files were analyzed and searched against Uniprot

human protein sequence database using Proteome Discoverer 1.4 (Thermo Fisher Scientific, USA). The parameters were set as follows: the protein modifications were carbamidomethylation (C) (fixed), oxidation (M) (variable), and myristyl on glycine (variable), Asn to Asp (variable); the enzyme specificity was set to trypsin; the maximum missed cleavages were set to 2; the precursor ion mass tolerance was set to 10 ppm, and MS/MS tolerance was 0.6 Da.

## Immunofluorescence analysis of subQ xenograft U87-CSC tumors

Double immunofluorescence staining was done as described [3]. Human-specific anti-DEspR mAb (5g12e8) and human-specific anti-Galectin-1 antibody (Abcam, cat#ab25138) were labeled with AlexaFluor(AF)-488 or AF568, and used at 1 μg/ml for anti-Galectin-1 and at 100 μg/ml for 5g12e8 mAbs on fixed, paraffin-embedded sections following antigen-retrieval. Digital photomicroscopy was done using a Zeiss Axioskop fluorescence microscope with auto-exposure settings.

## FACS analysis of U87-CSCs, MB-231-CSCs, H460-CSCs and Panc1-CSCs

U87-CSCs, MB-231-CSCs, H460-CSCs and Panc1-CSCs were incubated in ice-cold Hank's balanced salt solution (HBSS, Invitrogen, NY) plus 2 % FBS containing: (a) 10 μg/ml AF-568 labeled 7c5b2 mAb, or (b) 10 μg/ml AF-568 labeled IgG2b as isotype control. Duplicate samples were incubated for 20 min at 4 °C, washed, resuspended in 1 % FBS/HBSS, 1 % PFA, filtered and analyzed on an LSR-II-FACS instrument. Analysis was done using FloJo Flow Cytometry Analysis Software (http://www.FloJo.com).

## DEspR-inhibition of CSC-growth

DEspR-inhibition studies were performed as described [3]. CSCs (2000/well) were seeded in ultra-low attachment 96-well plate and treated with different blocking anti-hDEspR mAbs (5g12e8, 7c5b2 and 6g8g7) at 100 μg/ml, compared with control non-treated CSCs with six replicates for each. CSCs were cultured in optimal (5 % CO2, humidified incubator at 37 °C) non-adherent conditions. Anti-hDEspR (5g12e8, 7c5b2 and 6g8g7) mAbs were added at seeding, day-2 and day-4. Live and dead CSCs were counted using Trypan Blue on day-5.

## RNA sequence analysis

The NCBI Sequence Read Archive was searched on 7/17/2013 with a query sequence provided by the Genbank accession for "Homo sapiens dual endothelin-1(VEGFsp)/angiotensin II receptor (DEAR) mRNA, complete cds," gi|144,954,325|gb|EF212178.1|, against 727 sequencing runs (Additional file 5).

## Live cell imaging

Fluorescence and transmitted light images were acquired with a Zeiss LSM 710 Duo confocal microscope (Carl Zeiss, Thornwood NY). Excitation was from a 561 nm DPSS laser. A 63X 1.4 NA planapochromat oil immersion objective was used. The cells were in 35 mm dishes with coverglass bottoms. Emission was collected from 575–725 nm. Cells were maintained under physiological conditions using a Pecon stage-top incubation system which maintained 37° centigrade and 5 % CO2. Analysis was done using Image J (Image J, Wayne Rasband, NIH) with identical settings for brightness and contrast adjustments for all images. Live cell imaging with AF-568 labeled isotype IgG2b control was performed in identical conditions. At 60 min, cells were fixed briefly and mounted with Vectashield mounting medium with DAPI (Vector Laboratories, CA, Cat.# H-1200). Photomicroscopy was done using a Zeiss Axioskop fluorescence microscope (Carl Zeiss, Thornwood NY) with differential interference contrast (DIC).

## Statistical analysis

All data were analyzed for normality and descriptive statistics. The following statistical tests were performed using SigmaPlot 11.0 or PRISM 5: one-way analysis of variance (ANOVA) followed by Tukey multiple comparisons test (MCT) for CSC-growth inhibition experiments. A $P < 0.05$ was considered statistically significant.

## Additional files

**Additional file 1: Figure S1.** NCBI designation of Dear (alias DEspR) gene as pseudogene—updated May 2016.

**Additional file 2: Figure S2.** Sanger dideoxy-sequencing of DEspR spanning amino acid 14 (aa14) codon-TGG position within a Yamakawa compression-motif [Y-G-N1, 2-A-R]: DEspR 5′ T-G-G-A-A, shows two G's (GG), and downward slippage/compression of the two G's towards the 5′ T. A 3-nt long stem-loop structure spans the compression motif region in human DEspR but not in rat DEspR.

**Additional file 3: Figure S3.** Unedited RNA-Seq entries showing the questioned two G's with insertion A or TA document difficult sequencing region as seen in manual sequencing.

**Additional file 4: Table S1.** Protein profile pulled-down (Creative Proteomics). Proteins pulled-down with anti-DEspR mAb 5g12e8 from U87 CSC membranes.

**Additional file 5.** RNA-seq analysis. List of 727 sequencing runs searched with a DEspR query sequence in the NCBI Sequence Read Archive.

## Abbreviations
DEspR: dual endotheiln-1/VEGFsp receptor; VEGFsp: vascular endothelial growth factor signal peptide; CSC: cancer stem-like cell; ARMS: allele-specific amplification-refractory mutation system; ORF: open reading frame; FBXW7: F-box and WD repeat domain containing 7; mAb: monoclonal antibody; Gal-1: galectin-1; Gal-3: galectin-3; TMED10: transmembrane emp24 domain-containing protein 10; Rab-1b: ras-related protein Rab-1B; PNGase F: peptide-N-glycosidase F; MS: mass spectrometry.

## Authors' contributions
VLMH wrote the manuscript, co-supervised experimental design and data analysis, performed immunofluorescence analysis, co-supervised FACS analyses; MS performed RNA-seq analysis and revised manuscript; AMM performed Maxam–Gilbert sequencing, pull-down experiments and western blot analysis; GT performed pull-down experiments, western blot analysis, FACS analysis and DEspR inhibition of CSC-growth; KAP performed FACS analysis; KR and DJP performed mass spectrometry analysis; DJP advised on pull-down experiments and analysis; NR-O designed the experiments, supervised the studies, performed all statistical analyses and contributed to writing the manuscript. All authors read and approved the final manuscript.

## Author details
[1] Whitaker Cardiovascular Institute, Boston University School of Medicine, 700 Albany Street, Boston, MA 02118, USA. [2] Department of Medicine, Boston University School of Medicine, 700 Albany Street, Boston, MA 02118, USA. [3] Department of Pathology and Biomedical Engineering, Boston University, Boston, USA. [4] Cold Spring Harbor Laboratory, 1 Bungtown Road, Cold Spring Harbor, NY 11724, USA.

## Acknowledgements
We thank Francis J. Carr, Ph.D. for helpful discussions and manuscript editing; and Michael T. Kirber, Ph.D. for Live Cell Imaging in the Evans Department of Medicine Cell Imaging Core.

## Competing interests
Boston University has filed patents on DEspR for diagnostic and therapeutic agents with NRO and VLMH as co-inventors.

## Funding
This work was partially supported by Ignition Awards from BU Office of Technology Development, and CTSI Award from BU-Clinical and Translational Science Institute Grant NIH-U54TR001012. Some Mass Spectrometry analyses were performed at Cold Spring Harbor Laboratory Shared Resources, which are funded by the Cancer Center Support Grant 5P30CA045508. The funding body had no role in design, in the collection, analysis, and interpretation of data; in the writing of the manuscript; and in the decision to submit the manuscript for publication.

## References
1. NCBI.gov. DEAR dual endothelin-1(VEGFsp)/angiotensin II receptor pseudogene [Homo sapiens (human)]. http://www.ncbi.nlm.nih.gov/gene/102191832. Accessed 27 Aug 2015.
2. Glorioso N, Herrera VL, Bagamasbad P, Filigheddu F, Troffa C, Argiolas G, et al. Association of ATP1A1 and Dear SNP-haplotypes with essential hypertension: sex-specific and haplotype-specific effects. Circ Res. 2007;100:1522–9.
3. Herrera VL, Decano JL, Tan GA, Moran AM, Pasion KA, Matsubara Y, et al. DEspR roles in tumor vasculo-angiogenesis, invasiveness, CSC-survival and anoikis resistance: a 'common receptor coordinator' paradigm. PLoS ONE. 2014;9(1):e85821.
4. Meacham F, Boffelli D, Dhahbi J, Martin DIK, Singer M, Pachter L. Identification and correction of systematic error in high-throughput sequence data. BMC Bioinform. 2011;12:451.
5. Brodin J, Mild M, Hedskog C, Sherwood E, Leitner T, Andersson B, et al. PCR-induced transitions are the major source of error in cleaned up ultra-deep pyrosequencing data. PLoS One. 2013;8(7):e70388.
6. Pei B, Sisu C, Frankish A, Howald C, Habegger L, Mu XJ, et al. The GENCODE pseudogene resource. Genome Biol. 2012;13:R51.

7.  Kageyama Y, Kondo T, Hashimoto Y. Coding vs non-coding: translatability of short ORFs found in putative non-coding transcripts. Biochimie. 2011;93:1981–6.
8.  Uni-Prot.org. http://www.uniprot.org. Accessed 24 Oct 2015.
9.  Yamakawa H, Nakajim D, Ohara O. Identification of sequence motifs causing band compressions o human cDNA sequencing. DNA Res. 1996;3:81–6.
10. Camby I, Mercier ML, Lefranc F, Kiss R. Galectin-1: a small protein with major functions. Glycobiology. 2006;16:137R–57R.
11. Wong JJ, Ritchie W, Ebner OA, Selbach M, Wong JWH, Huang Y, et al. Orchestrated intron retention regulates normal granulocyte differentiation. Cell. 2013;154:583–95.
12. Wong JJ, Au AY, Ritchie W, Rasko JE. Intron retention in mRNA: no longer nonsense: known and putative roles of intron retention in normal and disease biology. Bioessays. 2016;38:41–9.
13. Levin JZ, Yassour M, Adiconis X, Nusbaum C, Thompson DA, Friedman N, et al. Comprehensive comparative analysis of strand-specific RNA sequencing methods. Nat Methods. 2010;7:709–15.
14. Faridani OR, McInerney GM, Gradin K, Good L. Specific ligation to double-stranded RNA for analysis of cellular RNA:RNA interactions. Nucleic Acids Res. 2008;36:e99.
15. Zhou H, Wang F, Wang Y, Ning Z, Hou W, Wright TG, et al. Improved recovery and identification of membrane proteins from rat hepatic cells using a centrifugal proteomic reactor. Mol Cell Proteomics. 2011;10:1–11.
16. Cao L, Clifton JG, Reutter W, Josic D. Mass spectrometry-based analysis of rat liver and hepatocellular carcinoma Morris hepatoma 7777 plasma membrane proteome. Anal Chem. 2013;85:8112–20.
17. Mirza SP, Halligan BD, Greene AS, Olivier M. Improved method for the analysis of membrane proteins by mass spectrometry. Physiol Genomics. 2007;30:89–94.
18. Almén MS, Nordström KJ, Fredriksson R, Schiöth HB. Mapping the human membrane proteome: a majority of the human membrane proteins can be classified according to function and evolutionary origin. BMC Biol. 2009;7:50–63.
19. Peng L, Kapp EA, McLauhlan D, Jordan TW. Characterization of the Asia Oceania human proteome organization membrane proteomics initiative standard using SDS-PAGE shotgun proteomics. Proteomics. 2011;11:4376–84.
20. Fagerberd L, Jonasson K, vonHeijne G, Uhlen M, Berglund L. Prediction of the human membrane proteome. Proteomics. 2010;10:1141–9.
21. Bensalem N, Masscheleyn S, Mozo J, Vallee B, Brouillard F, Trudel S, et al. High sensitivity identification of membrane proteins by MALDI TOF-MASS Spectrometry using polystyrene beads. J Proteome Res. 2007;6:1595–602.
22. Sanchez-Ruderisch H, Fischer C, Detjen KM, Welzel M, Wimmel A, Maing JC, et al. Tumor suppressor p16 INK4a: downregulation of galectin-3, an endogenous competitor of the pro-anoikis effector galectin-1, in a pancreatic carcinoma model. FEBS J. 2010;277:3552–63.
23. Maxam AM, Gilbert W. A new method for sequencing DNA. Proc Natl Acad Sci USA. 1977;74:560–4.
24. Ruiz-Opazo N, Hirayama K, Akimoto K, Herrera VL. Molecular characterization of a dual endothelin-1/angiotensin II receptor. Mol Med. 1998;4:96–108.

# 17

# Growth arrest specific gene 2 in tilapia (*Oreochromis niloticus*): molecular characterization and functional analysis under low-temperature stress

ChangGeng Yang[†], Fan Wu[†], Xing Lu, Ming Jiang, Wei Liu, Lijuan Yu, Juan Tian and Hua Wen[*]

## Abstract

**Background:** Growth arrest specific 2 (*gas2*) gene is a component of the microfilament system that plays a major role in the cell cycle, regulation of microfilaments, and cell morphology during apoptotic processes. However, little information is available on fish *gas2*. In this study, the tilapia (*Oreochromis niloticus*) *gas2* gene was cloned and characterized for the first time.

**Results:** The open reading frame was 1020 bp, encoding 340 amino acids; the 5′-untranslated region (UTR) was 140 bp and the 3′-UTR was 70 bp, with a poly (A) tail. The highest promoter activity occurred in the regulatory region (−3000 to −2400 bp). The Gas2-GFP fusion protein was distributed within the cytoplasm. Quantitative reverse transcription-polymerase chain reaction and western blot analyses revealed that *gas2* gene expression levels in the liver, muscle, and brain were clearly affected by low temperature stress. The results of *gas2* RNAi showed decreased expression of the *gas2* and P53 genes.

**Conclusion:** These results suggest that the tilapia *gas2* gene may be involved in low temperature stress-induced apoptosis.

**Keywords:** *Oreochromis niloticus*, Growth arrest specific gene 2, Functional characterization, Low-temperature stress

## Background

Growth and reproduction in fish are closely related to water temperature, light, gas pressure, and other climatic factors. Among them, ambient temperature is the most important factor affecting growth and development of fish [1–3]. In recent years, more and more attention has been paid to the molecular changes in fish responding to a low temperature environment [4–6]. Tilapia is a widely cultured warm water fish worldwide, with poor low temperature tolerance, and a growth temperature range of 16–38 °C. Therefore, culture is clearly subject to climate and temperature restrictions [7]. Studying the physiology,

biochemistry, and molecular mechanisms of tilapia under low temperature stress will help understand the effects of low temperature on fish and provide a theoretical basis for the tilapia breeding industry.

Transcriptome and digital gene expression (DGE) analyses were used in our previous study to analyze the gene expression changes in the liver of tilapia under low temperature stress [8]. As results, many differentially expressed genes involved in growth, development, immunity, apoptosis, and other physiological processes were detected. Among them, a gene that plays an important role in apoptosis, called growth arrest specific 2 gene (*gas2*), was identified. *Gas* genes are upregulated during serum starvation or cell contact inhibition in vitro [9, 10]. The protein encoded by *gas2* is a component of the microfilament system. It is highly conserved during evolution and plays a major role in the cell cycle, regulation

*Correspondence: wenhua.hb@163.com
[†]ChangGeng Yang and Fan Wu contributed equally to this work
Key Laboratory of Freshwater Biodiversity Conservation and Utilization of Ministry of Agriculture, Yangtze River Fisheries Research Institute, Chinese Academy of Fishery Sciences, Wuhan 430223, China

of microfilaments, and cell morphology during apoptotic processes, and regulation of calpain activity [11–14]. In this study, we cloned the tilapia *gas2* gene and analyzed structure and active region of its promoter, and subcellular localization of the protein. The *gas2* gene expression profile was analyzed in tilapia subjected to low-temperature stress. The relative changes observed will provide insight into the effect of low-temperature stress on tilapia.

## Methods

### Experimental fish

Tilapia was purchased from a fish hatchery in Guangxi Province, China and was transported to the experimental laboratory at the Yangtze River Fisheries Research Institute (Wuhan, Hubei Province, China). Their initial body weight was $140.0 \pm 10$ g. The fish were initially held for 2 weeks at 28 °C before commencing the experiments. Then, water temperature was increased to 30 °C for 1 week before they were subjected to a decrease in water temperature from 30 to 10 °C, which is near the lethal minimum [15], at a cooling at a rate of 1 °C/day. Dissolved oxygen was maintained at >5 mg/L with an air compressor, water pH was 7.2–7.5, and $NH_3$–N was $0.26 \pm 0.10$ mg/L.

### Sample preparation

Nine tissues (brain, gill, skin, muscle, heart, liver, eye, spleen, and intestines) were collected from each of three fish held at 30 °C. Three fish each from temperatures of 30, 25, 20, 15, and 10 °C were anesthetized, and samples of liver, brain, and muscle were collected, frozen in liquid nitrogen, and stored at −80 °C until use. Total RNA was extracted from these samples using Trizol reagent (Invitrogen, Carlsbad, CA, USA), according to the manufacturer's instructions. Liver, brain, and muscle tissues were ground in liquid nitrogen to extract total protein. Then, 300 µL of ice-cold lysis buffer (150 mM NaCl, 1.0% NP-40, 0.5% sodium deoxycholate, 0.1% sodium dodecyl sulfate, 50 mM Tris–HCl, pH 8.0, and protease inhibitors) were added to 5 mg tissue samples and maintained for 2 h at 4 °C with constant agitation. After a 20 min centrifugation at 12,000 rpm in a microcentrifuge at 4 °C, the tubes were placed on ice, the supernatant was aspirated to a fresh tube on ice, and the pellet was discarded.

### Cloning the tilapia *gas2* gene

Based on the DGE-tag sequences obtained and the reference tilapia genome data (https://www.ncbi.nlm.nih.gov/genome/197?genome_assembly_id=293496), four specific primers (Additional file 1: Table S1) were designed to perform the 5′ and 3′ rapid amplification of cDNA ends (RACE) procedure on this gene using the BD SMART™

RACE cDNA amplification kit (BD Biosciences/Clontech, Palo Alto, CA, USA) following the manufacturer's instructions.

Based on the cDNA sequences obtained and the reference tilapia genome data, two primers were designed to acquire the DNA promoter sequences (Additional file 1: Table S1).

### Sequence and phylogenetic analyses

The tilapia *gas2* motif was scanned using PROSITE (http://prosite.expasy.org/). The deduced amino acid sequence of tilapia *gas2* was submitted to the BLAST program (http://blast.www.ncbi.nlm.nih.gov) to search for counterpart sequences. Multiple sequence alignments were performed with the ClustalX 1.83 program. Then, an unrooted phylogenetic tree was constructed using the neighbor-joining algorithm in the MEGA 5.05 program, based on the sequence alignments and other Gas2 genes. The phylogenetic tree was tested for reliability by 1000 bootstrap replications. Putative transcription factor binding site motifs were detected by MatInspector software.

### Construction of a luciferase-reporter gene vector for the tilapia *gas2* gene promoter and pEGFP-N3-GAS2 gene expression vector

A promoter–deletion experiment was designed to monitor promoter activity of the 5′-flanking region. DNA fragments with a series of nested deletions were generated, and five DNA fragments were inserted into the pGL4.10 vector and named pGL4-1 (−3000 to 0 bp), pGL4-2 (−2400 to 0 bp), pGL4-3 (−1800 to 0 bp), pGL4-4 (−1200 to 0 bp), and pGL4-5 (−600 to 0 bp). pGL4.51 vector (Promega, E132A) with a CMV promoter was used as a reference (positive control), while pGL4.10 vector (Promega, E665A) without promoter was used as negative control.

The pEGFP-N3 expression plasmid was purchased from Invitrogen. The tilapia *gas2* open reading frame (ORF) was amplified by polymerase chain reaction (PCR), cloned into the pEGFP-N3 with the GAS-N-S1 and GAS-N-A1 primers (Additional file 1: Table S1), and named pEGFP-N3-GAS2.

### Construction of short hairpin RNA (shRNA) expression vectors

Three shRNA sequences were designed based on the *gas2* gene sequences. Three shRNA-expressing plasmids specifically targeting *gas2* (called shG1, shG2, and shG3) were constructed by GenePharma Corp. (Shanghai, China) using the pGPU6/Neo vector. Scrambled shRNA was used as a negative control. The shRNA sequences are shown in Additional file 2: Table S2.

## Cell culture, transient transfection, and detection of luciferase activity

The CHO-K1 cell line was purchased from the Institute of Biochemistry and Cell Biology, Shanghai Institutes for Biological Sciences, Chinese Academy of Sciences (Shanghai, China). CHO-K1 cells were cultured in F-12K medium containing 10% fetal bovine serum, 100 U/mL penicillin, and 100 U/mL streptomycin at 28 °C in 5% $CO_2$. The cells were seeded in 96-well plates for 24 h before transfection. When the cells were 90% confluent, an equivalent quantity of plasmid was transfected into the CHO-K1 cells following the transfection reagent instructions (Lipofectamine™ 2000; Invitrogen).

## Detection of luciferase activity

The cells were harvested after 48 h of transfection, and luciferase activity was detected on Molecular Device 5 using a Luciferase Reporter Gene Assay Kit (Beyotime, Shanghai, China), in accordance with the manufacturer's instructions. Three replicates were carried out for each sample.

## Culture of the tilapia brain cell line (TBC), sub-cellular localization, and tilapia gas2 RNAi

The TBC tilapia brain cell line (construction by the Yangtze River Fisheries Research Institute) was maintained in L15 medium (Hyclone, Logan, UT, USA) containing 20% fetal calf serum (JIBC, USA) and antibiotics. The flasks/plates were seeded at 50% confluency before transfection and were transfected into 6-well plates at constant numbers. The cells were maintained in medium without fetal calf serum or antibiotics prior to transfection. When the cells grew to 90% confluence, an equivalent quantity of plasmids (pEGFP-N3-GAS2 or shRNA) was transfected into the TBC cells following the transfection reagent instructions (Lipofectamine™ 2000, Invitrogen). The cells were stained for sub-cellular localization 24 h later with 1 mg/mL DAPI for 1 h away from light. All samples were examined under a Leica SP8 confocal laser scanning microscope (CLSM; Leica Microsystems,Bannockburn, IL, USA). The cells were harvested to carry on quantitative reverse transcription-PCR (qRT-PCR) for RNAi 48 h later.

## Quantitative RT-PCR analysis

qRT-PCR was conducted to determine the gas2 mRNA tissue distribution, the mRNA expression pattern under low-temperature stress, and expression of the P53 gene (Accession Number: XM_005463838). Expression of β-actin was used as the internal control. Total RNA was prepared with TRIzol reagent. cDNA was synthesized from each RNA sample (500 ng) using a PrimeScript® RT reagent kit (Takara Bio, Shiga, Japan), following the manufacturer's recommendations. qRT-PCR was conducted on an Applied Biosystems 7500 Real-Time PCR System with SYBR® Premix Ex Taq™ (Takara Bio). The primer pair GAS-R-S1 and GAS-R-A1, P53-R-S1 and P53-R-A1, and Actin-S1 and Actin-A1 was used to amplify the gas2, P53, and β-actin fragments, respectively (Additional file 1: Table S1).

Real-time PCR was carried out with 1 μL cDNA sample, 10 μL SYBR® Premix Ex Taq™, 0.4 μL ROX Reference Dye II, 0.4 μL PCR forward/reverse primers (10 mM), and 7.8 μL nuclease-free water. The thermocycling conditions for the reaction were as follows: 95 °C for 30 s, followed by 40 cycles consisting of 95 °C for 5 s, and 60 °C for 34 s. The reactions were carried out with three duplicates of each sample. The ratio changes in the target genes relative to the control gene (β-actin) were determined by the $2^{-\triangle\triangle CT}$ method.

## Recombinant expression and identification and preparation of the fusion protein antibody

The tilapia gas2 coding sequence was amplified using the GAS-P-S1 and GAS-P-A1 primers and sub-cloned into pET-30a to construct pET-30a-Gas. Then, expression of the His-tagged fusion protein was induced with pET-30a-Gas using 0.5 mM isopropyl-b-D-thiogalactopyranoside at 37 °C for 3–4 h. A Ni-NTA column was used to purify the tilapia Gas2 fusion protein, followed by sodium dodecyl sulfate–polyacrylamide gel electrophoresis (SDS-PAGE) and a western blot assay (with anti-His antibody) to detect the recombinant tilapia Gas2 protein. Anti-Gas2 from rabbit was prepared by China Wuhan ABclonal Biotech Co., Ltd. (Wuhan, China).

## SDS-PAGE and western blot assay

Protein concentrations were determined using the BCA method, separated on a 12% polyacrylamide gel under reducing conditions, and transferred to a polyvinylidene difluoride membrane. The membranes were blocked in blocking buffer (TBST, 5% skimmed milk in TBS containing 0.05% Tween-20) for 1 h and incubated overnight at 4 °C with primary rabbit antibodies against Gas2 or β-actin (1:1000; ABclonal) in TBST containing 1% skimmed milk. After washing with TBST (3 × 15 min), the membranes were incubated with goat anti-rabbit horseradish peroxidase-conjugated IgG (1:1000; Tiangen) for 1 h at room temperature. The reactive protein bands on the membrane were visualized using enhanced chemiluminescent reagents (Tiangen) and exposed in a darkroom. The expression intensities of the Gas2-specific bands were normalized against the β-actin bands.

### Statistical analysis

One-way analysis of variance (ANOVA) was used to analyze the log-transformed Ct values. When ANOVA identified differences among groups, Tukey's multiple-comparisons test (SAS Institute, Cary, NC, USA) was conducted to examine the differences among the treatments and values. A $p$ value <0.05 was considered significant.

## Results

### Cloning and sequence analysis of the tilapia gas2

We acquired the 1020-bp ORF of the tilapia *gas2* gene, encoding 340 amino acids based on the DGE-tag sequence. The molecular weight of the protein was determined to be 37.7 kDa, and the isoelectric point (pI) was 8.67. The 5′-UTR was 140 bp and the 3′-UTR was 70 bp with a poly (A) tail, according to 5′ and 3′ RACE cDNA amplification (Fig. 1). The sequence was submitted to Genbank (No. KY882138). The protein was identified as *gas2* and the calponin homology (CH) domain and *gas2*-related (GAR) domain consisted of 126 and 74 amino acid residues, respectively, according to a PROSITE analysis (Fig. 1). The closest homology was detected by multiple alignments of *gas2* deduced amino acid sequences with those of other species (Fig. 2).

### Phylogenetic analysis of tilapia gas2

A phylogenetic tree for the tilapia Gas2 protein and the other known Gas2 proteins was constructed by the neighbor-joining method. The list of species included is shown in Additional file 3: Figure S3. On the phylogenetic tree, tilapia Gas2 was distantly related with *Drosophila bipectinata* Gas2. And the tilapia Gas2 protein shared the closest relationship with *Neolamprologus brichardi* (Additional file 4: Figure S1).

### Cloning and promoter analysis of the tilapia gas2 5′-flanking region

3000 bp promoter sequence was cloned using a PCR method. As shown in Fig. 3, relative luciferase activity was detected from CHO-K1 cells transfected with the luciferase reporting assay vector. The luciferase activities from pGL4.51 (positive control), pGL4-1 (−3000 to 0 bp), and pGL4-2 (−2400 to 0 bp) were significantly higher than those in the negative control. Analysis of the promoter sequence in the region (−2400 to −3000 bp region) revealed that the main motifs of the sequence, e.g., Fox gene family, NeuroD, SOX/SRY and so on. (Fig. 3).

### Tilapia gas2 expression profile

qRT-PCR was employed to quantify the expression of tilapia *gas2* mRNA in different tilapia tissues from normal fish and in those experiencing low-temperature stress. Tilapia *gas2* mRNA was found in all nine tissues examined, with the highest levels in the spleen and liver (Fig. 4). The *gas2* expression changed in the muscle, brain and liver when tilapia was subjected to low temperature stress. The *gas2* expression levels at 10 and 15 °C were significantly higher in the liver compared with that at 30 °C. The gas2 expression increased first and then tended to decrease in muscle from fish exposed to decreasing temperature. The *gas2* expression was significantly higher at 15, 20, and 25 °C than that at 30 °C. Similarly, *gas2* expression in the brain was significantly higher at 10 and 20 °C than that at 30 °C. In contrast, expression was significantly lower in the brain at 25 °C than at 30 °C (Fig. 5).

### Tilapia Gas2 western blot assay

A western blot assay of Gas2 showed that Gas2 protein content in the brain did not change when temperature was decreased. Similar to *gas2* mRNA expression in muscle, Gas2 protein content was higher at 20 and 25 °C than that at 30 °C. However, Gas2 protein content in the liver was low at 30 °C and decreased as temperature was decreased, but protein content increased initially. Gas2 protein content was maximal at 10 °C (Fig. 6).

### Tilapia Gas2 subcellular localization

The wild-type green fluorescent protein (GFP) exhibited diffuse localization throughout the cell as seen on CLSM, whereas the transiently expressed Gas2-GFP fusion protein was distributed in a cytoplasmic network within the cell (Fig. 7).

### Tilapia gas2 RNAi expression

After transfecting the TBC and extracting RNA, the real-time PCR results showed that shG1 and shG3 effectively reduced *gas2* expression of the transduced cells and expression decreased by 1- to 3-fold. P53 expression levels also decreased 1- to 3-fold (Fig. 8).

## Discussion

Our DGE results (unpublished data) revealed that tilapia *gas2* expression increased significantly in the liver as temperature was decreased. We cloned the tilapia *gas2* gene. The full length tilapia *gas2* cDNA and the regulatory region were also amplified by 3′- and 5′-RACE and gene walking for further study. The predicted amino acid sequence of this gene was searched against the NCBI protein database, and a comparative analysis revealed high-sequence identity of the candidate protein with Gas2 of *Maylandia zebra*. The structural features of tilapia Gas2 were consistent with the well-known CH and GAR domains. The CH domain is a family of actin

ggtccacactggtttgtgtttttggaggctaacctgtgggtcggtgcaccgctgtgaaaa
gcatgaaaattgggaaccagcatctcacaggacgagtgagaggaaggtgcgcaccagcga

gctgaaccctcaggggaacc<u>ATG</u>TGCAGTTCCCTGAGCCCCAAACAGCCCAGCGGGCCCG
                        M  C  S  S  L  S  P  K  Q  P  S  G  P
GCCTGACAGACATGCAGCAGTACCAACAGTGGCTTTCCAGTCGACATGAGGCCAGCCTGC
G  L  T  D  M  Q  Q  Y  Q  Q  W  L  S  S  R  H  E  A  S  L
TGCCCATGAAGGAAGACCTGGCCCTGTGGCTCACTAATATTCTTGGTCTGGAAATCACTG
L  P  M  K  E  D  L  A  L  W  L  T  N  I  L  G  L  E  I  T
CTGAGAGCTTCATGGACCGCTTAGACAATGGCTTCCTGTTGTGCCAGCTGGCTGAGACAC
A  E  S  F  M  D  R  L  D  N  G  F  L  L  C  Q  L  A  E  T
TGCAGGAGAAGTTCAGGCAGAGTAATGGAGATCTGCCTACCCTTGGCAACAGCAAGAGAA
L  Q  E  K  F  R  Q  S  N  G  D  L  P  T  L  G  N  S  K  R
TTCCCAATCGGAGGATACCGTGTCGGCGGAGCGCACCATCCGGCTCTTTCTTTGCACGGG
I  P  N  R  R  I  P  C  R  R  S  A  P  S  G  S  F  F  A  R
ACAACACAGCCAACTTCCTGGCCTGGTGTCGTGAGGTCGGAGTTGGAGAAACATGTCTGT
D  N  T  A  N  F  L  A  W  C  R  E  V  G  V  G  E  T  C  L
TTGAATCAGAGGGTTTAGTTCTTCATAAGCAGCCACGAGAAGTGTGCCTTTGTCTTTTAG
F  E  S  E  G  L  V  L  H  K  Q  P  R  E  V  C  L  C  L  L
AGCTGGGCCGGATAGCATCACGGTACAATGTGGAGCCTCCTGGCCTTATAAAACTGGAGA
E  L  G  R  I  A  S  R  Y  N  V  E  P  P  G  L  I  K  L  E
AAGAGATTGAGCAAGAAGAGAAAGCGCCTCTACCTCCTCCGTCACCTGTATCTCCAGCTC
K  E  I  E  Q  E  E  K  A  P  L  P  P  P  S  P  V  S  P  A
AGCCTGTCCCTCCATCACCCTCTTCATCTCCTTCTCCATCCCCTTCCCCATCCCCATCCC
Q  P  V  P  P  S  P  S  S  S  P  S  P  S  P  S  P  S
CATCCCCAACTCCTACCAAAGTTACGTCTATCAAAAAAAGCACAGGGAAGCTGTTGGATG
P  S  P  T  P  T  K  V  T  S  I  K  K  S  T  G  K  L  L  D
ATGCCGTAAGACATATCGCCAATGATCCACCTTGCAGATGTGCAAATAAGTTCTGTGTAG
D  A  V  R  H  I  A  N  D  P  P  C  R  C  A  N  K  F  C  V
AGAGACAGTCGCAGGGTCGATATCGTGTTGGAGAGAAGATGCTCTTTATTCGGATGCTGC
E  R  Q  S  Q  G  R  Y  R  V  G  E  K  M  L  F  I  R  M  L
ACAATAAGCACGTGATGGTGCGAGTCGGTGGAGGCTGGGAGACCTTTGAGAGCTACCTGC
H  N  K  H  V  M  V  R  V  G  G  G  W  E  T  F  E  S  Y  L
TGAAACATGACCCATGCCGCATGCTCCAGATTTCCAGGGTGGAGGGCAAGATATCCCCCA
L  K  H  D  P  C  R  M  L  Q  I  S  R  V  E  G  K  I  S  P
TCAGCAGCAAGTCCCCTAATATCAAGGACCTCAGCCCTGACAGCTGCCTGGTGGTGGCAG
I  S  S  K  S  P  N  I  K  D  L  S  P  D  S  C  L  V  V  A

CGCACTACCGCAGCAAAAGG<u>TGA</u>agacatgagactgaaaaaaatgccatt<u>aataaa</u>agga
a  h  y  r  s  k  r  *
cagatcg<u><u>aaaaaaaaaaaaaaaaaaaaaaaaaaaaa</u></u>

**Fig. 1** The cDNA and predicted protein sequence of tilapia Gas2. The open reading frame sequences are shown in *capital letters*. The start and stop codons are marked by *box*. The AATAAA box is *underlined*, and the poly(A) region is *double-underlined*. The calponin homology (CH) domain (pos.: 34–159) and Gas2-related (GAR) domain (pos.: 227–300) were in *shadow*, respectively

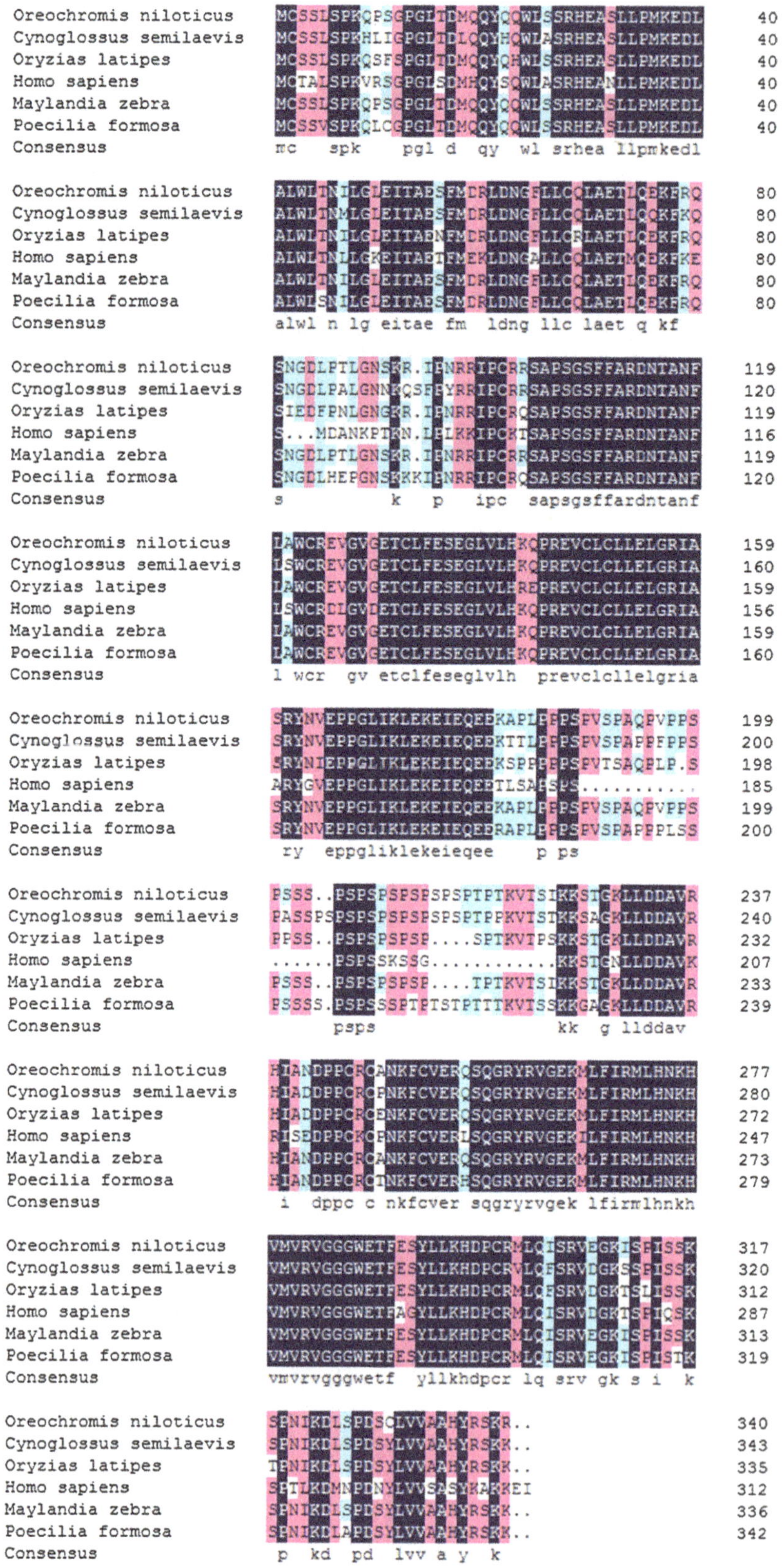

**Fig. 2** Multiple alignment of deduced amino acid sequences of Gas2 from *Cynoglossus semilaevis* (XP_008310260), *Oryzias latipes* (XP_004069628), *Homo sapiens* sequence (NP_005247), *Maylandia zebra* (XP_004575167), *Poecilia formosa* (XP_007578649)

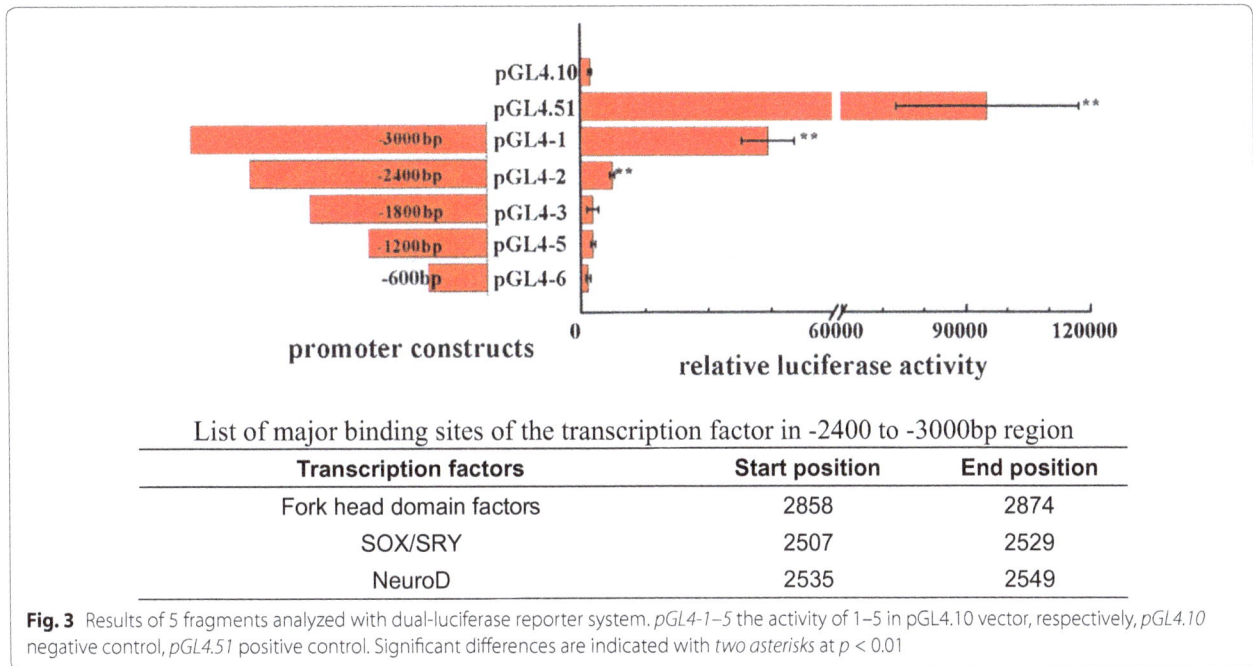

| List of major binding sites of the transcription factor in -2400 to -3000bp region | | |
|---|---|---|
| **Transcription factors** | **Start position** | **End position** |
| Fork head domain factors | 2858 | 2874 |
| SOX/SRY | 2507 | 2529 |
| NeuroD | 2535 | 2549 |

**Fig. 3** Results of 5 fragments analyzed with dual-luciferase reporter system. *pGL4-1–5* the activity of 1–5 in pGL4.10 vector, respectively, *pGL4.10* negative control, *pGL4.51* positive control. Significant differences are indicated with *two asterisks* at $p < 0.01$

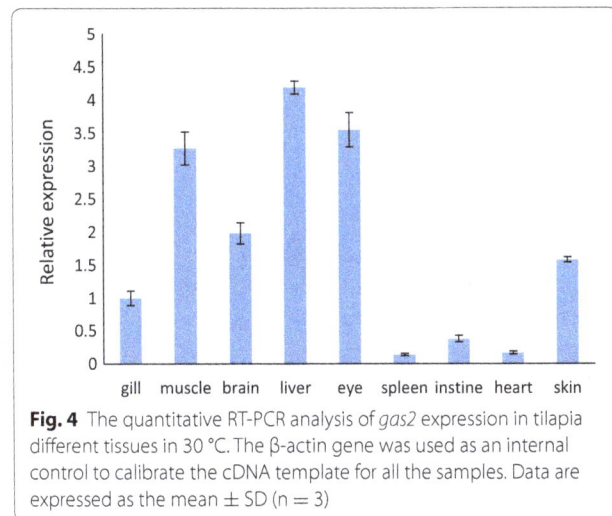

**Fig. 4** The quantitative RT-PCR analysis of *gas2* expression in tilapia different tissues in 30 °C. The β-actin gene was used as an internal control to calibrate the cDNA template for all the samples. Data are expressed as the mean ± SD (n = 3)

binding domains found in cytoskeletal and signal transduction proteins that directly links signal transduction molecules to the actin cytoskeleton via an association with F-actin [16, 17]. Most actin-binding proteins have two copies of the CH domain, but tilapia Gas2 had a single copy. The GAR domain is common in plakin and Gas2 family members. The GAR domain comprises about 57 amino acids and binds to microtubules [18]. Tilapia Gas2 had a GAR domain containing 73 amino acids. A Gas2 domain analysis revealed that it contained the CH and GAR domains, indicating that it has the potential to bind to microtubules.

The regulatory region (−3000 to −2400 bp) exhibited relatively higher promoter activity than other promoter regions according to a functional analysis of the tilapia *gas2* 5′-flanking region. In addition, several potential transcription factor binding sites were identified in the tilapia *gas2* positive regulatory region. The transcription factors binding to these sites are related to regulation of growth and development, cell cycle regulation, cell apoptosis, and immunoregulation. For example, transcription factors, such as the Fork head box, NeuroD, SOX/SRY.

Low-temperature stress not only can lead to cell swelling and membrane damage, resulting in cell necrosis [19], but can also trigger expression of a series of genes and complex physiological responses, including inhibition of cell proliferation, cell cycle changes, and cell apoptosis [6, 20–22]. Cell apoptosis or necrosis depends largely on the intensity of the low-temperature stress. Yao et al. [23] showed that the low and high temperature stress response mechanisms are similar in mussels and are dependent on caspase-3. They suggested a causal link between DNA damage at high and low temperatures and subsequent stress reactions, such as induced cell apoptosis. Low-temperature stress affects cell membrane fluidity and cell mass transport, leading to cell division, growth arrest, and apoptosis at the cellular level in tilapia [24]. Similarly, our previous DGE study reported that multiple genes are involved in apoptosis and change as temperature decreased, such as BCL2, E3 ligase, AXIN1, and *gas2* genes [8]. The *gas2* is a multifunctional gene involved in cell apoptosis. On

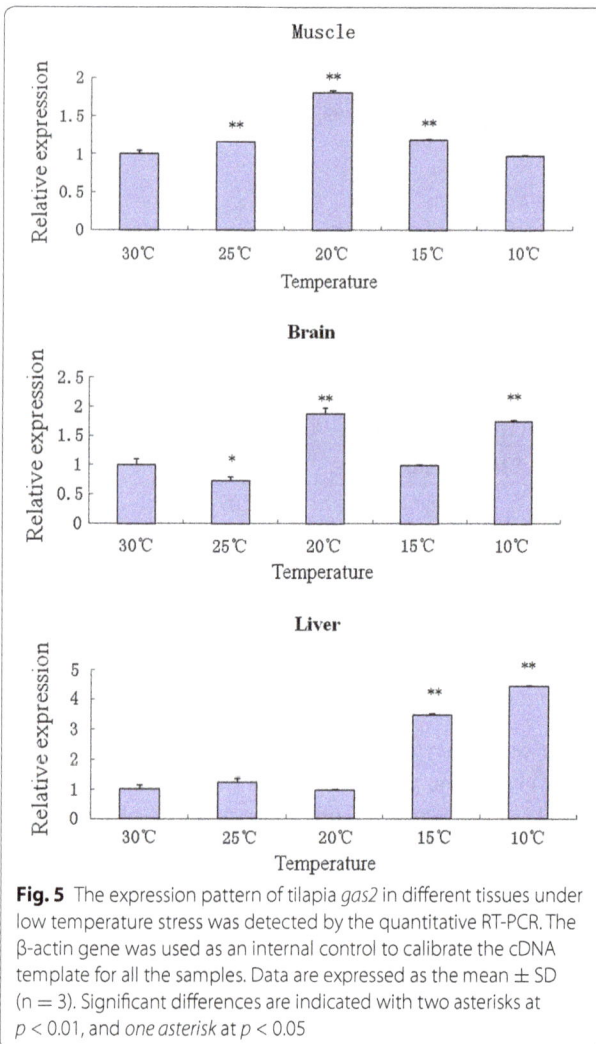

Fig. 5 The expression pattern of tilapia gas2 in different tissues under low temperature stress was detected by the quantitative RT-PCR. The β-actin gene was used as an internal control to calibrate the cDNA template for all the samples. Data are expressed as the mean ± SD (n = 3). Significant differences are indicated with two asterisks at p < 0.01, and *one asterisk* at p < 0.05

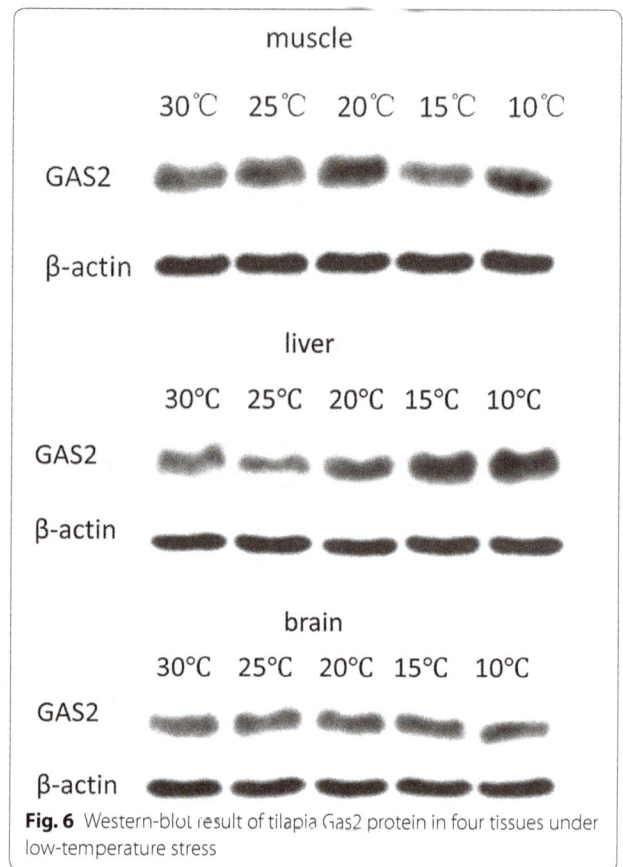

Fig. 6 Western-blot result of tilapia Gas2 protein in four tissues under low-temperature stress

the one hand, the *gas2* gene is a caspase-3 substrate that is cleaved by caspase-3 and is involved in the morphological changes of apoptotic cells [13, 25], on the other overexpressing *gas2* does not directly induce apoptosis but may increase sensitivity of cells to apoptotic signals [14]. Gas2 was shown to be a component of the microfilament network system [11, 26, 27], similarly, a network of Gas2 expression was identified in the cell cytoplasm in this study. Moreover, tissue distribution of Gas2 indicated that this gene was expressed ubiquitously, with the highest expression in the liver, indicating that liver might be one of the most important locations where Gas2 carry out physiological function. In addition, the changes in tilapia *gas2* mRNA and protein levels were detected under low-temperature stress by qRT-PCR and western blot analysis. Our early DGE experimental results in tilapia liver were validated by the present results, i.e., significantly higher expression of *gas2* was detected in the liver when temperature was dropped to

the lethal minimum of 10 °C than that at 30 °C. These results suggested that *gas2* synthesis was induced by low temperature, in the liver, indicating that the liver might be one of the most important tissues functioning in temperature stress response. Some studies have found that low-temperature stress can induce apoptosis of liver cells [28, 29]. Our finding of the higher expression of apoptosis-related gene (*gas2*) detected in the tilapia liver may also suggest the induction of liver cell apoptosis under low-temperature stress. The liver cells undergoing apoptosis may be related to the increasing stress sensitivity of these cells, and further results in a large effect on physiological functioning. This hypothesis may partly explain the reason for the poor low temperature tolerance of tilapia. The *gas2* RNAi results showed that shRNA was effective in reducing expression of the *gas2* gene in cells. Simultaneously, P53 gene expression also decreased. The inhibition of *gas2* gene expression may have reduced expression of the P53 gene. Some studies have shown that Gas2 affects stability of the P53 protein, and Gas2 binds to m-calpain in vivo and inhibits calpain activity, leading to increased P53 stability [14]. However, the reasons for these results are unclear and need further study.

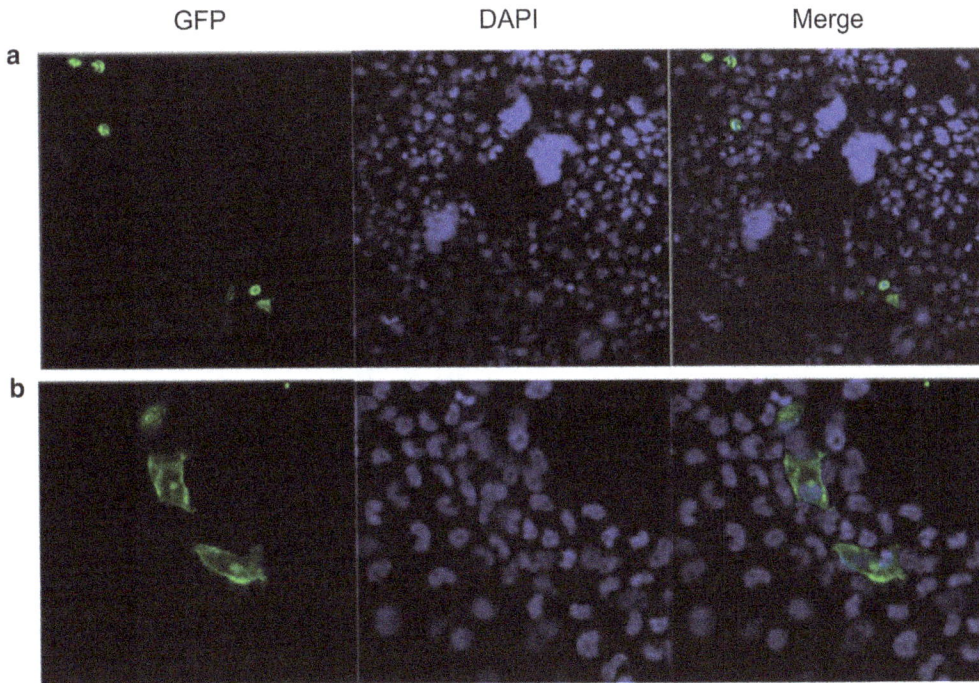

**Fig. 7** The subcellular location of tilapia Gas2 in TBC cells using a confocal laser endomicroscopy. **a** Localization of Gas2-GFP fusion green fluorescence proteins in the cytoplasm of TBC cells. **b** Control localization of GFP green fluorescence proteins

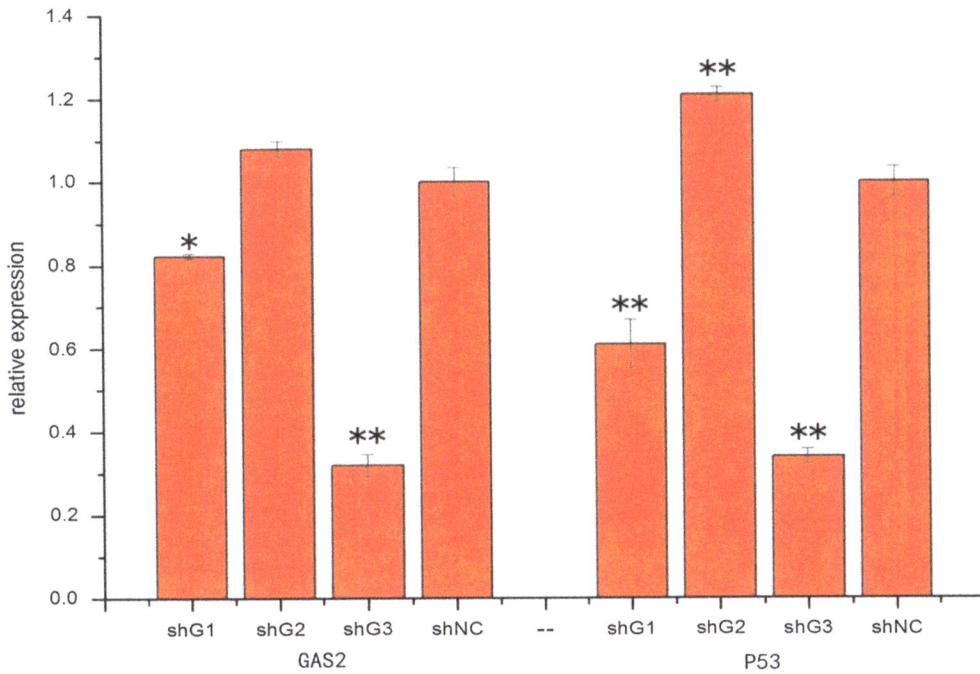

**Fig. 8** The expression pattern of *gas2* and P53 after shRNA transfection in TBC. GAS2: the relative expression of *gas2* gene after 48 h transfected shG1, shG2, shG3 and shNC. *P53* the relative expression of P53 gene after 48 h transfected shG1, shG2, shG3 and shNC. Data are expressed as the mean ± SD (n = 3). Significant differenced across medium injection were indicated with an *asterisk* at $p < 0.05$, and *two asterisk* at $p < 0.01$

## Conclusions

Here in this study, sequence cloning, phylogenetic analysis and functional promoter analysis were performed to better understand the characteristics of an apoptosis-related gene, tilapia *gas2*. Furthermore, qRT-PCR and western blot analyses indicated that *gas2* expression in multiple tilapia tissues (liver, muscle and brain) was significantly affected by low temperature stress. The regulated expression of this apoptosis-related gene revealed that low temperature may induce apoptosis of multiple tissues, partly explaining the sensitivity of tilapia to low-temperature on the molecular level.

## Additional files

**Additional file 1: Table S1.** Primers used in this study. The primers were used in cloning, vector construction and qRT-PCR.

**Additional file 2: Table S2.** The sequences of shRNA. The sequences information of shRNA were used in RNAi experiment.

**Additional file 3: Table S3.** List of species with their GenBank accession numbers. The species with GenBank accession numbers were used in phylogenetic analyses.

**Additional file 4: Table S4.** Phylogenetic tree of Gas2 amino acid sequences based on Neighbor-Joining (NJ) method. The bootstrap confidence values shown at the nodes of the tree are based on a 1000 bootstrap procedure, and the branch length scale in terms of genetic distance is indicated below the tree.

## Abbreviations

DGE: digital gene expression profiling; cDNA: complementary DNA; CHO-K1: Chinese hamster ovary K1 cell line; RNAi: RNA interference; DAPI: 4′,6-diamidino-2-phenylindole; IgG: immunoglobulin G; ORF: open reading frame.

## Authors' contributions

CGY, FW and HW had designed the study. CGY and FW performed cloning, expression assay, Western blot assay, sequence assays, as well as statistical analysis. FW, MJ and WL fed Nile tilapia and prepared the sample. CGY carried out recombinant expression. CGY and XL performed cell line culture, and data analysis. XL, LJY and JT performed transient transfection. All authors contributed to the writing of the manuscript. All authors read and approved the final manuscript.

## Acknowledgements

This aspect does not apply to the manuscript.

## Competing interests

The authors declare that they have no competing interests.

## Funding

This work was performed under the auspices of the National Natural Science Foundation of China (31402290) and the China Agriculture Research System (No. CARS-49).

## References

1. Guderley H. Metabolic responses to low temperature in fish muscle. Biol Rev. 2004;79(2):409–27.
2. Carginale V, Trinchella F, Capasso C, Scudiero R, Parisi E. Gene amplification and cold adaptation of pepsin in Antarctic fish. A possible strategy for food digestion at low temperature. Gene. 2004;336(2):195–205.
3. Ye CX, Wan F, Sun ZZ, Cheng CH, Ling RZ, Fan LF, Wang AL. Effect of phosphorus supplementation on cell viability, anti-oxidative capacity and comparative proteomic profiles of puffer fish (Takifugu obscurus) under low temperature stress. Aquaculture. 2016;452:200–8.
4. Long Y, Li LC, Li Q, He XZ, Cui ZB. Transcriptomic characterization of temperature stress responses in larval zebrafish. PLoS ONE. 2012;7(5):e37209.
5. Zerai DB, Fitzsimmons KM, Collier RJ. Transcriptional response of Delta-9-desaturase gene to acute and chronic cold stress in Nile tilapia *Oreochromis niloticus*. J World Aquacult Soc. 2010;41(5):800–6.
6. Chen Z, Cheng CH, Zhang J, Cao L, Chen L, Zhou L, Jin Y, Ye H, Deng C, Dai Z, et al. Transcriptomic and genomic evolution under constant cold in Antarctic notothenioid fish. Proc Natl Acad Sci USA. 2008;105(35):12944–9.
7. Cnaani A, Gall GAE, Hulata G. Cold tolerance of tilapia species and hybrids. Aquacult Int. 2000;8(4):289–98.
8. Yang C, Jiang M, Wen H, Tian J, Liu W, Wu F, Gou G. Analysis of differential gene expression under low-temperature stress in Nile tilapia (*Oreochromis niloticus*) using digital gene expression. Gene. 2015;564(2):134–40.
9. Manfioletti G, Ruaro M, Del Sal G, Philipson L, Schneider C. A growth arrest-specific (gas) gene codes for a membrane protein. Mol Cell Biol. 1990;10(6):2924–30.
10. Schneider C, King RM, Philipson L. Genes specifically expressed at growth arrest of mammalian cells. Cell. 1988;54(6):787–93.
11. Brancolini C, Schneider C. Gas2, a growth arrest-specific protein, is a component of the microfilament network system. J Cell Biol. 1992;117(6):1251–61.
12. Brancolini C, Marzinotto S, Schneider C. Susceptibility to p53 dependent apoptosis correlates with increased levels of Gas2 and Gas3 proteins. Cell Death Differ. 1997;4(3):247–53.
13. Lee KK, Tang MK, Yew DT, Chow PH, Yee SP, Schneider C, Brancolini C. gas2 is a multifunctional gene involved in the regulation of apoptosis and chondrogenesis in the developing mouse limb. Dev Biol. 1999;207(1):14–25.
14. Benetti R, Del Sal G, Monte M, Paroni G, Brancolini C, Schneider C. The death substrate Gas2 binds m-calpain and increases susceptibility to p53-dependent apoptosis. EMBO J. 2001;20(11):2702–14.
15. Li SS, Li CC, Dey M, Gagalac F, Dunham R. Cold tolerance of three strains of Nile tilapia, *Oreochromis niloticus* in China. Aquaculture. 2002;213(1):123–9.
16. Castresana J, Saraste M. Does Vav bind to F-actin through a CH domain? FEBS Lett. 1995;374(2):149–51.
17. Stradal T, Kranewitter W, Winder SJ, Gimona M. CH domains revisited. FEBS Lett. 1998;431(2):134–7.
18. Sun D, Leung CL, Liem RK. Characterization of the microtubule binding domain of microtubule actin crosslinking factor (MACF): identification of a novel group of microtubule associated proteins. J Cell Sci. 2001;114(Pt 1):161–72.
19. Hochachka PW. Defense strategies against hypoxia and hypothermia. Science. 1986;231(4735):234–41.
20. Nagle WA, Soloff BL, Moss AJ Jr, Henle KJ. Cultured Chinese hamster cells undergo apoptosis after exposure to cold but nonfreezing temperatures. Cryobiology. 1990;27(4):439–51.
21. Yi SX, Moore CW, Lee RE Jr. Rapid cold-hardening protects *Drosophila melanogaster* from cold-induced apoptosis. Apoptosis. 2007;12(7):1183–93.
22. Sonna LA, Fujita J, Gaffin SL, Lilly CM. Molecular biology of thermoregulation invited review: effects of heat and cold stress on mammalian gene expression. J Appl Physiol. 2002;92(4):1725–42.
23. Yao CL, Somero GN. The impact of acute temperature stress on hemocytes of invasive and native mussels (*Mytilus galloprovincialis* and *Mytilus californianus*): DNA damage, membrane integrity, apoptosis and signaling pathways. J Exp Biol. 2012;215(Pt 24):4267–77.

24. Los DA, Murata N. Membrane fluidity and its roles in the perception of environmental signals. Biochim Biophys Acta. 2004;1666(1–2):142–57.
25. Sgorbissa A, Benetti R, Marzinotto S, Schneider C, Brancolini C. Caspase-3 and caspase-7 but not caspase-6 cleave Gas2 in vitro: implications for microfilament reorganization during apoptosis. J Cell Sci. 1999;112(Pt 23):4475–82.
26. Zhang T, Dayanandan B, Rouiller I, Lawrence EJ, Mandato CA. Growth-arrest-specific protein 2 inhibits cell division in Xenopus embryos. PLoS ONE. 2011;6(9):e24698.
27. Zhu R, Mok MT, Kang W, Lau SS, Yip WK, Chen Y, Lai PB, Wong VW, To KF, Sung JJ. Truncated HBx-dependent silencing of GAS2 promotes hepatocarcinogenesis through deregulation of cell cycle, senescence and p53-mediated apoptosis. J Pathol. 2015;237(1):38–49.
28. Kerkweg U, Jacob M, De GH, Mannherz HG, Rauen U. Cold-induced apoptosis of rat liver endothelial cells: contribution of mitochondrial alterations. Transplantation. 2003;76(3):501–8.
29. Ibarz A, Martín-Pérez M, Blasco J, Bellido D, De OE, Fernández-Borràs J. Gilthead sea bream liver proteome altered at low temperatures by oxidative stress. Proteomics. 2010;10(5):963–75.

# Permissions

All chapters in this book were first published in MB, by BioMed Central; hereby published with permission under the Creative Commons Attribution License or equivalent. Every chapter published in this book has been scrutinized by our experts. Their significance has been extensively debated. The topics covered herein carry significant findings which will fuel the growth of the discipline. They may even be implemented as practical applications or may be referred to as a beginning point for another development.

The contributors of this book come from diverse backgrounds, making this book a truly international effort. This book will bring forth new frontiers with its revolutionizing research information and detailed analysis of the nascent developments around the world.

We would like to thank all the contributing authors for lending their expertise to make the book truly unique. They have played a crucial role in the development of this book. Without their invaluable contributions this book wouldn't have been possible. They have made vital efforts to compile up to date information on the varied aspects of this subject to make this book a valuable addition to the collection of many professionals and students.

This book was conceptualized with the vision of imparting up-to-date information and advanced data in this field. To ensure the same, a matchless editorial board was set up. Every individual on the board went through rigorous rounds of assessment to prove their worth. After which they invested a large part of their time researching and compiling the most relevant data for our readers.

The editorial board has been involved in producing this book since its inception. They have spent rigorous hours researching and exploring the diverse topics which have resulted in the successful publishing of this book. They have passed on their knowledge of decades through this book. To expedite this challenging task, the publisher supported the team at every step. A small team of assistant editors was also appointed to further simplify the editing procedure and attain best results for the readers.

Apart from the editorial board, the designing team has also invested a significant amount of their time in understanding the subject and creating the most relevant covers. They scrutinized every image to scout for the most suitable representation of the subject and create an appropriate cover for the book.

The publishing team has been an ardent support to the editorial, designing and production team. Their endless efforts to recruit the best for this project, has resulted in the accomplishment of this book. They are a veteran in the field of academics and their pool of knowledge is as vast as their experience in printing. Their expertise and guidance has proved useful at every step. Their uncompromising quality standards have made this book an exceptional effort. Their encouragement from time to time has been an inspiration for everyone.

The publisher and the editorial board hope that this book will prove to be a valuable piece of knowledge for researchers, students, practitioners and scholars across the globe.

# List of Contributors

Ana Raquel Soares, Noémia Fernandes, Marisa Reverendo, Gabriela M. R. Moura and Manuel A. S. Santos
Department of Medical Sciences and Institute for Biomedicine–iBiMED, University of Aveiro, 3810-193 Aveiro, Portugal

Hugo Rafael Araújo and José Luís Oliveira
IEETA, University of Aveiro, 3810-193 Aveiro, Portugal

Katherine Huang, Manoja B. K. Eswara, Zhaodong Xu, Tao Yu, Arthur Aubry, Izzy Livne-bar, Monika Sangwan, and Mohamad Ahmad
Lunenfeld Tanenbaum Research Institute, Mt Sinai Hospital, Toronto, ON, Canada

Mohamed Abou El Hassan
Lunenfeld Tanenbaum Research Institute, Mt Sinai Hospital, Toronto, ON, Canada
Clinical Chemistry Division, Provincial Laboratory Services, Queen Elizabeth Hospital, Charlottetown, PE, Canada
Department of Pathology, Faculty of Medicine, Dalhousie University, Halifax, NS, Canada

Rod Bremner
Lunenfeld Tanenbaum Research Institute, Mt Sinai Hospital, Toronto, ON, Canada
Department of Lab Medicine and Pathobiology, University of Toronto, Toronto, ON, Canada
Department of Ophthalmology and Vision Science, University of Toronto, Toronto, ON, Canada

Zuyao Ni
Lunenfeld Tanenbaum Research Institute, Mt Sinai Hospital, Toronto, ON, Canada
Donnelly Centre, University of Toronto, Toronto, ON, Canada

Fang Zhang and Zhenfa Ma
Key Laboratory of Swine Genetics and Breeding of Ministry of Agriculture & Key Laboratory of Agriculture Animal Genetics, Breeding and Reproduction of Ministry of Education, College of Animal Science, Huazhong Agricultural University, Wuhan 430070, Hubei, People's Republic of China

Zhuqing Ren and Yi Jin
Key Laboratory of Swine Genetics and Breeding of Ministry of Agriculture & Key Laboratory of Agriculture Animal Genetics, Breeding and Reproduction of Ministry of Education, College of Animal Science, Huazhong Agricultural University, Wuhan 430070, Hubei, People's Republic of China
The Cooperative Innovation Center for Sustainable Pig Production, Huazhong Agricultural University, Wuhan 430070, Hubei, People's Republic of China

Meijun Du and Liang Wang
Department of Pathology and MCW Cancer Center, Medical College of Wisconsin, Milwaukee, WI 53226, USA

Kang-Sheng Ma, Fen Li, Ying Liu, Ping-Zhuo Liang, Xue-Wei Chen and Xi-Wu Gao
China Agricultural University, Beijing, China

Jason A. Estep, Erin L. Sternburg, Gissell A. Sanchez and Fedor V. Karginov
Department of Cell Biology and Neuroscience, Institute for Integrative Genome Biology, University of California, Riverside, CA 92521, USA

Shubham Goyal, Takahiro Suzuki, Jing-Ru Li, Shiori Maeda, Mami Kishima, Hajime Nishimura, Yuri Shimizu and Harukazu Suzuki
Division of Genomic Technologies, RIKEN Center for Life Science Technologies, 1-7-22 Suehiro-cho, Tsurumi-ku, Yokohama, Kanagawa 230-0045, Japan

Puja Sinha and Manisha Sachan
Department of Biotechnology, Motilal Nehru National Institute of Technology, Allahabad 211004, India

Kiran Singh
Department of Molecular and Human Genetics, Banaras Hindu University, Varanasi, India

Sergio Vargas and Angelo Poliseno
Department of Earth and Environmental Sciences, Palaeontology & Geobiology, Ludwig-Maximilians-Universität München, Richard-Wagner Str. 10, 80333 Munich, Germany

**Gaurav G. Shimpi**
Department of Earth and Environmental Sciences,
Palaeontology & Geobiology, Ludwig-Maximilians-
Universität München, Richard-Wagner Str. 10,
80333 Munich, Germany
Marine Biotechnology and Ecology (MBE) Division,
CSIR-Central Salt and Marine Chemicals Research
Institute (CSMCRI), Bhavnagar, Gujarat 364002, India

**Gert Wörheide**
Department of Earth and Environmental Sciences,
Palaeontology & Geobiology, Ludwig-Maximilians-
Universität München, Richard-Wagner Str. 10,
80333 Munich, Germany
GeoBio-Center, Ludwig-Maximilians-Universität
München (LMU), Richard-Wagner Str. 10, Munich
80333, Germany
SNSB– Bavarian State Collections of Palaeontology
and Geology, Richard-Wagner Str. 10, 80333
Munich, Germany

**Lindsey Skrdlant and Ren-Jang Lin**
Department of Molecular and Cellular Biology, Irell
& Manella Graduate School of Biological Sciences,
Beckman Research Institute of the City of Hope,
Duarte, CA 91010, USA

**Jeremy M. Stark**
Department of Cancer Genetics and Epigenetics,
Irell & Manella Graduate School of Biological
Sciences, Beckman Research Institute of the City
of Hope, Duarte, CA 91010, USA

**Sanket Awate and Arrigo De Benedetti**
Department of Biochemistry and Molecular Biology,
Louisiana State University Health Sciences Center,
1501 Kings Highway, Shreveport, LA 71130, USA

**Isao Karube and Wataru Yoshida**
School of Bioscience and Biotechnology, Tokyo
University of Technology, 1404-1 Katakuramachi,
Hachioji, Tokyo 192-0982, Japan

**Daniyah H. Bay**
School of Bioscience and Biotechnology, Tokyo
University of Technology, 1404-1 Katakuramachi,
Hachioji, Tokyo 192-0982, Japan
Biology Department,Umm Al-Qura University,
Makkah, Kingdom of Saudi Arabia

**Annika Busch**
School of Bioscience and Biotechnology, Tokyo
University of Technology, 1404-1 Katakuramachi,
Hachioji, Tokyo 192-0982, Japan

Biosystems Technology, Institute of Applied Life
Sciences, Technical University of Applied Sciences
Wildau, Wildau, Germany

**Fred Lisdat**
Biosystems Technology, Institute of Applied Life
Sciences, Technical University of Applied Sciences
Wildau, Wildau, Germany

**Keisuke Iida**
Graduate School of Science and Engineering,
Saitama University, c/o Saitama Cancer Center,
Saitama, Japan

**Kazunori Ikebukuro and Kazuo Nagasawa**
Department of Biotechnology and Life Science,
Tokyo University of Agriculture and Technology,
Tokyo, Japan

**Lijian Yu and Michael R. Volkert**
Microbiological and Physiological Systems,
University of Massachusetts Medical School, 55
Lake Avenue North, Worcester, MA 01655, USA

**Mayuri Rege and Craig L. Peterson**
Program in Molecular Medicine, University of
Massachusetts Medical School, 373 Plantation
Street, Worcester, MA 01605, USA

**Sophia Pinz, Samy Unser and Anne Rascle**
Sophia Pinz and Samy Unser contributed equally to
this work Stat5 Signaling Research Group, Institute
of Immunology, University of Regensburg, 93053
Regensburg, Germany

**Jieying Qian, Naixia Chao, Chengtong Ma, Qicong
Chen, Jian Sun and Yaosheng Wu**
Key Laboratory of Biological Molecular Medicine
Research of Guangxi Higher Education, Department
of Biochemistry and Molecular Biology, Guangxi
Medical University, Nanning, Guangxi, People's
Republic of China

**Yong Liu**
Schools of Pharmacy, Guangdong Medical
University, Dongguan, Guangdong, People's
Republic of China

**Victoria L. M. Herrera, Ann Marie Moran,
Glaiza A. Tan, Khristine A. Pasion and Nelson
Ruiz-Opazo**
Whitaker Cardiovascular Institute, Boston
University School of Medicine, 700 Albany Street,
Boston, MA 02118, USA

Department of Medicine, Boston University School of Medicine, 700 Albany Street, Boston, MA 02118, USA

**Martin Steffen**
Department of Pathology and Biomedical Engineering, Boston University, Boston, USA

**Keith Rivera and Darryl J. Pappin**
Cold Spring Harbor Laboratory, 1 Bungtown Road, Cold Spring Harbor, NY 11724, USA

**ChangGeng Yang, Fan Wu, Xing Lu, Ming Jiang, Wei Liu, Lijuan Yu, Juan Tian and Hua Wen**
ChangGeng Yang and Fan Wu contributed equally to this work Key Laboratory of Freshwater Biodiversity Conservation and Utilization of Ministry of Agriculture, Yangtze River Fisheries Research Institute, Chinese Academy of Fishery Sciences, Wuhan 430223, China

# Index

www.ingramcontent.com/pod-product-compliance
Lightning Source LLC
Chambersburg PA
CBHW082029190326
41458CB00010B/3317